PROTEIN-METAL INTERACTIONS

ADVANCES IN EXPERIMENTAL MEDICINE AND BIOLOGY

Recent Volumes in this Series

Volume 40
METAL IONS IN BIOLOGICAL SYSTEMS: Studies of Some Biochemical and Environmental Problems
Edited by Sanat K. Dahr • 1973

Volume 41A
PURINE METABOLISM IN MAN: Enzymes and Metabolic Pathways
Edited by O. Sperling, A. De Vries, and J. B. Wyngaarden • 1974

Volume 41B
PURINE METABOLISM IN MAN: Biochemistry and Pharmacology of Uric Acid Metabolism
Edited by O. Sperling, A. De Vries, and J. B. Wyngaarden • 1974

Volume 42
IMMOBILIZED BIOCHEMICALS AND AFFINITY CHROMATOGRAPHY
Edited by R. B. Dunlap • 1974

Volume 43
ARTERIAL MESENCHYME AND ARTERIOSCLEROSIS
Edited by William D. Wagner and Thomas B. Clarkson • 1974

Volume 44
CONTROL OF GENE EXPRESSION
Edited by Alexander Kohn and Adam Shatkay • 1974

Volume 45
THE IMMUNOGLOBULIN A SYSTEM
Edited by Jiri Mestecky and Alexander R. Lawton • 1974

Volume 46
PARENTERAL NUTRITION IN INFANCY AND CHILDHOOD
Edited by Hans Henning Bode and Joseph B. Warshaw • 1974

Volume 47
CONTROLLED RELEASE OF BIOLOGICALLY ACTIVE AGENTS
Edited by A. C. Tanquary and R. E. Lacey • 1974

Volume 48
PROTEIN-METAL INTERACTIONS
Edited by Mendel Friedman • 1974

Volume 49
NUTRITION AND MALNUTRITION: Identification and Measurement
Edited by Alexander F. Roche and Frank Falkner • 1974

Volume 50
ION-SELECTIVE MICROELECTRODES
Edited by Herbert J. Berman and Normand C. Hebert • 1974

[American Chemical Society Symposium on

PROTEIN-METAL INTERACTIONS

Edited by

Mendel Friedman
Western Regional Research Laboratory
Agricultural Research Service
U. S. Department of Agriculture
Berkeley, California

PLENUM PRESS • NEW YORK AND LONDON

Library of Congress Cataloging in Publication Data

American Chemical Society Symposium on Protein–Metal Interactions, Chicago, 1973.
Protein–metal interactions.

(Advances in experimental medicine and biology, v. 48)
Includes bibliographies.
1. Metals in the body—Congresses. 2. Metals—Physiological effect—Congresses. 3. Proteins—Congresses. I. Friedman, Mendel, ed. II. American Chemical Society. III. Title. IV. Series. [DNLM: 1. Metals—Metabolism—Congresses. 2. Proteins—Metabolism—Congresses. W1AD559 v. 48 / QU55 S9864p 1973]
QP531.A42 1973 612′.0152 74-13406
ISBN 0-306-39048-5

Proceedings of the American Chemical Society Symposium on Protein–Metal Interactions held in Chicago, Illinois, August 27-29, 1973, with supplemental invited contributions.

A Division of Plenum Publishing Corporation
227 West 17th Street, New York, N.Y. 10011

United Kingdom edition published by Plenum Press, London
A Division of Plenum Publishing Company, Ltd.
4a Lower John Street, London W1R 3PD, England

Printed in the United States of America

Preface

Metal ions and proteins are ubiquitous. Therefore, not surprisingly, new protein-metal interactions continue to be discovered, and their importance is increasingly recognized in both physical and life sciences. Because the subject matter is so broad and affects so many disciplines, in organizing this Symposium, I sought participation of speakers with the broadest possible range of interests. Twenty-two accepted my invitation. To supplement the verbal presentations, the Proceedings include five closely related invited contributions.

The ideas expressed are those of the various authors and are not necessarily approved or rejected by any agency of the United States Government. No official recommendation concerning the subject matter or products discussed is implied in this book.

This book encompasses many aspects of this multifaceted field. Topics covered represent biochemical, immunochemical, bioorganic, biophysical, metabolic, nutritional, medical, physiological, toxicological, environmental, textile, and analytical interests. The discoveries and developments in any of these areas inevitably illumine others. I feel that a main objective of this Symposium, bringing together scientists with widely varied experiences yet with common interests in protein-metal interactions, so that new understanding and new ideas would result has been realized. I hope that the reader enjoys and benefits from reading about the fascinating interactions of metal ions and proteins as much as I did.

Although an adequate summary of the Symposium is not possible in a brief preface, I wish to express particular interest in the ideas reported by Professor Frieden: that the relative occurrence and participation of the various metals as essential elements in enzyme action and other life processes is an adaptive relationship to their relative abundance in the ocean. Undoubtedly, this adaptation is a continuing process. A more immediate practical concern voiced by D. K. Darrow and H. A. Schroeder that has received widespread publicity and debate is that children are highly susceptible to lead poisoning and that their exposure to lead nowadays comes mainly from automobile exhaust.

Of the invited contributions supplementing the Symposium, the paper by J. T. MacGregor and T. W. Clarkson deserves special mention. Dr. MacGregor collaborated with Dr. Clarkson, his former major professor, in this thorough review while the latter was out of the country dealing directly with an episode of mercury poisoning described in their paper. I believe their critical compilation of tissue distribution and toxicity of mercury compounds will greatly benefit the medical and other scientific communities in dealing with this useful but dangerous element.

I am confident that the Proceedings of the Symposium on "Protein-Metal Interactions" will be a valuable contribution to the literature. I am particularly grateful to Dr. I. A. Wolff, Chairman of the Protein Subdivision of the Division of Agricultural and Food Chemistry of the American Chemical Society, who invited me to organize this Symposium, to all contributors and participants for a well realized meeting of minds, and to R. N. Ubell, Editor-in-Chief, Plenum Publishing Company, for arranging publication as a volume in the series, ADVANCES IN EXPERIMENTAL MEDICINE AND BIOLOGY.

Mendel Friedman
Moraga, California
April 1974

Contents

THE EVOLUTION OF METALS AS ESSENTIAL ELEMENTS [with special reference to iron and copper]

Earl Frieden

Department of Chemistry

Florida State University

Tallahassee, Fla. 32306

The past few years have witnessed exciting progress in our understanding of the elements required for the growth and survival of the higher animal. As anticipated, these new discoveries have been exclusively among the trace elements, those elements which in minute quantities are essential for growth and development. As shown in Table I, after the discovery of the requirement for cobalt in 1935, there was a hiatus of about two decades before the essentiality of molybdenum (1953), chromium (1957) and selenium (1959) was confirmed. This was the end of an era in trace element research. From here on a major change in the research technique had to be devised. The animals, their food and their entire environment had to undergo a complete trace element decontamination, using special plastic houses, highly purified diets and filtered air. After a decade of painstaking research, Dr. Klaus Schwarz of the Veterans Administration and others added three new elements to the essential list: fluorine, tin and vanadium. In 1972, Dr. Edith Carlisle, University of California, Los Angeles, proved that silicon also was required for the growth and development of chicks. We now know that at least twenty-five of the 96 elements found on earth are required for some form of life (Figure 1; see also Table III).

Two-thirds of the lightest elements, 22 out of the first 34, are essential. These 22 elements plus molybdenum (no. 42), tin (no. 40) and iodine (no. 53) complete the list of 25 essential elements. It is an impressive fact that despite our current sophistication in biochemistry and molecular biology, there remain many fundamental, unanswered questions concerning the elements required for animal or plant life. Moreover, there has been little effort to relate the evolution of the dependence on specific elements to the past and present status of living organisms.

TABLE I

DISCOVERY OF TRACE ELEMENTS REQUIREMENTS OF ANIMALS

Iron . . . . .17th century	
Iodine 1850	Chatin, A.
Copper 1928	Hart, E.B., H. Steenbock, J. Waddell, and C.A. Elvehjem
Manganese . .1931	Kemmerer, A.R., and W.R. Todd
Zinc 1934	Todd, W.R., C.A. Elvehjem, and E.B. Hart
Cobalt 1935	Underwood, E.J., and J.F. Filmer Marston, H.R. Lines, E.W.
Molybdenum . .1953	deRenzo, E.C., E. Kaleita, P. Heytler, J.J. Oleson, B.L. Hutchings, and J.H. Williams Richert, D.A., and W.W. Westerfield
Selenium . . .1957	Schwarz, K., and C.M. Foltz
Chromium . . .1959	Schwarz, K., and W. Martz
Tin 1970	Schwarz, K., and D.B. Milne, and E. Vinyard
Vanadium . . .1971	Hopkins, L.L., and H.E. Mohr Schwarz, K., and D.B. Milne
Fluorine . . .1971	Schwarz, K., and D.B. Milne
Silicon . . .1972	Carlisle, E.M. Schwarz, K., and D.B. Milne

Adopted from K. Schwarz in "Trace Element Metabolism in Animals", C. F. Mills, Ed., E85 Livingstone, Edinburgh, p. 25 (1970). See this article for appropriate references.

In his classic text, The Fitness of the Environment, published in 1913, Lawrence J. Henderson concluded:

> The properties of matter and the course of cosmic evolution are now seen to be intimately related to the structure of the living being to its activities; they become, therefore, far more important in biology than has been previously suspected. For the whole evolutionary process, both

> cosmic and organic, is one, and the biologist may now rightly regard the universe in its very essence.

Despite the title of this far reaching work, Henderson put as much emphasis on the organism fitting to the environment through Darwinian evolution as on the milieu fitting the organism. Henderson wrote that the fitness of the environment was just one part of a reciprocal relationship of which the fitness of the organism is the other. This condition of life was expressed in the characteristics of water, carbonic acid and the compounds of carbon, hydrogen, and oxygen. These combined to provide "the best of all possible" environmnets for life. Henderson's basic approach is none the less applicable today and provides us with the starting point for our consideration of the basis for the selective dependence of life on the 25 natural elements. To fit their environment, living systems were involved in two major adaptations--first to the origin of life in the oceans and second to the utilization of oxygen in the atmosphere. These two adaptations have established the major guidelines for the requirement of the elements and the determination of the chemical structures accompanying the development of life on earth.

EVOLUTION OF LIFE IN THE SEA

It is widely accepted that life on earth first evolved in the sea or in saline tidal pools. This has led to an indelible imprint of the composition of the oceans in the chemistry of all cells and organisms. The oceans provided a convenient medium in which the primitive organism could float freely in a solution of the basic chemicals required to maintain life. For these primitive organisms the fluidity of a medium like water was indispensible. A denser matrix (e.g., solids) would have been too restrictive on the mobility of dissolved ions or metabolites. Finally, after millions of years, the metamorphic transition from aquatic to terrestrial existence began. This development was successful only if the transforming organisms were able to carry part of their marine environment with them. The distribution of salts in the blood and other body fluids reflects the oceanic nature of their evolutionary antecedents. The choice of sodium, potassium, calcium, magnesium and chloride ions for the major ionic environment of the cell, both inside and outside, stems from their availability in the primeval seas.

All this has been a widely recognized facet of comparative biochemistry (Prosser and Brown, 1961). What has not been fully appreciated is that the nature of the trace element requirement, particularly the metal ions, appears to have been limited by their availability in the early oceans. This is evident upon an examination of the concentrations in the oceans of the five bulk ions and the fourteen trace elements, (Table II; compare with Table III).

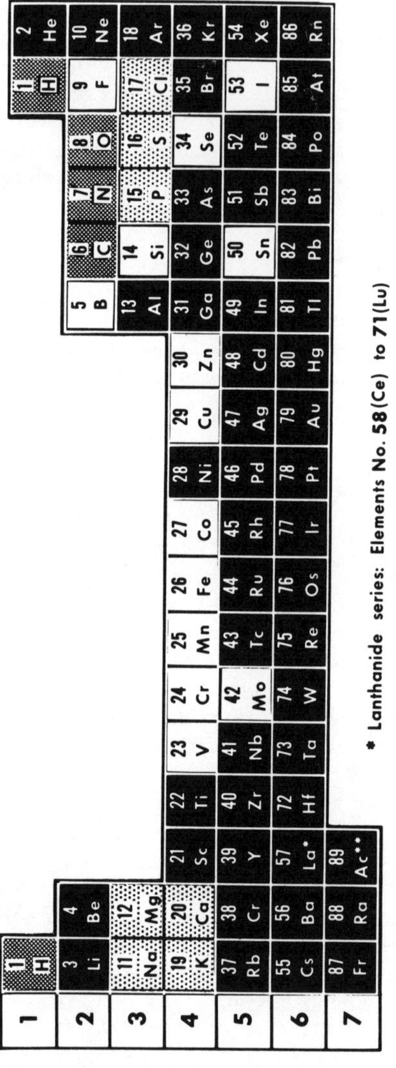

*** Lanthanide series: Elements No. 58(Ce) to 71(Lu)**

**** Actinide series: Elements No. 90(Th) to 103(Lw).**

ESSENTIAL LIFE ELEMENTS, 24 by the latest count, are clustered in the upper half of the periodic table. The elements are arranged according to their atomic number. The four most abundant elements that are found in living organisms (hydrogen, oxygen, carbon and nitrogen) are indicated by the grey cross-hatched areas. The seven next most common elements are shown in dotted areas. The 14 elements that are shown in white are needed only in traces.

Adapted from E. Frieden, Sci. Amer., July, 1972

Figure 1

Table II

Concentration of Some Crucial Elements in Sea Water

Element	Parts per Million	Element	Parts per Billion
K	380	Fe	10
Mg	1,350	Cu	3
Ca	400	Zn	10
Cl	19,000	Mn	2
Na	10,500	Mo	10
B	4.6	Co	.27
F	1.3	Se	.09
Si	3.0	Cr	.05
I	.06	V	2
		Sn	3

Taken from: E. D. Goldberg in The Sea (ed. by M. N. Hill), pp. 4-5, (1963), J. Wiley, New York.

The macro ions are present in concentrations 10^5 - 10^6 times those of the trace metals. The ten essential trace metals are present in 10 parts per billion or less. Three required non-metals, boron, fluorine, and silicon are present in 1-5 parts per million, but are still only trace elements in most systems, possibly due to the lack of the unique binding properties exhibited by these non-metals. Finally iodine proves to be an essential trace element to mainly non-marine forms. The special properties of these trace elements in both simple and complex organisms will be discussed later.

AEROBIC CELLS

Another major evolutionary adaptation met by living systems arises from changes in the atmosphere leading to an accumulation of and a dependence on oxygen. My colleague, Hans Gaffron (1960), of Florida State University, has summarized the major stages of biogenesis in terms of the transition from a highly reducing atmosphere to our present oxygen rich world. There was first an anaerobic era of excess hydrogen with ammonia, methane and water vapor and an accumulation of highly reduced organic substances. This was followed by a hydrogen poor period with the beginnings of inorganic or organic catalysis and traces of oxygen. In the third period, the anaerobic organisms deplete their useful reservoir of organic substances dooming the earth to essential dependence on photoreduction as the major initial source of energy. This leads to today's era of photosynthesis and ascendancy of cells dependent on oxidative energy, eliminating or driving the obligate anerobes "underground". Oxygen itself then becomes a major evolutionary force. As we shall see later the essentiality of numerous metals arises from their ability to promote the effective utilization of oxygen.

How environmental circumstances may determine whether an organism will survive such drastic changes as the step from anaerobic to aerobic conditions was shown in an actual laboratory model. In 1954, Jensen and Thofern found a micrococcus strain which grew well heterotrophically under anaerobic conditions. On contact with oxygen the cells die because they have neither an efficient mechanism for destroying the hydrogen peroxide which might be formed nor a respiratory system. But, when the nutrient medium contained a simple iron porphyrin, ordinary protohemin, these bacteria survived in air. The hemin was taken up into the cells and there combined with the right kind of protein to provide the bacteria with catalase and respiratory enzymes. The presence or absence of hemin in the surroundings at the time oxygen reaches the cells determines their fate.

THE THREE GROUPS OF ESSENTIAL ELEMENTS

The current state of knowledge about the 25 elements now recognized to be essential to the various types of organisms is summarized in Table III. In this table the elements are classified in three categories: (++) is used to describe the elements which have been shown to be essential to numerous representative animals in that group and are presumed to be required for life and survival by virtually every member of that group, (+) designates those elements for which a requirement has been demonstrated but the number of examples or evidence is so sparse as to limit any generalization of requirement (although the "unity" of nature strongly suggests a group requiremtent), and (-) indicates the lack of any definitive data supporting the essentiality of that element for the particular class of organism. This may be due to the absence of any experiments or to the failure to show a requirement in those experiments reported.

Each of the first group of six elements is required because they provide most of the atoms for the primary molecular building blocks for all forms of living matter: amino acids, sugars, fatty acids, purines, pyrimidines and their nucleotides. These molecules not only have independent biochemical roles but also are the respective constituents of the major categories of large molecules: proteins, cellulose, glycogen, starch, lipids and nucleic acids. Several of the 20 amino acids contain sulfur in addition to carbon, hydrogen, and oxygen. Phosphorus plays an important role in the nucleotides, such as adenosine triphosphate (ATP), which is central to the energetics of the cell. Adenosine triphosphate includes components that are also one of the four nucleotides needed to form

TABLE III

ESSENTIALITY OF THE ELEMENTS FOR LIVING ORGANISMS

Element	Bacteria	Algae	Fungi	Higher Plants	Inverte-brates	Verte-brates
H,C,N, O,P,S	++	++	++	++	++	++
K	++	++	++	++	++	++
Mg	++	++	++	++	++	++
Ca	+	++	+	++	++	++
Cl	++	++	-	++	++	++
Na	+	+	-	+	++	++
Fe	++	++	++	++	++	++
Cu	++	++	++	++	++	++
Zn	++	++	++	++	++	++
Mn	++	++	++	++	++	++
Mo	++	++	++	++	++	++
Co	++	+	+	+	++	++
Se	+	-	-	+	-	++
Cr	+	-	-	-	-	+
V	+	+	-	-	+	+
Sn	-	-	-	-	-	+
Ni	-	-	-	-	-	+
B	-	+	-	++	-	-
F	-	-	-	-	-	+
Si	-	+	-	+	-	+
I	-	+	-	+	+	++

KEY: (++) - Essential for several and, probably, all species in the group.
(+) - Essential for one or a few species in the group.
(-) - Not known to be essential for any species.

the double helix of deoxyribonucleic acid. Polysaccharides, such as cellulose, form the major structural component of plants and trees. Both sulphur and phosphorus are present in many of the small indispensible molecules called coenzymes. In bony animals phosphorus and calcium help to create strong supporting structures.

The second group of essential elements play a critical role in the electrochemical properties of cells from all forms of life. The principal cations are potassium and magnesium with a somewhat less versatile role for calcium and sodium. The principal anions or negative charges are provided by chloride, sulfate and phosphate and are widely distributed with an anomalous feature of fungi in not requiring chloride. These seven ions maintain the electrical neutrality of body fluids and cells and also play a part in maintaining the proper liquid volume of the blood and other fluid systems. Whereas the cell membrane serves as a physical barrier to the exchange of large molecules, it allows small molecules to pass relatively freely. The electrochemical functions of the anions and cations serve to maintain the appropriate relation of osmotic pressure and charge distribution on the two sides of the cell membrane. One of the striking features of the ion distribution is the specificity of these different ions. Cells are rich in potassium and magnesium, and the surrounding plasma is rich in sodium and calcium. It seems likely that the distribution of ions in the plasma of higher animals reflects the oceanic origin of their evolutionary antecedents.

The third and most numerous group of essential elements is those required only in very small, trace amounts. This latter fact in no way diminishes their great importance to the survival and proper functioning of all organisms. For the purpose of this discussion we are going to make a distinction within this group between the metals and the non-metals.*

The non-metals, boron, fluorine, silicon and iodine; will be mentioned briefly for several reasons. The first is that we don't know much about their requirement or biochemical role except for iodine. The indispensability of fluorine and silicon in the mammal has been reported in the last 18 months although the role of fluorine in preventing tooth decay was well known. Boron is required only in some plants and algae with no positive role in animals. Iodine, the heaviest required element presents a more interesting case. It is, of course, best known in the vertebrates as an essential part of the molecules which comprise the thyroid hormones, iodothyronine. These hormones play two probably related roles in the vertebrates (Frieden and Kent, 1974): (1) the initiation and control of metamorphosis and development in amphibia and in the higher vertebrates and (2) the adaptation of animals to homeothermy, which utilizes the calorigenic properties of these hormones. Yet the precise function of iodine in hormonal activity is still not understood. The role of iodine in invertebrates and plants is spotty and uncertain--there is no certainty of a group requirement for this element. The trace non-metals may be a more interesting group than imagined, but we will need much more information than is currently available.

---

*Boron and silicon are sometimes considered as metalloids, since they are at the borderline in the delineation of metals and non-metal properties.

The indispensible elements which most excite the bio-chemists are the trace metal ions, the most numerous category listed in Table III. They are especially interesting because of their widespread distribution, importance, and unique chemical properties. Five of these metal ions, Fe, Cu, Zn, Mn, Mo, appear to be generally required by all organisms. Cobalt will undoubtedly end up in this category. The other metals are required in such small amounts and the chances of contamination at this level are so great that highly refined experimental techniques were required to show their indispensibility. These metals include selenium, vanadium, chromium, and tin ions. Recently selenium has been identified as an essential component of an enzyme in the red blood cell, glutathione peroxidase, which is involved in the protection of the erythrocyte against the toxic, oxidative effects of hydrogen peroxide. There is growing evidence that nickel is also a required trace element. When we learn how to cope with the experimental conditions in the various media required for proper testing of different kinds of organisms, these four metals may prove to be needed by all cells. This experimental approach may also reveal additional new essential metals, the most likely being aluminum or germanium.

DEVELOPMENT OF CATALYSIS

What more can be said about the development of the need of cells for these 25 elements? The literature is replete with discussions of proteins, nucleic acids and informational macromolecules. Accordingly let us focus our attention on the development of the need for certain elements, in particular the metal ions. The six metals, iron, copper, zinc, manganese, molybdenum and cobalt, which are required by every major group of organisms have a reasonably well defined biological function. These six metals belong to a group known as the transition elements. The transition metals serve as essential constituents of a wide range of metalloproteins, usually enzymes, which participate in important oxidative, hydrolytic or transfer processes. Most of them undergo reversible changes in oxidation state which make them especially useful as intermediates in electron transfer or oxidation reactions. They also tend to form strong complexes with a variety of ligands of the type present in the side chains of proteins, and which gave extra stability to metalloprotein complexes.

How a typical catalytic activity of a transition metal enzyme might have evolved has been described by Melvin Calvin (1969) of the University of California (Berkeley). He pointed out the similarity between the enzymic reactions and the reactions known to the inorganic chemist. The model which he chose centers around the element iron, particularly in catalase, peroxidase and the cytochromes. Here a quantitative comparison was made between the ability of the simplest iron compounds to perform the same catalytic function as it has evolved in biological systems. Thus the accompanying (Figure 2)

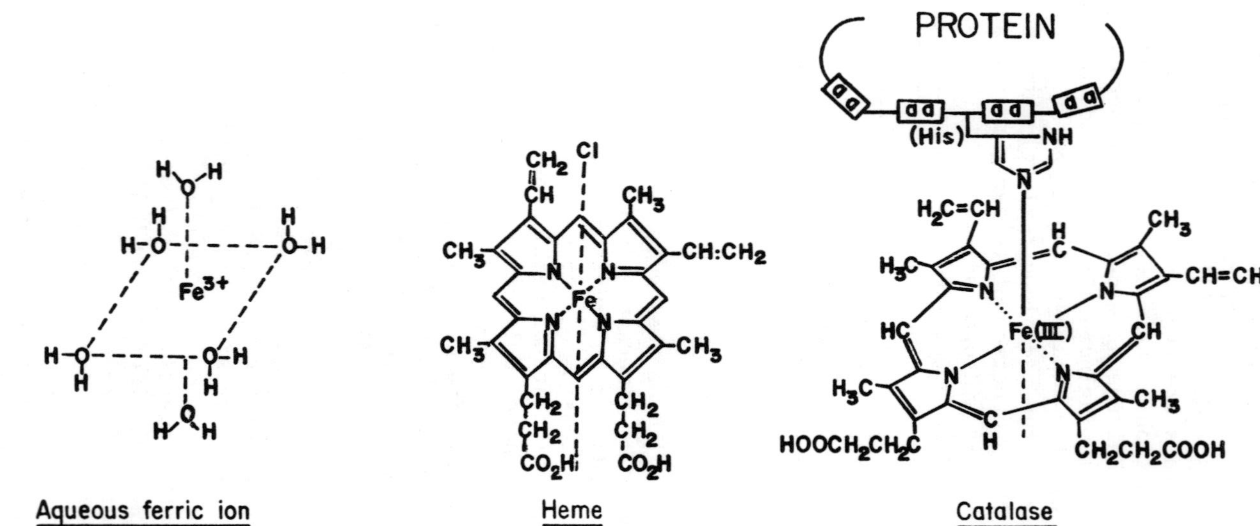

Figure 2. Evolution of a catalyst for the reaction $H_2O_2 \rightarrow H_2O + 1/2\ O_2$ as depicted by M. Calvin in <u>Chemical Evolution</u>, Oxford Univ. Press, 1969.

presents a comparison of the hydrated iron ion, an iron ion surrounded by protoporphyrin as in heme, and the iron containing heme group as incorporated into a protein surrounded by one or more coordinating groups, such as histidine, to form the catalase molecule. Ferric ion has a modest catalytic coefficient, about $10^{-5}mM^{-1}$ sec.$^{-1}$ at 0°. This number expresses the effeciency of the ferric ion in decomposing hydrogen peroxide. When ferric ion is incorporated into a porphyrin molecule in decomposing hydrogen peroxide increases by a factor of 10^3, to $10^{-2}mM^{-1}$ sec.$^{-1}$. Only if the heme group is integrated into a specific protein, as in catalase, does the catalytic efficiency go up to 10^5mM^{-1} sec.$^{-1}$, showing a remarkable increase by a factor of 10^7. Thus the efficiency of iron in catalase as a catalyst to decompose hydrogen peroxide has been enhanced by a factor of 10^{10}. These changes in the catalytic form of iron should not be imagined as having occurred in only three steps, but represent the kind of improvement in catalytic activity that must have developed during thousands of years of evolution to achieve the efficient metalloprotein catalysts of today.

Calvin (1969) points out another example of the utilization of metal ions for the development of superior catalysis. Again this involves reactions between a metal ion, cupric ion (Cu(II)), and molecular hydrogen. The expected products of this reaction are cuprous ion (Cu(I)) and oxidized hydrogen either as acid or water, depending upon the anion with which the cupric ion is associated. The rate of the reaction of molecular hydrogen with cupric ion is extremely slow, but it can be increased by the catalytic impetus of cuprous ion which is an effective catalyst for the reaction between hydrogen and cupric ion. Thus a system consisting of molecular hydrogen and cupric ion might remain unreacted for a long period until cuprous ion became available in the environment or until some other electron transfer process leading to the formation of cuprous ion; then this ion could serve as a catalyst for this electron transfer.

METALLOENZYMES, METALLOPROTEINS, AND METAL-ION ACTIVATED ENZYMES

Close to 1/3 of all the enzymes involve a metal ion as an essential participant (Vallee and Wacker, 1970). It is useful to think of these enzymes as comprising two major types, the metalloenzymes and the metal-ion activated enzymes, although the difference may be quantative rather than qualitative, representing a continuous spectrum in the relative binding between metal ion and protein.

The metalloenzymes contain a metal ion as an integral part of the molecule in a fixed ratio per molecule of protein. These metal ions are bound firmly to a limited number of specific sites on the protein; they show no evidence of dissociation under physiological conditions. Vigorous modification of the metalloenzyme is frequently required to remove the metal ion; this, in turn, results in the complete loss of enzyme activity. Under favorable conditions, the activity of the metal ion free protein (apoenzyme) can be restored by the addition of the original metal ion according to the reaction

$$\underset{\text{(inactive)}}{\text{apoenzyme + metal ion}} \rightarrow \underset{\text{(active)}}{\text{metalloenzyme}}$$

The addition of a different metal ion usually will not reconstitute the activity of the metalloenzyme except in a few rare instances. Thus the interaction between the active metal and the apoenzyme must be highly specific and unique. Metalloenzymes frequently involve metals which form the most stable chelates, particularly Fe(III), Cu(II), Zn(II) and Mn(II). A specific metal is used and reused for catalysis of a particular type of reaction--copper in oxidases which utilize oxygen, zinc in several dehydrogenases and hydrolases, iron-protoporphyrin in a variety of electron transfering enzymes and oxygenases. Selected examples of metalloenzymes are presented in Tables IV-VI.

The second category of enzymes which require metal ions are the metal-ion-activated enzymes. The metal ion is more loosely associated with the protein, but is essential for the maximum activity of the enzyme. For this group of enzymes, the metal may be removed from the protein during isolation and the metal-free protein may show some, but never maximal, activity. This greater dissociation suggests a lesser degree of specificity of the metal ion interaction although the protein moiety remains highly specific. The metals which have been found to form active metal-enzyme complexes include the six essential trace elements and the alkaline earth elements, magnesium, calcium, sodium and potassium. Some of these metal enzymes can be activated by several different metal ions which fulfill requirements for a particular ionic radius and stereochemistry. Thus, frequently there is some ambiguity as to which metal ion(s) function as activator(s) *in vivo*. The most effective activating metal ion(s) may not necessarily correspond to the ions actually controlling the enzyme in tissues. Some examples of important metal-ion activated enzymes are presented in Table IV.

The essentiality of metal ions is further reflected in a group of metallo-proteins for which a catalytic function has not been detected but which play an indispensible role in metabolism. The oxygen carrying proteins, hemoglobin, hemerythrin, myoglobin and hemocyanin and various metal transport or storage proteins such as transferrin, ferritin, and others are examples of this type of protein (Tables V, VI).

We now consider how metal ions may participate in the catalytic function of this large, specialized group of enzymes. Metal ions may function by playing an intimate role in the reaction mechanism or

TABLE IV

SOME EXAMPLES OF METALLOENZYMES AND METAL-ION ACTIVATED ENZYMES

[See Tables V & VI for Iron and Copper Enzymes]

Metal Ion	Enzyme	Biological Function
Zinc	Carbonic Anhydrase	CO_2 Formation
	Carboxypeptidase	Peptide hydrolysis
	Alcohol Dehyrogenase	Alcohol Metabolism
Cobalt (in corrinoid)	Methyl Malonyl-CoA Mutase	Fatty Acid Biosynthesis
	Glycerol Dehydrase	Glycerol Metabolism
	Ribonucleotide Reductase	Deoxynucleotide Biosynthesis
Moylboenum (with riboflavin)	Xanthine Oxidase	Purine Metabolism
	Nitrate Reductase	Nitrate Utilization
Manganese	Pyruvate Carboxylase	Carbohydrate Metabolism
Calcium	α-Amylase	Polysaccharide Hydrolysis
	Metal-Ion Activated Enzymes	
Zinc	Acylase	Hydrolysis
Magnesium	Hexokinase	Glucose Utilization
	Adenosine Triphosphatase	Phosphate Hydrolysis
	Carboxylase	Carbohydrate Metabolism
	DNAse	DNA breakdown

serving as a structural pre-requisite for the enzymes' activity. Many of the simplest enzymes, including certain metalloenzymes, consist of a single polypeptide chain which is highly folded and nearly spherical. Only a small fraction of these compact molecules are in the α-helical configuration. Their exterior surfaces have numerous hydrophilic or charged groups designed to promote solubility in water or dilute salt solutions. In their native conformation, many enzymes have a distinct cavity or cleft which is bounded by the locus of the protein involved in substrate binding and catalysis. The activity and specificity of the enzyme is determined to a large extent by the identity and sequence of the amino acids lining the active site.

For some of the metalloenzymes, the binding of the metal ion affects the electronic and stereochemical configuration of the active site. The most stable metal-protein complex will represent a compromise between the favored structural and bond features of the metal and the protein. The resulting metalloprotein may exist under an internal strain and therefore unusually reactive and show novel properties, achieving what Vallee and Williams (1968), have defined as an "entatic" state (a protein with a catalytic, active site). For metalloenzymes which contain copper, iron, or manganese, this activated condition may facilitate a change in the oxidation state of the metal ion. The best example is Cu(I) and Cu(II) which normally prefer tetrahedral and tetragonal geometries. Yet neither configuration may be attainable within the limitations of their binding site in the protein. This will produce distortion in the normal bonding of the metal ions and may tend to stabilize one oxidation state more than another. The usual effect on the copper proteins is to facilitate the reoxidation of the Cu(I) form to a blue Cu(II)-protein (Frieden, et. al. 1965).

Regardless of the structural function in oxidative enzymes, the specific role of some metal ions, usually iron or copper ion, is to serve as an intermediate electron carrier according to reaction sequences which is typical of the following:

$$AH_2 \searrow\!\!\swarrow Cu(II)_2\text{-Protein} \nwarrow\!\!\nearrow H_2O$$
$$A \swarrow\!\!\searrow Cu(I)_2\text{-Protein} \nearrow\!\!\searrow 1/2\ O_2$$

The oxidizable substrate reduces the metal ion-protein complex to a reduced metal-ion protein which is then reoxidized by molecular oxygen. In the absence of the electron carrier, the substrate AH_2 will not react directly with oxygen. A major advantage of this kind of intermediate step is that it permits the stepwise oxidation of energy rich biological substrates, enabling the diversion of the energy derived from oxidation to the formation of energy rich compounds (see also Figure 4).

In either the metalloenzymes or the metal-ion activated enzymes, the metal ion may serve to lock the geometry of the active site so that only certain substrates can be accommodated. This is shown in Figure 3 a and b in which the metal ion is responsible for maintaining the tertiary and quartenary structures of an enzyme protein. Thus the metal may serve or maintain the protein. Thus the metal may serve or maintain the proper relation between regions of the molecule on the same polypeptide chain (Figure 3a) or to keep several chains together as in Figure 3b. Removal of the metal ion causes denaturation or dissociation into subunits and the disintegration of the active site. An example of this type of metal ion is the zinc containing dehydrogenases in which the metal ion performs both a catalytic and a structural role.

Another illustration of primarily a structural effect occurs when the metal ion is bound at an allosteric site (Figure 3c), one that is not at the active center. Despite its distance from the active site, the metal ion can modify substrate binding or reactivity by inducing electronic displacements or conformational changes at the active site.

One of the earliest ideas for the function of a metal ion was the bridge theory proposed by Hellerman in 1937. The metal ion promotes catalysis by serving as a bridge from the substrate to the active site of the enzyme to form an active intermediate as diagrammed in Figure 3d. This is believed to explain the activation of the hydrolytic enzyme, arginase, by Mn(II) and other divalent metal ions. When a co-enzyme or prosthetic group is also present, the metal ion may bind both the substrate and coenzyme to the active site (Figure 3e,f). The coenzymes, pyridoxal phosphate, thiamine pyrophosphate, flavin adenine dinucleotide etc., frequently involve a metal ion which serves as a complexing agent for the substrate, or coenzyme, or more likely, the metal ion could bind the substrate to a coenzyme-protein complex (Figure 3f).

IRON AND COPPER

For reasons which will be outlined later, we conclude that iron and copper must be central to the development of our requirement for all metal ions. Indeed, it has proved to be convenient to identify major periods in the development of our contemporary civilization over the past 8,000 years as the copper and iron ages. Yet living cells have been under the domination of a combined iron and copper age which had its beginnings with the advent of aerobic cells about one billion years ago. Thus, iron and copper have become so firmly entrenched in modern biochemical thinking that it is difficult to imagine life in a cell without these two metals. It is quite certain that their essentiality to every form of life from plants to man arises from their role as prosthetic groups in a number of important enzymes and proteins. These two metals have become preeminent because of their

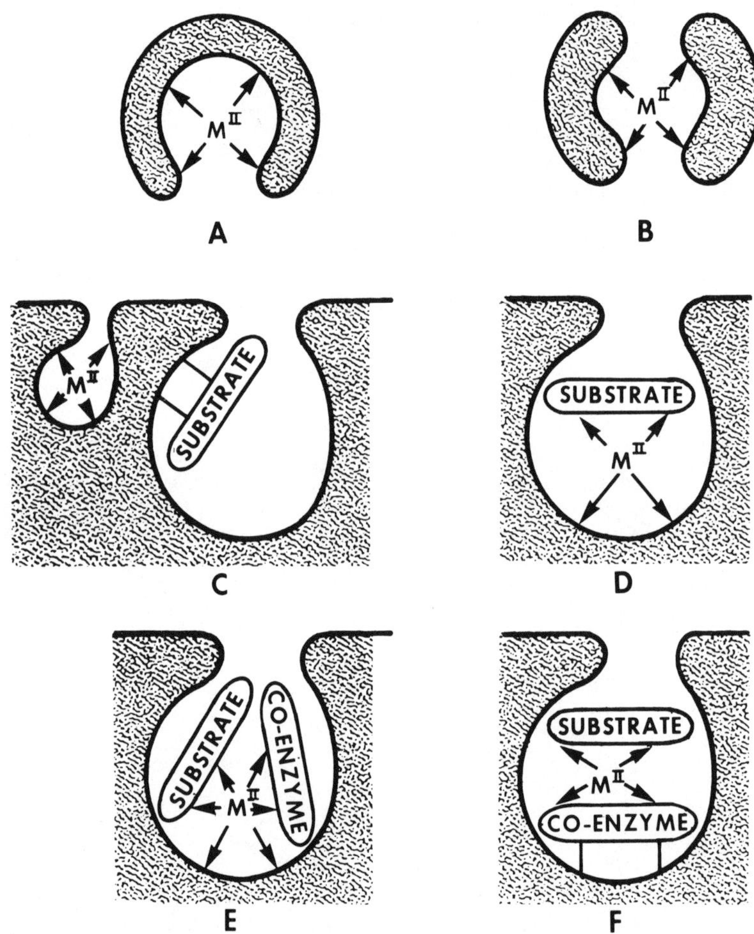

Figure 3. Possible functions of metal ions in enzyme systems. [Modified from Butler (1971).]

direct involvement in the adaptation to aerobicity and their role as terminal oxidases and respiratory proteins (Vallee and Wacker (1970)). If the biochemical evolution of these two metals is considered, a number of important relationships and correlations arise which have not been described previously [see later discussion related to Figure 4].

No eukaryotes and only a few prokaryotes, e.g. some species of lactic acid bacteria, are capable of survival without iron or copper. Iron, in particular, plays a major role in the biogeochemical history of the earth. Iron and nickel make up the core of the earth and since iron occurs most commonly in meteorites, it is probably universally distributed throughout the solar system. The whole earth is 35% iron and its crust is 5% iron. In the reducing atmosphere of the primitive earth, there was a need for the development of ligating and transporting mechanisms for the very insoluble ferric hydroxide compounds so that iron could play its vital role in electron transfer and oxidation to pave the way for the aerobic revolution that must have started about 10^9 years ago. J. B. Neilands (1972) of the Univ. of California has examined the possibility that the procaryotes with most primitive iron metabolism might well be the direct descendents of the primordial cell since only certain anaerobic and lactic acid bacteria do not contain heme compounds. He cites the ferrodoxins and rubredoxins, iron-sulfur compounds found in anaerobic bacteria, as representing the most primordial iron complexing system. For electron transport, oxygenation and hydroxylation, the evolution of iron compounds proceeded from sulfur to nitrogen ligands culminating in the heme proteins and highly specialized non-heme proteins we know today. For transport and storage, the iron ligands were selectively modified by stepwise transition to a mixed sulfur-nitrogen and sulfur-oxygen and, finally to an all oxygen type of coordination characteristic of ferritin, transferrin and the low molecular weight non-protein iron binding siderochromes.

A concommitant development of copper ligands may have been overwhelmed by the ubiquity of iron in the oceans and on the earth. The only copper-porphyrin equivalent to the heme group is found as a brilliant red pigment in the wings of the rare African bird, the touraco (Frieden, 1963). The fact that uniquely useful copper proteins developed at all may be attributed to the superior chelating properties of cupric ion. The heme biosynthetic machinery, however, was too entrenched to permit competition with iron for the protopor phyrin structure. The magnesium porphyrin derivatives in chlorophyll and the cobaltous porphyrin structures in the corrinoids (vitamin B_{12}) represent highly specialized adaptations for which iron may have been chemically unsuitable.

The importance of both iron and copper are reflected in the broad spectrum of useful metalloproteins (Tables V and VI) which have evolved to most effectively utilize the versatility of these two

TABLE V

MAJOR TYPES OF IRON ENZYMES AND PROTEINS

Heme Proteins	Non-Heme Proteins
Hemoglobin-O_2 Carrier	Hemerythrin-O_2 Carrier (Invert)
Myoglobin-Muscle O_2 Carrier	Transferrin-Fe Transport
Hydroperoxidases-Catalase, Peroxidases	Ferritin-Iron Storage
Cytochromes-Electron Transport	Ferrodoxin-Electron/Transport
Cytochrome Oxidase-Terminal Oxidase	Nitrogenase-N_2 Fixation
Tryptophan Oxygenase-Trp Oxidation	Succ. Dehydrog. Aconitase-Krebs Cycle
Xanthine Oxidase-Purine Metabolism	Ribonucleotide Reductase
	Lipoxidase

TABLE VI

SOME EXAMPLES OF COPPER-PROTEINS

Protein; Enzyme	Source; Function
Hemocyanin	O_2 carrier in invertebrates
Azurin	Electron carrier in bacteria
Plastocyanin	Electron carrier in plants
Tyrosinase	Phenol oxidation
Ceruloplasmin	Cu transport in serum; ferroxidase
Superoxide Dismutase	Protection against superoxide
Lysine Oxidase	Cross linking of collagen and elastin
Galactose Oxidase	Sugar metabolism; contains pyridoxal
Ascorbate Oxidase	Terminal Oxidase in some plants
Cytochrome Oxidase	Terminal Oxidase in most cells)

metal ions as electron and oxygen carriers and for oxygenation, hydroxylation and other crucial metabolic processes. The remaining discussion will concern some of the interrelations between these two key metals.

PROTECTION FROM SUPEROXIDE AND PEROXIDE

A most fundamental relation between copper and iron metallo-enzymes may be deduced from the recent observations of I. Fridovich

TABLE VII

ENZYMIC PROTECTION AGAINST SUPEROXIDE AND PEROXIDE

1. Superoxide Dismutase Systems - Cu-Zn or Mn-Zn

$$2O_2^- + 2H^+ \longrightarrow O_2 + H_2O_2$$

2. Catalase - Heme

$$2H_2O_2 \longrightarrow 2H_2O + O_2$$

3. Peroxidase Systems - Heme

$$H_2O_2 + AH_2 \longrightarrow 2H_2O + A$$

(1972), J. M. McCord and their coworkers at Duke Univ. They have found that a well known ubiquitous group of copper proteins, including erythrocuprein, hepatocuprein, etc., possess a unique enzymic activity, superoxide dismutase. The reaction catalyzed is the dismutation of peroxide ion to molecular oxygen and hydrogen peroxide (Table VI). This furnishes a convenient method of disposing of a highly reactive intermediate, the superoxide ion, O_2^-. The superoxide dimutase system is coupled with a group of heme (iron-proto-porphyrin) enzymes that decompose hydrogen peroxide either to oxygen and water as in the catalase systems, or with the aid of some hydrogen donor molecule (AH_2) to water and a dehydrogenated product (A) (Table VII). These two types of enzymes provide a disposal route for the two principal toxic by-products of oxygen reduction, superoxide ion and peroxide ion. There is no dearth of biochemical reactions which produce hydrogen peroxide. The reactions which produce superoxide ion have been much more difficult to identify, but the expanding list includes enzymic reactions such as milk xanthine oxidase, rabbit liver aldehyde oxidase, rabbit liver dihydroorotate dehydrogenase, pig liver diamine oxidase, the autoxidation of hemoglobin and myoglobin and probably many other one-electron transfers. It turns out that this protection is a pre-requisite for the adaptation of living cells to the utilization of oxygen. It represents an indispensible link between the two groups of metalloproteins, a copper enzyme with superoxide dismutase activity and a heme enzyme with catalase or peroxidase activity. It should be added that all the superoxide dismutases contain zinc which appears to be important in structure but not in function. A few bacterial and mitochondrial mutases contain manganese instead of copper and lack zinc (Fridovich, 1972). Fridovich has recently suggested that the cupro-zinc superoxide dismutases and the Mn-superoxide dismutases evolved independently in protoeukaryotes and prokaryotes respectively. (This conforms to proposals that mitochondria evolved out of an endosymbiotic relationship between these proteukaryotes and anaerobic prokaryotes).

The presence of these enzymes and the development of life in oxygen has been correlated by McCord et. al. (1971), Fridovich and coworkers (see review of Fridovich, 1972). The earth's atmosphere has evolved from a highly reducing environment to its current oxygen-rich status. It was first recognized by Pasteur that certain bacteria grow best in oxygen-free atmosphere and will not survive in oxygen. It has been found that these strict anaerobes show no superoxide dismutase and, generally, no catalase activity. Virtually all aerobic organisms, particularly those containing cytochrome systems, were found to contain both types of enzymes, superoxide dismutase and catalase and/or peroxidase. The evolution of this team of copper and iron enzymes appears to have been a pre-requisite for the development of life in oxygen as we find it today. There remains a special group of aerotolerant anerobes which survive exposure to air and metabolize oxygen to a limited extent but do not contain cytochrome systems. These microorganisms possess superoxide dimutase activity but no catalase activity.

OXYGEN CARRYING PROTEINS

Blood has always been associated with animal life and, if for no other reason, iron and copper metalloproteins would be forever linked in nature by their occurrence in blood as the oxygen carrying chromo-proteins. The color of these pigments, red for the iron proteins, attracted early attention and they are the oldest recognized metalloproteins. The most unbiquitous respiratory pigments are of the hemoglobin type, a protein attached to a heme group which consists of a ferrous-protoporphyrin complex. Hemoglobin is found in plants, in many invertebrates, and in all vertebrates. Myoglobin is a smaller oxygen carrying hemo-protein which exists exclusively in muscle. A limited class of invertebrates, the sipunculid worms, contain hemerythrin, a non-heme iron containing protein. A larger group of invertebrates, in the four classes Cephalaopoda, Crustacea, Gastropoda and Xiphosura have as their oxygen carrying pigment a beautiful blue copper protein known as hemocyanin. Thus the ability to serve as respiratory proteins in nature appears to be the exclusive property of the copper and iron proteins (Prosser and Brown, 1961).

Why has no other essential metal ion or other type of respiratory protein developed to satisfy this important function, principally for the animal kingdom? For a possible answer to this question, we should examine what a protein needs to succeed in this activity. It needs to be able to form a stable dissociable complex with the highly reactive molecule, O_2. Transition metals excel in this capacity; few other chemical groups can do this. In fact, all efforts to devise other physiologically compatible model oxygen carriers have failed to date. The compounds that come closest to emulating oxygen binding properties of hemoglobin, hemerythrin, and hemocyanin contain cobalt and other transition metal ion derivatives. This limits our consideration to metals such as Co, Mn, Mo, and Zn. None of these metal ions match Cu(II) in their ability to form stable chelates with amino acids, peptides or proteins. The well known Irving-Williams series on the relative ability of divalent metal ions of the first transition series to interact with nitrogen donor ligands is Cu(II) > Zn(II) > Ni(II) > Co(II) > Fe(II) > Mn(II). In sea water cobalt is less than one-tenth as concentrated as copper and is therefore much less likely to have been utilized in the primitive oceans. Manganous ions are less stable against further oxidation and are much poorer chelators. Though plentiful, zinc does not have some of the key chemical characteristics of the other transition metals, e.g. zinc ion has a complete set of 3*d* electrons.

ADAPTATION OF IRON AND COPPER UTILIZING SYSTEMS

But probably the most compelling reason favoring iron and copper mechanisms is related to our previous discussion about the development of copper and iron enzymes as protectants against toxic oxygen by-products. It seems a certainty that the development of superoxide dismutase and catalase-peroxidase systems for the survival of aerobes preceeded the later evolution of the respiratory pigments and enzymes as depicted in Figure 4. Therefore these cells already had the

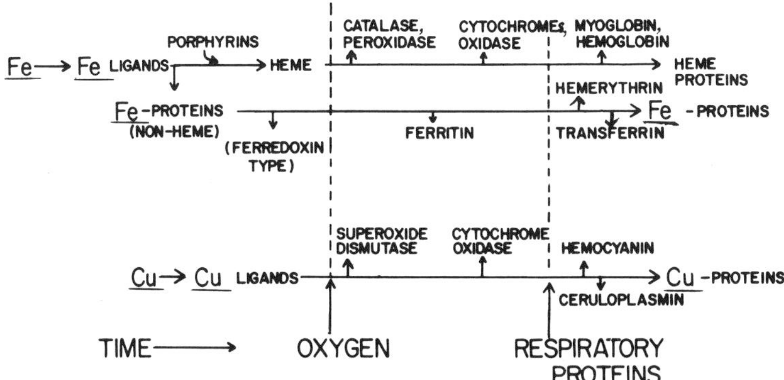

Figure 4. Diagram depicting the evolutionary sequence and the development of the hemeproteins, the non-heme iron proteins and the copper proteins. The horizontal axis is essentially a time line with the advent of oxygen era and the respiratory proteins indicated.

necessary machinery to utilize copper and iron ions and to convert iron into heme for the biosynthesis of whatever metalloprotein was useful. For iron, this ultimately led to the development of specialized storage and transport proteins, the ferritins, the transferrins, and the mucosal iron binding proteins. For copper, there was a corresponding development of the serum copper transport and iron mobilizing protein, ceruloplasmin (ferroxidase) and copper binding proteins in the liver and other tissues. It was obviously most convenient to use these two metal ions for any new task for which they were capable, including the formation of the oxygen carrying proteins-hemoglobin, hemocyanin and hemerythrin. It should be emphasized that we do not impose any common genetic origin on the protein moieties of these metalloproteins, although this is not excluded. The genetic history of the two pairs of polypeptide chains which comprise the vertebrate hemoglobins has been described by Ingram (1961).

CUPRO-FERRO PROTEINS

If a biochemist is asked to identify the _one_ enzyme which is most vital to all forms of life, he would probably name cyrochrome C oxidase. This is the enzyme, found in all aerobic cells, which introduces oxygen into the oxidative machinery that produces the energy which we need for physical activity and biochemical synthesis. When cytochrome oxidase action is blocked as in cyanide poisoning, all cellular activity grinds to a quick halt, indeed in only several minutes. This enzyme may be regarded as the ultimate in the integration of the function of iron and copper in biological systems. Here in a single molecule, we combine the talents of iron and copper ions to bind oxygen, reduce it with electrons from other cytochromes in the hydrogen electron transport chain and, finally, to convert the reduced oxygen to water. [The cytochrome portion of the electron transport system is shown schematically in Figure 5]. Yet it is a frustrating fact that this key enzyme has proved to be extremely difficult to isolate, purify and otherwise successfully study. We do know that each molecule weighs about 140,000 daltons and has two copper ions and two heme groups which are required for its function (Chance et al., 1971).

A chemically similar enzyme, tryptophan-2,3-dioxygenase, catalyzes the insertion of molecular oxygen into the pyrrole ring of tryptophan yielding N-formylkynurenine. It also has a widespread distribution and has been purified to homogeneity from rat liver and the microorganism, _Pseudomonas acidovorans_. In a molecule weighing 167,000 daltons, it has been found to have two atoms of Cu(I) and two molecules of heme (Brady et.al., 1968).

Finally, a recently discovered enzyme (Cavallini et.al., 1968) cysteamine oxygenase, has been reported to contain one atom each of iron, copper and zinc ions in a molecule weighing 100,000 daltons.

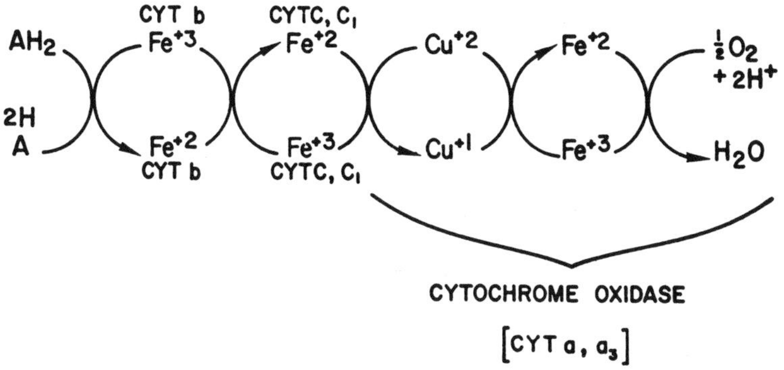

Figure 5. Diagram illustrating a portion of the cytochrome electron transport system of mitochondria. The arrows show the direction of the flow of electrons. The sequence of Cu → Fe → O_2 in cytochrome oxidase is probable but not certain.

PLANT ENZYMES

Plants have not escaped the influence of the copper-iron relationship. Many plants have a small but significant oxidative respiration which utilizes the cytochrome c-cytochrome oxidase system. This can be regarded as a vestige from the same development as the primitive aerobic cell. However, iron and copper electron carriers are also involved in the photo-induced electron flow in photosynthetic tissues (Bishop, 1971). It is now believed that photosynthesis in green plants proceeds by the cooperation of two photochemical systems, each catalyzed by light absorbed through a specific pigment system. These two photosystems, I and II (Figure 6) are interrelated to their own cooperative electron transport system which directs the sequential transfer of electrons derived from water to a reduction of $NADP^+$. While all the components of this system may not be identified, at least three cytochromes (heme) and one ferridoxin (non-heme iron) are involved. Plastocyanin, a small copper protein is localized in the reducing side of cytochrome f. The initial transfer of electrons from water requires manganous ions, and magnesium is an essential component of the chlorophyll molecule located in both pigment matrices.

BIOSYNTHESIS OF CONNECTIVE TISSUE PROTEINS

Somewhere in the development of complex organisms from their single cell precursors, it became advantageous for cells to aggregate to form tissues and, ultimately, organs. The development of extra-

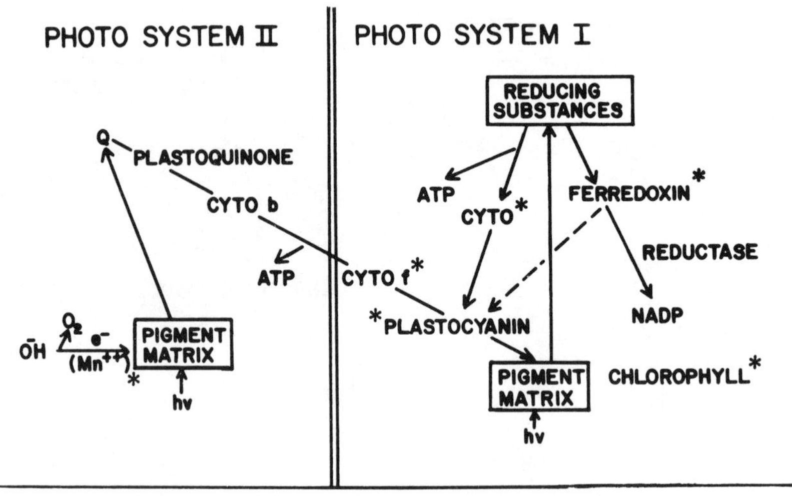

Figure 6. A generalized scheme for the photoinduced electron flow in photosynthetic tissue. Direction of electron flow is indicated by the arrows. The metal ion systems are indicated by asterisks. [Adapted from N. Bishop, Ann. Rev. Biol. 1971, p. 199.]

cellular attachments was greatly enhanced by the presence of hardy macromolecules of the connective tissue of the type now identified as collagen and elastin. Recent work on the biosynthesis of these connective tissue components has revealed a direct role for iron and copper enzymes and a probable role for manganous and zinc ions, as well.

Collagen is synthesized by a series of sequential steps involving assembly of ribosomes of a proline- and lysine-rich polypeptide precursor called "protocollagen". The hydroxylation of appropriate proline and lysine residues in proto-collagen to hydroxyproline and hydroxylysine occur before the collagen is extruded into the extracellular matrix. These hydroxylation steps require iron as Fe(II) with a maximum effect at 10 micromolar; no other metal ion will replace iron (Procker, 1971). The synthesis of complete collagen can be blocked by an iron chelator, 1 millimolar α,α'-dipyridyl. The hydroxyamino acids confer additional stability to the collagen molecule by increased prospects for intermolecular hydrogen bonding and for cross linking involving sugar groups.

Since copper deficiency has been shown to lead to aneurysms and soft bones in experimental animals, a role of copper in promoting the tensile strength of the fibrous proteins, collagen and elastin, has long been suspected. It has now been established that a copper enzyme, lysine oxidase, is the determining factor in providing chemical cross linking of polypeptide chains to stabilize these structures (Chou et. al., 1969). The most important cross-linkage occurs between four lysyl residues which form a pyridinium ring, joining neighboring peptide chains. These cross-links were sufficiently stable to be isolated after the hydrolysis of collagen yielding the cyclic isomers, desmosine (Figure 7) and isodesmosine. Without lysine oxidase and its induced crosslinkages, soluble collagen and elastin molecules with less tensile strength are formed.

The skeletal and postural defects in manganese deficiency in numerous mammals have been related to a reduction in tissue mucopolysaccharide content. These observations have been accounted for by the recognition of the requirement for Mn(II) by many glycosyltransferase enzymes, including those needed for chondroitin sulfate synthesis (Leach, 1971). The two critical manganese dependent enzymes recognized so far are: (1) a polymerase system which is responsible for polysaccharide chain elongation and (2) a galactosyl transferase system whose products are involved in attaching oligosaccharides to the protein portion of the glycoprotein.

The specific role of zinc in connective tissue metabolism is much less certain. Zinc has a beneficial effect on wound healing which probably involves connective tissue metabolism. In zinc deficiency, many of the active bone cells that are remote from the circulation seem to differentiate abnormally (Westmoreland, 1971).

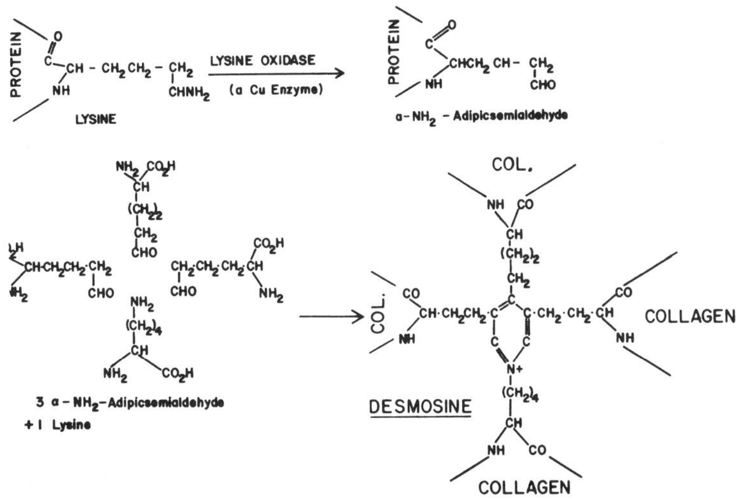

Figure 7. Formation of desmosine crosslinks in collagen initiated by the copper enzyme, lysine oxidase.

CERULOPLASMIN, IRON CYCLES AND IRON METABOLISM

In this last section we will consider the connection between copper compounds, particularly the serum copper protein ceruloplasmin, and iron metabolism. This subject is discussed further in a latter chapter in this volume on the role of ferroxidase and ferrireductase in iron metabolism Frieden and Osaki (1974) (this volume). The essentiality of copper in the mammal was first recognized in 1928 by Hart, Steenbock and Elvehjem, Univ. of Wisconsin, who reported that the copper deficient experimental animal became seriously anemic. This anemia has now been traced to a low plasma copper and a reduced level of the catalytic serum copper protein, ceruloplasmin, now known to be a molecular link between copper and iron metabolism. Ceruloplasmin appears to control the rate of iron uptake by transferrin, the iron transport protein in serum which delivers iron to tissues which need it, for example, the maturing erythrocyte for the biosynthesis of hemoglobin. In our laboratory at Florida State Univ. we have been able to relate this iron mobilizing property to the ferroxidase activity of ceruloplasmin, it's ability to catalyze the oxidation of ferrous to ferric iron so that transferrin can bind the ferric ion and transport it to iron receptor sites wherever they may be (see review by Frieden, 1971). The importance of these interconversions of ferrous to ferric ion in iron metabolism can be appreciated best by a consideration of the important ferrous to ferric cycles which seem to be involved in iron metabolism (Figure 8) (Frieden, 1973). We have already emphasized this in regards to their function in the vital hemoproteins, such as the cytochromes or the non-heme iron enzymes.

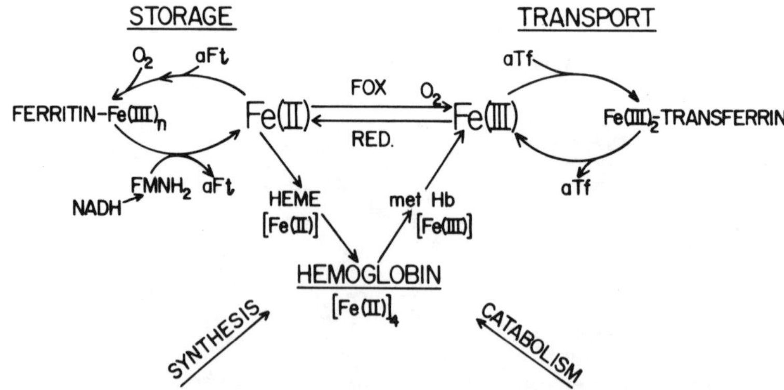

Figure 8. The ferrous to ferric cycles in iron metabolism and their importance in the storage, transport, biosynthesis, and catabolism of iron. A key influence is exerted by the ferroxidases (Fox I, Fox II), copper enzymes which catalyze the oxidation of Fe(II) to Fe(III) and the ferrireductases. [Diagram modified from Frieden (1973).]

One of the focal points of iron metabolism is the ferroxidase reaction in which the oxidation of ferrous ion to ferric ion is catalyzed by ceruloplasmin (ferroxidase) and other ferroxidases. When this occurs in the serum, the ferric ion is quickly trapped as the iron-free form of transferrin, apotransferrin. On the basis of the calculations, it was proposed that the spontaneous oxidation of ferrous to ferric ion was inadequate to account for the need for iron in the form of ferri-transferrin, thus requiring the ferroxidase activity supplied by ceruloplasmin (Osaki et al., 1971). The rate of iron metabolization reached a maximum at 0.2 μM, less than one-tenth the normal level of plasma ceruloplasmin, 2 μM. Confirmation _in vivo_ has been provided by a research group at the University of Utah (Ragan et al., 1969; Roeser et al., 1970) which showed that ceruloplasmin, but not Cu(II), was responsible for the immediate mobilization of serum iron in the copper deficient pigs and rats.

Serum iron is principally derived from iron stored as ferritin in the reticulo-endothelial cells of the liver. Ferritin consists of a polyhydroxy iron polymer with a molecular weight of about 150,000 surrounded by a protein coat of molecular weight 465,000. Its formation is now known to involve a Fe(II)-apoferritin complex followed by an enhanced oxidation of Fe(II) to Ferri-ferritin (Macara et al., 1972; Bryce and Crichton, 1973). The release of iron from ferritin also has been a subject of continuing interest. Recently we explored the cytoplasm of bovine liver cells for a reductive ferritin-iron release mechanism and found a microsomal fraction of vertebrate liver which contained a ferritin reductase system which releases iron from ferritin as ferrous ion (Osaki et al., 1974). This system requires the vitamin derivatives nicotinamide dinucleotide and flavin mononucleotide (Sirivech et al., 1974).

Since iron as ferric ion exhibits limited solubility in aqueous media and tends to polymerize at physiological pH's, it is maintained in a highly mobile state by its interaction with apotransferrin to form ferri-transferrin, constituting the transport form of iron. Ferri-transferrin then reacts directly with reticulocytes to provide the iron necessary for hemoglobin biosynthesis and along with ferritin provides iron to other tissues for all iron compounds. The iron is probably released as ferrous ion or undergoes an appropriate reduction in preparation for its incorporation into hemoglobin in the developing reticulocyte.

The release of transferrin or ferritin iron to or inside the reticulocytes provides Fe(II) for the next significant step in the biosynthesis of the iron containing proteins, of which hemoglobin is by far the most important, comprising almost two-thirds of the body iron. Presumably the production of other iron proteins including myoglobin, the cytochromes etc., would also use similar sources of iron. For the biosynthesis of hemoglobin and myoglobin, iron must be inserted into a protoporphyrin nucleus exclusively as Fe(II) to form protoheme IX, the heme component of hemoglobin and myoglobin. This is accomplished with the aid of ferrochelatase, an enzyme found in soluble fractions of vertebrate liver and erythrocytes. Studies of the ferrochelatase from rat liver reveal a high selectivity for its metal ion requirement - only Fe(II), Co(II) and traces of Mn(II) are incorporated into porphyrin; all other metal ions were inert or inhibitory (Labbe and Hubbard, 1961).

For an active protein, the hemoglobin molecule is extremely stable and probably survives the lifetime of the red cell. Eventually the enzymic machinery, which maintains the integrity of the erythrocyte, ages and cell dies. The iron in hemoglobin becomes more susceptible to oxidation, resulting in the formation of methemoglobin which cannot carry oxygen. Surrounded by the dying cell, the methemoglobin molecule is thus programmed for destruction after capture by the spleen, the globin portion to its constituent amino acids, the protoporphyrin to the bile pigments and the iron for reuse in the biosynthesis of new iron-containing proteins. The oxidation state of the emergent iron is not known, but in the presence of the numerous reducing agents of the red cell and the plasma, e.g. reduced glutathione, ascorbic acid, it is likely that much of the iron emerges as Fe(II), requiring the action of ferroxidase for oxidation, prior to transport as ferri-transferrin.

Thus iron utilizing systems have evolved in which iron must go through the ferrous state to be mobilized or to be integrated into the most prevalent iron compound, hemoglobin. But for storage as ferritin or for transport as transferrin, iron ultimately must be converted to the ferric state. At virtually every known step of iron metabolism including storage, transport, biosynthesis and degradation, the ferrous to ferric cycles play a major role in the multi-metabolic pathways of the most important metal constituent in all living systems. By virtue of its ferroxidase activity the serum copper enzymes, ceruloplasmin, exerts its influence on the metabolism of iron by facilitating the ferrous to ferric cycle. Ceruloplasmin and iron mobilization represent one of the most recent examples of the evolution of copper-iron relationships in the vertebrates.

CONCLUDING SUMMARY

Life on earth has developed a selective dependence on 25 of the naturally occurring chemical elements. This has involved two major adaptations--first to the origin of life in the sea and, second, to the utilization of oxygen. The essential elements fall into three groups: (1) Six elements (H,O,C,N,S,P) which comprise the primary building blocks for all cells, (2) five elements (K,Na,Mg,Ca,Cl) which satisfy the ionic requirements for cells, and (3) the fourteen remaining elements which are required only in trace amounts for a variety of improtant metabolic activities. Ten of these trace elements are essential metals, but only six (Fe,Cu,Zn,Mn,Mo, and Co) are known to be required by all types of organisms. These six transition metals serve as essential components of a wide variety of metalloproteins, usually enzymes, which participate in important hydrolytic, oxidative or transfer biological processes. Thus, almost one-third of the enzymes are metalloenzymes or metal-ion-activated enzymes. The former contain a metal ion as integral part of the molecule. In the latter, the metal ion is more loosely associated with the protein, but essential for maximum catalytic activity. The metal ions may function by playing an intimate role in the catalytic reaction mechanism or by serving as a structural pre-requisite for the enzymes' activity. There are also a group of metalloproteins without any known catalytic activity but which serve, indispensibly, as oxygen carrying proteins, or as metal transport or storage proteins.

Central to the development of all aerobic life are the metals iron and copper, which have become closely associated over a billion years of biochemical evolution, as depticted in Figure 4. These two elements became preeminent among the trace metals because of their terminal oxidases, respiratory proteins, and for other more specialized metabolic reactions. One of their primary roles has involved the protection of the aerobic cell from the highly toxic oxygen byproducts, superoxide ion and hydrogen peroxide. This took the form of the copper enzyme, superoxide dismutase, and the heme enzymes, catalase and peroxidase. The success of the aerobes was accompanied by the development of more sophisticated iron and copper enzymes including the cytochromes, the cupro-ferri (heme) enzyme, cytochrome oxidase, and electron transfer reactions in plants. With the increasing complexity of organisms, the cellular machinery utilizing iron and copper was adapted for the production of the oxygen carrying proteins, hemoglobins, hemerythrins and hemocyanins. These later stages of evolution were accompanied by a parallel development in biosynthetic enzymes associated with connective tissue and other more specific processes. A recent example of the continuing association of iron and copper in vertebrates is the role of the serum copper protein, ceruloplasmin, and its ferroxidase activity, in the mobilization of iron in preparation for the biosynthesis of hemoglobin.

REFERENCES

Brady, F., Monaco, M. E., Forman, H. J., Schutz, G. and Feigelson, P. (1972). J. Biol. Chem., 247,7915.

Bryce, C. F. A. and Crichton, R. R. (1973). Biochem. J., 133, 301.

Butler, E., (1971). Physiology and Biochemistry of the Domestic Fowl, vol. 2 (ed. by D. J. Bell and B. M. Freeman) Acad. Press N. Y.

Calvin, M. (1969). Chemical Evolution, Oxford Univ. Press, N. Y.

Carlisle, E. (1972). Science, 178, 619.

Cavallini, D., Supre, S., Scandurra, R., Graziani, M. T. and Cotta-Rasmusino, F. (1968). Europ. J. Biochem., 4, 209.

Chance, B., Yonetani, T. and Mildvan, A. S. (Eds.) (1971). Probes of structure and function of macromolecules and membranes, 2, Academic Press, N. Y., p. 575.
Chou, W. S., Savage, J. E. and O'Dell, B. L. (1969). J. Biol. Chem., 244, 5785.
Frieden, E. and Kent, A. B. (1974). in press.
Frieden, E. (1973). Nutr. Reviews, 31, 41.
Frieden, E. (1972). Sci. Amer., 227, 52.
Frieden, E. (1971). Adv. in Chem. Series, Bioinorg. Chem., Am. Chem. Soc.
Frieden, E. (1963). in Horizons in Biochemistry (Ed. by M. Kasha and B. Pullman).
Frieden, E. and Kent, A. B. (1974). Int. Embyrological Congress, Sorrento, Italy.
Frieden, E. and Osaki, S. (1974). This volume.
Frieden, E., Osaki, S. And Kobayashi, H. (1965). J. Gen. Physiol., 49, 213.
Gaffron, H. (1960). Persepctives Biol. Med. III, 163.
Goldberg, E. D. (1963). in The Sea (ed. by M. N. Hill), p. 4-5, J. Wiley, N. Y.
Hart. E. B., Steen bock, H. B., Waddell, J. and Elvehjem, C. A., (1928). J. Biol. Chem.
Hellerman, L. (1937). Physiol. Rev., 17, 454.
Henderson, L. (1913). The Fitness of the Environment, N. Y.
Ingram, V. M. (1961). Nature, 189, 704.
von Jenson, J., Thofern, E. (1954). Zeit. F. Naturforsch, 9b, 596.
Labbe, R. F., and Hubbard, N., (1961). Biochim. Biophys. Acta. 52, 130.
Leach, R. M., Jr. (1971). Fed. Proc., 30, 991.
Macara, T. G., Hoy, T. G. and Harrison, P. M. (1972). Biochem. J., 126, 151.
McCord, J. M., Keele, B. B. and Fridovich, I. (1971). Proc. Nat. Acad. Sci. U. S. 68, 1024.
Neilands, J. B. (1972). in Structure and Bonding, 11, Springer-Berlag, N. Y., p. 145.
Osaki, S., Johnson, D. and Frieden, E. (1971). J. Biol Chem., 246, 2746.
Osaki, S., Johnson, D. and Frieden, E. (1966). J. Biol. Chem., 241, 2746.
Osaki, S., Sirivech, S. and Frieden, E. (1974). in press.
Prosser, C. L. and Brown, F. A. (1961). in: Comp. Animal Physiol., W. B. Saunders, Philadelphia, Pa., p. 198.
Prockop, D. J. (1971). Fed. Proc., 30, 984.
Ragan, H. A., Nacht, S., Lee, G. R., Bishop, C. R. and Cartwright, G. E. (1969). Am. J. Physiol., 217, 1320.
Roeser, H. P., Lee, G. R., Nacht, S. and Cartwright, G. E. (1970). J. Clin. Invest., 49, 2408.
Schwarz, K. (1974). TEMA-II Symposium, in press.
Sirivech, S., Frieden, E. and Osaki, S. (1974). In press.

Vallee, B. and Wacker, W. E. C. (1970). Metalloproteins in "The Proteins," V, (Ed. H. Neurath), Academic Press, N. Y.

Vallee, B. and Williams, R. J. P. (1968). Proc. Nat. Acad. Sci. U.S. 59, 498.

Westmoreland, N. (1971). Fed. Proc., 30, 1001.

2

THE FUNCTIONAL ROLES OF METALS IN METALLOENZYMES

J. F. Riordan and B. L. Vallee

Biophysics Research Laboratory

Department of Biological Chemistry

Harvard Medical School, Boston, Ma. 02115

There is now widespread appreciation for the fact that overall and local changes in protein conformation are critical to the mechanism of enzymatic catalysis. Yet, it has been difficult to document the occurrence and judge the significance of such changes, since methods for the analysis of protein structure often require conditions which cannot be compared easily and directly with those designed for determining function. This is true even while the number,resolving power and sophistication of suitable approaches has increased rapidly.

The importance of such structural changes to the thermodynamics of the enzyme-substrate complex is universally recognized, but the role, if any, of local conformational changes in facilitating catalysis is unresolved. Nor is it known how much of the enzyme's chemical reactivity, essential for the process, exists *prior to* the formation of the enzyme-substrate complex.

Efforts to delineate structural changes accompanying enzyme action have prompted a search for new approaches designed to meet the requirements of simultaneous high speed and resolving power which would discern individual steps critical to catalysis. In the following we will briefly review some of the structural and functional characteristics of metalloenzymes and indicate a number of approaches which we have employed to deal with these problems.

Functional and Structural Effects of Metals in Monomeric Metalloenzymes. Distinctive catalytic roles for metals have been discerned largely by inspection of enzymes composed of a single polypeptide chain while structural roles have been recognized primarily through studies of those having quaternary structure. Thus, the metal atom is known to play a distinct catalytic role in the monomeric carboxypeptidases A (Vallee and Riordan, 1968) and B (Wintersberger _et al_., 1965) or the carbonic anhydrases (Keilin and Mann, 1940). Much less is known about the effects of metals on stabilization of tertiary structure (Rosenberg, 1960). A zinc atom serves in catalysis of certain single chain bacterial enzymes, e.g., the neutral protease from _B. subtilis_ or thermolysin from _B. thermoproteolyticus_, but, in addition, calcium atoms apparently stabilize their secondary and/or tertiary structure (Latt _et al_., 1969; Matsubara and Feder, 1970).

Metal atoms at the active site of enzymes can be removed and restored reversibly, and different metals can be substituted for those found in the native state. The resultant spectrochemical and functional consequences can be examined without major interference from structural alterations which might complicate interpretation of the data. Similar lines of investigation have been extended successfully to multichain enzymes. However, in these instances owing to the existence of subunit structure there are problems including variations in metal content, oxidation of thiol groups, instability of proteins, difficulty of removal and reconstitution of metals and the existence of isoenzymes. Yet, investigations of multichain metalloenzymes have extended an appreciation of the biological function of metals, as well as their roles in protein structure (Ulmer, 1970).

Functional and Structural Effects of Metal Atoms in Polymeric Metalloenzymes. Mediation of the association and dissociation of monomers and polymers by metal ions are usually monitored by means of concurrent metal analyses and ultracentrifugation, either in the presence of or after treatment with chelating, denaturing, or chemical modifying agents or after changing pH, temperature or protein concentration.

In this fashion metals have been shown to affect the degree of association of a number of metalloproteins (Table I). In instances where the state of polymerization correlates directly with catalysis, the metal can assume an additional, distinctive role. However, in other metal ion dependent polymerizing systems, biological consequences of subunit interactions have not yet been discerned, and the significance of the metal remains to be established (Ulmer and Vallee, 1971).

Different metal ions can serve different roles in the same enzyme. Thus, in _B. subtilis_ amylase two calcium atoms are essential for enzymatic activity but a zinc atom serves to dimerize the protein though it does not affect function (Vallee _et al_., 1959). Zinc is present only in the regulatory but not the catalytic sub-

TABLE I
REPRESENTATIVE EXAMPLES OF METALLOENZYMES

Enzyme	Metal	Cofactor	M.W.	Stoichiometry metal and coenzyme	Source
			CALCIUM		
Amylase	Ca Zn	-- None	-- 50,000	1 Ca 0.5 Zn	*Bacillus subtilis*
Pseudomonas protease	Ca	None	48,000	1-2 Ca	*Pseudomonas aeruginosa*
			COPPER		
Plastocyanin	Cu	None	21,000	2 Cu	Spinach, *Chlorella Chenopodium*
Azurin	Cu	None	14,600	1 Cu	*Bordetella pertussis*
Rhus blue protein (stellacyanin)	Cu	None	16,800	1 Cu	*Rhus vernicifera*
Monoamine oxidase	Cu	PLP	170,000	1 Cu	Bovine plasma
D-Galactose oxidase	Cu	None	75,000	1 Cu	*Dactylium dendroides*
Uricase	Cu	None	12,000	1 Cu	Mammalian liver

TABLE I continued...

Tyrosinase (polyphenyloxidase)	Cu	None	119,000	4 Cu	Mushroom
Dopamine-β-hydroxylase	Cu	Ascorbic acid	290,000	2 Cu	Bovine adrenal medulla
Laccase	Cu	None	120,000 141,000	6 Cu	*Rhus vernicifera*
Ascorbic acid oxidase	Cu	None	140,000	8 Cu	Squash, cucumber
Cytochrome oxidase	Cu	Heme	Unknown	1 Cu/heme	Bovine heart
			IRON		
Ferredoxin (clostridial)	Fe	None	6,000	7 Fe	*Clostridium pasteurianum acidiurici*
Spinach ferredoxin	Fe	None	11,600	2 Fe	Spinach leaves
Adrenodoxin (adrenal ferredoxin)	Fe	None	13,000	2 Fe	Hog adrenal mitochondria
Putida redoxin	Fe	None	12,000	2 Fe	*Pseudomonas putida*
Rubredoxin	Fe	None	5,900	1 Fe	*Clostridium pasteurianum*

TABLE I continued...

Rubredoxin	Fe	None	---	1 Fe	*Micrococcus aerogenese*
Metapyrocatechase	Fe	None	140,000	3 Fe	*Pseudomonas arvilla*
Pyrocatechase	Fe	None	95,000	2 Fe	*Pseudomonas arvilla*
Protocatechuate 3,4-dioxygenase	Fe	None	700,000	8 Fe	*Pseudomonas aeruginosa*
Agavain	Fe	None	52,000	1 Fe	Sisal extract
NADH-dehydrogenase	Fe	Flavine nucleotide	80,000	4 Fe	Hog heart
Succinate dehydrogenase	Fe	Flavine nucleotide	200,000	4 Fe	Bovine heart
Succinate dehydrogenase	Fe	Flavine nucleotide	Unknown	8 Fe/ flavine	Hog heart
Succinate dehydrogenase	Fe	Flavine nucleotide	200,000	4 Fe/ 1 flavine	Yeast
Dihydroorotate dehydrogenase	Fe	FAD, FMN	115,000	4 Fe 2 FAD	*Zymobacterium oroticum*
Aldehyde oxidase	Fe Mo	FAD	300,000	8 Fe, 2 Mo, 2 FAD	Hog liver

TABLE I continued...

Xanthine oxidase	Fe Mo	FAD	300,000	8 Fe, 2 Mo, 2 FAD	Bovine Milk
			MANGANESE		
Pyruvate carboxylase	Mn	Biotin, ATP, acetyl COA, Mg^{2+}	655,000	4 Mn 4 biotin	Chicken liver
			MOLYBDENUM		
Nitrate reductase	Mo	FAD	Unknown	Unknown	_Neurospora crassa_
			ZINC		
Carbonic anhydrase	Zn	None	30,000	1 Zn	Bovine erythrocytes
Carboxypeptidase A	Zn	None	34,300	1 Zn	Bovine pancreas
Carboxypeptidase B	Zn	None	34,300	1 Zn	Porcine & bovine pancreas
Neutral protease	Zn	None	44,700	1 Zn, 3 Ca	_B. subtilis_
Neutral protease	Zn	None	40,000	1 Zn	_B. megaterium_
Alcohol dehydrogenase	Zn	NAD	150,000	4 Zn	Yeast

TABLE I continued...

Alcohol dehydrogenase	Zn	NAD	80,000	4 Zn, 2 NAD	Equine liver
D-lactic cytochrome reductase	Zn	FAD	50,000	4-6 Zn	Yeast
Alkaline phosphatase	Zn	None	89,000	4 Zn	*E. coli*
Aldolase	Zn	None	65,000 75,000	1 Zn	Yeast
Dipeptidase	Zn	None	47,200	1 Zn	Hog kidney
Leucine aminopeptidase	Zn	None	300,000	4-6 Zn	Porcine kidney
Leucine aminopeptidase	Zn	None	320,000	12 Zn	Bovine lens
Thermolysin	Zn	None	37,500	1 Zn, 4 Ca	*B. thermoproteolyticus*
Leucostoma peptidase A	Zn	None	22,500	1 Zn, 2 Ca	Cottonmouth moccasin
Neutral protease	Zn	None	63,000	1 Zn, Ca	*B. cereus*
Aspartate transcarbamylase	Zn	None	310,000	6 Zn	*E. coli*

units of aspartate transcarbamylase and it does not appear to affect function directly (Rosenbusch and Weber, 1971; Nelback et al., 1972). The role of metal ions in regulatory processes is only now beginning to be explored, but will likely emerge to be an important means by which metals can exert biological effects. A few examples will further document the role of metals in subunit interactions.

Yeast alcohol dehydrogenase (YADH), a tetramer of molecular weight 150,000, is inhibited instantaneously by 1,10-phenanthroline through the formation of a protein-metal-chelate mixed complex. Inhibition is reversed on dilution. On prolonged exposure of the enzyme to the inhibitor, however, zinc is removed, and the enzyme dissociates into four subunits with consequent loss of catalytic activity. Both the dissociation and the inhibition are prevented by NADH (Kägi, 1960).

Horse liver alcohol dehydrogenase contains 4 gram atoms of zinc per molecular weight of 80,000. The zinc atoms involved in function appear to differ from those which stabilize quaternary structure. Two zinc atoms are essential for catalytic activity, react readily with chelating agents such as 1,10-phenanthroline and sodium diethyldithiocarbamate, and exchange easily with ^{65}Zn but do not seem to participate in subunit interactions. The other two zinc atoms do not react with the same chelating agents and do not exchange with ^{65}Zn under the same conditions as those that participate directly in catalysis. However, carboxymethylation of the active center sulfhydryl groups selectively labilizes these chemically unreactive zinc atoms. In 8 M urea, the protein dissociates into two subunits and, upon removal of this remaining fraction of zinc, into two additional fragments (Drum et al., 1967; Pho and Bethune, 1972). Cobalt and cadmium can selectively replace either one or the other or both of these two types of native zinc atoms generating new metalloalcohol dehydrogenases with spectral and enzymatic properties characteristic of these metals and even forming "metal-hybrid-enzymes" (Drum and Vallee, 1970b).

Metals can also change the secondary and tertiary structures of polymeric proteins without dissociating their quaternary structure. Glutamine synthetase of E. coli is a metal-enzyme complex that undergoes a functionally significant modulation of secondary and tertiary conformation (Stadtman et al., 1968). The enzyme consists of 12 identical subunits of molecular weight 50,000 arranged as two superimposed hexagons. It binds from 1 to 4 gram atoms of Mn^{2+} per monomer with varying stability constants. EDTA removes Mn^{2+} resulting in loss of catalytic activity, a slight decrease in the sedimentation constant, an increase in intrinsic viscosity, but without dissociating the subunits. However, the apoenzyme is more susceptible both to denaturation and to chemical modification and is easily dissociated by exposure to urea, alkaline pH or organic mercurials. Readdition of metals to the polymeric apoenzyme restores activity in a time-dependent fashion, beginning with a rapid exposure of hydrophobic groups, while sulfhydryl side chains are "buried", and the environment of aromatic residues slowly alters (Stadtman et al., 1968; Shapiro and Ginsberg, 1968). Neither removal nor restoration of the metal appears to affect the arrangement of subunits; it seems likely that the metal mediates small changes in secondary and tertiary structure.

Metals as Probes of Enzyme Mechanisms. Rapid advances in techniques for the purification of proteins and analysis of metals have greatly facilitated both the recognition of metalloproteins as discrete entities and their characterization. As a consequence, it has become clear that metalloproteins play critical roles in many vital areas of biology, serving in storage, transport and in virtually all aspects of enzymatic catalysis, and they may well participate in other biological areas in a manner which is not yet apparent.

This field offers new opportunities for a merger of the experimental approaches of organic, physical, inorganic and biochemistry, providing new avenues for interaction between otherwise unrelated interests (Dessy et al., 1971).

Many highly purified metalloproteins are now available in quantity to permit structural, spectral, magnetic and thermodynamic studies which could not be performed easily a decade ago, owing to lack of material, suitable instrumentation or both. Simultaneously, the understanding of the chemistry of metal complex ions has increased greatly leading to an appreciation of the role of model systems in studying mechanisms of metalloenzyme action.

A significant effort has addressed the proposition that the mechanism of action of (nonmetallo)enzymes may be understood ultimately in terms of the physical and organic chemistry underlying the exceptional chemical reactivity of those amino acid side chains involved in catalytic function (Vallee and Riordan, 1969; Shaw, 1970). Similarly, the biological role of metals may become obvious as a result of an application of the chemical basis for apparent differences between the properties of well-defined complex metal ions and those of metalloenzymes. The information on the role of metal ions in stabilization of protein structure may be expected to increase owing to structural studies by indirect methods, e.g., deuterium and tritium exchange, or by direct methods, e.g., x-ray diffraction. The ultimate elucidation of enzyme catalysis, however, requires additionally that the functional characteristics of an enzyme be related to its structural features through information gained while the enzymic reaction is actually in progress. This requires physical-chemical means to probe features of enzyme action which are syncatalytic i.e., occurring synchronous with catalysis. The properties of metalloenzymes are exceptionally well suited to that purpose. For example, the physical-chemical characteristics of colored metal atoms present in such enzymes allow probing of their protein environments by spectral techniques. Iron, copper and cobalt complex ions display properties quite different from those of metalloenzymes containing these same metals. Many other physical-chemical characteristics of metals in metalloenzymes are unusual when compared to those of well-defined model coordination complexes. On the other hand, the same features of the vast majority of _non-enzymatic metalloproteins_ are similar to those of such models (Vallee and Williams, 1968a,b).

The characteristic physicochemical properties resulting from the interaction of metals with simple models and, presumably, protein ligands arise from three sources: properties of the metal altered by the ligands, properties of the ligand altered by the metal and specific _de novo_ properties of the resultant complex. The environment could perhaps influence each of these, though, at this juncture, there are few means to differentiate between the possibilities. However, metal atoms might be regarded as intrinsic probes of their protein environment.

It has been difficult to identify the metal binding ligands of metalloenzymes. Cysteinyl, tyrosyl and histidyl residues have been implicated most often. But until now chemical means alone have been inadequate to define metal binding sites of metalloenzymes unequivocally. The characteristics of chelates, in which the mode of metal coordination is known quite precisely have been compared with those of metalloenzymes, to predict the nature of the metal ligands in the latter. However, bidentate complexes, which have been studied most thoroughly, are inadequate as models since recent evidence indicates that metals in metalloenzymes are more likely coordinated through at least three ligands (Lindskog, 1963; Dennard and Williams, 1966; Peisach _et al_., 1967; Piras and Vallee, 1967; Gould and Ehrenberg, 1968; Lipscomb _et al_., 1968) and possibly even more (Riordan and Vallee, 1969).

Few multidentate complex ions suitable for appropriate comparisons with metalloenzymes have been studied thus far. In addition, in the three-dimensional protein structure, ligands may acquire unusual characteristics, perhaps as a result of conformational or neighboring group effects. Further, the protein environment of metal binding sites, perhaps comprising hydrophobic characteristics, may differ distinctly from that of metal complex ions employed for comparison.

The unusual spectra of metalloenzymes disappear when the enzyme is denatured; they then resemble those of model coordination complexes. If denaturation is reversible, the original spectra can be restored. These and other facts have led to the proposition that the mode of metal coordination in metalloenzymes is closely related to their activities and could reflect the thermodynamic conditions of a site which is prepared for catalysis and which might thus be recognizable in the absence of substrate. We have described such a catalytically poised state, intrinsic to the active site of an enzyme, as "entatic", literally in a thermodynamically stretched state or under tension (Vallee and Williams, 1968a,b; Vallee and Wacker, 1970; Vallee *et al.*, 1971; Williams, 1971).

Not all metals are good probes of their structural environment in the sense, e.g., that they have convenient spectral properties. However, in cases where the metal displays poor probe characteristics, e.g., in zinc metalloenzymes, it has often been possible to replace zinc with cobalt. Cobalt-substituted zinc enzymes display visible spectra and also exhibit enzymic activity. Using cobalt as a probe, its environment can be examined in the enzyme alone, during and after formation of the enzyme-substrate complex, during the catalytic process, and finally, when in equilibrium with products. Table II shows the spectral properties of some cobalt substituted zinc metalloenzymes. The value of magnetic circular dichroic and of EPR studies of such systems has been pointed out recently (Kaden *et al.*, 1972; Kennedy *et al.*, 1972).

Work performed in our laboratory on carboxypeptidase A, a single-chain enzyme, and on alcohol dehydrogenase of horse liver, a multichain enzyme may serve for purposes of illustration. In the former the metal seems largely functional and in the latter they appear to play structural as well as functional roles.

Cobalt Carboxypeptidase A. The catalytically essential zinc atom in a number of metallopeptidases can be replaced with cobalt to give active products but with more favorable probe properties. Carboxypeptidase A of pancreas has been studied most extensively in this regard (Neurath *et al.*, 1970; Lipscomb *et al.*, 1970; Vallee *et al.*, 1970). The spectra of cobalt carboxypeptidase A are indicative of low symmetry of the cobalt coordination site (Latt and Vallee, 1971).

The visible spectrum of cobalt carboxypeptidase has a shoulder at about 500 nm and maxima at 555 and 572 nm, both with absorptivities slightly greater than 150 (Figure 1). Two infrared bands are centered at 940 nm and 1570 nm ($\varepsilon = 20$). The location of the high energy band at 940 nm is unusual, since the near infrared absorption bands of divalent cobalt complex ions are generally found at longer wavelengths; their position in cobalt spectra can often probe the metal environment more effectively than can the visible bands. In this case the 940 nm maximum suggests that cobalt interacts more strongly with the ligands of the metalloenzyme than with those of model complex ions.

TABLE II

SPECTRAL PARAMETERS OF Co(II) COMPLEX IONS AND Co(II) SUBSTITUTED METALLOENZYMES

Compound	Absorption λ, nm (Absorptivity, M^{-1} cm^{-1})	Ellipticity λ, nm ($[\Theta]^{25}$x10^{-4})	Notes
1. Simple Co(II) complexes			
$[Co(H_2O)_6]^{2+}$	510 (∿5) 1200 (∿2)		Octahedral
$[CoCl_4]^{2-}$	685 (700), complex 1700 (100), complex		Tetrahedral
$[Co(OH)_4]^{2-}$	600 (∿150), complex 1400 (∿50), complex		Tetrahedral
$[Co(Et_4dien)]Cl_2$[a]	520 660 950		Distortion from trigonal bipyramidal
$[Co(Tren\ Me)_6Cl]Cl$[b]	500 (120) 625 (80) 800 (∿30) doublet 1750 (∿30)		
2. Co(II) enzymes			
Co(II) carbonic anhydrase	510 (280) 615 (300) 900 (30)	460 (30) 610 (30)	Low symmetry
	550 (380) 640 (280) 1250 (90)	550 (30)	
Co(II) alkaline phosphatase	510 (280) 610 (210) sh	470 (20) 575 (-8)	Low symmetry
	555 (350) 640 (250)	520 (20)	
Co(II) carboxypeptidase	500 sh 555 (∿150) 950 (∿25)	500 sh 538 (-5)	Distorted tetrahedral
	572 (∿150)		
Co(II) yeast alcohol[c] dehydrogenase	620 (∿800) 657 (∿800) 710 (∿600)		Low symmetry

TABLE II continued...

Co(II) neutral protease	475 sh 525 (sh)	 555 (50-100)	500 sh	550 (-8)	Low symmetry
Co(II) yeast aldolase	490 sh	530 (150)	485 (24)	530 (30)	Low symmetry
Co(II) yeast enolase[d/]	490 (sh)	540 (30-40) 575 (sh)	490 sh	540 (-5)	Low symmetry

[a/] In organic solvents.

[b/] In organic solvents.

[c/] Enzyme contained both cobalt and zinc.

[d/] Metal-enzyme complex.

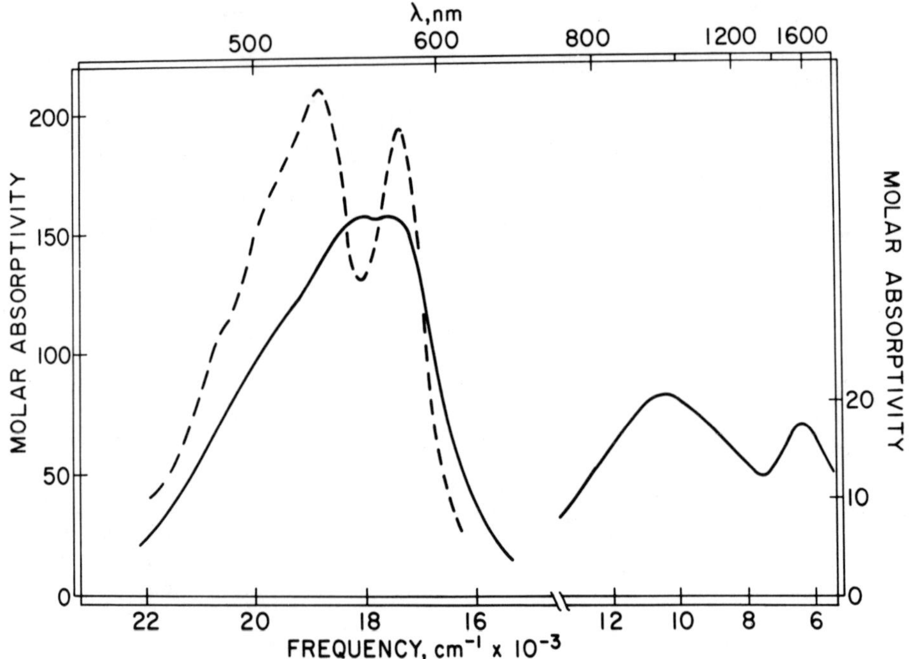

Figure 1. Absorption spectra of cobalt carboxypeptidase for (a) visible region. The enzyme, 1 mM, was dissolved in 1 M NaCl, 0.005 M Tris-Cl, pH 7.1, 20° (—— ——). Another enzyme solution, about 3 mM, was diluted with glycerol to 45% v/v and cooled to 4.2°K (- - - -) for spectral measurements. (b) Near-infrared region. Apocarboxypeptidase, 1.5 mM, was dissolved in 1 M NaCl, 0.005 M [D] Tris-Cl, D_2O, pH 7.2. The sample cuvet contained 1.5 mM enzyme plus $Co(SO_4)$ in D_2O buffer to yield a final total cobalt concentration of 2.0 mM, and, hence, a 0.5 mM excess of free Co(II) ions; the reference cell contained 1.5 mM apoenzyme, brought to volume with buffer.

Spectral absorption measurements at liquid helium temperature, 4.2°K, increase resolution and reveal two new shoulders and a shift of the maximum at 555 to 532 nm while that at 572 nm remains the same. Absorptivity is not reduced significantly mitigating against vibronic interactions as significant factors in the generation of spectral intensity. A gradual increase of the temperature to 25°C reverses the spectral changes. In a magnetic field of 47,000 gauss the absorption band at 572 nm becomes optically active.

Circular dichroic spectra also reflect the geometry of the metal binding site and may encompass information on the influence of vicinal factors on the cobalt atom in the enzyme. A negative band at 538 nm and the shoulder near 500 nm probably correspond to the low wavelength maximum and shoulder of the absorption band, enhanced in the spectrum observed at 4.2°K.

The EPR spectrum of Co(II) carboxypeptidase A is broad and quite featureless and can be described by these approximate g-values -- g_1 ~5.6, g_2 ~4.4 and g_3 ~2.7. These do not differ greatly from those derived from the spectra of distorted 4-coordinate Co(II) complex ions (Kennedy _et al._, 1972). Such an enzyme spectrum would be consistent with a structure in which Co(II) forms three strong bonds to the protein and one bond (or indeed none at all) of a different character (Vallee and Williams, 1968).

Agents which affect enzymatic activity might also alter the spectra if these were to reflect catalytic potential. Indeed, a large number of inhibitors bring about major shifts in the absorption and circular dichroic spectra of cobalt carboxypeptidase (Figure 2). Further spectral details are observed as a function of the structure and concentration of inhibitors added. The patterns of these spectral shifts are characteristic for inhibitors with different functional groups (Latt and Vallee, 1971).

The effects of substrates on the circular dichroic spectrum of cobalt carboxypeptidase yield particularly interesting information. The circular dichroic spectrum of the cobalt carboxypeptidase·glycyl-L-tyrosine complex differs from that both of the enzyme itself and from that of most of the other enzyme-inhibitor complexes examined. The extremum of the negative Cotton effect of the cobalt enzyme at 537 nm shifts to 555 nm with an inversion of sign, while the molar ellipticity increases from -500 to +2000, a four-fold change in magnitude (Figure 2). Both the inversion of the sign and the marked enhancement of the spectrum are striking and suggest major electronic rearrangements around the cobalt atom upon interaction with substrates. This impression is reinforced by the fact that in the absorption spectrum the band at 940 nm splits into two, located at ∿850 nm and ∿1150 nm, respectively, while that at 1570 nm is shifted to 1420 nm.

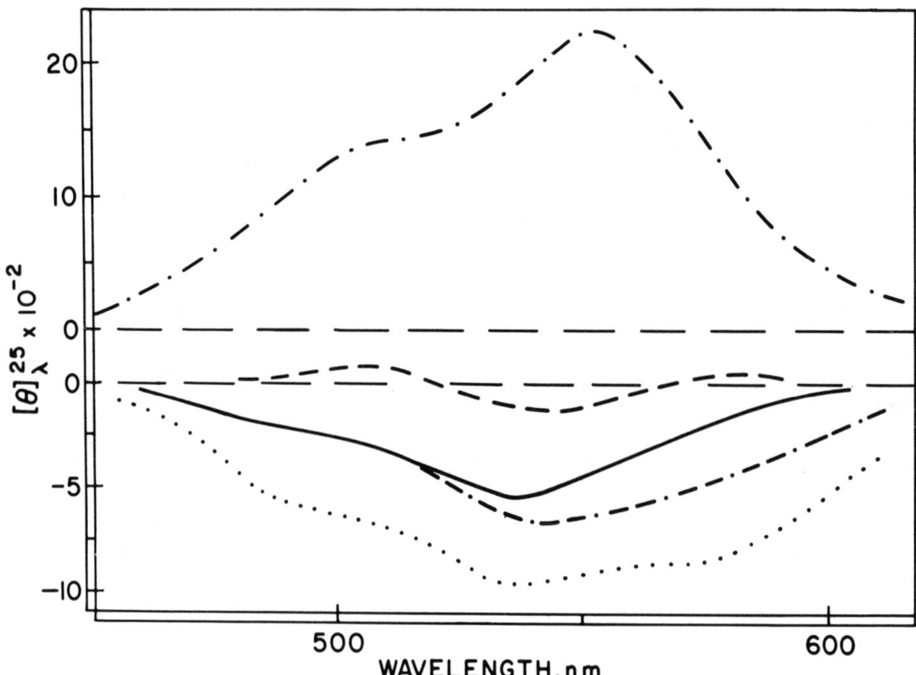

Figure 2. Effect of glycyl-L-tyrosine, β-phenylpropionate, and L-phenylalanine on the circular dichroic spectrum of cobalt carboxypeptidase (———). The enzyme was dissolved in 1 M NaCl, 0.005 M Tris-Cl, pH 7.1. Spectra of the enzyme were also obtained in the presence of glycyl-L-tyrosine, 10 mM (-·-·-·-·), L-phenylalanine, 9 mM (— — —), and β-phenylpropionate, 2 mM (--·--·), or 9 mM (····).

The absorption and circular dichroic spectra of cobalt carboxypeptidase jointly indicate that in the enzyme the catalytically active metal occupies an irregular, asymmetric, tetrahedral site. While delineating constraints on the nature of the metal-site geometry, the spectra also suggest that the outer cobalt electrons are delocalized, indicating strong interactions with the protein. These changes could reflect directional forces generated at the active site by the asymmetry around the cobalt atom, magnified and shifted upon enzyme-substrate interaction. In this instance, cobalt serves simultaneously as an integral part of the catalytic apparatus and as a spectral probe, perhaps capable of providing new insight into enzyme action (Vallee and Latt, 1970; Latt and Vallee, 1971).

Cobalt Procarboxypeptidase. The unusual spectra of the catalytically essential, chromophoric cobalt atom of cobalt carboxypeptidase suggesting irregular coordination geometries and low symmetry of the cobalt site (Vallee and Williams, 1968b), might serve to identify unsuspected catalytic properties. Toward this end the spectral features of cobalt procarboxypeptidase have been examined.

The entatic site hypothesis would suggest that a metalloprotein, thought to be enzymatically inert but exhibiting an entatic spectrum, actually has catalytic potential. Procarboxypeptidase was examined to test this hypothesis and its predictive value. Bovine carboxypeptidase A, molecular weight, 34,600 and its precursor, procarboxypeptidase, molecular weight 87,000 contain a single zinc atom which can be replaced with cobalt.

Zymogen activation has been thought to involve conformational changes that might either affect the catalytic site or reveal, form or complete the substrate-binding site--or all of these. The spectrochemical properties of the cobalt procarboxypeptidase-carboxypeptidase pair are virtually identical and, based on this, the zymogen has been shown to exhibit substantial peptidase activity, as great or greater than that of the native enzyme. The visible, IR, CD and MCD spectra of cobalt procarboxypeptidase are virtually identical to those of cobalt carboxypeptidase in all respects. Similarly, the perturbation of these spectra by inhibitors, such as β-phenylpropionate, and by substrates, such as glycyl-L-tyrosine, are also the same as those of cobalt carboxypeptidase. The absorption and CD spectra suggested that in both the zymogen and the enzyme the metal is poised to participate in the catalytic process.

N-haloacylated aromatic amino acids are peptide substrates of carboxypeptidase, minimal in size and complexity but with the specificity requirements of the enzyme, and cobalt procarboxypeptidase catalyzes their hydrolysis as readily as does the native enzyme. Both cobalt procarboxypeptidase and the native zinc enzyme exhibit Michaelis-Menten kinetics toward chloro- and dichloroacetyl-phenylalanine. Further, their pH optima, k_{cat} and K_m values are all virtually the same. Both β-phenylpropionate and β-phenyllactate are competitive inhibitors with identical apparent K_i values. Perturbation spectra of the cobalt zymogen-enzyme pair constitute another link between their spectral and catalytic features of the enzyme-zymogen pair. In this instance the entatic site hypothesis led to the detection of spectra characteristic of a catalytically active system in a protein thought to be inert. We are unaware of previous instances in which, based on consideration of symmetry and geometry of metal coordination, metal spectra have served as guides to the existence of activity in a metalloprotein. Such entatic spectra may turn out to be indicative of unsuspected catalytic properties of other metalloproteins.

Liver Alcohol Dehydrogenase. Similar approaches have led to the successful complete replacement of the zinc atoms of horse liver ADH either with cobalt or with cadmium, resulting in enzymatically active enzymes exhibiting specific activities and spectral properties characteristic of these metals and not seen in the native enzyme (Drum and Vallee, 1970b). Co-LADH has a broad absorption band in the near-ultraviolet region centered at 340 nm (ε = 6500) associated with two positive and three negative circular dichroic bands between 300 nm and 450 nm. Absorption maxima also occur at 655 nm (ε = 1330), 730 nm (ε = 800), and in the near infrared between 1000 and 1800 nm (ε = 270-540) (Figure 3). A small negative ellipticity band is centered at 620 nm.

Cd-LADH exhibits an intense absorption band centered at 245 nm, which may assist in identifying the ligands of the proteins binding the metal since its molar absorptivity (ε = 10,200) is close to that of 14,000 reported for the cadmium mercaptide chromophores of metallothionein.

The circular dichroic spectrum of Co-LADH is richer in detail than that of any other comparable cobalt enzyme studied thus far. It may be analyzed on one hand in terms of protein structure, as reflected in the intrinsic and side-chain Cotton effects of the protein, and of superimposed extrinsic Cotton effects resulting from either binding of NADH or of 1,10-phenanthroline (OP).

These metal substitutions also perturb the side-chain Cotton effects of LADH. Between 250 and 300 nm the circular dichroic spectra of these metallodehydrogenases exhibit remarkable fine structure, characteristic of each. The circular dichroic difference spectrum of Cd-LADH vs. Zn-LADH below 270 nm exhibits a positive ellipticity band which may well correspond to the maximum in the absorption spectrum at 245 nm. In the region of maximum absorption due to aromatic amino acid residues, the ellipticity of the Co-LADH exceeds that of both Zn- and Cd-LADH (Drum and Vallee, 1970b).

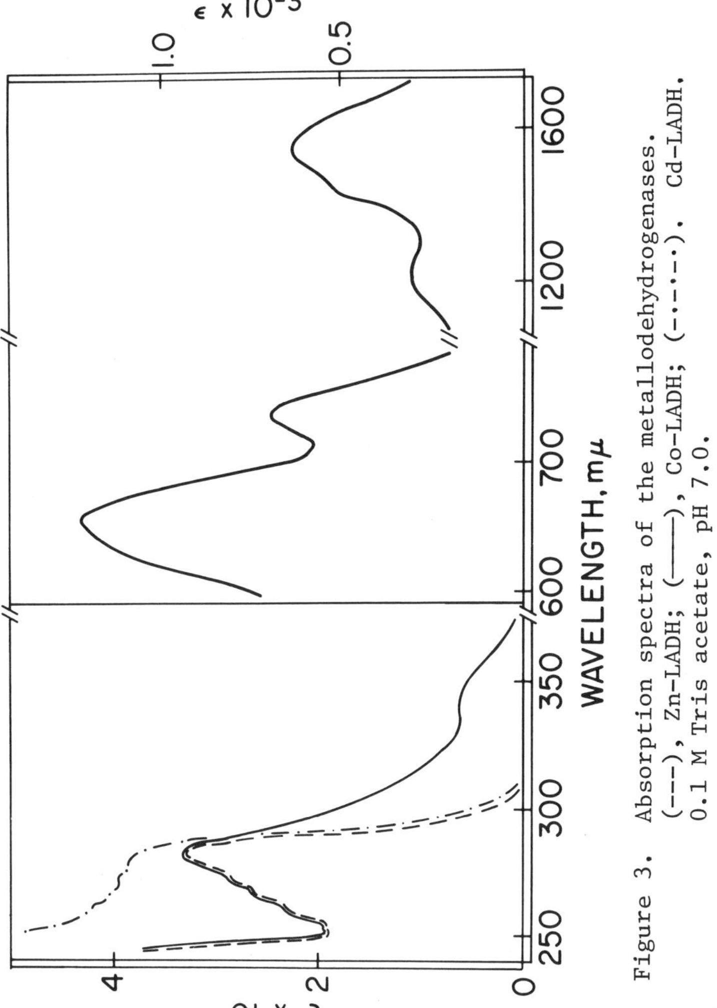

Figure 3. Absorption spectra of the metallodehydrogenases. (---), Zn-LADH; (——), Co-LADH; (-·-·-·). Cd-LADH. 0.1 M Tris acetate, pH 7.0.

The effects of the metal substitutions in LADH are reflected also in the optical properties of extrinsic chromophores, bound asymmetrically. Both the reduced coenzyme (NADH) and 1,10-phenanthroline (OP) bind to Zn-, Co-, and Cd-LADH while generating extrinsic Cotton effects, distinctive for each, with magnitudes which related quantitatively to binary complex formation (Figure 4).

Similarly, the amplitude of the circular dichroic band of the Cd-LADH-OP complex, centered at 271 nm, markedly exceeds that of Zn-LADH-OP, while that of the Co-LADH-OP complex is considerably smaller. Thus the rotational strengths of these mixed complexes afford sensitive probes of the identity of the metal substituent at the active site.

Through work of this type it has become apparent that some of the metals present in certain metalloenzymes are critical to the catalytic step while others seem to serve primarily in protein structure (Drum *et al.*, 1967; Drum and Vallee, 1970a). It would appear that two such discrete roles for metal atoms are maintained when cobalt or cadmium, for example, are substituted in Zn-LADH. Both the cobalt and cadmium enzymes bind 2 moles of OP, suggesting that, like the native zinc enzyme, only two of the metal atoms are accessible to this agent. Selective replacement either of the "functional" or of the "structural" metal atoms with other metals, resulting in "metal-hybrid enzymes," can now be documented by employing the unusual absorption and circular dichroic properties discussed, which augment the criteria available in the past.

Cobalt as a Spectral Probe for Other Metalloenzymes. A similar approach using the probe properties of metals has also been employed for other proteolytic enzymes. Thus far, only a small number of well-characterized proteolytic enzymes are known to contain a metal, generally zinc in the native state (Vallee and Wacker, 1970; Vallee and Latt, 1970; Vallee *et al.*, 1971). The spectral and enzymatic characteristics of cobalt carboxypeptidase have been discussed in detail but analogous studies in other similar systems have also become possible recently.

The zinc of both thermolysin and neutral protease can be replaced by cobalt and other metals to yield active enzymes and these derivatives exhibit characteristic spectra (Vallee and Latt, 1970). In addition, the spectral characteristics of carbonic anhydrase, alkaline phosphatase of E. coli and yeast aldolase have been studied intensively by replacing their native zinc atoms with cobalt.

Even though the total number of cobalt enzymes examined in this manner thus far is small, certain patterns are becoming evident. The absorption spectra of many of these cobalt substituted

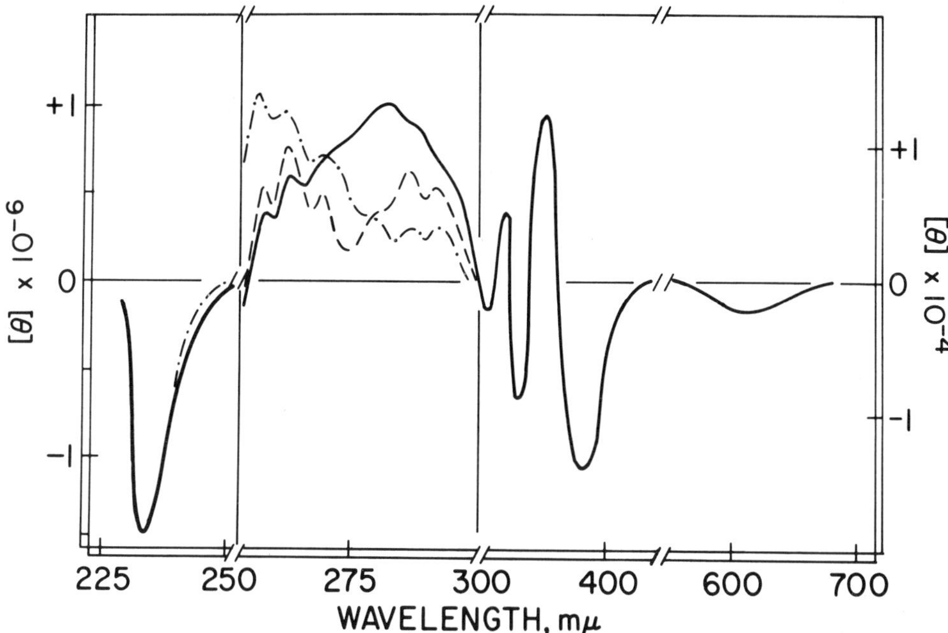

Figure 4. Circular dichroic spectra of the metallodehydrogenases. (---), Zn-LADH; (——), Co-LADH; (-·-·-·), Cd-LADH. The symbol [θ] denotes molecular ellipticity based on the concentration and molecular weight of LADH. 0.1 M Tris acetate, pH 7.0.

enzymes are quite similar and many have optically active bands of similar magnitude and position. Even though it is premature to interpret the spectral details, it is still tempting to speculate that similarities, when they exist, might reflect common functional characteristics. Conceivably similarities and differences of spectra could imply specific mechanisms.

SUMMARY

While early studies of metalloenzymes have revealed the specific participation of metal ions in enzymatic catalysis, it has become increasingly apparent that some metalloenzymes may contain discrete categories of metal atoms, critical on one hand to the catalytic step and, on the other, to protein structure. These different roles can be discerned by inspection of both single chain and multichain enzymes. In some single chain mammalian enzymes, e.g., carboxypeptidase A and B or carbonic anhydrase, the metal atom plays largely a catalytic role, and stabilization of tertiary structure has not as yet been demonstrated decisively. In certain single chain bacterial enzymes, e.g., the neutral protease from *B. subtilis* or thermolysin from *B. thermoproteolyticus*, a zinc atom serves in catalysis, but, in addition, calcium atoms apparently stabilize secondary and/or tertiary structure. In single chain enzymes, the removal and restoration of catalytically active metal atoms is accomplished relatively easily, but in multichain enzymes, this process has proven more difficult, suggesting that some of the integral metal atoms may affect quaternary structure. These two fundamentally discrete roles for metal atoms can be distinguished by various physicochemical approaches. Selective replacement of one or the other of these two types of native metal atoms with other metals becomes possible, resulting in "hybrid-metal-enzymes." Thus, in horse liver alcohol dehydrogenase, containing 3.4 to 3.6 gram atoms of zinc per molecular weight of 80,000, the zinc atoms involved in function can be differentiated from those which participate in structure. Only two zinc atoms seem essential for catalytic activity; they are readily accessible to the ambient medium as gauged by their relative reactivity toward chelating agents such as 1,10-phenanthroline, α,α'-dipyridyl and sodium diethyldithiocarbamate as well as by their relative ease of isotope exchange. They do not appear to participate in subunit interactions but can be exchanged for cobalt or cadmium resulting in enzymatically active species. The balance of the zinc atoms are quite unreactive or "buried," do not react easily with the same chelating agents, or exchange with isotopic zinc under the same conditions as "active site" zinc atoms. However, the chemically "unreactive" metal atoms are selectively labilized by carboxymethylation of the enzyme and appear to participate in stabilizing the quaternary structure: the protein dissociates into subunits in 8 M urea and upon removal of this fraction of "structural" zinc atoms. Substitution of cobalt and cadmium for zinc generates new alcohol dehydrogenases exhibiting spectral and enzymatic properties, characteristic of these metals. In an analogous fashion, E. coli alkaline phosphatase also contains discrete zinc atoms which participate in catalysis and which can be differentiated from the other "structural" zinc atoms in the molecule. Substitution of cobalt for zinc yields an enzymatically active enzyme which exhibits its characteristic enzymatic and spectral properties. Different metal ions may also serve different roles in the same enzyme, e.g., the two calcium atoms in B. subtilis amylase are enzymatically active, while a zinc atom dimerizes the protein without affecting function. Zinc has also been shown to be present only in the regulatory subunits of aspartate transcarbamylase while a metal atom has not been found so far in the catalytic subunits. Replacement of zinc by metals with good probe properties, e.g., cobalt or manganese, afford opportunities for the exploration of the environment of these various functional and structural species by a number of spectroscopic approaches.

ACKNOWLEDGEMENTS

This work was supported by Grants-in-Aid GM-15003 and GM-02123 from the National Institutes of Health, of the Department of Health, Education and Welfare.

REFERENCES

Dennard, A. E. and Williams, R. J. P. (1966). Transition Metal Chemistry, Carlin, R. L., Ed., New York, N. Y., Marcel Dekker, Vol. 2, 115.

Dessy, R., Dillard, J. and Taylor, L., Eds. (1971). Bioinorganic Chemistry, Washington, D. C., American Chemical Society.

Drum, D. E., Harrison, J. H., IV, Li, T.-K., Bethune, J. L. and Vallee, B. L. (1967). Proc. Natl. Acad. Sci. U.S. 57:1434.

Drum, D. E., Li, T.-K. and Vallee, B. L. (1969). Biochemistry 8: 3783.

Drum, D. E. and Vallee, B. L. (1970a). Biochemistry 9:4078.

Drum, D. E. and Vallee, B. L. (1970b). Biochem. Biophys. Res. Commun. 41:33.

Gould, D. C. and Ehrenberg, A. N. (1968). Symposium on Hemocyanin in Physiology and Biochemistry of Haemocyanins, Ghiretti, F., Ed., New York, Academic Press, 27.

Hunter, F. E., Jr. and Lowry, O. N. (1956). Pharmacol. Rev. 8:89.

Kaden, T. A., Holmquist, B. and Vallee, B. L. (1972). Biochem. Biophys. Res. Commun. 46:1564.

Kägi, J. H. R. and Vallee, B. L. (1960). J. Biol. Chem. 235:3188.

Keilin, D. and Mann, T. (1940). Biochemical J. 34:1163.

Latt, S. A. and Vallee, B. L. (1971). Biochemistry 10:4263.

Li, T.-K. and Theorell, H. (1969). Acta Chem. Scand. 28:892.
Li, T.-K. and Vallee, B. L. (1972). Modern Nutrition in Health and Disease, 5th Edition, Goodhart, R. S. and Shils, M. E., Eds., Philadelphia, Lea & Febiger, p. 372.
Lindskog, S. (1963). J. Biol. Chem. 283:945.
Lipscomb, W. N., Hartsuck, J. A. Reek, G. N., Jr., Quiocho, F. A., Bethge, P. H., Ludwig, M. L., Steitz, T. A., Muirhead, H. and Coppola, J. C. (1968). Brookhaven Symp. Biol. 21:24.
Lipscomb, W. N., Reeke, G. N., Jr., Hartsuck, J. A., Quiocho, F. A. and Bethge, P. H. (1970). Phil. Trans. Roy. Soc. Lond. B257:177.
Matsubara, H. and Feder, J. (1970). The Enzymes, 3rd ed., Vol. 3, Boyer, P. D., Ed., New York, Academic Press, p. 721.
Nelbach, M. E., Pigiet, V. P., Gerhart, J. C. and Schachman, H. C. (1972). Biochemistry 11:315.
Neurath, H., Bradshaw, R. A., Petra, P. H. and Walsh, K. A. (1970). Phil. Trans. Roy. Soc. Lond. B257:159.
Peisach, J., Levine, W. G. and Blumberg, W. E. (1967). J. Biol. Chem. 242:2847.
Pho, D. B. and Bethune, J. L. (1972). Biochem. Biophys. Res. Commun. 47:419.
Piras, R. and Vallee, B. L. (1967). Biochemistry 6:348.
Rosenberg, A. (1960). Arkiv. Kemi 17:25.
Rosenbusch, J. P. and Weber, K. (1971). Proc. Natl. Acad. Sci. 68:1019.
Scrutton, M. C., Wu, C. W. and Goldthwait, D. A. (1971). Proc. Natl. Acad. Sci. U.S. 68:2497.
Shapiro, B. M. and Ginsberg, A. (1968). Biochemistry 7:2153.
Slater, J. P., Mildvan, A. S. and Loeb, L. A. (1971). Biochem. Biophys. Res. Commun. 44:37.
Shaw, E. (1970). Physiol. Revs. 50:244.
Stadtman, E. R., Shapiro, B. M., Ginsburg, A., Kingdon, D. S. and Denton, M. D. (1968). Brookhaven Symp. Biol. 21:378.
Thompson, R. H. S. and Cummings, J. N. (1970). Biochemical Disorders in Human Disease, 3rd ed., Thompson, R. H. S. and Wooton, I. D. P., Eds., New York, Academic Press, 12, p. 431.
Ulmer, D. D. (1970). Effects of Metals on Cells, Subcellular Elements and Macromolecules, Coleman, J. R. and Miller, M. W., Eds., Springfield, Ill., Charles C. Thomas, p. 11.
Ulmer, D. D. and Vallee, B. L. (1971). Bioinorganic Chemistry, Advances in Chemistry Series 100, Dessy, R., Dillard, J. and Taylor, L., Eds., Washington, D. C., American Chemical Society, p. 187.
Vallee, B. L. (1955). Advances in Protein Chemistry, Vol. 10, Anson, M. L., Bailey, V. and Edsall, J. T., Eds., New York, Academic Press, p. 317.
Vallee, B. L. (1961). Federation Proc. 20(3):71.
Vallee, B. L. (1971). Federation Proc. 20, Suppl. 10:71.
Vallee, B. L. and Coombs, T. L. (1959). J. Biol. Chem. 234:2615.

Vallee, B. L. and Latt, S. A. (1970). Structure-Function Relationships of Proteolytic Enzymes, Desnuelle, P., Neurath, H. and Ottesen, M., Eds., Copenhagen, Munksgaard, p. 144.

Vallee, B. L. and Riordan, J. F. (1968). Brookhaven Symp. Biol. 21:91.

Vallee, B. L. and Riordan, J. F. (1969). Ann. Rev. Biochem. 38:733.

Vallee, B. L., Riordan, J. F., Auld, D. S. and Latt, S. A. (1970). Phil. Trans. Roy. Soc. Lond. B257:215.

Vallee, B. L., Riordan, J. F., Johansen, J. T. and Livingston, D. M. (1971). Cold Spring Harbor Symp. Quant. Biol. 36:517.

Vallee, B. L., Stein, E. A., Summerwell, W. N. and Fischer, E. H. (1959). J. Biol. Chem. 234:2901.

Vallee, B. L. and Ulmer, D. D. (1972). Ann. Rev. Biochem. 41:91.

Vallee, B. L. and Wacker, W. E. C. (1970). Metalloproteins in The Proteins, 2nd ed., Neurath, H.,Ed., New York, Academic Press, Vol. V.

Vallee, B. L. and Williams, R. J. P. (1968a). Chem. in Brit. 4:397.

Vallee, B. L. and Williams, R. J. P. (1968b). Proc. Natl. Acad. Sci. U.S. 59:498.

Warburg, O. (1949). Heavy Metal Prosthetic Groups in Enzyme Action. Oxford, Clarendon Press.

Williams, R. J. P. (1967). Endeavour 26:96.

Williams, R.J.P. (1968). Chem. in Brit. 4:397.

Williams, R.J.P. (1971). Cold Spring Harbor Symp. Quant. Biol. 36: 53.

Wintersberger, E., Neurath, H., Coombs, T.L. and Vallee, B.L. (1965). Biochemistry 4:1526.

3

STUDIES ON CARBOXYPEPTIDASE A

E. T. Kaiser, Tai-Wah Chan, and Junghun Suh

Department of Chemistry, University of Chicago,

Chicago, Illinois 60637

In this article the main theme to be discussed will be our continuing studies on the mechanism of action of the zinc-containing metalloenzyme bovine carboxypeptidase A. A summary of earlier mechanistic studies in our laboratory on carboxypeptidase A was published recently (Kaiser and Kaiser, 1972). Except for some of the essential highlights of that article much of the material covered in the present article is new.

The nature of the ligands binding the active site zinc ion in carboxypeptidase A has been elucidated by a combination of X-ray and peptide sequencing studies (Lipscomb _et al._, 1969, Bradshaw _et al._, 1969). Three ligands from the protein, two histidines (His-69 and His-196), and a glutamate residue (Glu-72), bind the metal ion. In addition, a fourth ligand to the zinc ion, a water molecule is present. Because the carboxylate group of the glutamate residue bound to the zinc ion carries a mononegative charge, the zinc ion bears a single positive net charge. Three main species of carboxypeptidase A exist, differing at their amino termini (Sampath-Kumar _et al._, 1964). Carboxypeptidase A_α, the species on which the X-ray work has been done (Lipscomb et al., 1966) is 307 amino acids in length, carboxypeptidase A_β is 305 amino acids in length, and carboxypeptidase A_γ, the species on which most of the mechanistic and kinetic studies have been performed contains 300 amino acids.

Carboxypeptidase A catalyzes the hydrolysis of the peptide or ester bonds of N-acyl-α-amino acids and O-acyl α-hydroxy acids, respectively, adjacent to the terminal free carboxyl groups of these substrates. By analogy to the analysis of Bender and Kézdy of structure-reactivity relationships in chymotrypsin-catalyzed reactions (Bender and Kézdy, 1965), we have been testing the form of the Taft equation (Taft, 1956) shown in equation 1 for its applicability to the analysis of reactivity patterns in carboxypeptidase A-catalyzed ester and peptide hydrolysis (Kaiser and Kaiser, 1972). Here ρ_X^* is the reaction constant, σ_X^* is the Taft constituent constant for the group X, and S_{R_1} and S_{R_2} are factors which account for the influence of the substituents R_1 and R_2 on substrate reactivity. It must be emphasized here that this discussion of substrate reactivity in carboxypeptidase-catalyzed reactions is limited to those simple substrates in which R_1 contains no amide linkage because more complex substrates often show substrate activation or inhibition behavior, complicating the analysis of their reactivity (Lumry *et al.*, 1951; Kaiser *et al.*, 1964; McClure *et al.*, 1964; Kaiser *et al.*, 1965; Awazu *et al.*, 1967).

$$\log \frac{[k_{cat}/K_{m_{app}} \text{ for } R_1R_2X]}{[k_{cat}/K_{m_{app}} \text{ for reference compound}]} \tag{1}$$

$$= \rho_X^* \sigma_X^* + S_{R_1} + S_{R_2}$$

Comparison of the ratio of the rate constants for the carboxypeptidase A catalyzed hydrolysis of esters and peptides with the corresponding ratio for the hydroxide ion catalyzed reactions has indicated that the effect of variations in the group X on $k_{cat}/K_{m_{app}}$ appears only to reflect electronic influences of X and does not involve productive binding of X to the enzyme (Kaiser, 1970). Comparison of the reactivities of cinnamoyl, furylacryloyl, and p-nitrocinnamoyl derivatives of L-β-phenyllactate or of L-mandelate toward carboxypeptidase with the reactivities toward hydroxide ion of a series of ethyl esters containing the same acyl groups indicated that R_1 also exerts only an electronic influence on the rate of the enzymatic reaction (Kaiser and Kaiser, 1972).

Because these conclusions were based on limited rate data, we have been pursuing the extension of our reactivity analysis to establish the bounds within which it applies. Most recently, we have studied the hydrolysis of a series of p-substituted O-(trans-cinnamoyl)-L-β-phenyllactates. The results we obtained for the nonenzymatic and enzymatic hydrolyses of these compounds are summarized in Table 1. In Figure 1 the relationship between log k_{cat}/K_m and log k_{OH^-} is illustrated. The points for the various substituted derivatives fall close to the line drawn, but the unsubstituted species gives a result which falls considerably below the line. In other words, the unsubstituted ester is substantially less reactive to carboxypeptidase action than might be expected from a simple linear correlation between the enzymatic and hydroxide ion catalyzed reactions. However, in a gross sense, at least if one considers that reactivity to hydroxide ion catalysis reflects essentially the electronic influence of the substituents, then our data suggest that there is very little enzymatic specificity exerted by carboxypeptidase A in catalyzing the hydrolysis of the p-substituted O-(trans-cinnamoyl)-L-β-phenyllactates.

TABLE 1

Carboxypeptidase- and Hydroxide-Catalyzed Hydrolysis of Acyl-Substituted O-(trans-Cinnamoyl)-L-β-Phenyllactates

Substituent	k_{cat} (sec^{-1})	$10^4 K_m$ (M)	$10^{-6} k_{cat}/K_m$ ($M^{-1}sec^{-1}$)	$10^2 k_{OH}$ ($M^{-1}sec^{-1}$)
p-NO_2	205	1.14 ± 0.2	1.77 ± 0.03	1.64 ± 0.02
p-CN	228	1.14 ± 0.01	1.95 ± 0.02	1.58 ± 0.02
p-Cl	144	1.36 ± 0.2	1.01 ± 0.02	0.542 ± 0.003
p-H	57.3	1.64 ± 0.01	0.348 ± 0.002	0.405 ± 0.001
p-MeO	64.8	2.08 ± 0.02	0.301 ± 0.04	0.19 ± 0.001

In the enzyme-catalyzed reactions the substrate concentrations employed ranged from 3×10^{-5} to 1.34×10^{-3} M. The reactions were run in 0.5 M NaCl, 0.05 M Tris-HCl at pH 7.5 and 25.0°. The range of OH^- concentrations used was 0.014 to 0.04 M, and $\mu = 0.55$ in the hydroxide ion-catalyzed runs.

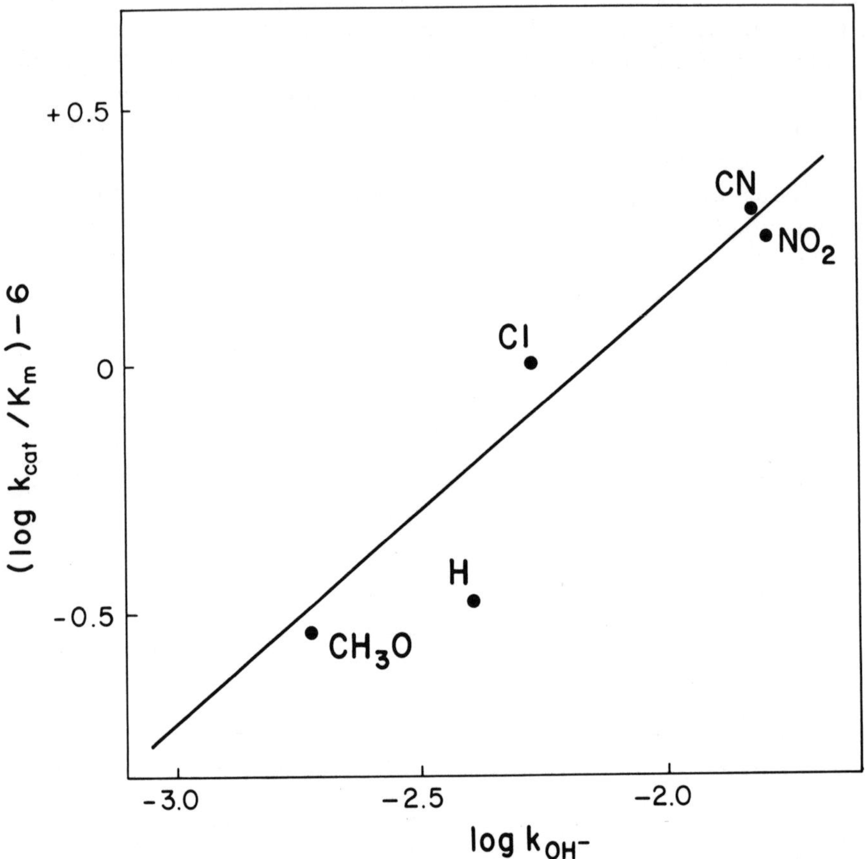

Figure 1. The relationship between log k_{cat}/K_m and log k_{OH^-} is illustrated.

In Figure 2, a plot is shown of the dependence of the logarithms of various kinetic parameters for the hydrolysis of the p-substituted cinnamoyl esters in terms of the values of σ^{o} (Taft). Equations 2-4 summarize the relationships found and the correlation coefficients obtained. The significant deviation of the point for the unsubstituted compound from the straight line drawn in the k_{OH^-} vs σ^{o} plot suggests that at least part of the deviation seen for this compound in the plot given in Figure 1 is due to its unexpectedly high alkaline lability. It should be pointed out in any case that the log k_{OH^-} values correlate slightly better with Jaffe σ values (Jaffe, 1953) than with the σ^{o} values as shown in equation 5. This is not true, however, for the kinetic parameters for the enzyme-catalyzed reaction.

$$\log k_{cat}/K_m = (0.89 \pm 0.14)\,\sigma^{o} + (5.62 \pm 0.07) \quad \text{correlation coefficient} = 0.96 \tag{2}$$

$$\log k_{cat} = (0.64 \pm 0.13)\,\sigma^{o} + (1.87 \pm 0.07) \quad \text{correlation coefficient} = 0.94 \tag{3}$$

$$\log k_{OH^-} = (0.95 \pm 0.12)\,\sigma^{o} - (0.73 \pm 0.06) \quad \text{correlation coefficient} = 0.98 \tag{4}$$

$$\log k_{OH^-} = (0.92 \pm 0.07)\,\sigma - (0.67 \pm 0.04) \quad \text{correlation coefficient} = 0.991 \tag{5}$$

From a rather limited comparison of nonenzymatic and enzymatic data, using as models various ester derivatives with L-β-phenyllactate and L-mandelate as the alcohol leaving groups (Kaiser and Kaiser, 1972), the argument has been made that carboxypeptidase A shows specificity for the R_2 moiety, in contrast to the findings for variations in the X and R_1 groups.

Figure 3 summarizes the picture developed by us (Kaiser and Kaiser, 1972) for the interaction of carboxypeptidase A with simple ester and peptide substrates where the plane represents the surface of the enzyme, a, represents the binding site for the substrate carboxyl group, b, the binding site for R_2, and c, the catalytic site. The interaction of the carboxylate groups with site a is postulated in view of the stringent specificity requirements of carboxypeptidase A for a free terminal carboxyl group in the substrate. While the interactions shown in Figure 3 are ones which are required from the reactivity patterns of simple substrates with carboxypeptidase (Waldschmidt–Leitz,1931;

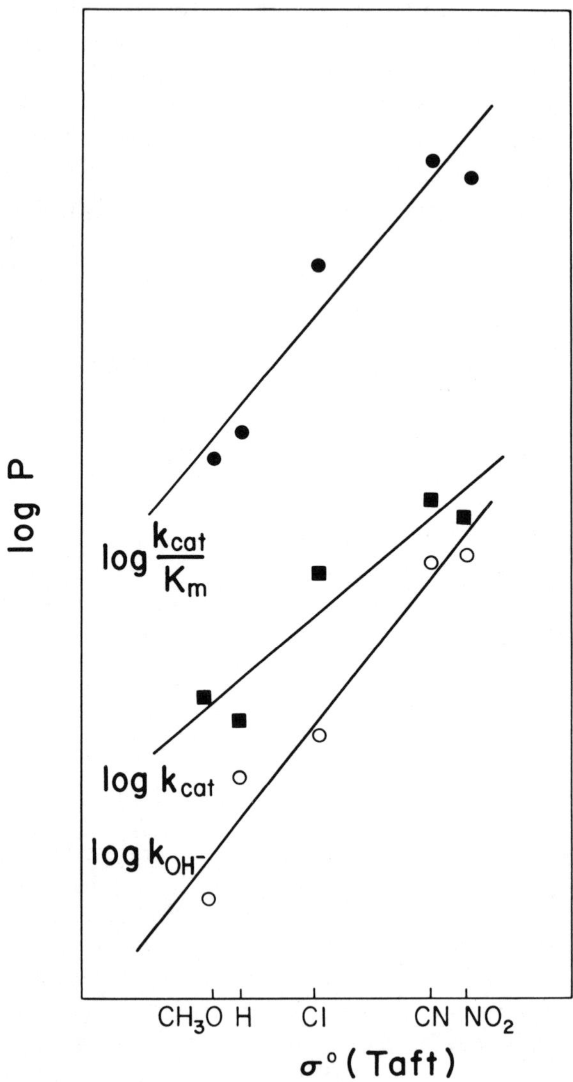

Figure 2. The dependence of the logarithms of various kinetic parameters in terms of σ^0 (Taft) is shown for the p-substituted cinnamoyl esters.

Hofmann and Bergmann, 1940) there clearly may be additional important interactions even for simple substrates. In particular, observations that removal of the zinc ion from carboxypeptidase A causes a drastic diminution of the enzyme's esterase and peptidase activities (Coleman and Vallee, 1962; Felber et al., 1962), have been interpreted to indicate that zinc ion participates directly in the catalytic process. An alternative possibility that these findings do not eliminate is that the active site zinc ion merely acts to hold the enzyme's peptide backbone in a geometry such that the reactive functional groups are positioned properly for catalysis to occur. In the case of some measurements in this laboratory with freshly prepared apoenzyme derived from a species first identified as a discrete form of carboxypeptidase, carboxypeptidase A_{δ}, but later shown to be a mixture containing mainly carboxypeptidase A_{β} (Petra et al., 1971) small but definite esterase activity was detected in assays with O-(trans-cinnamoyl)-L-β-phenyllactate (Glovsky, 1972). Despite careful attempts to avoid the problem, the residual esterase activity may have been due to metal ion contamination of the apoenzyme solution. However, the possibility that under certain conditions apocarboxypeptidase can be catalytically active may be worth exploring further.

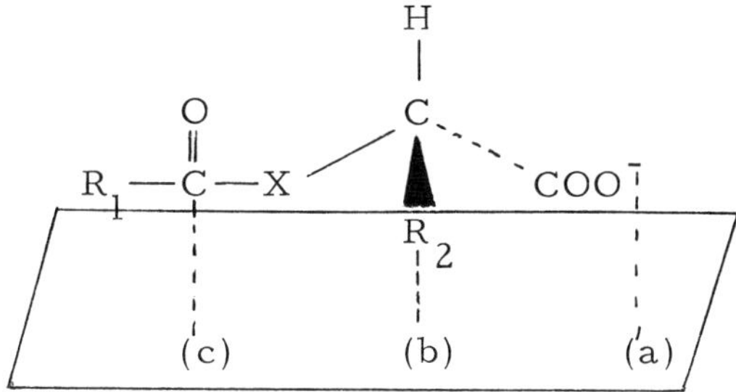

Figure 3. The interaction of carboxypeptidase with simple substrates is illustrated.

In Figure 4a, the interactions of glycyl-L-tyrosine (Gly-Tyr) with carboxypeptidase as seen by X-ray crystallographic study (Lipscomb et al., 1969) are illustrated. In one interaction the C-terminal side chain of the substrate is inserted into a pocket in the enzyme, thereby displacing several water molecules. No specific interactions of the C-terminal side chain of Gly-Tyr with residues of the protein are clearly dominant. In

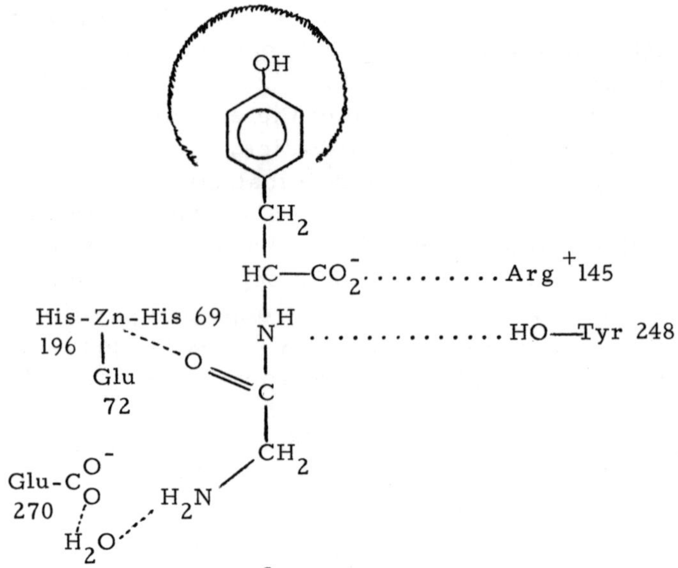

Figure 4a. Specific binding interactions of Gly-Tyr with carboxypeptidase A as deduced from X-ray crystallographic analysis are shown.

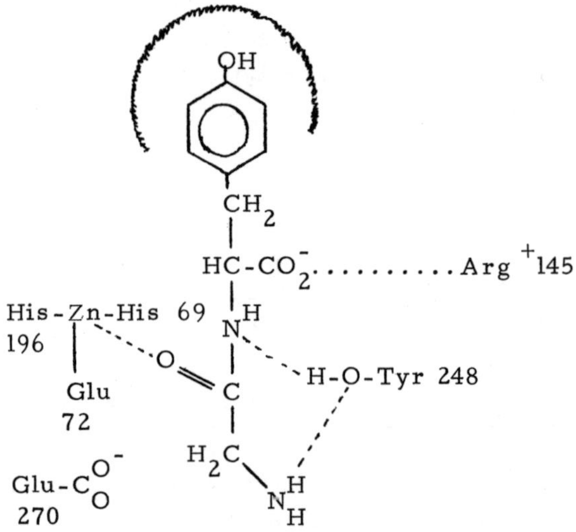

Figure 4b. The presumed productive binding of Gly-Tyr to carboxypeptidase A is illustrated (Lipscomb, 1970).

another interaction the C-terminal carboxylate group of the substrate forms a salt link with the guanidinium group of Arg 145. In a further interaction the carbonyl oxygen of the substrate's susceptible peptide bond replaces the water molecule which is present as one of the four ligands to the zinc ion. Finally, in an interaction possible only with the dipeptides which have a free terminal amino group, Glu 270 binds through a water molecule to the free amino group of Gly-Tyr. This last interaction has been proposed (Lipscomb _et al._, 1969; Hartsuck and Lipscomb, 1971) to be responsible for the very slow rate of cleavage of Gly-Tyr by carboxypeptidase A. From the results of model building studies with longer substrates (Hartsuck and Lipscomb, 1971) it has been argued that the first three interactions mentioned above are characteristic of a productive complex but the last is not. In Figure 4b the postulated productive mode of binding of Gly-Tyr is shown.

The principal points which arise from the comparison Lipscomb and his co-workers (Lipscomb _et al._, 1970) have made of the three dimensional structure of carboxypeptidase A with the complex of the enzyme with Gly-Tyr are that the only parts of the enzyme near enough to the peptide bond of the substrate to be involved directly in catalysis are Glu 270, Tyr 248 and the active site zinc ion. The only other group of the protein within 3 Å of a functional group of the substrate is Arg 145 which binds the substrate carboxylate function.

In terms of the picture developed in Figure 3 from the structure-reactivity analysis, the following assignments are consistent with the X-ray data: site a, Arg 145; site b, the enzyme pocket; and site c, the carboxylate group of Glu 270. In addition to the minimal enzyme-substrate interactions required by the structure-reactivity correlations, the X-ray results indicate that at least for peptides there is an interaction between the reactive carbonyl group of the substrate and the active site zinc ion. Also, the NH group of the scissile peptide bond is in close proximity to the phenolic hydroxyl of Tyr 248.

One of the principal aspects of enzyme reactivity which must be considered when mechanisms are developed is the pH dependency of the catalytic process. Significant differences have been observed in the pH dependencies of the rates of carboxypeptidase-catalyzed hydrolysis of various synthetic substrates. Dipeptides and the nonspecific ester substrate _O_-acetyl-L-mandelate show one type of behavior whereas tripeptides and specific ester substrates like O-(*trans*-cinnamoyl)-L-β-phenyllactate show another. For example, the bell-shaped pH-k_{cat}/K_m profile for the hydrolysis of O-acetyl-L-mandelate was postulated to reflect the ionization of two functional groups in the enzyme, one with a pK_a of 6.9 and the other with a pK of 7.5 (Carson and Kaiser, 1966). These pK values were not far from those determined from the corresponding curves for the peptides N-carbobenzoxyglycyl-L-alaninate (6.3, 8.2) and N-carbobenzoxyglycyl-L-glutamate (6.3, 8.4) (Slobin and Carpenter, 1966). The pH-rate data for dipeptides must be viewed with caution in general, however, since the hydrolytic behavior of these substrates is often complex and this could markedly affect the interpretation of the observed pK values. In agreement with X-ray and chemical modification studies (to be discussed later) it has been hypothesized (Kaiser and Kaiser, 1972) that the pK seen on the acid side (pK_1 6.3-6.9) represents the ionization of the free carboxyl group of Glu 270, while the higher pK seen in these curves (pK_2 7.5-8.4) may perhaps reflect the ionization of tyrosine 248, an assignment which is also reasonable in terms of the X-ray diffraction and chemical modification studies.

The broader pH-k_{cat}/K_m curves observed in the hydrolysis of the specific ester substrate O-(*trans*-cinnamoyl)-L-phenyllactate (Hall *et al.*, 1969) and tripeptides like N-carbobenzoxyglycylglycyl-L-phenylalaninate (Auld and Vallee, 1970) were interpreted to indicate that two ionizing groups on the enzyme are involved in these reactions, one with a pK of 6.5 and the other with a pK of 9-9.4. While the lower pK probably corresponds to the one seen for O-acetyl-L-mandelate and the dipeptides, it has been suggested that the higher pK for specific ester substrates and tripeptides could reflect the ionization of the water ligand bound to the active site zinc ion (Kaiser and Kaiser, 1972;Glovsky *et al.*, 1972). Besides the X-ray diffraction studies on native carboxypeptidase and its complexes with synthetic substrates (Lipscomb *et al.*, 1969; Lipscomb, 1970), nmr studies with manganese carboxypeptidase on enzymatically active species in which Mn^{++} has replaced the active site Zn^{++} ion indicate that this water ligand is excluded by the incoming peptide or ester substrate (Shulman *et al.*, 1966; Navon *et al.*, 1968).

While the acetylation of tyrosine residues in carboxypeptidase A eliminates enzymatic activity toward the nonspecific ester substrate O-acetyl-L-mandelate (Lipscomb *et al.*, 1968), a result consistent with the assignment of the pK 7.5 ionization seen in the native enzyme-catalyzed hydrolysis of this compound, such modification of the enzyme has little effect on the reaction of the specific substrate O-(*trans*-cinnamoyl)-L-β-phenyllactate (Glovsky *et al.*, 1972). This provides strong evidence that the presence of a proton in the phenolic group of Tyr 248 is not required for the action of carboxypeptidase A on the latter substrate.

Some support for the assignment of the pK value which is ≥ 9 to the ionization of the active site metal ion-water complex has been obtained from an examination of the pH dependency of the hydrolysis of O-(*trans*-cinnamoyl)-L-β-phenyllactate by manganese carboxypeptidase (Glovsky *et al.*, 1972). Analysis of the pH dependency of k_{cat}/K_m in the case of this manganese enzyme catalyzed reaction revealed that the alkaline pK (pK_2) was 9.3. It is of interest to note that nmr dispersion experiments indicate that there is no proton ionization from the hydration shell of manganese carboxypeptidase A in the pH range from 8 to 9 (Lipscomb, personal communication), and it thus appears from these experiments that the pK_a value for the ionization of a water ligand bound to Mn^{++} in the enzyme must be above 9. Although the implication of our analysis that the ionization constants for active site Zn^{++} and Mn^{++}-water complexes in the metallocarboxypeptidases may be very similar is surprising, no adequate inorganic models exist with which these results might be compared. In terms of the data available, the pK_2 assignments made seem reasonable.

Returning briefly to the assignment of the free carboxyl group of Glu 270 as the enzymatic group with pK_a 6.5 ± 0.4 seen in the carboxypeptidase A catalyzed hydrolysis of the synthetic substrates which have been discussed, there is good support from chemical modification for this identification. In particular, the pH dependency of the modification of carboxypeptidase A with *N*-ethyl-5-phenylisoxazolium-3'-sulfonate (Petra, 1971) or *N*-bromoacetyl-*N*-methylphenylalaninate (Hass and Neurath, 1971), both of which attack glutamate 270 with concomitant abolishment of enzymatic activity toward esters and peptides, was consistent with the assignment made. The rate of inactivation of the enzyme with these inhibitors showed a dependence on the ionization of a functional group in the enzyme with a pK of approximately 7.

Two very reasonable modes by which the carboxylate group of Glu 270 might act are as a nucleophilic catalyst and as a general base catalyst. The structure-reactivity analysis presented earlier fits a picture in which the carboxylate group acts as a nucleophilic catalyst. However, it has seemed worthwhile to examine kinetic solvent isotope effects on carboxypeptidase A-catalyzed reactions in the hope that they might shed further light on the pathway by which catalysis occurs.

In the interpretation of kinetic results a change in solvent from H_2O to D_2O can give rise to several complicating effects which include possible conformational changes in the protein, differences in the solvation of the transition state for reaction or changes in the activity of nucleophiles in the enzyme (Jencks, 1963). These complications as well as the complexity of the enzymatic reactions which may proceed through several intermediate stages must be kept in mind and necessitate a cautious approach in reaching mechanistic conclusions from the kinetic solvent isotope effect data.

When the deuterium oxide kinetic solvent isotope effect in the carboxypeptidase A-catalyzed hydrolysis of the peptide substrate N-(N-benzoylglycyl)-L-phenylalaninate was compared to that measured for the specific ester substrate O-(*trans*-cinnamoyl)-L-phenyllactate, a significant difference was seen. For the carboxypeptidase A-catalyzed hydrolysis of the dipeptide no kinetic solvent isotope effect on the value of k_{cat}/K_m was found, and the effect on k_{cat} $(k_{cat}^{H_2O}/k_{cat}^{D_2O} = 1.33 \pm 0.15)$ is small compared with the effect expected for a reaction in which proton transfer occurs in a rate determining step(Kaiser and Kaiser, 1972). As can be seen from Table 2, a substantially larger kinetic solvent isotope effect was observed for k_{cat}/K_m in the enzymatic hydrolysis of the specific ester substrate O-(*trans*-cinnamoyl)-L-β-phenyllactate, and the

$$k_{cat}^{H_2O}/k_{cat}^{D_2O}$$

ratio was close to 2. The other esters for which values are shown in the Table exhibit smaller kinetic solvent isotope effects with respect to the parameters k_{cat}/K_m than does O-(*trans*-cinnamoyl)-L-β-phenyllactate. Certain points should be noted with regard to the differences between the various esters. Apparently, *p*-nitro-substitution in the acyl group reduces the kinetic solvent isotope effect seen in the case of the parameter

k_{cat}/K_m in a given series. Furthermore, the change in the alcohol leaving group from L-β-phenyllactate, a specific leaving group, to L-mandelate, a nonspecific one, causes a significant diminution in the kinetic solvent isotope effect measured. The kinetic solvent isotope effects on k_{cat} and k_{cat}/K_m in the case of the nonspecific ester O-(trans-p-nitrocinnamoyl)-L-mandelate, thus, are negligible. This would seem to indicate that with this ester as with the peptide substrate, a catalytic step involving proton transfer is not important kinetically.

To accommodate the findings discussed, the mechanism of Figure 5 has been proposed. According to this mechanism, the carboxylate group of Glu 270 acts as a nucleophile attacking the carbonyl group of the hydrolytically susceptible ester (or peptide) bond in the substrate. The zinc ion at the active site of the enzyme may function to orient the carbonyl group of the substrate and possibly to polarize it, facilitating attack by Glu 270. A tetrahedral intermediate is formed as a result of the attack of the carboxylate function of Glu 270 on the carbonyl group of the substrate and it breaks down to give an anhydride species with concomitant formation of the α-hydroxy acid (α-amino acid)product from the C-terminal portion of the substrate. At the end, the anhydride species decomposes, regenerating the free enzyme.

The mechanism proposed accommodates the facts which we have discussed, with particular regard to the esterase function of carboxypeptidase. For example, the examination of structure-activity relationships for carboxypeptidase A substrates has led to conclusions which are consistent with the mechanism of Figure 5. Specifically, the similarity found for the effects of R_1 and X on the carboxypeptidase A and hydroxide ion-catalyzed hydrolysis of simple synthetic substrates is consistent with the underlying hypothesis of Figure 5 that the mechanism by which carboxypeptidase A-catalyzed hydrolysis reactions occur is a nucleophilic one, involving the formation of a tetrahedral intermediate. The assignments discussed above of the pK values seen in the hydrolysis of O-(trans-cinnamoyl)-L-β-phenyllactate to the ionization of the carboxyl group of Glu 270 and to that of the water ligand bound to the active site zinc ion are in accord with the proposed mechanism. In the mechanism the reactive state of the enzyme is suggested to be one in which the free carboxyl group Glu 270 is in its anionic form and the water ligand bound to the active site zinc ion is unionized.

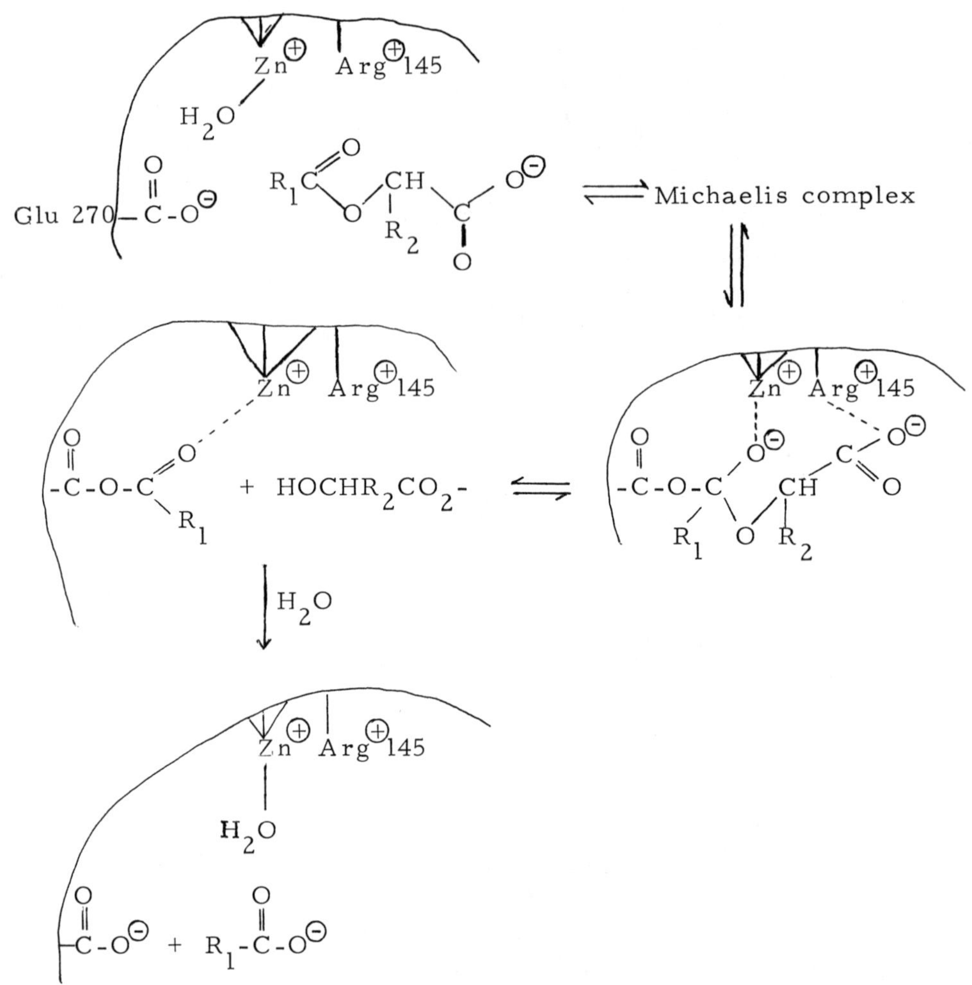

Figure 5. --Proposed mechanism of action of carboxypeptidase A as an esterase and a peptidase. The -O- in an ester is replaced by an -NH- group in a peptide. A single positive charge is shown for the zinc ion because the other positive charge is neutralized by a carboxylate ligand from the enzyme. For those substrates where modification of tyrosine residues appears to alter carboxypeptidase reactivity significantly, it is still unclear whether the binding or catalytic steps are affected.

To accommodate the absence of a deuterium oxide kinetic solvent isotope effect on k_{cat}/K_m in the carboxypeptidase A catalyzed hydrolysis of the dipeptide N-(N-benzoylglycyl)-L-phenylalaninate and of the tripeptide N-(N-benzoylglycyl)-glycyl-L-phenylalaninate (Auld and Vallee, 1970), the rate limiting step in these reactions has been proposed to be the formation of an anhydride intermediate according to the mechanism of Figure 5. Because the deuterium oxide kinetic solvent isotope effects on the parameters k_{cat}/K_m and k_{cat} for the carboxypeptidase A-catalyzed hydrolysis of the specific ester O-(*trans*-cinnamoyl)-L-β-phenyllactate are close to 2 the rate-controlling step in this reaction has been postulated to be the breakdown of an anhydride intermediate. In terms of this analysis, proceeding from top to bottom through the list of esters in Table 2, the decrease in the deuterium oxide kinetic solvent isotope effect seen is considered to reflect a gradual changeover from rate-controlling anhydride breakdown to rate-controlling anhydride formation. Our conclusion reached earlier that proton transfer is not involved in a catalytically important step in the reaction of O-(*trans*-p-nitrocinnamoyl)-L-mandelate is consistent with the postulate that anhydride formation is rate-controlling in the case of this nonspecific ester substrate just as it is for the di- and tripeptides.

Recently, in a continuation of our efforts to test the proposed mechanism of Figure 5 we have investigated the reactions of thiolester substrates with carboxypeptidase A. Specifically, we have examined the carboxypeptidase A-catalyzed hydrolysis of S-(*trans*-cinnamoyl)-α-mercapto-β-phenylpropionate over a wide pH range. Perhaps the most remarkable observation we have made with this thiolester is that both of its enantiomers react with the enzyme. This finding stands in marked contrast with the situation seen for the oxygen ester analog, O-(*trans*-cinnamoyl)-L-β-phenyllactate, where only the L-isomer is reactive. The (-)-isomer of the thiolester substrate is the more reactive one, exhibiting a bell-shaped pH dependency for the kinetic parameter k_{cat}/K_m. Two groups on the enzyme with pK_a values, respectively, of 6.2 (pK_{a_1}) and 9.0 (pK_{a_2}) are involved in the hydrolysis of the (-)-isomer, and the $(k_{cat}/K_m)_{lim}$ value found is $(1.02 \pm 0.03) \times 10^3\ M^{-1}sec^{-1}$. The pK values are reminiscent of the ones found for the reactions of carboxypeptidase A with oxygen ester analogs (Hall *et al.*, 1969; Glovsky *et al.*, 1972; Kaiser and Kaiser, 1972). The (+)-isomer of the thiolester reacts with carboxypeptidase at a respectable rate

TABLE 2

Kinetic Solvent Isotope Effects in Carboxypeptidase-Catalyzed Hydrolysis Reactions

Substrate	$\frac{(k_{cat})^{H_2O}}{(k_{cat})^{D_2O}}$	$\frac{(k_{cat}/K_m)^{H_2O}}{(k_{cat}/K_m)^{D_2O}}$	$\frac{(K_m)^{H_2O}}{(K_m)^{D_2O}}$
O-(trans-cinnamoyl)-L-β-phenyllactate	1.95 ± 0.12	1.78 ± 0.18	1.11 ± 0.17
O-(trans-p-nitro-cinnamoyl)-L-β-phenyllactate	1.51 ± 0.004	1.49 ± 0.07	0.98 ± 0.05
O-(trans-cinnamoyl)-L-mandelate	1.56 ± 0.002	1.13 ± 0.01	1.36 ± 0.01
O-(trans-p-nitro-cinnamoyl)-L-mandelate	1.09 ± 0.002	1.01 ± 0.02	1.04 ± 0.02
N-(N-benzoylglycyl)-L-phenylalaninate	1.33 ± 0.15	0.86 ± 0.34	1.55 ± 0.47

The buffer solutions contained 0.5 M NaCl and 0.05 M Tris. The pH of the H_2O buffer solutions was 7.50 ± 0.02, and the pD of the D_2O buffer solutions was 8.05 ± 0.02. Kinetic runs were performed at 25.0°.

although it is considerably lower than that seen for the (-)-isomer; $(k_{cat}/K_m)_{lim}$ = (1.66 ± 0.11) x 10^2 $M^{-1}sec^{-1}$. Besides the finding that both thiolester enantiomers react with carboxypeptidase A, the very surprising observation has been made from pH dependency studies on the (+)-isomer that the pK_{a_1} value is 8.2 for the reaction of carboxypeptidase with this slowly hydrolyzed compound. The value of pK_{a_2} is 9.6, a figure somewhat (but perhaps not significantly) higher than the pK_{a_2} values seen for the (-)-isomer and the oxygen ester analogs. Despite the very substantial difference between the pK_{a_1} value measured in the carboxypeptidase A-catalyzed hydrolysis of the (+)-isomer of the thiolester and those for the other substrates, it may be that the mechanism of reaction of this substrate is similar to that for the other substrates, although an additional ionizing group with pK_a = 8.2 interferes with the kinetic observation of the group of lower pK_a normally seen. Alternatively, the mechanism of reaction of the slowly hydrolyzed thiolester may be quite different from that for other ester substrates. A possible interpretation of the pH dependency results obtained with the (+)-isomer is illustrated in Figure 6. According to this mechanism, the catalytically active species would be one in which zinc hydroxide acts to attack the carbonyl group of the thiolester substrates with the possible assistance of the hydroxyl group of Tyr-248, functioning as an electrophilic catalyst. If this mechanism were to hold, it would mean that the interpretation of the ionization constants measured in the enzyme-catalyzed hydrolysis of the (+)-isomer is different than that which we normally have considered for the action of carboxypeptidase A on synthetic substrates. Although the mechanism given in Figure 6 requires that the group of lower pK (perhaps Tyr-248) must be in its unionized form for reaction to take place and that the group of higher pK, possibly the zinc-water complex at the active site of the enzyme, must be in the ionized form (zinc hydroxide), this reversal of the more usual interpretation of the kinetically measured pK_a values is perfectly consistent with the pH dependency of k_{cat}/K_m observed for the hydrolysis of the (+)-isomer. It must be emphasized, however, that this mechanism is an extremely speculative one, and in many ways it is unattractive to postulate that two isomers of the same substrate react by different pathways at the enzyme active site. It will certainly be necessary to establish firmly the relationship between the action of carboxypeptidase on the (+)-isomer of S-(trans-cinnamoyl)-α-mercapto-β-phenylpropionate and on the other substrates of the enzyme, and we are therefore carrying

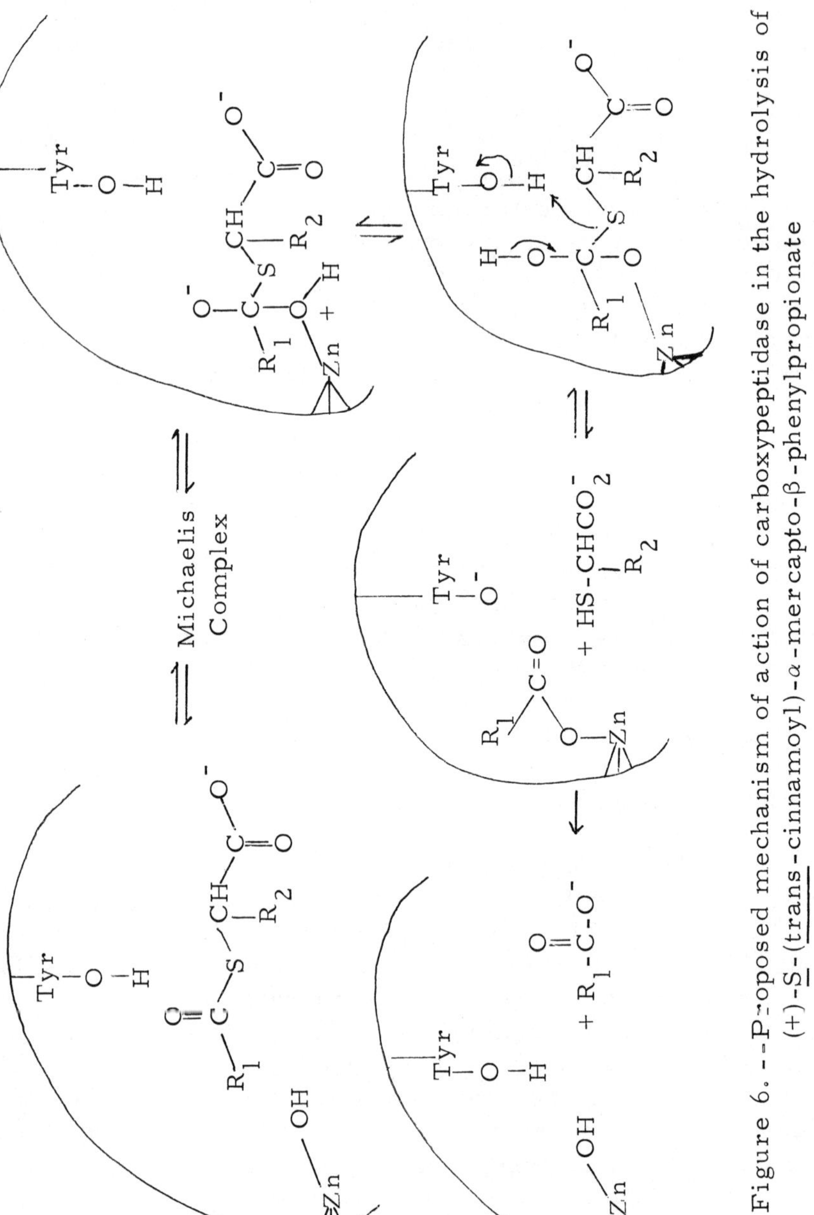

Figure 6. --Proposed mechanism of action of carboxypeptidase in the hydrolysis of (+)-S-(trans-cinnamoyl)-α-mercapto-β-phenylpropionate

out kinetic experiments with the enzyme modified at various functional groups. Additional kinetic work with different metallocarboxypeptidases would also be desirable.

Besides the need for clarification of the way in which the slowly hydrolyzed thiolester isomer reacts with the enzyme, it is of substantial interest to determine the reasons for the considerably decreased rate of hydrolysis of even the rapidly hydrolyzed isomer relative to the hydrolysis of the corresponding oxygen ester analog. It is possible that the decreased rates of reaction of the thiolesters as compared to oxygen esters reflect the less effective electrophilic catalysis of thiolester hydrolysis by the zinc ion at the active site of the enzyme. This explanation seems quite reasonable since the thiolesters appear to be substantially less reactive to electrophilic catalysis in their hydrolytic reactions compared to the behavior of the oxygen analogs, as evidenced by kinetic measurements in acid solution. From the standpoint of the catalytic action of nucleophilic groups on the enzyme such as the carboxylate residue of Glu 270, it seems unlikely that the thiolesters would be more slowly hydrolyzed than the oxygen esters; indeed, one might expect that the oxygen esters could be more reactive. For instance, the second order rate constant for the hydroxide ion catalyzed hydrolysis of O-(trans-cinnamoyl)-L-β-phenyllactate is $(4.04 \pm 0.02) \times 10^{2}$ $M^{-1}sec^{-1}$ while the corresponding rate constant for the alkaline hydrolysis of S-(trans-cinnamoyl)-α-mercapto-β-phenylpropionate is $(2.02 \pm 0.02) \times 10^{2}$ $M^{-1}sec^{-1}$ at 25°. It must be pointed out, however, that at the present time the possibility cannot be discounted that thiolesters simply bind to carboxypeptidase A in a very different manner than oxygen esters and/or that the difference in size between the oxygen atom in the reactive bond of the thiolesters is highly significant in determining the extent of productive binding of the substrate.

From the foregoing discussion it should be apparent that much remains to be done to establish the pathway of the reaction of carboxypeptidase A with ester and peptide substrates. It would be highly desirable to find direct evidence for the postulated formation of the mixed anhydride intermediate of Figure 5. Furthermore, the role of the metal ion at the active site of the enzyme in the catalytic process is perhaps the least clearly understood aspect of the enzyme's action. Indeed, the recent observation that a species proposed to be Co(III)-carboxypeptidase A catalyzed the hydrolysis of esters although it did not react with peptide substrates has raised the important question whether in the action of the native zinc enzyme on ester substrates the substrate carbonyl group can enter the inner coordination sphere of the metal ion as suggested by the scheme of Figure 5 (Kang et al., 1972). The results with the new metal ion modified enzyme leave some doubt whether the metal ion is in fact directly involved in the catalytic process at all or whether it may not simply function to hold the requisite enzyme functional groups in a geometry suitable for productive binding and reaction with the substrate. The investigation of carboxypeptidase action remains a fertile and challenging field.

Acknowledgment

The support of this research by grants from the National Institute of Arthritis, Metabolism and Digestive Diseases is gratefully acknowledged.

References

Auld, D. S. and Vallee, B. L. (1970), Biochemistry, 9, 4352.

Awazu, S., Carson, F. W., Hall, P. L. and Kaiser, E. T. (1967), J. Amer. Chem. Soc., 89, 3627.

Bender, M. L. and Kézdy, F. J. (1965), Ann. Rev. Biochem., 34, 49.

Bradshaw, R. A., Ericsson, L. H., Walsh, K. A. and Neurath, H. (1969), Proc. Natl. Acad. Sci., U.S., 63, 1389.

Carson, F. W. and Kaiser, E. T. (1966), J. Amer. Chem. Soc., 88, 1212.

Coleman, J. E. and Vallee, B. L. (1963), Biochemistry, 2, 1460.

Felber, J. P., Coombs, T. L. and Vallee, B. L. (1962), Biochemistry, 1, 231.

Glovsky, J. (1972), Ph.D. Thesis, University of Chicago.

Glovsky, J., Hall, P. L. and Kaiser, E. T. (1972), Biochem. Biophys. Res. Commun., 47, 244.

Hall, P. L., Kaiser, B. L. and Kaiser, E. T. (1969), J. Amer. Chem. Soc., 91, 485.

Hartsuck, J. A. and Lipscomb, W. N. (1971) in "The Enzymes," Boyer, P. D., editor, Academic Press, New York, p. 1-56.

Hass, G. M. and Neurath, H. (1971), Biochemistry, 10, 3535, 3541.

Hofmann, K. and Bergmann, M. (1940), J. Biol. Chem., 134, 225.

Jaffe, H. H. (1953), Chem. Rev., 53, 191.

Jencks, W. P. (1963), Ann. Rev. Biochem., 32, 657.

Kaiser, B. L. (1970), Ph.D. Thesis, University of Chicago.

Kaiser, E. T., Awazu, S. and Carson, F. W. (1964), Biochem. Biophys. Res. Commun., 21, 444.

Kaiser, E. T. and Carson, F. W. (1965), Biochem. Biophys. Res. Commun., 18, 457.

Kaiser, E. T. and Kaiser, B. L. (1972), Accounts Chem. Res., 5, 219.

Kang, E. P., Storm, C. B. and Carson, F. W. (1972), Biochem. Biophys. Res. Commun., 49, 621.

Lipscomb, W. N., Coppola, J. C., Hartsuck, J. A., Ludwig, M. L., Muirhead, H., Searl, J. and Steitz, T. A. (1966), J. Mol. Biol., 19, 423.

Lipscomb, W. N., Hartsuck, J. A., Reeke, G. N., Jr., Quiocho, F. A., Bethge, P.H., Ludwig, M. L., Steitz, T. A., Muirhead, H. and Coppola, J. C. (1968), Brookhaven Symp. Biol., 21, 24.

Lipscomb, W. N., Hartsuck, J. A., Reeke, G. N. and Quiocho, F. (1969), Proc. Natl. Acad. Sci. U.S., 64, 28.

Lipscomb, W. N. (1970), Accounts Chem. Res., 3, 81.

Lipscomb, W. N., Reeke, G. N., Jr., Quiocho, F. A. and Bethge, P. (1970), Phil. Trans. Roy. Soc. London, Ser. B, 251, 177.

Lumry, R., Smith, E. L. and Glantz, R. R. (1951), J. Amer. Chem. Soc., 73, 4330.

McClure, W. O., Neurath, H. and Walsh, K. A. (1964), Biochemistry, 3, 1897.

Navon, G., Shulman, R. G., Wyluda, B. J. and Yamane, T. (1970), J. Mol. Biol., 51, 15.

Petra, P. H. (1971), Biochemistry, 10, 3163.

Petra, P. H., Hermodson, M. A., Walsh, K. A. and Neurath, H. (1971), Biochemistry, 10, 4023.

Sampath-Kumar, K. S. V., Clegg, J. B. and Walsh, K. A. (1964), Biochemistry, 3, 1728.

Shulman, R. G., Navon, G., Wyluda, B. J., Douglass, D. C. and Yamane, T. (1966), Proc. Natl. Acad. Sci. U.S., 56, 39.

Slobin, L. I. and Carpenter, F. H. (1966), Biochemistry, 5, 499.

Taft, R. W., Jr., in "Steric Effects in Organic Chemistry," (1956), M. S. Newman, Ed., Wiley, New York, N.Y., p. 556.

Waldschmidt-Leitz, E., (1931), Physiol. Rev., 11, 358.

4

ZINC-WOOL KERATIN REACTIONS IN NONAQUEOUS SOLVENTS

N. H. Koenig and Mendel Friedman

Western Regional Research Laboratory

Agricultural Research Service

U.S. Department of Agriculture

Berkeley, California 94710, U.S.A.

We have been studying chemical treatments of keratins, especially wool textiles, in order to improve processing methods and products for the benefit of wool growers, manufacturers, and consumers. Reactions of wool with zinc compounds in water have been reported (e.g., McPhee, 1965, and Habib et al., 1973), but data are lacking for similar studies with nonaqueous solvents. Zinc has an active role in certain proteins such as carboxypeptidase A, carbonic anhydrase, and alkaline phosphatase (Coleman, 1971; cf this volume: Kaiser et al.; Riordan and Vallee). This review summarizes our work with several reactions of wool with zinc salts in organic solvents, particularly dimethylformamide (DMF).

Three types of reactions are described: (1) surface deposition of polymers from 4-vinylpyridine and zinc chloride (Koenig and Friedman, 1972); (2) internal chemical modification by cylic acid anhydrides followed by zinc acetate (Koenig and Friedman, 1972a); and (3) internal modification of wool (without pretreatment) with zinc acetate (Koenig, et al., 1974). In the first reaction, the polymer is formed from 4-vinylpyridine by coordination with zinc chloride. Deposits are formed within the wool after evaporating a volatile solvent, in which the materials have been applied. The other reactions, internal modification with zinc acetate with or without anhydride pretreatment, are done in a high-boiling solvent. The solvent is a dipolar aprotic swelling agent, such as DMF or dimethyl sulfoxide (DMSO).

SURFACE POLYMER DEPOSITION VERSUS INTERNAL MODIFICATION

Wool can be modified in two general ways: surface deposition or internal chemical modification (Whitfield and Wasley, 1964). One type of surface modification is described more fully in the section of 4-vinylpyridine-zinc chloride polymers. In this case an adherent deposit forms on all wool surfaces, but is intimately intermixed throughout.

The second type, chemical modification, requires penetration into the internal keratin structure. Wool keratin consists of polypeptide chains with attached reactive groups such as amino, hydroxyl, amide, sulfhydryl, phenolic, guanidino, imidazole, and carboxyl. The amino acid chains of the wool molecule are held together laterally by electrostatic bonds, hydrogen bonds, hydrophobic interactions, dipole attractions and disulfide bonds. The disulfide (cystine) bonds crosslink the wool, giving a giant molecule that is insoluble. Therefore, some molecules that react with soluble proteins will not modify wool because they are too large or of unsuitable polarity to penetrate the wool molecule. The presence of internal structural differences such as crystalline and noncrystalline regions adds further complexity to wool modification.

Usually, only very small molecules can easily get into the close-knit wool structure. However, in the presence of liquids that swell wool, medium size molecules can enter also. In recent years, a few special nonaqueous liquids, particularly DMF (Koenig, 1962) and DMSO (Koenig, 1961), have been found suitable for internal chemical modification of wool. Moreover, the course of many chemical reactions is profoundly affected by dipolar aprotic solvents (Parker, 1969) such as DMF or DMSO (Friedman, 1967).

SOLVENT EFFECTS IN PROTEIN REACTIONS

Protein-solvent interactions may be illustrated with DMSO, in which the following two main factors influence the course of reaction (Friedman, 1968; Friedman and Koenig, 1971).

(1) Preferential solvation of positive charges by DMSO leaves negative charges and other nucleophilic centers free from destabilizing influences of positive charges and ion pair aggregation.

(2) The strong dipole of the sulfoxide group in DMSO acts as a hydrogen-bond acceptor and alters acid-base equilibria, nucleophilic and electrophilic reactivities of reactants, and stabilities of transition states.

These factors imply that chemical reactivities of protein functional groups should be subject to strong DMSO-solvent effects for the following reasons:

(A) Proteins are polyelectrolytes in which the unequal distribution of positive and negative charges except at the isoelectric point depends on the amino acid composition of each particular protein.

(B) Hydrogen-bonding and hydrophobic interactions of peptide bonds and side chain groups are responsible for conformations that proteins assume in solution; in turn, the geometric details, helicity and folding influence chemical reactivities of functional groups.

(C) The degree of ionization of protein functional groups, which determines the concentration of each nucleophilic species at any given pH, and the inherent nucleophilic reactivities of protein functional groups should depend on the dielectric constant, dipole moment, and hydrogen-bonding abilities of their solvent environment.

(D) At room temperature, the effects of DMSO on wool are subtle, primarily to swell the wool, enhance reactivities of basic functional groups, and facilitate sulfhydryl-disulfide interchange. In the range 80-120°C, DMSO is a powerful medium for accomplishing chemical modification. At the same time, unless suitable reagents are present to modify side chains in wool, hot DMSO alone can collapse the helical structure, resulting in marked shortening of wool fibers.

Although DMSO exerts powerful pharmacological effects on animal organisms, the chemical basis of this biological activity is still largely unknown. We suggest that wool is a useful model for studying the molecular mechanism by which DMSO alters properties of hair, skin, and related body proteins. Zinc is an essential trace mineral in nutrition, is required for protein synthesis, and has been studied in hair (Hambidge et al., 1972).

EXPERIMENTAL PROCEDURE

The fabric was a scoured, undyed, worsted flannel, 6.5 oz/yd^2, cut into circles (8.0 cm, 1.2g). These were exhaustively extracted with trichloroethylene followed by ethanol. The wool was dried before treatment, except for anhydride-pretreated wool, which was at laboratory moisture content. In general, reagents and solvents were of good commercial grade. 4-Vinylpyridine was Eastman's[1]

---

[1]Reference to a company or product name does not imply approval or recommendation of the product by the U.S. Department of Agriculture to the exclusion of others that may be suitable.

practical grade, bp 53-55°C. Zinc acetate was the hydrate $Zn(C_2H_3O_2)_2 \cdot 2H_2O$, the Baker Analyzed reagent. Succinic anhydride was Eastman's highest grade.

Weighed samples were treated in Petri dishes (Koenig, 1961). A solution of 4-vinylpyridine and zinc chloride in acetone was added to the wool in a Petri dish. The sample was heated in an oven to evaporate the acetone and complete the reaction. Other samples were pretreated with succinic anhydride or citraconic anhydride in DMF as previously described (Koenig, 1965). Samples of native wool (no pretreatment) were treated similarly with zinc salts in DMF. All treated samples were then rubbed gently in hot 2-butanone, extracted overnight with ethanol in a Soxhlet apparatus, oven-dried, and reweighed.

4-VINYLPYRIDINE-ZINC CHLORIDE POLYMERS

Reaction Conditions. A durable, microscopically rough surface deposit forms on wool and other substrates when they are heated with 4-vinylpyridine and zinc chloride applied in acetone. No deposit forms when the reagents are applied singly. Since zinc chloride is known to enhance polymerization of vinylpyridine (Tazuke and Okamura, 1967), the deposite is probably poly(4-vinylpyridine) with coordinately bound zinc. We believe that this or analogous treatments may be useful for modifying various fibrous substrates or for preparing metal ion adsorbents.

TABLE I

Treatment of Wool with Varying Amounts of 4-Vinylpyridine and Zinc Chloride in Acetone[a]

4-Vinylpyridine g	Zinc Chloride g	Weight Gain %
0.5	0.5	27
0.2	0.5	24
0.5	0.2	3
0.0	0.5	3

[a]1.2 g dry wool (8 cm circle), 10 ml acetone, 30 min, 120°C.

The relative influence of 4-vinylpyridine and zinc chloride in wool treatments is illustrated in Table I. Both compounds are necessary; moreover, a higher ratio of zinc chloride gives a higher weight gain.

Surface Deposits on Fibrous Substrates. To see whether the reaction is dependent on wool as a substrate, other fibrous materials were treated similarly (Table II). The weight gains are nearly equal for four materials of varied chemical structure. These weight gains are a little greater than the amount of 4-vinylpyridine used. These results suggest that the reaction does not depend on the number or type of particular reaction sites in the substrates. Since the treatment is effective with various substrates, it has potentially wide application.

Microscopic examination (Figure 1) shows extensive deposits on all external fiber surfaces. Non-uniform pieces of reaction product adhere to the fibers, here pulled apart to show this clearly. These observations suggest that an insoluble product forms on all available surfaces and within free spaces of the substrate mass.

Nature of Reaction. 4-Vinylpyridine has a basic nitrogen atom that strongly coordinates various transition metal ions including zinc (II), copper (II), cobalt (II) (Agnew and Brown, 1971), and mercury (II) (Friedman and Waiss, 1972). Crystalline complexes with a ratio of 2 moles of 4-vinylpyridine per mole of zinc chloride have been isolated. In solution, a mixture of 2:1 and 1:1 complexes can exist (Tazuke and Okamura, 1967). These complexes are intermediates in the polymerization of 4-vinylpyridine.

We suppose that heating the substrate in the presence of 4-vinylpyridine and zinc chloride forms a surface or interstitial polymer on and around the substrate. The pendant pyridine groups

TABLE II

Application of 4-Vinylpyridine and Zinc Chloride to Various Substrates[a]

Material	Gross Structure	Weight Gain %
Wool	Worsted flannel	56
Leather	Oil-tanned chamois	52
Cotton	Woven towel	48
Polyester	Double knit	56

[a] 0.5-0.6 ml 4-vinylpyridine, 1.0-1.5 g zinc chloride.

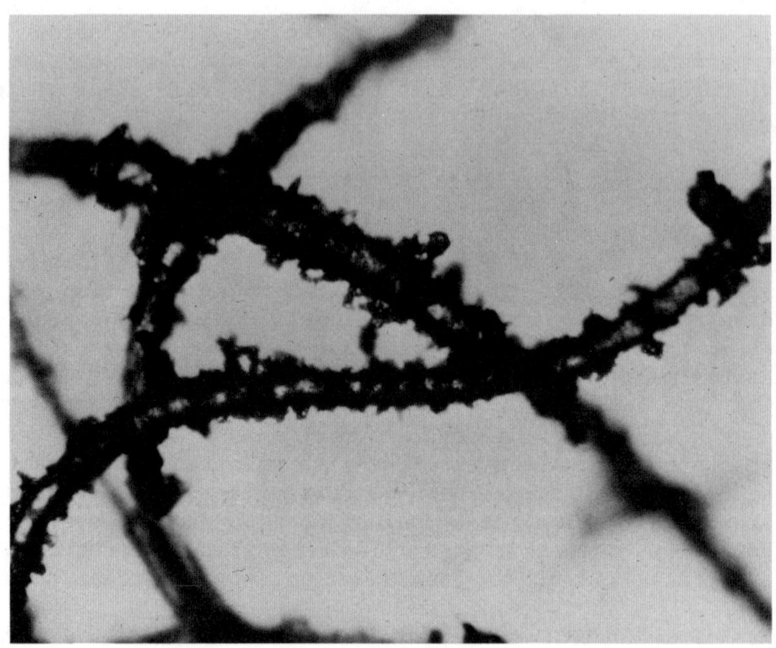

Figure 1. Wool fibers treated with 4-vinylpyridine-zinc chloride.

of the polymer may be free, or may coordinate through nitrogen with a zinc chloride molecule. Since zinc chloride can coordinate two ligands, it may also be bonded to wool or other suitable substrates. Possibly, under favorable stoichiometric and steric conditions, the zinc chloride may coordinate two pyridine rings to crosslink the polymer.

Elemental analyses indicate that treated wool samples contain substantial amounts of zinc and chlorine but with no easily identified stoichiometry. Ultraviolet spectroscopic analyses of hydrolysates of treated wool show the characteristic pyridine absorption at 255 nm.

ANHYDRIDE PRETREATMENT

Modification of New Carboxyl Groups by Zinc Acetate. When wool is chemically modified by cyclic acid anhydrides, the reaction introduces new carboxyl groups (Koenig, 1965; Whitfield and Friedman, 1973), which are potential sites for additional modification. We shall discuss primarily the reaction of zinc acetate in DMF with

TABLE III

Modification by Zinc Acetate[a] After Succinic Anhydride[b]

Time hr	Temp °C	Additional Weight Gain, %
0.25	110	10
0.50	110	15
1.0	110	15
1.5	110	16
1.0	25	0
3.0	25	4

[a] Succinic anhydride modified wool, 1.0 g zinc acetate, 6 ml DMF.

[b] Replicate samples, 1.2 g dry wool, 1.0 g succinic anhydride, 6 ml DMF, 0.75 hr, 110°C, 15% weight gain.

succinic anhydride modified wool. The treatment is designed to neutralize and crosslink the many carboxyl groups introduced by succinic anhydride.

Table III shows the effects of time and temperature on weight gain due to zinc adduct by succinic anhydride modified wool. The reaction with zinc acetate leveled off at 15% weight gain in about 30 min at 110°C. At room temperature reaction was much slower, as usual for wool treatments in DMF (Koenig, 1962; 1965).

Nature of Reaction. Reaction with zinc acetate, already shown by weight increase, markedly decreases felting shrinkage (Table IV). Surprisingly, analogous zinc chloride treatment gave no weight gain or shrinkage protection. One possible explanation is that acetate ion combines with the proton of the carboxyl group to form acetic acid, which is volatilized by heating, leaving zinc-carboxylate groups. Binding of zinc to carboxylate occurs in the enzyme carboxypeptidase A (Quiocho and Lipscomb, 1971). Moreover, since zinc readily coordinates various ligands, the zinc atoms are probably bound to other wool sites by secondary valences. Accordingly, acetate ion may assist in replacing a carboxyl group proton by a zinc atom, which may also coordinate other ligand groups in wool.

To test this hypothesis, we also treated succinic anhydride modified wool with cadmium acetate, a related compound since cadmium is in the same Periodic Table column as zinc. In agreement with our hypothesis, we observed a weight gain of the same order of

TABLE IV

Treatment with Various Salts After Succinic Anhydride Modification[a]

Salt[b]	Additional Weight Gain,%	Area Shrinkage[c] %
Zinc acetate	14	3
Zinc chloride	0	39
Cadmium acetate	19	20
Calcium acetate	1	37

[a]Replicate samples, 1.2 g dry wool, 1.0 g succinic anhydride, 6 ml DMF, 0.75 hr, 110°C, 15% weight gain.

[b]Succinic anhydride modified wool, 1.0 g salt, 6 ml DMF, 0.75 hr, 110°C.

[c]Untreated wool, 37-43%. Milled in an Accelerotor at 1780 rpm for 2.0 min with 200 ml of 0.5% sodium oleate solution at 40°C.

magnitude as with zinc acetate, although shrinkage protection was lower (Table IV). By contrast, another divalent cation acetate, calcium acetate, did not react. This also supports our hypothesis because calcium differs greatly from zinc in its coordination behavior.

Mechanism of Shrink Resistance. Table V shows the Accelerotor shrinkage of various modified samples. Although larger anhydrides

TABLE V

Accelerotor Shrinkage of Modified Wool Fabrics

Treatment	Weight Gain %	Area Shrinkage[a] %
Succinic anhydride only	17	44
Succinic anhydride followed by zinc acetate[b]	14[c]	3
Citraconic anhydride only	11	43
Citraconic anhydride followed by zinc acetate[b]	13[c]	3

[a]Untreated wool, 37-48%. Accelerotor milled as in Table IV.

[b]Anhydride modified wool, 1.0 g zinc acetate, 6 ml DMF, 0.75 hr, 110°C

[c]In addition to anhydride weight gain.

impart shrinkage protection (Koenig, 1965), wool modified only by succinic or citraconic anhydride was not protected. However, subsequent zinc acetate treatment produced nearly shrinkproof wool. Both steps are needed. Zinc acetate treatment of unmodified wool does not protect against shrinkage. Although Accelerotor tests show that the two-step treatment gives excellent shrinkage protection, preliminary data indicate less protection under household laundering conditions. The longer laundering tests possible allow time for diffusion wherein zinc is replaced by protons.

The question arises as to the mechanism of increased shrink resistance. The effect may relate to increased electrostatic repulsion caused by converting carboxyl groups to a cationic derivative (Stigter, 1971). Additional factors may contribute, e.g., crosslinking of carboxylate or other wool groups by zinc atoms and coordination of the zinc atoms to other nearby ligands—amino, imidazole, guanidino, phenolic groups.

COMPARISON OF ZINC SALTS

Solvent Effects with Zinc Acetate. The preceding section described wool made shrink resistant with succinic anhydride followed by zinc acetate, a two-step process with DMF as solvent in both steps. While running controls without anhydride pretreatment, we noted an unexpectedly high weight gain for native wool treated with zinc acetate. Consequently, we compared properties of zinc derivatives of native wool and wool enriched in carboxyl groups in more detail. Although anhydride pretreatment is needed for shrink resistance, wool treated with zinc acetate alone is of interest because it has improved insect resistance.

Chemical modification of wool with zinc acetate is feasible in several polar organic swelling solvents (Table VI). It is possible

TABLE VI

Modification by Zinc Acetate in Nonaqueous Solvents[a]

Solvent	Time hr	Weight Gain %
DMF	0.75	14
DMSO	0.50	12
1-Methyl-2-pyrrolidinone	0.75	17
Ethylene glycol	0.75	16

[a]1.2 g dry wool, 2.0 g zinc acetate, 6 ml solvent, 110°C.

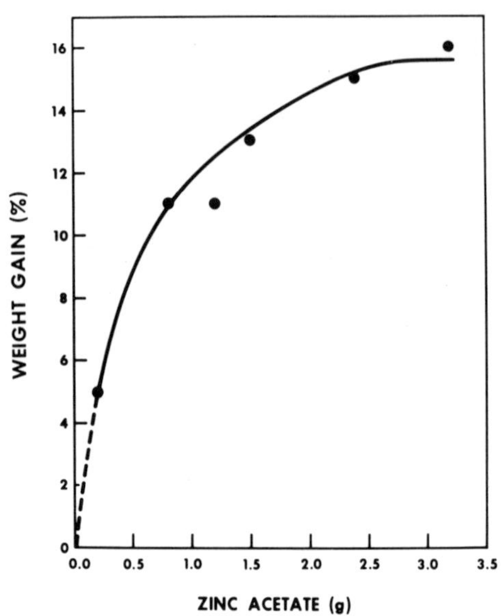

Figure 2. Effect of zinc acetate concentration on weight gain. 1.2 g dry wool, 6 ml DMF, 45 min 110°C.

that zinc acetate reacts with ethylene glycol, since this is a protic solvent, but there is also a high degree of reaction with wool. The low solubility of zinc acetate in less polar organic solvents is a restrictive factor, as is the poor swelling of wool by such solvents.

Evidence for Internal Chemical Modification. Figure 2 relates weight gain to concentration of zinc acetate. The curve rises sharply initially, then more gradually until it apparently levels off at about 16%. The curve could not be extended because zinc acetate has limited solubility. There appears to be a limit to the amount of zinc acetate that reacts with a given weight of wool, despite a large excess of reagent at higher concentrations. Consequently, the data suggest that the reagent penetrates the wool structure and reacts with all accessible sites. Internal chemical modification in DMF has been discussed for other reagents, isocyanates (Koenig, 1962), and anhydrides (Koenig, 1965).

The effect of temperature on reaction rate was studied also, with 1.5 g zinc acetate in 6 ml DMF (Table VII). Reaction was fast at 110°C., reaching 12% weight gain in 1/2 hr and leveling off at 13%. Reaction was slower at 80°C., but had a similar maximum weight gain (12%) in 1 hr. There was a tremendous rate decrease when the temperature was lowered to 26°C., only 1% weight gain in 2 hr, just as occurs for other modifications in DMF (Koenig, 1962, 1965).

TABLE VII

Rate of Wool Modification by Zinc Acetate at Various Temperatures[a]

Temperature °C	Reaction Time hr	Weight Gain[b] %
26	2.0	1
26	24	9
26	168	11
80	0.5	8
80	1.0	12
80	2.0	12
110	0.50	12
110	0.75	13
110	1.0	12
110	1.5	13

[a]1.2 g dry wool, 1.5 g zinc acetate, 6 ml DMF.

[b]Values based on duplicate runs.

However, the same maximum was approached, 11% weight gain in 168 hr, as at higher temperatures. Accordingly, the same limiting weight gain at different temperatures (Table VII), and at high concentrations (Figure 2) both suggest that the wool is chemically modified.

The zinc acetate may be neutralizing wool carboxyl groups, displacing a proton and forming zinc-carboxylate ionic linkages. Since zinc cations also form complexes with donor groups (ligands) (Cotton and Wilkinson, 1966), zinc may also be bound by coordinate (semi-ionic) linkages to wool ligands such as amino, imidazole, guanidino and phenolic groups. We presume that binding of the zinc cation and associated anion is by ionic or semi-ionic (coordinate) linkages rather than normal covalent bonds. Ionic bonds, based on electrical attraction, should be broken by water more easily than the covalent bonds usually formed in durably chemically modified wool. Ionic character accords with preliminary data indicating the treatment has low stability toward water washing. A preliminary experiment indicates that subsequent treatment with aqueous formaldehyde improves water durability of the zinc acetate step.

Zinc Acetate Versus Zinc Halides. Comparison of zinc salts in Table VIII shows that zinc acetate (13% weight gain) is much more reactive than the two zinc halides (5% weight gain). The ratio of weight gains for the different salts is in line with the ratio of zinc analyses, which suggest that the wool also contains some added anion. The contrast between zinc acetate and zinc chloride is even

TABLE VIII

Comparison of Zinc Acetate and Zinc Halides[a]

Salt	0.75 hr Weight Gain,%	0.75 hr Zinc in Wool, %[b]	1.5 hr Weight Gain,%	1.5 hr Zinc in Wool, %[b]
Zinc acetate	13	7.6	13	9.2
Zinc chloride	5	3.7	5	3.9
Zinc bromide	5	3.2	6	3.8

[a] 1.2 g dry wool, 1.5 g salt, 6 ml DMF, 110°C, based on duplicate runs.

[b] Determined by X-ray fluorescence spectroscopy.

greater if wool is pretreated with succinic anhydride. In this case, the relative values were 14% additional weight gain for zinc acetate and 0% for zinc chloride (Koenig and Friedman, 1972a).

To explain why zinc chloride reacts less with anhydride-treated wool than with native wool, we presume that zinc chloride can react with amino (basic) groups only, whereas zinc acetate can react with both amino and carboxyl groups. Unlike the chloride, the acetate can form a slightly ionized volatile product, acetic acid, loss of which may be the driving force for neutralizing carboxyl groups. Reaction with succinic anhydride decreased the number of available amino groups, forming amide side chains which do not bind zinc salts, and increases the number of carboxyl groups (Koenig, 1965). The reactive amino groups are thus replaced with an equal number of reactive carboxyl side chains. Since zinc acetate reacts with both amino and carboxyl groups, the weight gain of zinc acetate is essentially the same for native wool and anhydride-treated wool. On the other hand, anhydride treatment greatly reduces the weight gain of zinc chloride, because it removes amino groups that could react with zinc chloride and forms new carboxyl groups that do not react with zinc chloride.

The weight gain when zinc salts react with native wool would tend to decrease in the order of anion weight - bromide, acetate, chloride - to the extent that anions remain in the dry products. Another possible factor in weight gain relates to acid strength. Reaction of amino groups with zinc chloride can form HCl, which would compete for remaining amino groups. Because HCl is a strong acid, amino-hydrochloride formed by zinc chloride could be more stable than amino-acetate formed by zinc acetate. Accordingly, zinc chloride would react with fewer amino groups than zinc acetate and the weight gain with zinc chloride would be less.

CONCLUDING REMARKS

Wool, cotton, polyester, and leather were durably modified by 4-vinylpyridine and zinc chloride applied in acetone. The pendant pyridine rings of the surface deposit, probably poly(4-vinylpyridine), may be free or coordinated through nitrogen and zinc, which may also be bonded to the substrate.

Wool fabrics treated with 4-vinylpyridine and zinc chloride feel reasonably soft and flexible at weight gains up to about 20% but not at higher levels. A potential benefit of the treatment is the decreased laundering shrinkage. No shrinkage occurred when a sample at 24% weight gain was milled in an Accelerotor as in Table IV, but for 6.0 min. A wool control sample milled simultaneously had 41% area shrinkage.

Modification of wool by cyclic acid anhydrides such as succinic or citraconic anhydride introduces new carboxyl groups. Subsequent reaction of the modified wool with zinc or cadmium acetate in hot DMF further increases fabric weight by over 15% and retards felting shrinkage. Under similar conditions, neither zinc chloride nor calcium acetate is effective. A possible explanation of these effects is offered.

Native wool (no anhydride pretreatment) also reacts with zinc acetate in hot DMF. After extraction with 2-butanone and ethanol, the fabric gains up to 16% in weight. Concentration and temperature studies suggest there is internal chemical modification of the wool. Relative weight gains for wool and wool pretreated with succinic anhydride suggest that zinc acetate can react with both amino and carboxyl groups, whereas zinc chloride can react with amino groups only.

Unlike wool pretreated with succinic anhydride, treatment with zinc acetate alone does not impart shrink resistance; hence soil should be removed by drycleaning. Fastness to drycleaning is expected, because no significant weight loss occurred when treated samples were extracted with drycleaning-type solvents. Treated samples were soft, flexible and slightly yellowed. At higher weight gain, the fabric was slightly harsh, with some treatment shrinkage.

Increased concern with toxicity of finishing chemicals has prompted our interest in chemicals considered safe, such as zinc salts. The zinc acetate treated fabric was moderately resistant to attack by carpet beetle larvae.

ACKNOWLEDGEMENTS

We are indebted to Miss Linda Brown, Mrs. Iva Ferrel, Miss Diane Gardner, Mrs. Mary Muir, and Mr. Eddie Marshall for technical assistance, and to Dr. Wilfred Ward for a critical manuscript review.

REFERENCES

Agnew, N. H. and Brown, M. E. (1971). Solid-state polymerization of vinylpyridine coordination compounds. J. Polymer Sci. A-1, 9:2561-2574.

Coleman, J. E. (1971). Metal ions in enzymatic catalysis, in Progress in Bioorganic Chemistry, Volume 1, edited by Kaiser, E. T. and Kezdy, F. J., Wiley-Interscience, New York.

Cotton, F. A. and Wilkinson, G. (1966). Advanced Inorganic Chemistry, 2nd Ed., Interscience, New York, pp. 604, 610.

Friedman, M. (1967). Solvent effects in reaction of amino groups in amino acids, peptides, and proteins with α,β-unsaturated compounds. J. American Chem. Soc. 89:4709-4713.

Friedman, M. (1968). Solvents effects in reactions of protein functional groups, Quarterly Repts. Sulfur Chem. 2:125-144.

Friedman, M. and Koenig, N. H. (1971). Effect of dimethyl sulfoxide on chemical and physical properties of wool. Textile Res. J. 40:605-609.

Friedman, M. and Waiss, A. C. (1972). Mercury uptake by selected agricultural products and by-products. Environ. Sci. Technol. 6:457-458.

Habib, F. K., Parkin, M. and Whewell, C. S. (1973). The action of zinc and cadmium amines on wool. J. Textile Inst. 64:112.

Hambidge, K. M., Hambidge, C., Jacobs, M., and Baum, J. D. (1972). Low levels of zinc in hair, anorexia, poor growth, and hypogeusia in children. Pediat. Res. 6:868-874.

Koenig, N. H. (1961). Isocyanate modification of wool in dimethyl sulfoxide. Textile Res. J. 31:592-596.

Koenig, N. H. (1962). Modification of wool in dimethylformamide with mono- and diisocyanates. Textile Res. J. 32:117-122.

Koenig, N. H. (1965). Wool modification with acid anhydrides in dimethylformamide. Textile Res. J. 35:708-715.

Koenig, N. H. and Friedman, M. (1972). Surface modification of wool and other fibrous materials by 4-vinylpyridine and zinc chloride. Textile Res. J. 42:319-320.

Koenig, N. H. and Friedman, M. (1972a). Reaction of zinc acetate with wool carboxyl groups derived from cyclic acid anhydrides. Textile Res. J. 42:646-647.

Koenig, N. H., Muir, M. W. and Friedman, M. (1974). Reaction of wool with zinc acetate in dimethylformamide. Accepted by Textile Res. J.

McPhee, J. R. (1965). Reaction of wool with sodium sulfite and Zn^{++}, Textile Res. J. 35:383-384.

Parker, A. J. (1969). Protic-dipolar aprotic solvent effects on rates of biomolecular reactions. Chem. Rev. 69:1-32.
Quiocho, F. A. and Lipscomb, W. N. (1971). Carboxypeptidase A: A protein and an enzyme, Adv. Protein Chem. 25:1-78.
Stigter, D. (1971). On a correlation between the surface chemistry and the felting behavior of wool, J. American Oil Chem. Soc. 48:340-343.
Tazuke, S. and Okamura, S. (1967). Effects of metal salts on polymerization. Part III. Radical polymerizabilities and infrared spectra of vinylpyridines complexed with zinc and cadmium salts, J. Polymer Sci. A-1,5:1083-1099.
Whitfield, R. E. and Friedman, M. (1973). Flame-resistant wool. III. Chemical modification of wool with chlorendic and related halo-organic acid anhydrides, Textile Chem. Colorist,5:76-78.
Whitfield, R. E. and Wasley, W. L. (1964). Reactions of proteins, in Chemical Reactions of Polymers, edited by Fettes, E. M., Interscience, New York.

5

HEME-PROTEIN-LIGAND INTERACTIONS

J. B. Fox, Jr., M. Dymicky, and A. E. Wasserman

Eastern Regional Research Center, Agricultural Research Service, U.S. Department of Agriculture, Phila., Pa. 19118

Recent studies have suggested the use of nitrogenous heterocyclic compounds as color forming reagents in cured meats (Tarladgis, 1967; Howard et al., 1973). During screening in frankfurters of a variety of such compounds, including a series of substituted pyridines, we observed the production of a variety of colors. The principal colors were: purple, produced by strong electron acceptors in the 4 position on the pyridine nucleus; orange, produced by methylpyridine derivatives; and red or pink, produced by all other substituents in the 3 or 4 position. These results led us to a more detailed study of the spectra of these pigments, using bovine myoglobin in pH 5.5 buffer which is approximately the pH of meat.

NATURE OF THE HEMOCHROME

In order to form the hemochrome, it is necessary to denature the protein. The heme of myoglobin is partly buried in the surface of the protein (Figure 1), resulting in steric hindrance of ligand binding to the heme, an effect first observed by St. George and Pauling (1951). The cleft in the protein which contains the heme is too narrow to admit the pyridine nucleus and the protein must be partly denatured. Sodium lauryl sulfate (SLS) not only denatures the protein, producing a soluble hemochrome (Howard et al., 1973) but also solubilizes the heme (Simplicio, 1972). The nature of the SLS-denatured globin hemochrome has been defined only at pH 8.0, by Van den Oord and Wesdorp (1969) who found that the binding of SLS to myoglobin at pH 8.0 required 17 moles of SLS to completely denature a mole of pigment. Our results of the titration of metmyoglobin with SLS at pH 5.5 are shown in Figure 2. The initial precipitation reaction was linear with SLS concentration and was complete at a ratio of 15 moles of SLS/mole of myoglobin. Upon continued addition of SLS, the pigment was resolubilized. The sigmoid shape of the curve implies an equilibrium process which began at a 28 molar excess and was completed at a 55 molar (0.1% SLS) excess. Katz et al. (1973) recently reported that a point of maximum volume decrease of myoglobin occurs at a molar ratio of 60:1 (SLS:Mb), which is close to the value we have observed for the resolubilization of the pigment. Assuming the protein is still part of the resolubilized hemochrome, the total molecular weight of the micelle is about 32,000 D, composed of 18,000 D of myoglobin and 14,000 D of SLS. We assumed the globin was part of the micelle because: 1. The spectrum of the SLS-denatured globin hemochrome has absorption maxima at 530 and 570 nm at both pH 5.5 and pH 8.0, typical of the diimidazolyl hemochrome (Van den Oord and Wesdorp, 1969). Only the protein can furnish the imidazole groups, probably from the F8 (proximal) and E7 (distal) histidine residues. 2. Only 0.1% SLS was required at pH 5.5 to resolubilize 0.050 mM metmyoglobin (MetMb), whereas Simplicio (1972) found it required 1.0% SLS to solubilize hemin. The difference we attribute to the solubilizing effect of the protein.

FORMATION OF THE HEMOCHROMES

Solutions containing 0.025 mM MetMb and 0.1% SLS in $\Gamma/2 = 50$ mM acetate buffer, pH 5.5, were used to study the kinetics of formation and spectra of the substituted pyridine hemochromes. Ligand concentration varied from 0.1 to 25 mM. The reference was a buffer solution and temperature was controlled to 20.0° ± 0.1 in a Cary 14 spectrophotometer. The reaction was started with excess $Na_2S_2O_4$. The spectra of three typical reaction products are shown in Figure 3. The two major bands at ca. 530 and 555-560 nm are due to porphyrin ring π electron excitations and were common to all substituted pyridine hemochromes. Absorption bands at ca. 480 nm were typical of the methyl substituted pyridine hemochromes, whereas bands in the 600-750 nm region were typical of the hemochromes formed from pyridine derivatives with strong electron acceptor groups in the 4 position. These bands are due to charge transfer between the iron and the ligand (Brill and Williams, 1961).

KINETICS

Akoyunoglou et al. (1963) had difficulties obtaining kinetic and equilibrium data for the formation of nitrogenous base-heme complexes, indicating that more than a simple equilibrium between heme and ligand was involved. We had similar difficulties determining the rates of formation of these complexes as well as the

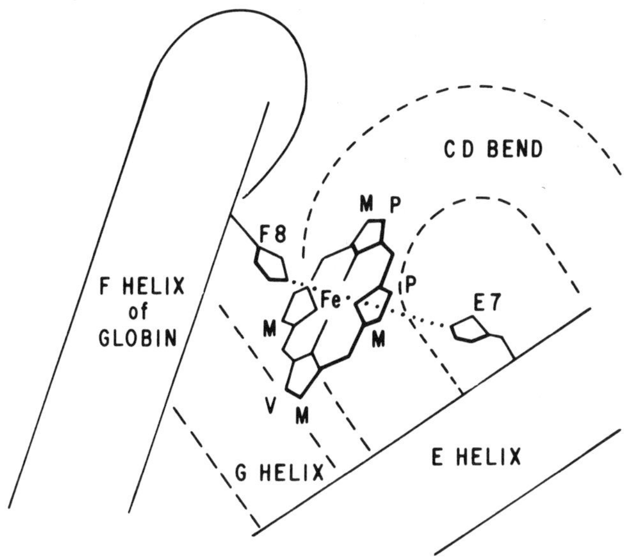

Figure 1. Position of the heme in myoglobin.

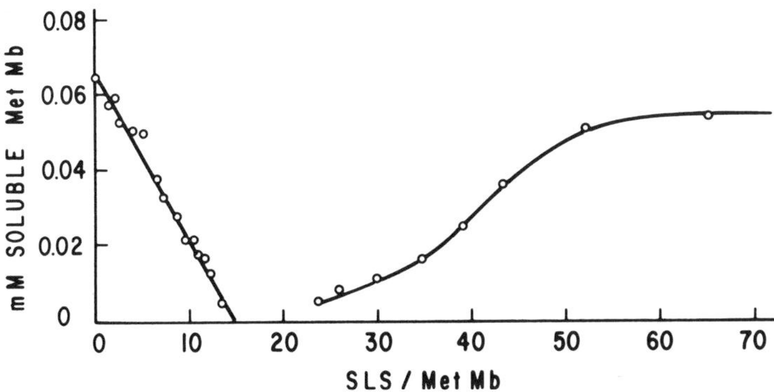

Figure 2. Precipitation and resolubilization of metmyoglobin by sodium lauryl sulfate.

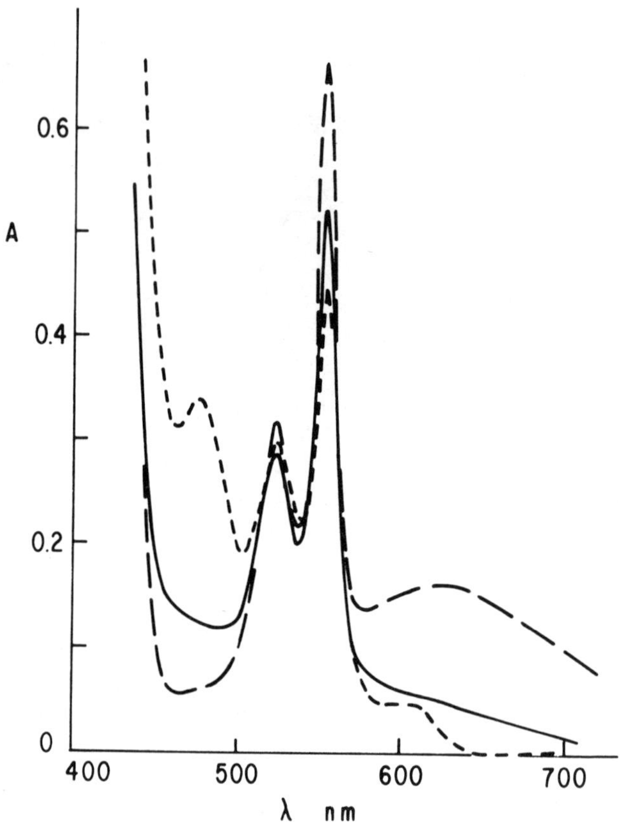

Figure 3. Substituted pyridine SLS-denatured globin hemochromes. ——, 3-acetylpyridine (rose), — —, 4-carboxymethylpyridine (purple), and ----, 3-methylpyridine (orange).

order of the reaction. With large molar excesses of substituted pyridines, the reaction was first order with respect to the heme pigment concentration. These first order rate constants were dependent on the substituted pyridine concentration, but the dependence was greater than third order. If pyridine concentration was less than 1-2 mM, no complex formation was observed, whereas at 25 mM the reaction went too fast to follow. The process of binding 1 or 2 molecules of ligand to the heme is apparently a complex multiple ligand reaction with the SLS-denatured globin heme micelle.

At ligand concentrations of 25 mM and higher, the reaction took place in two stages. The first stage was completed in seconds, and was characterized by the formation of the typical absorption spectra, with the λ_{max} lying between 554 and 555 nm for all substituted pyridines studied. The second stage took up to an hour to complete. It was characterized by the bathochromic shift of the absorption maxima from the common starting point to final values ranging from 557 to 567 nm, each value characteristic of the particular pyridine derivative.

This reaction sequence suggested three alternative explanations: 1. The initial rapid reaction resulted in an imidazolyl-pyridyl heme complex, probably through replacement of either the proximal or distal histidine, followed by a slow replacement of the imidazole of a second histidine by another molecule of pyridine derivative; 2. The initial reaction forms a dipyridine hemochrome, followed by a slow interaction between the ligand and the micelle. Changing the character of the micelle would alter its interaction with the peripheral groups of the heme, resulting in a spectral shift of the absorption bands of the porphyrin as suggested by Caughey et al. (1966); 3. There were multiple forms of the micelle which reacted at different rates with the pyridine. Our evidence indicates that any or all of the explanations may apply, depending on the pyridine derivatives and/or the method of forming the hemochrome.

Alternative 1. Mono to Dipyridine Hemochrome

All of the hemochromes had essentially the same initial wavelength of maximum absorption. This suggests that the hemochrome π electron structure was basically that of a pyridine complex, modified by an imidazolyl group in the other ligand position. The single pyridine ligand may have affected the fine positioning of the heme peak, but the latter effect was not detectable. With the formation of the dipyridine hemochrome, the π electron interaction should be much stronger, resulting in a more pronounced shift.

Substituents on the pyridine ring changed the position of the λ_{max} of the hemochrome, and also affect the positions of the ultraviolet absorption maxima of pyridine, which are the π electron excitation bands of the pyridine ring. To determine if a relationship existed between the λ_{max}, the hemochrome λ_{max} were plotted as a function of the substituted pyridine λ_{max} (Figure 4). The substituents on the pyridine nucleus are differentiated as either electron donors or acceptors. Falk et al. (1966) suggested a positive correlation based on the electron donating character of the substituent, but the intermixing of the two types of substituents in the figure shows that such a correlation is not strictly

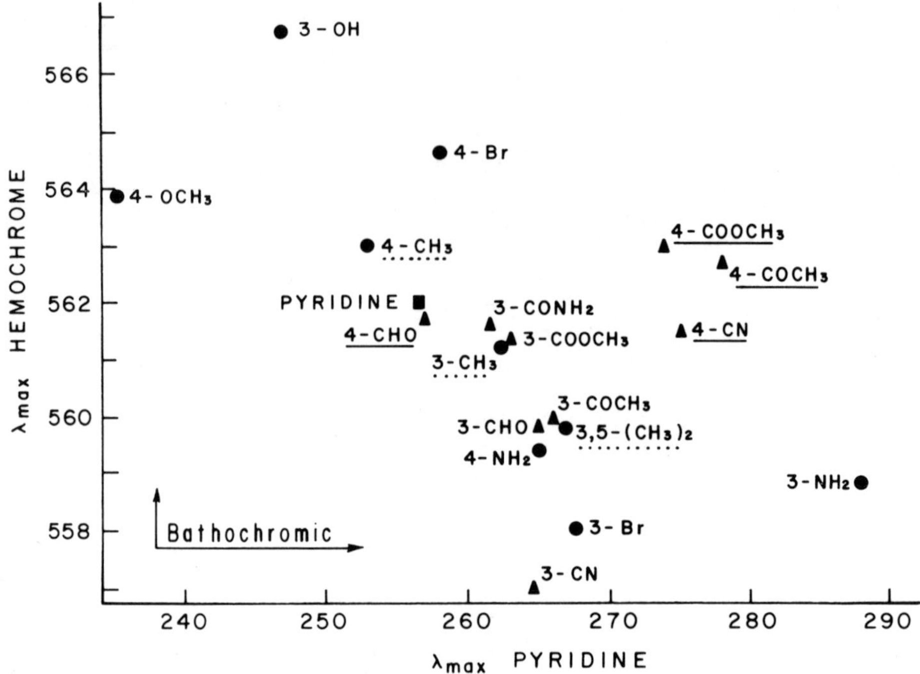

Figure 4. Wavelength of maximum absorption of substituted pyridine hemochromes as a function of the wavelength of maximum absorption of the substituted pyridine. , pyridine; 0, donor and Δ, acceptor substituents. _____, purple; ...., orange. For the data, r=53, significant at the 1% level.

correct. However, acceptor and donor substituents are also classified as either bathochromic or hypsochromic in their effect on the ultraviolet spectra of the pyridine ring, and it is this characteristic of the substituents that is the basis of the relationship in the plot. The correlation coefficient for the regression is 0.53, which corresponds to a probability of 0.01 (Snedecor, 1946). There was no statistical correlation between λ_{max} of the hemochrome and either the stability constant of the pyridine hemochrome, as suggested by Falk et al. (1966) or the acid dissociation constant of the pyridine derivative. The spectral correlation may be interpreted in terms of the basic electron structure of the complex. Substituuents on the pyridine ring change the energy levels of both the bonding and antibonding orbitals of the pπ electrons. If the substituent raises the bonding orbital energy level, the effect will

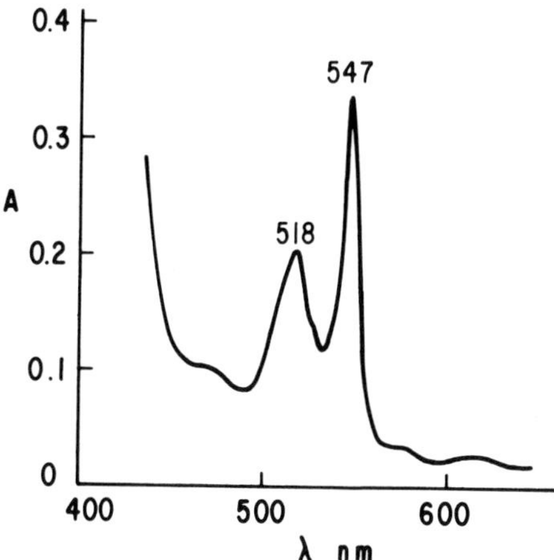

Figure 5. Spectra of the gelatinous liquid from fresh ham cooked either in air or nitrogen at 95°.

be transferred to the dπ orbitals of the iron with which the pyridine has formed π bonds (Pauling, 1949; Falk et al., 1966). The π electrons of the porphyrin ring also form bonds with the dπ orbitals of the iron. Energy transfer between the dπ electrons of the iron would raise the bonding energy levels of the porphyrin ring. Since the antibonding orbitals of the porphyrin are not affected, the energy gap between the bonding and antibonding orbitals would be decreased and the λ_{max} shifted towards longer wavelengths. Since the correlation is negative, the increase in the bonding orbital energy levels of the pyridine due to substituents must be accompanied by a correspondingly greater increase in the antibonding orbital energy levels.

These energy level shifts are the result of the inductive and conjugative (hyperconjugative) effects of the substituent. If these energy level values were known for the various derivatives, it seems likely that a better relationship than wavelength* could be established for the data of Figure 4. Chandra and Basu (1959) have calculated the values of the hyperconjugative and inductive effects for 3- and 4-methylpyridine and 3,5-dimethylpyridine. Their results showed that the wavelength shifts were linear with either effect. Since our results are also linear for these three derivatives, no specific correlation to either hyperconjugative or inductive effects can be made.

Alternative 2. Micelle Interaction and Color

Such an interaction was first suggested by Caughey et al. (1966). We have not been able to derive any direct evidence from our studies on this point. However, we have made an observation which bears on the question of detergent-heme interactions. The drip from several ham samples cooked at 95° all had an orange-pink color and the typical spectra of a pyridine hemochrome (Figure 5). Tappel (1957) suggested that the pigment of cooked fresh meats is that of niacinamide hemochrome, partly from the reflectance spectral characteristics and partly because niacinamide is the only nitrogen-containing aromatic ring compound occurring in meat in sufficient quantities to form the complex.

The unusual feature of the spectrum of Figure 5 is that the wavelength maxima occur at 518 and 547 nm, some 10 and 15 nm towards shorter wavelengths than the corresponding maxima of the SLS-denatured globin hemochrome (Figure 4). The heme micelle cannot be the same for the two pigments, and if the ham drip pigment is the niacinamide complex, the difference in spectra must be due to the micelle difference.

Alternative 3. Multiple Reacting Forms

The third alternative was supported by the spectral changes that occurred during the formation of the 3-hydroxypyridine hemochrome (Figure 6). The typical peak at about 555 nm appeared quickly in the first stage, but during the second stage, instead

*The relation between the spectra of the pyridine derivative and the hemochromes formed from them is more accurately expressed as a function of wave number as it is the latter which directly related to the energy of the orbitals involved. During the review of this manuscript, Dr. D. Quimby of E.R.R.C. pointed out that the variability of the data of Figure 4 could be considerably reduced by treating the data of the 3 and 4 position substituted pyridines separately.

of the peak shifting to a longer wavelength, a new peak was formed at 567 nm, both peaks showing simultaneously. Since the 567 nm peak was produced slower, and disappeared faster than the 555 nm peak, it could not have been in equilibrium with the latter and must have been produced de novo. This situation can occur only if there are two different reacting forms of the heme micelle, one of which reacts faster with the pyridine than does the other.

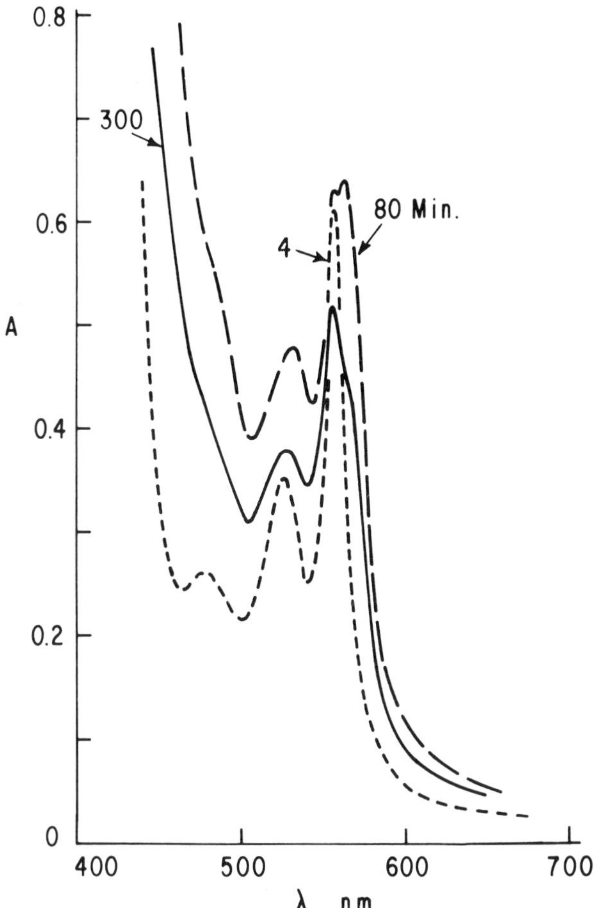

Figure 6. Spectra of the mixture of 3-hydroxypyridine and SLS-denatured globin hemochrome during the slow state of the reaction.

HUE OF THE PYRIDINE HEMOCHROMES

The differences in the λ_{max} of the various pyridine hemochromes that have just been discussed are not sufficient to account for the observed variations in hue of the hemochromes. Hemochromes in Figure 4 were either orange (dotted underline), purple (solid underline) or various shades of red, inclining to a rose (bluish-red) tint. The orange pigments had strong charge transfer bands in the 480 nm region which, by reducing the amount of blue light transmitted, produces the orange hue. Conversely, the charge transfer bands at 600 to 750, by reducing the amount of red light transmitted, produce a purple hue.

To explain the position of these bands the simple relationship of electron donor or acceptor may be invoked. The electron donating methyl groups raise the energy levels of the antibonding orbitals of the π electrons of the pyridine ring. This increases the energy difference of the transfer of a charge from the $d\pi$ orbitals of the iron to the antibonding π orbitals of the pyridine ring. The absorption bands for this transfer therefore appear at shorter wavelengths. The effect is also observed in the fine positioning of the charge transfer bands of the hemochromes. The data of Table I show that the lower the energy of the π electron excitation of the pyridine ring, the lower the energy of charge transfer. Electron accepting substituents would have the opposite

TABLE I

Absorption Maxima of Various Pyridine Derivatives and the Corresponding Hemochromes

Substituent	λ_{max}, nm Pyridine	Hemochrome
4-methyl	253.0	ca. 472
3-methyl	262.5	477
3,5-dimethyl	267.0	482
4-formyl	257	643
4-carboxymethyl	274	657
4-cyano	275	678
4-acetyl	278	740

effect. Increasing the positive charge on ferric complexes would also have the opposite effect, as suggested by Brill and Williams (1961). In these complexes, the charge transfer is from the ligand to the iron. Raising the energy levels of the pyridine pπ bonding orbitals decreases the energy gap between them and the antibonding orbitals of the iron, decreasing the energy gap and shifting the absorption bands towards longer wavelengths.

SUMMARY

A negative correlation has been found to exist between the λ_{max} of various substituted pyridines and the hemochromes formed from them, and we have developed a qualitative theory to explain the phenomenon. Spectral variations in the hemochromes have been observed which are attributed to variations in the heme micelle, either total composition of the micelle, as in the case of the ham exudate, or micro variations in one micelle as in the case of the 3-hydroxypyridine hemochrome.

ACKNOWLEDGMENTS

The authors wish to acknowledge the technical assistance of Mr. E. M. DiEgidio.

Reference to brand or firm name does not constitute endorsement by the U.S. Department of Agriculture over others of a similar nature not mentioned.

REFERENCES

Akoyunoglou, J.-H.A., Olcott, H.S. and Brown, W.D.(1963). Biochemistry 2:1033.

Brill, A.S. and Williams, R.J.P.(1961). Biochem. J. 78:246.

Caughey, W.S., Fujimoto, W.Y. and Johnson, B.P.(1966). Biochemistry 5:3830.

Chandra, A.K. and Basu, S.(1959). J. Chem. Soc. (London), 1623.

Falk, J.E., Phillips, J.N. and Magnusson, E.A.(1966). Nature 212: 1531.

Howard, A., Duffy, P., Else, K. and Brown, W.D.(1973). J. Agr. Food Chem. 21:894.

Katz, S., Miller, J.E. and Beall, J.A.(1973). Biochemistry 12:710.

Pauling, L.(1949). in "Haemoglobin", Eds., Roughton, F.J.W. and Kendrew, J.C., Butterworths Scientific Publs., London, p.60.

St. George, R.C.C. and Pauling, L.(1951). Science 114:629.

Simplicio, J.(1972). Biochemistry 11:2525.

Snedecor, G.W.(1946). "Statistical Methods", 4th Ed., Iowa State College Press, Ames. p. 149.
Tappel, A.L.(1957). Food Res. 22:404.
Tarladgis, B.G.(1967). U.S. Patent No. 3,360,381.
Van den Oord, A.H.A. and Wesdorp, J.J.(1969). European J. Biochem. 8:263.

6

^{13}C NMR STUDIES OF THE INTERACTION OF Hb AND CARBONIC ANHYDRASE WITH $^{13}CO_2$

Frank R. N. Gurd, Jon S. Morrow, Philip Keim, Ronald B. Visscher, and Robert Marshall, Dept. of Chemistry, Indiana Univ., Bloomington, Indiana 47401

This report deals with the prospects for ^{13}C NMR in defining the interactions of CO_2 and its hydrated derivatives with small molecules and various proteins. The importance of bicarbonate ion as a buffer component is well recognized. The insolubility of many bicarbonate and carbonate salts is a central fact around which most people interested in metal-protein interactions try to work. The most common strategy is to remove or at least to avoid adding bicarbonate while setting up a metal-peptide or metal-protein study. The availability of ^{13}C-enriched CO_2 and bicarbonate and of NMR spectrometers capable of studying the ^{13}C signals makes it convenient to reexamine the roles of these important interacting species in certain metal-protein systems of unquestionable importance.

ROLES OF CO_2 AND BICARBONATE

The best known biological roles of CO_2 and the bicarbonate-carbonate pair are in CO_2-fixation, decarboxylation reactions, CO_2-binding to hemoglobin, and the interconversion process catalyzed by carbonic anhydrase. The last two of these have been studied by ^{13}C NMR and are dealt with below (Patterson and Ettinger, 1960; Matwiyoff and Needham, 1972; Morrow et al., 1973; Koenig et al., 1973).

The roles of CO_2 and its derivatives in systems of interest to those attending this Symposium are conveniently discussed in terms of the scheme in Figure 1. The top part of the figure shows the routes of conversion of CO_2 to bicarbonate and carbonate. The proton capture and release involved in the interconversion between bicarbonate and carbonate is fast, so that either is available as

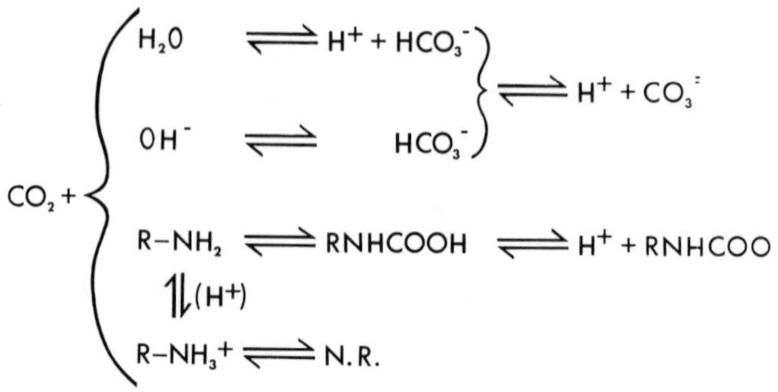

$H_2O \rightleftharpoons H^+ + HCO_3^-$
$\rightleftharpoons H^+ + CO_3^=$
$OH^- \rightleftharpoons HCO_3^-$
CO_2 +
$R-NH_2 \rightleftharpoons RNHCOOH \rightleftharpoons H^+ + RNHCOO$
(H^+)
$R-NH_3^+ \rightleftharpoons$ N.R.

Figure 1. Reactions available to CO_2 in aqueous solutions of amines. The uncatalyzed hydration of CO_2 is relatively slow, while the rest are quite fast.

a reactive species in the presence of the other. The interconversion is fast enough on the NMR time scale that the two species are represented by a single resonance in ^{13}C-NMR. The chemical shift of the single resonance falls at a weighted average value under these conditions of rapid exchange (Bovey, 1969). The result is the relationship shown in Figure 2 connecting the observed resonance

12.0
11.0
10.0
9.0
8.0
pH
Bicarb. Titration
μ = 0.46 M T = 32°
24
28
32
ppm

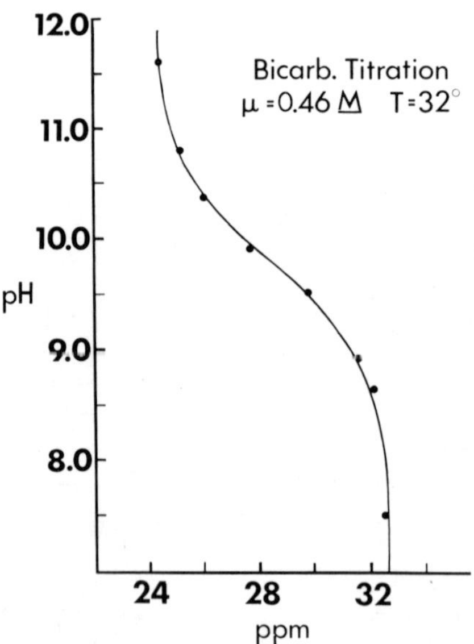

Figure 2. Chemical shift of bicarbonate-carbonate ^{13}C NMR resonance as a function of pH, referenced to external CS_2.

position with the pH of the solution according to the theoretical Henderson-Hasselbalch equation. The observed chemical shift falls between the acid and base limits of 32.63 ± 0.12 ppm and 24.38 ± 0.13 ppm upfield of external CS_2, respectively. An obvious consequence of the mobility of this equilibrium is that the total concentration of bicarbonate and carbonate can act as their sum in any but the most rapid processes. Because of this fact and the nature of the NMR averaging it will be convenient in what follows to speak of the bicarbonate-carbonate component as an entity in contrast with various other forms of CO_2 derivatives.

In the presence of carbonic anhydrase, the Zn-containing enzyme of remarkably high turnover number (Lindskog and Coleman, 1973), the interconversion of CO_2 and the bicarbonate-carbonate component is very rapid. The ^{13}C NMR observation of this effect is discussed below. Of great importance to us is the obverse, that in the absence of carbonic anhydrase a mixture of CO_2 and bicarbonate-carbonate will equilibrate slowly in terms of chemical and physiological processes. This implies that the two forms will to some extent diffuse separately. For our purposes it means that both the CO_2 and the bicarbonate-carbonate forms are separately available to enter the metal-protein interaction systems that are the central concern of this Symposium.

Carbamino Derivatives. The lower part of Figure 1 shows the dominant form of reaction of CO_2 with an amine group in a peptide or protein. The reaction is analogous with that of cyanate with an amine in that the reactive species are uncharged (Johncock et al., 1958; Garner et al., 1973). The form $RNHCOO^-$ is a carbamate. The requirement for the basic form of the amine and for the acidic form, CO_2, limits the stoichiometry of the reaction as shown in some detail below. Because of the difference in pK, an α-amino group will compete for the CO_2 rather well with an ϵ-amino group. The reaction in both directions has an appreciable activation energy and is characterized by finite rate constants (Caplow, 1968).

With this sketch as background it is appropriate to ask what CO_2 will mean to us as people concerned with ionic interactions of proteins. Apart from having potential solubility in a non-polar region of a protein, CO_2 itself can form carbamino compounds with amino groups, and quite possibly with imidazole and sulfhydryl groups as well. The well established carbamino formation with amino groups illustrates two subsidiary consequences. First, the conversion of the amino group to the carbamino form removes potential metal binding electron donor groups from equilibrium and in this sense competes with the metal-protein interaction as it would be observed in the absence of CO_2. Second, the carbamino formation not only changes the charge state at the amino group by substituting a predominant negative charge for a neutral or positive function, but it also presents a new cation binding site. In this sense the carbamino formation removes transition metal binding sites and introduces oxygen ligands that may well show generally lower affinity for transition metals but enhanced affinity for, say, alkaline earth metal ions. Steric consequences of the change must not be overlooked, nor must the consequences of changes in hydrogen bonding patterns.

The meaning of the bicarbonate and carbonate to the metal-protein system can be considered in two ways. First, these ions generally interact strongly with metal ions with obvious competitive effects to be anticipated. A special case of particular interest is that of the ternary interaction with transferrin (Warner and Weber, 1953). Second, these ions have a potential for binding to proteins as ions in their own right. Hydrogen bonding or strong ionic interactions of carboxyl groups in proteins are well documented. It seems most probable that bicarbonate or carbonate binding will become widely recognized in future.

The classical means of distinguishing between the various forms of CO_2 and its derivatives has depended on manometric or potentiometric determination of CO_2 pressure, convertible to a value for dissolved CO_2 concentration, determination of total "carbonates" of all forms by acid displacement, and pH determination on the original equilibrium mixture. This set of measurements can yield a value for the total carbamino form. The computation is by difference and is subject to error if bicarbonate or carbonate as such are bound to the protein. Some success has been achieved with a strategy in which unbound bicarbonate and carbonate are collectively removed by rapid gel filtration at high pH, with the idea that dissociation of true carbamino compounds will be relatively slow. The "total CO_2" carried out of the gel with the protein is at that point in carbamino form, and is deemed to correspond to the original carbamino content plus the previously free CO_2 that had been rapidly taken up by the protein at high pH (Rossi-Bernardi et al., 1969).

Carbamino Amino Acids. The potential usefulness of ^{13}C NMR in defining the carbamino derivatives of amino acids is shown in the following paragraphs. The proton-decoupled spectrum in Figure 3 (Morrow et al., 1973) shows the pattern of glycine containing the natural abundance of ^{13}C equilibrated at pH 8.90 with $^{13}CO_2$ in which the mole fraction C^{13} was 0.09. The chemical shifts are expressed as parts per million, ppm, upfield of external CS_2. The small resonance in the figure at 126.20 ppm represents an added internal dioxane standard. The spectrum shows glycine in both the free and carbamino forms. The most upfield resonance at 150.9 ppm represents the C^{α} of free glycine, the smaller nearby resonance at 147.1 ppm represents C^{α} of the carbaminoglycine. Other evidence establishes that the relative intensities of these two resonances

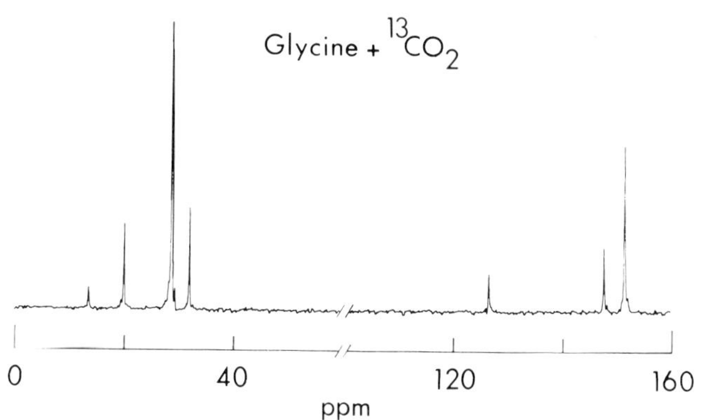

Figure 3. ^{13}C NMR spectrum of 3.0 M glycine equilibrated with 1.0 M total carbonate at pH 8.90. Recycle time was 10 sec, 232 transients.

are meaningful and that the unchanged glycine is the dominant form in this experiment. The two most downfield resonances at 19.6 and 13.3 ppm, respectively represent the carboxyl carbons of the unchanged and the carbamino forms of glycine. The actual adduct enriched in $^{13}CO_2$ is represented by the dominant resonance at 28.4 ppm. Immediately upfield of the adduct resonance is the resonance of the bicarbonate-carbonate component (Morrow et al., 1973).

The spectrum in Figure 3 shows that addition of the CO_2 moiety to the amino group causes a change in chemical shift of the C^α also attached to the nitrogen atom, and that the effect extends further to the carboxyl carbon nucleus of the glycine. The magnitude of the change is referable primarily to the conversion to the uncharged imino form of the amine (Gurd et al., 1971) and secondarily to the inductive effect of the added carboxyl group itself. In the range of pH where carbamino compounds are formed readily with glycine the adduct shows little sensitivity to pH changes, indicating that the pK for dissociation of the proton from the carbamino adduct(cf. Figure 1) is relatively low (Roughton and Rossi-Bernardi, 1966).

Several strong arguments can be developed to show that the enriched adduct is indeed bound to the N atom of the glycine. A particularly direct one is illustrated in Figure 4 (Morrow et al., 1974). The upper part shows the spectrum of glycine enriched with ^{15}N, in which the upfield ^{13}C signal of C^α is split by coupling with the nitrogen isotope of spin $\frac{1}{2}$. The lower part of the figure shows the results of equilibration with $^{13}CO_2$. The upfield part of the spectrum now shows the C^α of the adduct as the smaller composite peak, under the influence of the spin state of the ^{15}N nucleus. The splittings of the carboxyl resonances occupying the

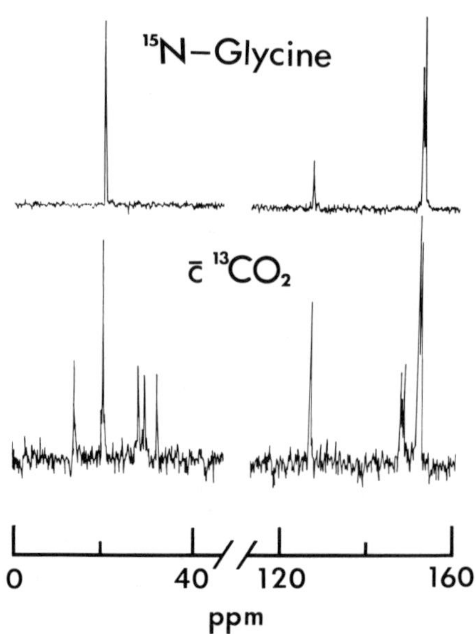

Figure 4. ^{13}C NMR spectrum of 1.0 M ^{15}N-Glycine. Top is free form, and bottom is equilibrated with 1.0 M total carbonates, pH 8.90. Recycle time was 10 sec; 128 transients.

most downfield positions in the spectrum are small. Near 28.4 ppm can be seen the twin resonances of the $^{13}CO_2$ adduct to the ^{15}N-glycine, showing the influence of the ^{15}N nucleus in splitting the ^{13}C signal. At 31.8 ppm appears the free bicarbonate-carbonate signal which is of course not itself subject to the splitting. The patterns of the splittings confirm clearly the expected structure of the carbamino derivative of glycine.

The basis for the incompletely stoichiometric formation of the carbamino compounds was sketched above. An example of the dependence of carbamino formation on pH for a given total concentration of reactants is shown in Figure 5. The results fit reasonably well with previous manometric analyses (Stadie and O'Brien, 1936). Some approximations used in the present concentration estimates can be avoided by procedures now being used which will be published separately. Discrepancies with the results in Figure 5 are small, however. The failure to obtain clear predominance of a given component in an equilibrium is a familiar experience for a coordination chemist.

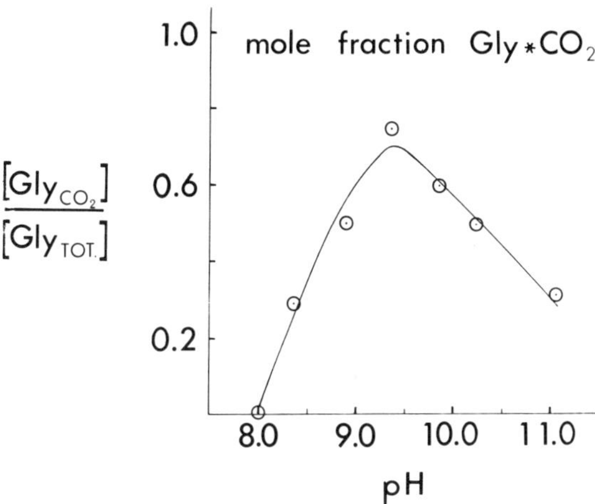

Figure 5. Mole fraction carbamino glycine as function of pH as measured by ^{13}C NMR. Total glycine was 0.5 M, total carbonates 0.5 M, temperature was 29° to 30°.

Before turning to the results with proteins an illustration of the sensitivity of ^{13}C NMR to conformational effects may be introduced economically. The point is illustrated in Figure 6,

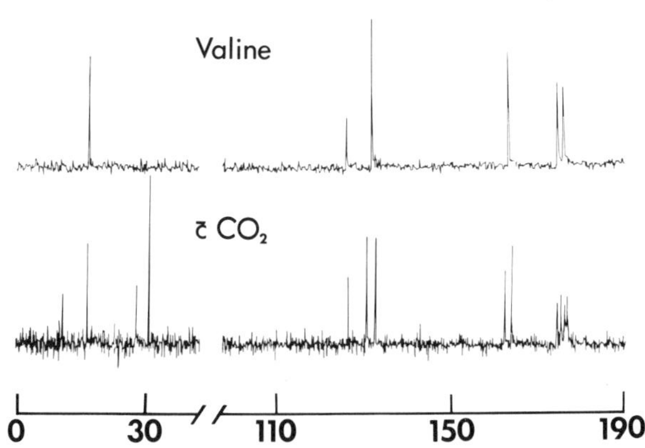

Figure 6. ^{13}C NMR spectrum of 0.50 M valine, free in top spectrum, and equilibrated with 0.75 M total carbonates in lower spectrum; pH = 8.93.

which shows the spectra of valine and of valine equilibrated with $^{13}CO_2$. The valine spectrum shows, from left to right, resonances of the carboxyl carbon, dioxane, C^{α}, C^{β} and the pair $C^{\gamma 1}$ and $C^{\gamma 2}$. Note that the two methyl groups are not equivalent because they sense the optical isomerism of the L-valine invested in C^{α} (Gurd et al., 1971). The lower spectrum shows the two components just as in Figure 3 for glycine. Of special interest is the observation of the four small peaks representing the two forms of the methyl pairs. A close analysis shows that the carbamino form has a larger splitting between the two methyl groups (Morrow et al., 1974). This observation indicates that the two methyl groups experience different influences from the added $-CO_2^-$ group, analogous to previous observations on substituted isopropyl groups (Kroschwitz et al., 1969).

INTERACTION OF CO_2 WITH HEMOGLOBIN

The best recognized case of carbamino formation involving a protein is that of hemoglobin (Henriques, O. M., 1928; Roughton, 1964). It has long been recognized that the presence of a ligand such as O_2 or CO bound to the heme iron affects the ease of formation of carbamino compounds probably involving terminal amino groups (Kilmartin and Rossi-Bernardi, 1971). The effect is known to be reciprocal, so that the bound CO_2 behaves as an allosteric effector (Roughton, F. J. W., 1964).

The effect of ligand state on the pattern of carbamino formation is illustrated in Figure 7. This figure shows the spectra of human hemoglobin in the deoxy-, carbon monoxy-, and methemoglobin forms. The pH values were 7.44, 6.98, and 6.78, respectively, and the range of total carbonates 10 to 18 mM. The pCO_2 varied between 20 and 60 mmHg. The mole fraction of ^{13}C was 0.89. Each spectrum shows the bicarbonate-carbonate resonance at approximately 32.7 ppm upfield of CS_2. The presence of significant dissolved CO_2 is shown by the resonances at 68.3 ppm. A clear carbamino resonance peak at 30.0 ppm is visible in the deoxyhemoglobin spectrum. Corresponding resonances are barely seen in the other two cases.

More detail may be seen in the spectra shown in Figure 8. The upper part shows the spectrum of deoxyhemoglobin A_0 in the presence of $^{13}CO_2$, and the lower part that of the carbonmonoxy derivative under similar conditions. The conditions include a total carbonate level of 50 mM and pCO_2 of 70 to 50 mmHg, at mole fraction ^{13}C of 0.87. The bicarbonate-carbonate resonance is so strong in each case that the tracing goes off the figure and is flanked on either side by spinning side bands. Under each set of conditions the most prominent carbamino resonance is at 30.0 ppm and probably corresponds to the terminal derivative of the α-chain

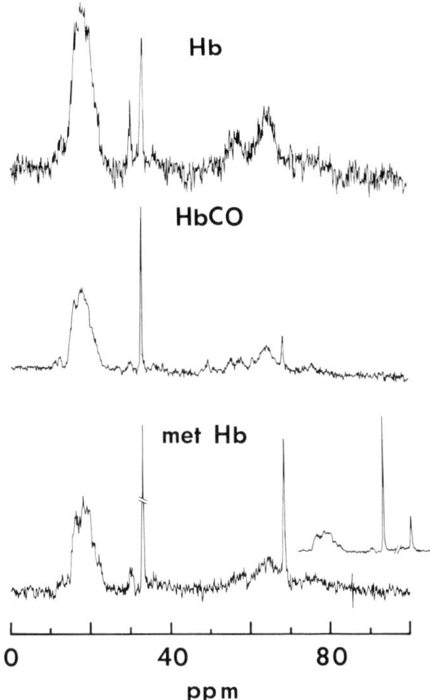

Figure 7. 15.1 MHz ^{13}C NMR spectra of various states of human hemoglobin (A_o) equilibrated with $^{13}CO_2$. See text for details.

of the human hemoglobin (Morrow and Gurd, in preparation). The prominent resonance on the downfield side seen in the deoxyhemoglobin spectrum at 29.4 ppm probably is ascribable to the β-chain derivative. It is much less pronounced in the liganded form shown in the lower part of the figure. The smaller resonance components become more pronounced at higher pH (Morrow et al., 1973).

The assignment of the resonance at 30.0 ppm receives support from the results shown in Figure 9. Here the two major components of the hemoglobin of the white leghorn chicken were separated and tested for carbamino compound formation. A normal pattern is shown for the hemoglobin A_{II}, but not for the A_I which appears to contain acetylated terminal amino groups in the α-chains. These measurements were made at pH 7.28 to 7.00.

The results sampled here have been extended (Morrow and Gurd, in preparation) to quantitate the formation of various components as a function of pH. It is possible by these means to define formation constants and the pK values of the amino groups involved.

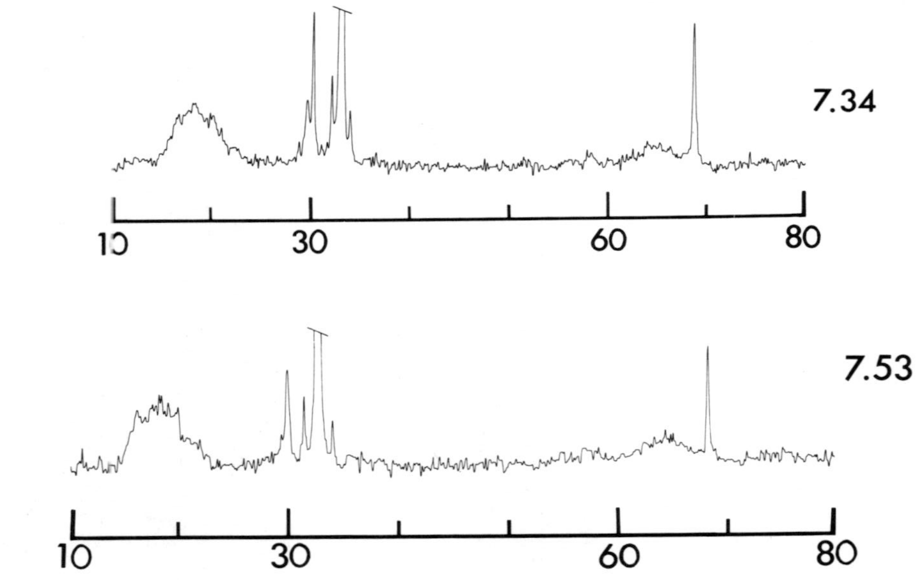

Figure 8. 25.2 MHz ^{13}C NMR spectra of human hemoglobin (A_o) at slightly higher pH and carbonate levels than in figure 7. (Top) Deoxy Hb, pH 7.34, 8192 scans. (Bottom) HbCO, pH 7.53, 8.92 scans, see text.

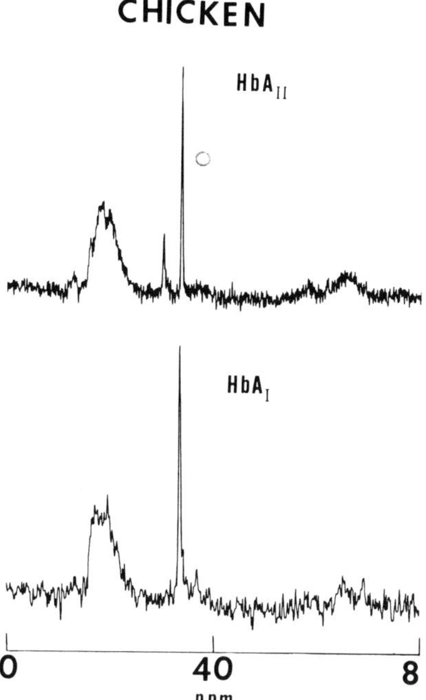

Figure 9. 15.1 MHz ^{13}C NMR spectra of the two major hemoglobin components of the white leghorn chicken, in the deoxy state. pCO_2 was 20 mmHg in both cases. The N-terminal of the α-chains in HbA_I is reported to be acetylated (Matsuda, et al., 1964).

The directness of the ^{13}C NMR method makes it a good complement to the skillful application of more classical techniques by Rossi-Bernardi and others (Rossi-Bernardi, et al., 1969). The subject should not be left without acknowledging the prior, independent work of Matwiyoff and Needham (1972).

INTERACTION OF $^{13}CO_2$ WITH CARBONIC ANHYDRASE

Matwiyoff and Needham (1972) and Koenig et al. (1973) have studied the interaction of $^{13}CO_2$ with carbonic anhydrase independently of some observations made in this laboratory (Morrow et al., 1973). As has been made clear above both CO_2 and bicarbonate-carbonate components, the substrates of the carbonic anhydrase reaction, may be observed readily by ^{13}C NMR. If the interchange of forms, called "exchange" in NMR terminology, is sufficiently rapid there will be a tendency of the individual signals to change in chemical shift position in the direction of merging. In the limit

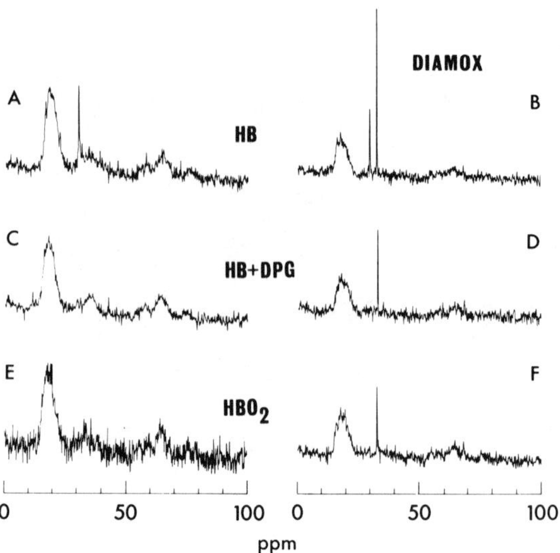

Figure 10. 15.1 MHz ^{13}C NMR spectra of various states of hemoglobin preparations equilibrated with $^{13}CO_2$ in the presence and absence of acetazolamide. For each spectrum 8192 accumulations were obtained with 1.36 sec recycle time at 33-34°. (A) Hb; (B) Hb + acetazolamide; (C) Hb + 2,3-DPG; (D) Hb + 2,3-DPG + acetazolamide; (E) HbO_2 + 2,3 DPS; (F) HbO_2 + 2,3-DPS + acetazolamide.

of fast exchange a weighted average shift may be seen, as a single sharp resonance exemplified in Figure 2. Relatively slow exchange rates will yield a variety of partially merged forms in which the individual resonances characteristically approach each other as the chemical shifts move away from the individual extreme values. Characteristically, also, the two still separate resonances are each broadened (Bovey, 1969).

Figure 10, taken from Morrow et al. (1973), shows three spectra on either side. Reading down on either side these spectra deal with $^{13}CO_2$ equilibrated, respectively, with deoxyhemoglobin, deoxyhemoglobin in the presence of ample 2,3-diphosphoglycerate, and lastly oxyhemoglobin. On the left side are shown the preparations as isolated from hemolysates without special purification, and on the right side after the addition of the carbonic anhydrase inhibitor acetazolamide under the name Diamox^{R}. The dramatic effect of the carbonic anhydrase in causing signal broadening of the bicarbonate-carbonate resonance is obvious. The results in Figure

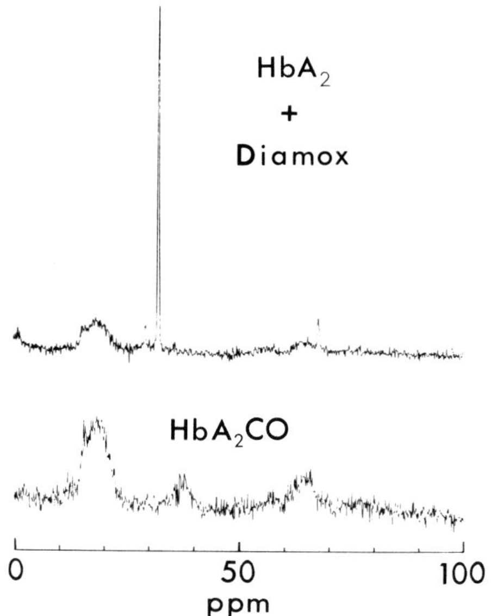

Figure 11. 15.1 MHz ^{13}C NMR spectra of human hemoglobin component HbA_2 ($\alpha_2\delta_2$), equilibrated with $^{13}CO_2$ near physiological conditions. (Top) HbA_2, with 0.1 mM acetazolamide, pH = 7.08. (Bottom) HbA_2CO, $\overline{pH}$ 7.16. The broadness of the bicarbonate resonance near 33 ppm is due to the presence of small amounts of active carbonic anhydrase.

10 have the further points of interest that the 2,3-diphosphoglycerate inhibits the carbamino derivative formation even under conditions where this is probably primarily resident on α-chains, and that the oxyhemoglobin does likewise. The latter observation correlates with the other liganded ferrohemoglobin example, the carbonmonoxy form shown in Figures 7 and 8.

Figure 11 shows another experiment in which the minor human hemoglobin A_2 component ($\alpha_2\delta_2$) was isolated (Dozy et al., 1968). The isolation procedure leaves considerable contamination of the hemoglobin with carbonic anhydrase. The figure shows first that the hemoglobin A_2 is similar to A_o in forming a discernible carbamino derivative more easily in the deoxy- than in the carbonmonoxy-form, and second that the absence of carbonic anhydrase inhibitor causes both broadening and shift of the substrate resonances.

CONCLUSION

The attention given to carbamino derivative formation in hemoglobin should not obscure the generality of the reaction. The relative ease of making observations by ^{13}C NMR has made it possible to observe the reaction with numerous peptides and with sperm whale myoglobin as well as with a wide variety of hemoglobins (Morrow et al., 1973, 1974). It is possible, in fact, that allosterically controlled structural factors are responsible under certain conditions for restricting rather than enhancing the reactivity of the terminal amino groups in hemoglobin.

A freely accessible terminal amino group in a protein may be expected to undergo reaction to a considerable extent under physiological conditions (Stadie and O'Brien, 1937; Morrow et al., 1974). For example, given a pKa for the amino group of 7.66 with a reasonable corresponding value of pKc of 4.74, where

$$Kc = \frac{[R\text{-}NHCOO^-][H^+]}{[R\text{-}NH_2][CO_2]}$$

then at pH 7.34 and pCO_2 equal to 40 mmHg, approximately 25% of the amino group will be in the $R\text{-}NHCOO^-$ form. In a system where there is allosteric enhancement the consequences of formation of the derivative may be magnified (Roughton, 1964).

An authoritative assessment of the importance of CO_2 and bicarbonate in physiology and biochemistry has been published recently by Hastings (1970). The review by Professor Hastings sets out clearly the history of the discovery of the carbamino derivatives of hemoglobin. The review goes on to demonstrate the breadth of the implications of CO_2 metabolism. In terms of the whole subject, Hastings (1970) reiterates a long-standing observation of his that "The body is more interested in maintaining a constant CO_2 tension of the arterial blood than in maintaining a constant pH". A further quotation is apposite: "As the years have worn on and the evidence for the desire of the body to maintain a constant CO_2 tension has been ignored, I have come to the conclusion that it is easy to accept homeostasis with respect to pH because all enzyme activity and enzyme reactions are pH sensitive. But because we have been without a biochemical reason, it is not so easy to accept homeostasis with respect to CO_2 tension, except insofar as changes in it bring about changes in pH indirectly."

Hastings (1970) further draws attention to some examples in which CO_2 concentration and bicarbonate concentration appear separately to control patterns of metabolism in liver cells. The effects are so striking that one is tempted to speculate on regulatory roles for each of these species at the level of key enzymes in the pathways of glycogen synthesis and long-chain fatty acid synthesis, among others.

To invert the argument, it is reasonable to suppose that the CO_2 concentration will affect all proteins and peptides that are not specifically protected at the amino-terminus. Such protected termini are found very frequently (Needleman, 1970). In this regard, one should again note the parallel with the potentiality for transition metal binding (Stadie and O'Brien, 1937; Shearer et al., 1967).

Clearly the development of the ^{13}C NMR technique will hasten the recognition of more roles for CO_2 in biochemistry. Because of the great enrichment that is readily attained the method has enough sensitivity for many purposes. Several forms of CO_2 may be observed concurrently. Distinct carbamino derivatives can be distinguished. Quantitation is possible from integrated intensities. The method is applicable to whole cell suspensions at the risk of some loss of sensitivity and resolution (Matwiyoff and Needham, 1972; Morrow et al., 1973).

ACKNOWLEDGMENTS

The advice and counsel of Professor Adam Allerhand concerning the NMR work and the use of the ^{15}N enrichment is gratefully acknowledged. The advice and encouragement of Professors A. Baird Hastings and John T. Edsall has been extremely helpful. Professor Frank Zeller provided facilities for working with chickens. This work was supported by PHS Grants HL-14680 and HL-05556. This is the 58th paper in a series dealing with coordination complexes and catalytic properties of proteins and related substances.

REFERENCES

Bovey, F. A. (1969) Nuclear Magnetic Resonance Spectroscopy (Academic Press, New York), pp. 187-188.

Caplow, M. (1969) in Carbon Dioxide: Chemical, Biochemical and Physiological Aspects, eds. Forster, R. E., Edsall, J. T., Otis, A. G., and Roughton, F. J. W. (NASA SP-188, Washington, D. C.), pp. 47-52.

Dozy, A. M., Kleihaver, E. F., and Huisman, T. H. J. (1968) J. Chromatography 32, 723-727.

Garner, M. H., Garner, W. H., and Gurd, F. R. N. (1973) J. Biol. Chem. 248, 5451-5455.

Gurd, F. R. N., Lawson, P. J., Cochran, D. W., and Wenkert, E. (1971) J. Biol. Chem. 246, 3725-3730.

Henriques, O. M. (1928) Biochem. Z. 200, 1-24.

Johncock, P., Kohnstam, G., and Speight, D. (1958) J. Chem. Soc. (London) 2544-2551.

Kilmartin, J. V., and Rossi-Bernardi, L. (1971) Biochem. J. 124, 31-45.

Koenig, S. H., Brown, R. D., Needham, T. E., and Matwiyoff, N. A. (1973) Biochem. Biophys. Res. Comm. 53, 624-630.

Kroschwitz, J. I., Winokur, M., Reich, H. J., and Roberts, J. D. (1969) J. Am. Chem. Soc. 91, 5927-5928.

Lindskog, S., and Coleman, J. E. (1973) Proc. Nat. Acad. Sci. USA 70, 2505-2508.

Matsuda, G., Maita, T., and Nakagina, M. (1964) J. Biochem. (Tokyo) 56, 490-491.

Matwiyoff, N. A. and Needham, T. E. (1972) Biochem. Biophys. Res. Commun. 49, 1158-1164.

Morrow, J. S., Keim, P., and Gurd, F. R. N. (1974) in preparation.

Morrow, J. S., Keim, P., Visscher, R. B., Marshall, R. C., and Gurd, F. R. N. (1973) Proc. Nat. Acad. Sci. USA 70, 1414-1418.

Needleman, S. B., ed. (1970) Protein Sequence Determination, Springer-Verlag, New York, pp. 82-83.

Patterson, A., Jr., and Ettinger, R. (1960) Z. Elektrochem. 64, 98-110.

Rossi-Bernardi, L., Pace, M., Roughton, F. J. W., and Van Kempen, L. (1969) in Carbon Dioxide: Chemical, Biochemical and Physiological Aspects, eds. Forster, R. E., Edsall, J. T. Otis, A. G., and Roughton, F. J. W. (NASA SP-188, Washington, D. C.), pp. 65-71.

Roughton, F. J. W. (1964) in Handbook of Physiology, Section 3: Respiration, eds. Fenn, W. O. and Rahn, N. (American Physiological Society, Washington, D. C.), Vol. 1, p. 767.

Roughton, F. J. W., and Rossi-Bernardi, L. (1966) Proc. Roy. Soc. (London) B164, 381-400.

Shearer, W. T., Bradshaw, R. A., Gurd, F. R. N., and Peters, T. (1967) J. Biol. Chem. 242, 5451-5459.

Stadie, W. C., and O'Brien, H. (1936) J. Biol. Chem. 112, 723-758.

Stadie, W. C. and O'Brien, H. (1937) J. Biol. Chem. 117, 439-470.

Warner, R. C., and Weber, I. (1953) J. Amer. Chem. Soc. 75, 5094-5101.

7

THE ANION-BINDING FUNCTIONS OF TRANSFERRIN

P. Aisen, A. Leibman, and R.A. Pinkowitz

Departments of Biophysics and Medicine
Albert Einstein College of Medicine
Bronx, New York

Although the presence in blood serum of iron which was not part of heme, and was neither ultrafiltrable nor dialyzable, was recognized almost 50 years ago (Katz, 1970), the precise physico-chemical nature of this iron-bearing component was established only after the development of reliable procedures for the fractionation of blood proteins. In 1945 Holmberg and Laurell in Sweden established the existence of an iron-transporting protein in blood which they punningly named transferrin (Holmberg and Laurell, 1945, 1947). Almost simultaneously Schade and Caroline (1946) in the United States demonstrated that the bacteriostatic activity of blood plasma was in part due to a protein with an enormous avidity for iron which they therefore named siderophilin. These investigators further established the remarkable similarity of the iron-complexing properties of the blood protein to those of conalbumin, a protein present in egg white (1944). Because of this similarity, the designation ovotransferrin for the egg protein is currently gaining widespread usage (Feeney and Komatsu, 1966). Finally, a "red protein" from milk has also been shown to display the same type of iron-binding activity as transferrin and conalbumin, and hence has been given the name lactoferrin (Masson and Heremans, 1967, 1968). This is something of a misnomer, since lactoferrin occurs in a variety of physiological fluids as well as in leucocytes. Together, serum transferrin, conalbumin (ovotransferrin) and lactoferrin comprise the class of proteins often designated as "the transferrins" (Feeney and Komatsu , 1966).

The need for iron carriers like the transferrins in complex organisms is readily apparent. At the pH, oxygen tension and ionic composition of most physiological fluids the thermodynamically favored state of iron is Fe(III). Taking the solubility product of $Fe(OH)_3$ as $10^{-36}M^4$, it follows that the equilibrium concentration of free ferric ion cannot exceed $10^{-14}M$. In man, the daily turnover of hemoglobin iron approximates 30 mg or 5×10^{-4} moles. If sufficient iron for the biosynthesis of hemoglobin is to reach the bone marrow, an effective iron transport system must be available. The transferrin molecule has the central role in this system.

Possibly an even more important function than simple transport of iron is carried out by serum transferrin, however. The protein is capable of a specific interaction with receptors on the surface of the hemoglobin-synthesizing reticulocyte, and possibly with other iron-requiring cells (Jandl *et al*, 1959) so that iron is given up only to those cells with a special need for it. The importance of this regulatory function of transferrin may be seen when it is realized that excess iron, like other heavy metals, is toxic. Indeed, acute iron poisoning is one of the commoner types of poisoning in childhood. Furthermore, several cases have been reported in the clinical literature of atransferrinemia, a condition characterized by a genetic inability to synthesize transferrin (Heilmeyer, 1964; Goya *et al*, 1972). Patients with this disorder ultimately succumb to the paradoxical combination of iron deficiency anemia and chronic iron overload of most of the body's tissues. Nothing could more dramatically illustrate the role of transferrin in maintaining iron in soluble form in the circulation, and selectively delivering it to cells with specific requirements for iron.

In order for transferrin to carry out these functions, it is necessary on the one hand that it bind iron sufficiently tightly so as to resist hydrolysis at physiological pH, but yet reversibly so that it can deliver iron to the immature red blood cell, or reticulocyte, for the biosynthesis of hemoglobin. The reversible nature of the binding has been shown experimentally by studies in which transferrin was shown to persist in the circulation with a half-life of about 10 days, while the half-life of iron bound to transferrin is only 40 minutes (Katz, 1961). It is also possible, in *in vivo* experiments, to show that transferrin can be recycled after it has given up its iron to the reticulocyte (Fletcher and Huehns, 1967). The binding of iron to the protein is so strong that simple dissociation of iron from protein to cell is excluded on kinetic grounds (Aisen and Leibman, 1968), and a specific mechanism by which the reticulocyte induces release of iron from transferrin must exist. The dual problem thus posed is how can transferrin bind iron sufficiently tightly so as to resist hemolysis (the apparent stability constant for iron binding in physiological fluids is of the order of $10^{24}M^{-1}$), yet yield its metal to cells with specific requirements for iron.

The first clue to the elucidation of these questions was

provided almost as soon as transferrin became available in form sufficiently pure for study. Schade and his coworkers (1949) demonstrated that the characteristic salmon-pink color of the iron-transferrin complex did not develop in the absence of carbon dioxide in one of its forms, possibly bicarbonate. This was confirmed several years later in a detailed quantitative study of iron binding to conalbumin by Warner and Weber (1953). Using carbonic anhydrase, these workers showed that a hydrated form of carbon dioxide was involved in the binding of metal ions by conalbumin, and on the basis of electrophoretic mobility measurements of the protein in its metal-free and metal-complexed forms suggested that the active form was the bicarbonate ion. Recently, it has been argued that the carbonate ion may be the bound form of carbon dioxide (Bates and Schlabach, 1973), but for what follows in this paper this is not a critical problem. Our concern will be with the role of anion binding by transferrin and the interactions - chemical and physiological - between the anion and metal binding functions of the protein.

Properties of the Transferrins. Each of the transferrins so far studied is a single-chain protein with a molecular weight in the range 75-80,000. The molecule has two specific metal-binding sites which are similar, but perhaps not quite identical (Aasa, 1972; Price and Gibson, 1972; Aisen, Lang and Woodworth, 1972). These sites will accept Fe(III), Cu(II), Zn(II), Ga(III), Mn(III), Cr(III), Tb(III), Eu(III), Er(III), and Ho(III), and perhaps other ions as well (Feeney and Komatsu, 1966; Aisen, Aasa and Redfield, 1969; Luk, 1971). Of these, Fe(III) is by far the most tightly bound, and capable of displacing other metal ions from the specific sites (Aisen, Aasa and Redfield, 1969). In order for Fe(III) to bind to transferrin, however, it appears that a suitable anion must be concomitantly bound. Unless an anion acceptable to the protein is available, specific binding of Fe(III) can not be demonstrated, at least by EPR spectroscopy (Price and Gibson, 1972). Furthermore, the specific anion-binding site is activated only when a metal ion is bound. Apotransferrin will not tightly bind bicarbonate or other anions (Aisen, Aasa, Redfield, 1969; Aisen, Pinkowitz and Leibman, 1973). Thus, the anion- and metal-binding functions of transferrin may be described as showing an extraordinary degree of positive cooperativity.

For all cases so far studied, the stoichiometries of metal ion and anion binding by transferrin are identical (Warner and Weber, 1953; Schade and Reinhart, 1966; Masson and Heremans, 1968; Aisen, Aasa and Redfield, 1969; Aisen and Leibman, 1973). The protein can accept up to two metal ions, and for each metal ion bound an anion is also bound. This precise stiochiometry, taken with the strong interdependence of metal ion and anion binding, have led us to study the anion binding properties of transferrin in some detail.

Spectroscopic Studies of the Effects of Anion-binding on Fe(III) binding. Ordinarily, bicarbonate occupies the anion binding site of transferrin. When care is taken to exclude bicarbonate, however, a variety of other anions are capable of occupying this site to form ternary complexes of Fe(III), transferrin and anion (Aisen *et al*, 1967). Most of these complexes give rise to optical spectra which closely resemble the spectrum of native Fe-transferrin-bicarbonate. In order to demonstrate their existence, and study them quantitatively, it is necessary to supplement the usual spectrophotometric methods with electron paramagnetic resonance (EPR) spectroscopy. EPR is particularly useful in the study of paramagnetic protein-metal complexes because of its sensitivity to the immediate environment of the bound metal ion.

Static magnetic susceptibility measurements have shown that iron specifically bound to transferrin is in the high spin (S=5/2) ferric (d^5) state (Aisen, Aasa and Redfield, 1969). The phenomenological spin Hamiltonian describing the EPR spectra of high spin Fe(III) in sites of less than cubic symmetry is characterized by two zero-field splitting parameters, D and E. The ratio of D to E, commonly designated λ, reflects the symmetry of the crystal field about the central metal ion. An axial distortion from cubic symmetry is represented by the D term, while a planar or rhombic distortion is expressed in the E term. It is always possible to choose a coordinate system such that $0 \leq \lambda$ 1/3. The familiar spin Hamiltonian is then written as:

$$H = \beta \underline{H}.g.\underline{S} + \underline{D}\,\hat{S}_z^{\,2} - 1/3S(S+1) + \underline{E}(\hat{S}_x^{\,2} - \hat{S}_y^{\,2})$$

zeeman axial rhombic

The relative magnitudes of these three terms then determine the fields at which resonances are observed for the spin 5/2 manifold. Quartic terms due to a cubic field are usually very much smaller than the quadratic terms and are customarily neglected because of the number of parameters they introduce. When the Zeeman term is substantially larger than the others, resonances will occur near $g' = 2$, where g' is defined as $\beta H/h\nu$. This situation, of importance in a number of Mn(II)-proteins, does not occur in the transferrins at X- or K- band frequencies. When the Zeeman term is small compared to the zero-field terms, the symmetry parameter λ determines the g' values of the resonances. If the iron site shows axial symmetry and the D term is largest so that λ is near zero, resonances in a frozen solution are seen near $\underline{g}' = 6$ and $g' = 2$. The former arise from ferric ions oriented so that the symmetry axis of the zero-field tensor is perpendicular to the applied magnetic field, while the latter resonance line originates from parallel orientations. When λ is close to 1/3, and D is much larger than the Zeeman energy, then all orientations of ferric ions will give rise to resonances near $g' = 4.3$. The symmetry of the Fe(III) binding sites of the transferrins, and therefore the EPR spectra to which they give rise, depend strongly on the occupant of the anion-binding site (Pinkowitz, R.A. and Aisen, P., 1972).

The EPR spectrum of the physiological Fe-transferrin-bicarbonate complex, in purified form and as it occurs in fresh blood serum, are shown in Fig. 1. The resonance lines are centered about $\underline{g}' = 4.3$, but the splittings indicate that λ is not exactly 1/3. Computer solutions of the spin Hamiltonian for the S = 5/2 manifold indicate that a value for λ of 0.315 best fits theory to experiment (Aasa, 1970). In Fig. 2 are shown the resonances observed from a variety of other Fe-transferrin-anion complexes. The nitrilotriacetate complex, shown in Fig. 2A is very sharp, indicating that the $\underline{g}'$ values are very nearly isotropic and λ is close to 1/3. The thioglycollate complex, in Fig. 2B, shows a distinct splitting which is, however, less pronounced than for the bicarbonate form. Accordingly, λ must be slightly larger than for the bicarbonate case; a value of 0.32 is probably a good choice. On going to the malonate and oxalate cases, in Fig. 2C and 2D, a very pronounced change is seen in the spectra. Resonances are now seen near $g' = 5.7$ and $g' = 6.8$ (a small fraction of the transferrin is still present in the bicarbonate form, accounting for the resonances near $g' = 4.3$). Clearly, the iron sites have become more axial. Although careful computer analyses of these spectra have not been undertaken, rough calculations indicate that λ must be less than 0.1.

Substituting nitrilotriacetate for bicarbonate has still another effect on the EPR parameters. By measuring the intensity of the EPR signal as a function of temperature between 1.3°K and 4.2°K it is possible to estimate the magnitude of D in the spin Hamiltonian. For the bicarbonate forms of all Fe(III)-transferrins so far studied D falls between 0.31 cm^{-1} and 0.36 cm^{-1}. Both the symmetry and the strength of the crystal field about the iron atom, then, are profoundly affected by the occupancy of the anion binding site. Whether this is due to a change in ligands coordinated to the metal, or to a change in their geometry, or both, is not yet clear. It is not even known whether the anion is itself a ligand of the metal. But, since the metal-binding properties of transferrin are so profoundly influenced by the associated anion-binding functions, it is attractive to suppose that the interaction between the anion and metal-binding sites has direct physiological relevance.

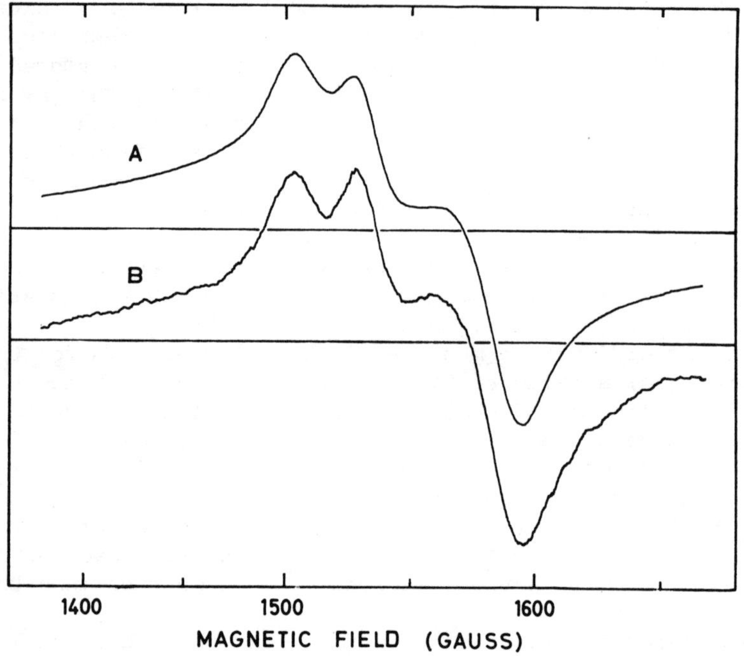

Fig. 1. EPR spectra obtained at 90° K of: A, Fe(III)-transferrin-bicarbonate isolated from Cohn fraction IV-7; and B, freshly drawn normal human serum. The modulation amplitude is 10 gauss, the microwave frequency is 9.208 GH_z, and the microwave power is 20 milliwatts. (From Aisen *et al*, 1967).

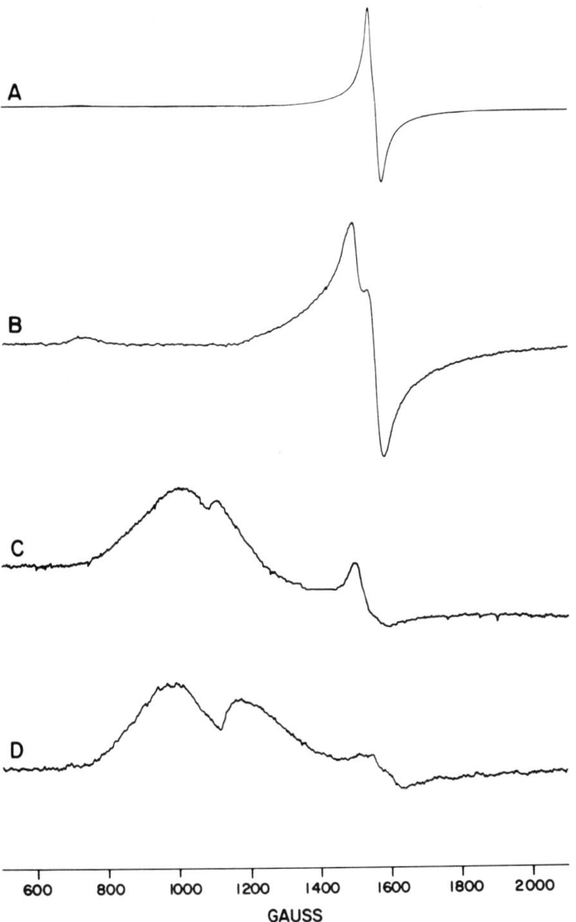

Fig. 2. Ternary complexes of Fe(III), transferrin, and various anions. A, Fe(III)-transferrin-nitrilotriacetate; B, Fe(III)-transferrin-thioglycollate; C, Fe(III)-transferrin-malonate; and D, Fe(III)-transferrin-oxalate.

The Exchangeability of Bicarbonate Specifically Bound to Transferrin. Since the metal and anion binding functions of transferrin are strongly linked with each other, studies of the exchangeability of transferrin-bound bicarbonate were undertaken to gain understanding of some of the factors which affect the stability of the anion-protein bond. Transferrin specifically labelled with $H^{14}CO_3^-$ was prepared by displacing nitrilotriacetate from the ternary Fe-transferrin-nitrilotriacetate complex, prepared in the absence of all forms of CO_2 (Aisen *et al*, 1967), with ^{14}C-labelled bicarbonate (Aisen *et al*, 1973). Satisfactory preparations were also obtained by anerobically adding labelled bicarbonate to a CO_2-free mixture of $Fe(NH_4)_2(SO_4)_2$ and apotransferrin. In both cases unbound labelled bicarbonate was removed by exhaustive dialysis against a working buffer of 0.1 M KCl-0.05 M HEPES (N-2-hydroxyethylpiperazine-N'-2-ethanesulfonate) at pH 7.4. It was also possible to isolate labelled monoferric transferrin by isoelectric focussing from transferrin only 30% saturated with iron (Aisen, Lang and Woodworth, 1972). This procedure depends on the change in net charge of the transferrin molecule - and therefore a change in isoelectric point - for each Fe(III) ion that is bound (Aisen, Leibman and Reich, 1966).

The exchange of transferrin-bound labelled bicarbonate with bicarbonate free in solution was followed by observing the disappearance of radioactivity from the protein using the Crowe-Englander dialysis apparatus. In the absence of anions other than those of the working buffer, the half-time for bicarbonate exchange at 25°C was approximately 20 days. Because of this slowness, it was not feasible to determine whether the exchange process obeyed simple first-order kinetics, and detailed studies were not attempted. However, when a metal-complexing anion capable of occupying the specific anion-binding site of transferrin was present, the exchange rate increased markedly, and kinetic patterns could be deduced. For instance, millimolar nitrilotriacetate decreased the half-time for bicarbonate exchange in monoferric transferrin to 7 hours (Fig. 3), and the exchange conformed to a simple first order reaction. More interesting was the exchange promoted by nitrilotriacetate in diferric transferrin. In this case, the kinetic behavior was not simple first order. It could be well described, however, by the sum of 2 first-order exchange reactions, each accounting for half of the radioactivity at zero time (Fig. 3). Qualitatively similar results were obtained with other complexing agents capable of specifically binding to transferrin, such as citrate and thioglycollate, although the actual kinetic constants varied for each agent. Indeed, the bicarbonate ion itself was an effective promoter of bicarbonate exchange in transferrin (Aisen *et al*, 1973). A likely mechanism, then, to account for these observations is:

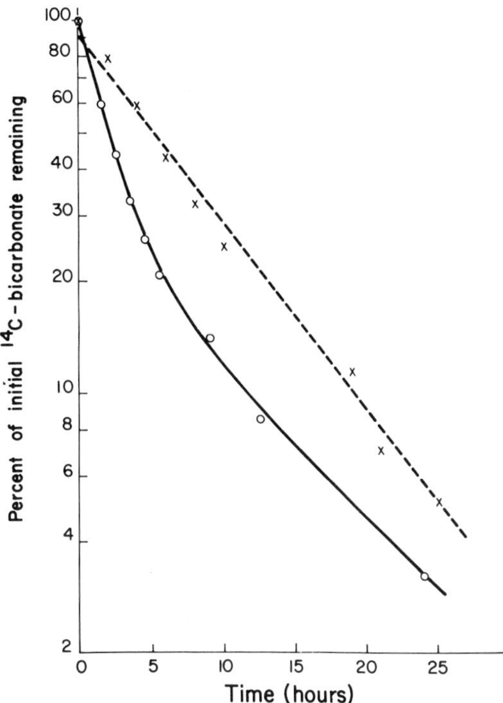

Fig. 3. Exchange of transferrin-bound [^{14}C] bicarbonate at ambient pCO_2 and in the presence of 10^{-3}M nitrilotriacetate. (X), monoferric transferrin; (o), diferric transferrin. The dashed line is a plot of the function F = 100% [exp(-0.093t)] and the solid line is a plot of the function F = 50% [exp(-0.17t) + exp(-0.022t)]. (From Aisen et al, 1973).

$$\text{Fe-transferrin-H}^{14}\text{CO}_3^- + \text{anion} \rightarrow \text{Fe-transferrin-anion} + \text{H}^{14}\text{CO}_3^-$$

$$\text{Fe-transferrin-anion-} + \text{HCO}_3^- \rightarrow \text{Fe-transferrin-HCO}_3 + \text{anion}$$

The simple first-order exchange in monoferric transferrin is a finding which suggests that the two metal-binding sites of the protein are not distinguishable by the intrinsic exchange behavior of their associated anion-binding functions. However, when both metal-binding sites are filled, the anion-binding sites are no longer identical in their exchange behavior. One site now exchanges more rapidly, and the other more slowly, than in the monoferric protein. Since the binding of Fe(III) to transferrin is known to be random, both sites should be equally populated and any intrinsic difference between the two sites of the protein should be evident in the monoferric preparation. Transferrin with one atom

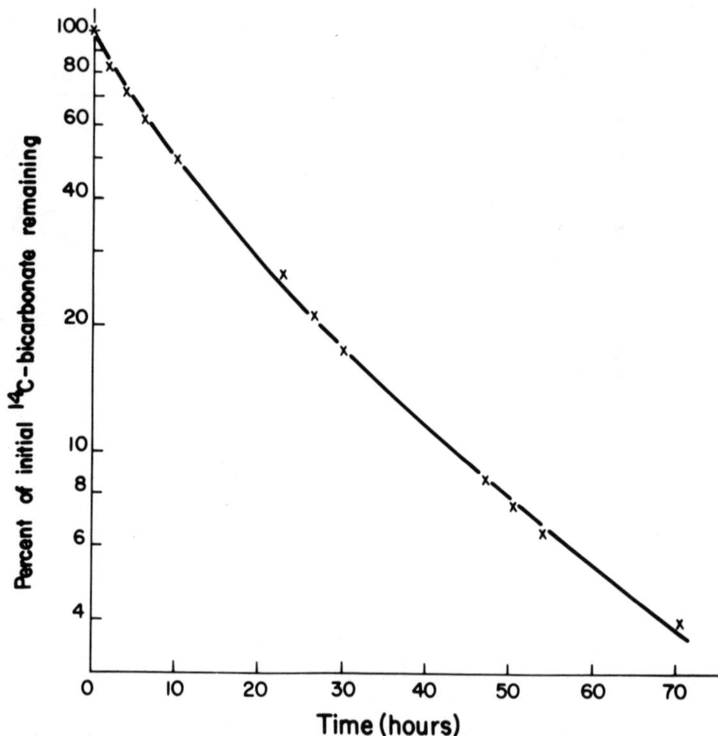

Fig. 4. Bicarbonate exchange from monolabelled transferrin brought to 90% saturation with iron and unlabelled bicarbonate. The experimental points are represented by x, and the solid line is the function $F = 45\% [\exp(-0.12t) + \exp(-0.036t)] + 10\% [\exp(-0.062t)]$. (From Aisen _et al_, 1973).

bound shows a simple exponential loss of the radioactivity in its exchangeable bicarbonate, so that it seems reasonable to suppose that there is no _intrinsic_ difference between the sites. But because the exchange behavior of diferric transferrin requires two exponential terms for its description, we feel that a site-site interaction must exist. This hypothesis is strengthened by an experiment in which labelled monoferric transferrin was brought to 95% saturation with unlabelled Fe(III) and unlabelled bicarbonate. This monolabelled diferric transferrin now showed the complex exchange behavior characteristic of fully labelled diferric transferrin (Fig. 4).

There may be a physiological relevance of this site-site interaction. Several years ago Fletcher and Huehns reported observations which pointed to a difference between the two binding sites of transferrin in their capacity to donate iron to the reticulocyte for the biosynthesis of hemoglobin (Fletcher and Huehns, 1967; 1968). The basis for this Fletcher-Huehns effect is not known. If, however, the bicarbonate-binding site of the protein is involved in its interaction with the reticulocyte, then the site-site interaction we have observed may help explain a physiological difference between the sites. The iron atom associated with the faster exchanging site might be more readily available for donation to the reticulocyte. Thoughts such as this have led us to try to study the role of the anion-binding site of transferrin in its interaction with the reticulocyte.

Bicarbonate and the Transferrin-Reticulocyte Interaction. The role of transferrin in providing iron for the synthesis of hemoglobin by the reticulocyte has been known for some time (Jandl et al, 1959), but the mechanism by which iron is released from protein to cell is still only poorly understood. That a specific mechanism must exist is implied by two well-established facts: (1) the binding of iron to transferrin is so tight under physiological conditions that spontaneous dissociation of iron from the protein is excluded, and (2) when iron is taken up by the reticulocyte from transferrin the protein itself is conserved and recycled in the iron transport chain (Katz, 1961). Because of the interdependence of the anion- and metal-binding functions of the protein, investigations were undertaken to determine whether the anion-binding site played any part in the mechanism or iron uptake by reticulocytes (Aisen and Leibman, 1972).

To do this, comparisons were made of the behavior of Fe-transferrin-bicarbonate and Fe-transferrin-oxalate. The oxalate form of the protein was chosen because it is relatively easy to prepare and comparatively stable, reverting to the bicarbonate form only over the course of hours or days even when bicarbonate is available to displace the specifically-bound oxalate (Aisen et al, 1967), as it must be in a physiological experiment. In our first studies, the ability of rabbit reticulocytes to take up ^{59}Fe from the two transferrin complexes was measured. After an incubation of 1 hour, 5.0 μg of Fe per ml of reticulocytes was taken up from the bicarbonate complex, while only 1.75 μg of Fe was taken up from the oxalate complex (Fig. 5). Since the oxalate complex was, of necessity, prepared in a physiological buffer, it contained about 20% of the Fe bound to the bicarbonate form of transferrin. This may have accounted for some of the observed uptake from this preparation, so that the true uptake of iron from the oxalate complex is even less than that measured. Once taken up by the reticulocyte, the iron from either form of the protein is readily incorporated into heme.

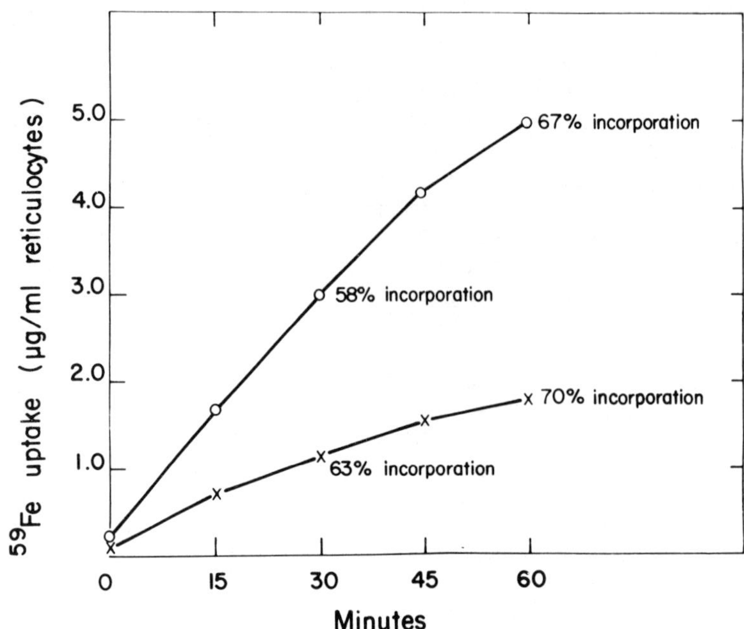

Fig. 5. Uptake of iron by reticulocytes from ^{59}Fe(III)-transferrin-oxalate (x—x) and ^{59}Fe(III)-transferrin-bicarbonate (o—o). The percentage figures refer to the fraction of ^{59}Fe activity bound to reticulocytes and recovered in heme isolated from washed cells. (From Aisen and Leibman, 1973).

When reticulocytes are incubated with an equimolar mixture of Fe-transferrin-bicarbonate and Fe-transferrin-oxalate, the uptake or iron from the bicarbonate form of the protein is depressed (Fig. 6). A control experiment showed that this effect could not have been due to inhibition of reticulocyte function by free oxalate. It seemed reasonable to suppose, therefore, that the oxalate and bicarbonate forms of transferrin competed on an equal basis for the reticulocyte receptors. The oxalate complex, when occupying a receptor site, prevented an Fe-transferrin-bicarbonate molecule from taking its place there, but could not itself function as an iron donor. Further evidence for this hypothesis was provided by measuring the actual binding of ^{125}I-labelled proteins to reticulocytes (Fig. 7). No appreciable difference was detected in the binding behavior of Fe-transferrin-oxalate and Fe-transferrin-bicarbonate. We felt, therefore, that the relative inability of the reticulocyte to utilize iron from the oxalate complex was due to an inability to transfer iron from protein already bound to the reticulocyte. The suggestion was offered that the postulated iron-releasing mechanism of the reticulocyte involved an attack on the

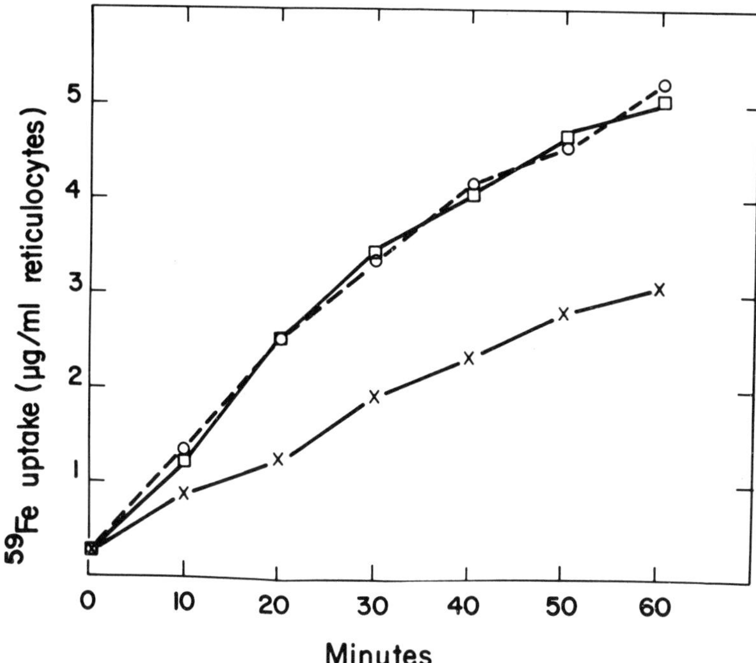

Fig. 6. Uptake of iron from ^{59}Fe(III)-transferrin-bicarbonate in the presence of unlabelled Fe(III)-transferrin-oxalate (x— x). Control experiments are also shown with ^{59}Fe(III)-transferrin-bicarbonate alone (o— o) and ^{59}Fe(III)-transferrin-bicarbonate plus a 2-fold molar excess of potassium oxalate. (From Aisen and Leibman, 1973.)

anion-binding site of the protein. When bicarbonate occupies the site, it is removed by the reticulocyte following which the iron-protein bond is readily broken. When the site is occupied by oxalate, the reticulocyte is unable to remove this anion, and the iron protein bonds remain stable. Admittedly, our work is far from completed, but the hypothesis fits the available facts and to us seems attractive and worthy of further study.

Summary and Conclusions. Transferrin has long been known and studied for its remarkable metal-binding properties. More recently, the interrelation between metal-binding and specific anion-binding

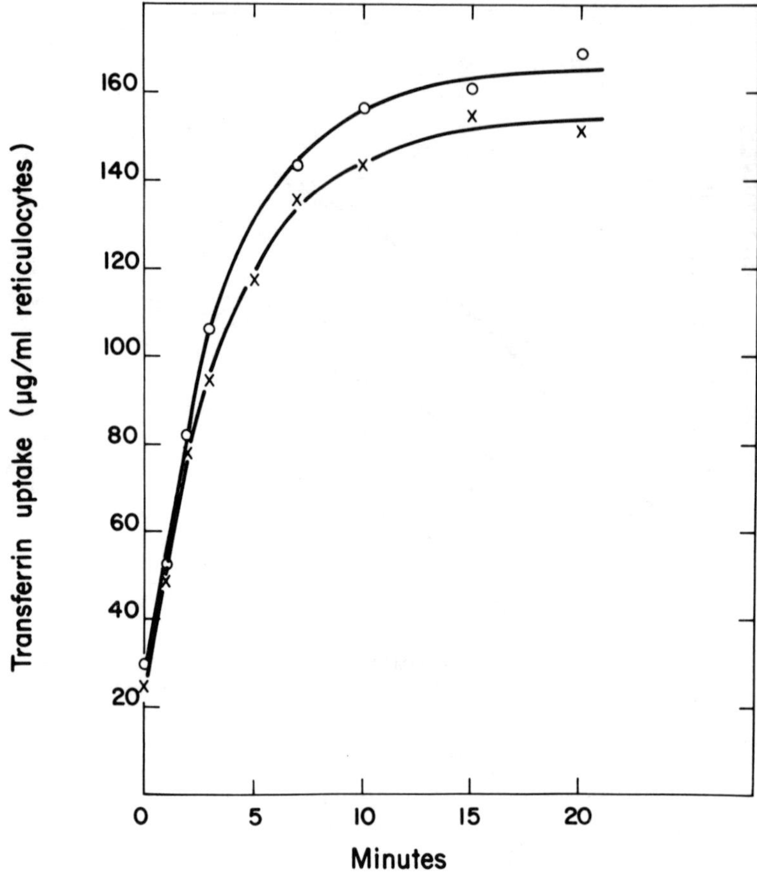

Fig. 7. Binding of Fe(III)-transferrin-bicarbonate (o— o) and Fe(III)-transferrin-oxalate (x— x), each labelled with ^{125}I, to reticulocytes. (From Aisen and Leibman, 1973).

has also come to be appreciated. The cooperativity between the anion- and the metal-binding functions is so strong that neither function appears to exist in the absence of the other. A site-site interaction is also displayed by the two anion-binding sites of the transferrin molecule. It may be that these cooperative interactions lie at the heart of the mechanism by which the reticulocyte removes iron from the protein.

REFERENCES

Aasa, R. (1970). Powder line shapes in the EPR spectra of high-spin ferric complexes. J. Chem. Phys. 52: 3919-3930.

Aasa, R. (1972). Re-interpretation of the electron paramagnetic resonance spectra of transferrin. Biochem. Biophys. Res. Comm. 49: 806-812.

Aisen, P., Leibman, A., and Reich, H.A. (1966). Studies on the binding of iron to transferrin and conalbumin. J. Biol. Chem. 241: 1666-1671.

Aisen, P., Aasa, R., Malmstrom, B.G., and Vanngard, T. (1967). Bicarbonate and the binding of iron to transferrin. J. Biol. Chem. 242: 2484-2490.

Aisen, P., and Leibman, A. (1968). Citrate-mediated exchange of Fe^{3+} among transferrin molecules. Biochim. Biophys. Res. Comm. 32: 220-226.

Aisen, P., Aasa, R., and Redfield, A.G. (1969). The chromium, manganese and cobalt complexes of transferrin. J. Biol. Chem. 244: 4628-4633.

Aisen, P., Lang, G., and Woodworth, R.C. (1972). Spectroscopic evidence for a difference between the binding sites of conalbumin. J. Biol. Chem. 248: 649-653.

Aisen, P., and Leibman, A. (1972). The role of the anion-binding site of transferrin in its interaction with the reticulocyte. Biochim. Biophys. Acta 304: 797-804.

Aisen, P., Leibman,A., Pinkowitz, R.A., and Pollack, S. (1973). Exchangeability of bicarbonate specifically bound to transferrin. Biochemistry 12: 3679-3684.

Aisen, P., Pinkowitz, R.A., and Leibman,A. (1973). Spectroscopic and other studies of the anion-binding sites of transferrin. Annals N.Y. Acad. Sci., in press.

Bates, G.W., and Schlabach, M.R. (1973). A study of the anion binding site of transferrin. FEBS Letters 33: 289-292.

Feeney, R.E., and Komatsu, S.K. (1966). The transferrins. Structure and Bonding 1: 149-206.

Fletcher, J., and Huehns, E.R. (1967). Significance of the binding of iron by transferrin. Nature 215: 584-586.

Fletcher, J., and Huehns, E.R. (1968). Function of transferrin. Nature 218: 1211-1214.

Goya, N., Miyazake, S., Kodate, S., and Ushio, B. (1972). A family of congenital atransferrinemia. Blood 40: 239-245.

Heilmeyer, L. (1964). Human hyposideraemia. In "Iron Metabolism: An International Symposium" (F. Gross, ed). Springer-Verlag, Berlin, 201-213.

Holmberg, C.G., and Laurell, C.-B. (1945). Studies on the capacity of the serum to bind iron. A contribution to our knowledge of the regulation of serum iron. Acta Physiol. Scand. 10: 307-319.

Holmberg, C.G., and Laurell, C.-B. (1947). Investigations in serum copper: I. Nature of serum copper and its relation to the iron-binding protein in human serum. Acta Chem. Scand. 1: 944-950.

Jandl, J.H., Inman, I.K., Simmons, R.L., and Allen, D.W. (1959). Transfer of iron from serum iron-binding protein to human reticulocytes. J. Clin. Invest. 38: 161-185.

Katz, J.H. (1961). Iron and protein kinetics studies by means of doubly labeled human crystalline transferrin. J. Clin. Invest. 40: 2143-2152.

Katz, J.H. (1970). Transferrin and its functions in the regulation of iron metabolism. In "Regulation of Hematopoiesis" (A.S. Gordon, ed), Appleton-Century-Crofts, New York, 539-577).

Luk, C.K. (1971). Study of the nature of the metal-binding sites and estimate of the distance between the metal-binding sites in transferrin using trivalent lanthanide ions as fluorescent probes. Biochemistry 10: 2838-2843.

Masson, P.L., and Heremans, J.F. (1967). Studies of lactoferrin, the iron-binding protein of secretion. Protides of the Biological Fluids. Proc. Colloq. Bruges 14: 115-124.

Masson, P.L., and Heremans, J.F. (1968). Metal-combining properties of human lactoferrin (red milk protein). I. The involvement of bicarbonate in the reaction. Europ. J. Biochem. 6: 579-584.

Pinkowitz, R.A., and Aisen, P. (1972). Zero-field splittings of iron complexes of the transferrins. J. Biol. Chem. 247: 7830-7834.

Price, E.M., and Gibson, J.F. (1972). A re-interpretation of bicarbonate-free ferric transferrin E.P.R. spectra. Biochim. Biophys. Res. Comm. 46: 646-651.

Price, E.M., and Gibson, J.F. (1972). Electron paramagnetic resonance evidence for a distinction between the two iron-binding sites in transferrin and in conalbumin. J. Biol. Chem. 247: 8031-8035.

Schade, A.L., and Caroline, L. (1944). Raw hen egg white and the role of iron inhibition of _Shigella dysenteriae_, _Staphylococcus aureus_, _Escherichia coli_ and _Saccharomyces cervisiae_. Science 100: 14-15.

Schade, A.L., and Reinhart, R.W. (1966). Carbon dioxide in the iron and copper siderophilin complexes. Protides of the Biological Fluids. Proc. Colloq. Bruges. 14: 75-81.

Warner, R.C., and Weber, I. (1953). The metal-combining properties of conalbumin. J. Amer. Chem. Soc. 75: 5094-5101.

8

LACTOFERRIN CONFORMATION AND METAL BINDING PROPERTIES

R. M. Parry, Jr. and E. M. Brown*

Eastern Regional Research Center, Agricultural Research Service, U.S. Department of Agriculture, Phila., Pa. 19118

Lactoferrin† is an iron binding protein found in milk and other mammalian secretions. It has properties similar to the serum transferrins and to ovotransferrin of avian egg-white (reviewed by Feeney and Komatsu, 1966). These proteins specifically bind two moles of ferric ion per mole of protein. For each Fe(III) bound, one molecule of bicarbonate is incorporated into the complex (Masson and Heremans, 1968). All of these Fe(III) binding proteins have been found to have similar amino acid compositions, optical spectra and electron paramagnetic resonance spectra.

Two physiological roles have been proposed for the transferrin-like proteins. The primary role for serum transferrin is believed to be transport of iron to the reticulocytes for incorporation into the hemoglobin of red blood cells. *In vivo*, where the iron binding sites generally are not saturated, the protein may act also as a bacteriostatic agent. The same two functions have been proposed for ovotransferrin, but with the bacteriostatic function considered to be the more important. Evidence for the bacteriostatic function of lactoferrin has been provided by Bullen et al. (1972), but while direct evidence for an iron transport function is lacking, it has not been disproved.

Studies by Welty et al. (1972) on the lactoferrin concentration in the cow mammary gland during involution and mastitis infection indicated that lactoferrin rapidly increases under these

---

*National Research Council Postdoctoral Research Fellow.
†Lactoferrin refers to the metal-free protein; iron(III)lactoferrin is the protein saturated with 2 moles ferric ion per mole protein.

conditions. Lactoferrin appears to be unique among the whey proteins in that its concentration changes roughly parallel the periods of resistance to infection in the mammary gland, so that its concentration is very low in normal milk whereas it is much higher in the lacteal secretion of the non-producing gland.

The iron binding proteins such as lactoferrin, ovotransferrin and transferrin are characteristically composed of a single polypeptide chain (Mann et al., 1970; Querinjean et al., 1971) and the body of evidence accumulated to date does not support the hypothesis that a repeating sequence occurs in the proteins (Bezkorovainy and Grohlich, 1973; Elleman and Williams, 1970). The proteins have been reported to be between 70,000 and 90,000 molecular weight. Several laboratories have found molecular weights near 76,000 for bovine lactoferrin (Castellino et al., 1971), human lactoferrin (Querinjean et al., 1971), ovotransferrin (Greene and Feeney, 1968), and human transferrin (Mann et al., 1970). Other workers have shown 86,000 molecular weights for bovine lactoferrin (Groves, 1960) and ovotransferrin (Fuller and Briggs, 1956; Elleman and Williams, 1970). The iron binding proteins are also characterized as having many intramolecular disulfide bonds, no free sulfhydryls, but sufficient bound carbohydrate to put them in the class of glycoproteins (Feeney and Komatsu, 1966).

The most apparent difference between these metal-binding proteins is their relative affinity for iron and the conditions necessary for _in vitro_ removal of iron. The lactoferrins (bovine and human milk) apparently have a higher affinity for iron than ovotransferrin and transferrin (Aisen and Leibman, 1972). Furthermore, removal of the iron from lactoferrin requires exposure to pH 2, while metal-free forms of transferrin and ovotransferrin can be prepared at pH 4. Exposing the protein to pH 2 causes a distinct conformational change in transferrin and ovotransferrin (Feeney and Komatsu, 1966; Phelps and Cann, 1956). Dielectric dispersion measurements reported by Rosseneu-Motreff (1971) have shown that the saturation of transferrin with iron modifies the parameters of the ellipsoidal shape so that the molecule expands slightly, becoming more spherical after metal incorporation. Thus, the protein conformation appears to play a role in the metal complexation mechanism and needs to be considered in an explanation of the binding.

Studies on the amino acids involved in the metal binding site have shown that the more important residues include tyrosines and nitrogen ligands from either histidine or tryptophan residues. Warner and Weber (1953) first postulated that the tyrosyl phenolic groups were involved in this complex because of the release of three protons for each Fe(III) bound to transferrin. More recently, Michaud (1968) and Luk (1971) have found, by pH difference spectroscopy experiments, that four tyrosyl residues are involved in the two metal binding sites of transferrin and ovotransferrin. The EPR studies of Aisen and Liebman (1972) have shown that at least one nitrogen plays a role at each metal binding site in lactoferrin and transferrin. The Mössbauer studies of Spartalian and Öosterhuis (1973) indicate that there are two or three nitrogen ligands for each iron in transferrin. At this writing, experimental methodology has not yielded any definitive evidence for which residue(s) contribute the nitrogen ligands.

Luk's (1971) novel studies using the fluorescent Tb(III)-Fe-(III) transferrin complex indicated that the two binding sites are about 43 Å apart. Recent reports by Aisen et al. (1973, see also his work in this volume) and Aasa (1972) have given the first evidence that the two metal-combining sites are nonequivalent in ovotransferrin and that the protein does not bind metals in a simple random fashion.

We have undertaken this study of bovine lactoferrin to obtain further information about its iron binding sites and the protein conformational changes which occur on metal complexation.

ISOLATION METHOD

Lactoferrin from bovine milk has not been studied as extensively as transferrin and ovotransferrin, probably because of its low concentration in normal milk. It has been reported (Groves, 1960) that lactoferrin concentration may be as low as 20 mg/l. of milk. Our experience with isolation of the protein from milk indicated considerable variation in the amount of lactoferrin present. It appeared that the level of lactoferrin present may be directly proportional to the level of other minor whey proteins such as bovine serum albumin and immune globulin proteins. Higher amounts of these proteins are often found in late lactation milk or in milk from a cow with a mastitis infection of the mammary gland. Higher levels of these proteins are also found when the mammary gland secretes colostrum immediately after calving. The lactoferrin concentration is at least ten-fold higher in colostrum but rapidly decreases in the following three to four days as the gland begins to produce normal milk.

The fluid expressed from the mammary gland of non-lactating cows was found to be the richest source of lactoferrin. This prelacteal secretion yielded purified protein in gram quanities (2 to 4 g/l. of fluid). A further advantage in using prelacteal secretion is that the lactoperoxidase concentration in this fluid is negligible. This enzyme, which has a molecular weight and an isoelectric point similar to lactoferrin, is a particularly difficult contaminant to remove from preparations. Samples of lactoferrin prepared from milk or colostrum, in some cases, would be sufficiently contaminated with lactoperoxidase after purification to render them of questionable value in characterization studies.

The purification procedure is essentially that described by Groves (1965). Some modifications were made to speed this isolation. The nitrilotriacetate complex of iron(III) was added to most preparations at the beginning stages of purification so that all lactoferrin was saturated. Isolation from milk was begun by warming the milk to 30°, adjusting the pH to 4.6 to precipitate the casein, readjusting the whey to pH 6.5 and absorption of the positively charged proteins on Amberlite IRC-50 (approximately 50 g/l.). After stirring several hours at 4° the resin was recovered, washed with water and the proteins eluted with 0.2 M phosphate, pH 6.5. Solutions were dialyzed and concentrated by ultrafiltration.

The initial stages of the purification from colostrum differed slightly from that used for milk. First day post-partum colostrum was obtained and stored frozen until used; the viscous liquid was diluted with two volumes water per volume colostrum, the casein precipitated and the supernatant containing iron(III)lactoferrin was fractionated as described below.

Preliminary treatment of the prelacteal secretion with iron caused the viscous opaque fluid to turn distinctly pink. Addition of ammonium sulfate to 50% saturation precipitated unwanted protein. Solid ammonium sulfate was added to the supernatant to make it 75% saturated at 2°, a pink precipitate containing iron(III)lactoferrin was recovered and the precipitate was analyzed.

After these preliminary treatments for the three mammary secretions, all preparations were pumped through diethylaminoethyl (DEAE) cellulose columns, preequilibrated with 0.005 M phosphate, pH 8.2. This treatment effectively adsorbed most contaminating whey proteins and allowed the lactoferrin to pass through unadsorbed. The effluent was concentrated by ultrafiltration and chromatographed on a phosphocellulose column equilibrated with 0.1 M phosphate, pH 6.0 (2 x 40 cm) using a step-wise gradient of increasing pH and phosphate buffer concentration. Lactoferrin was eluted on the addition of 0.2 M phosphate - 0.1 M NaCl, pH 7.5, dialyzed and concentrated if necessary. Some iron was apparently removed from the protein during this chromatography and iron was again added to the preparation with the excess being removed by gel filtration on Sephadex G-25.

Metal-free lactoferrin was prepared according to the method of Johanson (1960) by titrating the solution to pH 2, whereupon the solution became colorless. The free iron was then adsorbed on Dowex 50-12X resin. Rapid titration of the protein back to pH 6.5 gave good spectroscopic-quality lactoferrin solutions. This method leaves not more than 0.1 mole of iron per mole of lactoferrin. Iron was determined colorimetrically as the 1,10-phenanthroline complex after protein digestion with sulfuric acid and iron reduction with hydroxylamine hydrochloride (Snell and Snell, 1959).

MATERIALS AND METHODS

Sedimentation velocity experiments were carried out in a Spinco Model E ultracentrifuge with schlieren optics. Normally 15 mm, 4° sector, Kel F centerpieces were used, but protein concentrations below 2 g/l. were studied with 30 mm centerpieces. Stock protein solutions for all ultracentrifuge experiments were passed through Sephadex G-25, previously equilibrated with the appropriate solvent, and then dialyzed 18-24 hr against that solvent. Dilutions of the protein stock solution were made with the dialyzate.

Sedimentation equilibrium experiments were carried out using the Yphantis method (1964). The ultracentrifuge was equipped with interference optics, focused at the 2/3 plane of the cell, and photographs made on II G spectroscopic glass plates. The sample cell was modified for external loading as described by Ansevin et al. (1970). Fringe measurements were made on a Nikon comparator and a pre-blank and post-blank measured to insure no change in cell distortion during the experiment. Commonly the final speed chosen for the run was 20,410 RPM. At the start of the experiment the sample was run at 2 ω^2 for 1/2 hr and 2/3 ω^2 for 1/2 hr before setting the final speed. Under these conditions, equilibrium was reached within 18 hr. The molecular weights were calculated according to the method of Roark and Yphantis (1969) using Dr. Roark's computer program.

Extinction coefficients were determined for lactoferrin in water using optical density measurements on a Zeiss PMQ II spectrophotometer. The protein was dried to a constant weight *in vacuo* at 50°, usually for about 48 hr.

Partial specific volume measurements were made at 25° on a Paar densimeter in water. Solvent densities were also measured during the time of the ultracentrifuge experiments. Solvent viscosities were taken from the International Critical Tables.

Gel electrophoresis did not prove to be a reliable test for purity of lactoferrin, particularly with regard to lactoperoxidase contamination. The critical test of purity used for this study was based upon the visible and ultraviolet spectra because of the distinctive Soret band of lactoperoxidase. Absorbance ratios of A_{280}/A_{465} = 27 to 28 and A_{410}/A_{465} = 0.8 to 0.85 were considered acceptable values for purified lactoferrin.

Long-term storage has been a continuing problem with lactoferrin. Storage as the frozen 2% protein solution either with or without metal for two months or more caused the formation of turbidity which increased with time. Lyophilization also produced turbid solutions when the protein was redissolved, but the instability did not depend on the length of storage. We chose the lyophilization method for iron(III)lactoferrin which was stored desiccated at 4°, removing the turbidity with Millipore filters (3 μ pore size). Metal-free protein was prepared immediately before use.

Ultra Pure urea and guanidine hydrochloride were purchased from Schwarz-Mann Biochemicals. Mercaptoacetic acid (Aldrich) was freshly distilled immediately before use. Other chemicals were the best available grades and were used without further purification. Glassware was cleaned with chromic acid-sulfuric acid cleaning solution, rinsed with water, then soaked in concentrated nitric acid and rinsed with copious amounts of water to avoid heavy metal contamination. The water used in this study was doubly deionized and was the best available with respect to conductivity, metal contamination, fluorescence, and absorbance. It compared favorably with water which was deionized and then glass distilled.

Cobalt-, chromium-, and manganese-protein complexes were prepared by the method of Aisen et al. (1969), except that chromic chloride rather than chromous chloride was used for the formation of Cr(III)lactoferrin.

For spectrophotometric experiments, solutions were 0.1 molar in KCl in addition to the specified buffers or perturbants. Neutral, aqueous solutions were either unbuffered at pH 6.5-7.5, or buffered at pH 7.0 with 0.005 M N-(2-hydroxyethyl piperazine)-N'-2-ethane sulfonic acid (HEPES). No differences were noted in the presence or absence of the buffer. Adjustments of pH were made by dilution with appropriate amounts of KCl-KOH or KCl-HCl mixtures.

Absorption spectra were obtained with a Cary Model 14 recording spectrophotometer and single wavelength absorbances with a Zeiss PMQ-II. Beer's law was shown to hold over all practical concentrations. Difference spectra involving only pH changes were obtained using matched 1 cm cuvettes. The effects of urea, guanidine and mercaptoethanol on the difference spectrum were obtained using tandem 1 cm cells so that protein and solvent concentrations could be matched.

Solvent perturbation difference spectra were obtained using a minimum of two different concentrations of iron(III)lactoferrin in aqueous solution, lactoferrin in aqueous solution and in 8 M urea, and 8 M urea solution of lactoferrin which had been disulfide cleaved with mercaptoacetic acid by the method of Herskovits and

Laskowski (1962). Concentrations were selected so that A_{280} was between 1.0 and 2.2. The perturbants used were sucrose, ethylene glycol, glycerol, and dimethylsulfoxide in aqueous and 8 M urea solutions with a final perturbant concentration of 20%. The data were obtained and analyzed using the method of Herskovits and Sorensen (1968a,b). Spectra were recorded with the Cary 14 at 5 Å per second, dynode setting = 3, slit control = 25; slit width was not more than 0.7 mm at wavelengths greater than 250 nm.

The circular dichroism (C.D.) spectra were recorded using a Jasco Model ORD/UV/CD-5 or a Cary 60 recording spectropolarimeter with a Model 6002 circular dichroism attachment. Protein concentrations and optical path lengths were adjusted to keep the photomultiplier voltage below 900 volts. C.D. data in the 600-300 nm region is expressed as the mean molar ellipticity [θ] deg cm^2 per decimole of iron bound, and below 300 nm in terms of protein concentration as the mean molar ellipticity $[\theta]_{MRW}$, using 113 as the mean residue weight.

Fluorescence measurements were made on an Aminco-Bowman spectrophotofluorometer with excitation at 280 nm and emission 345-360 nm.

RESULTS AND DISCUSSION

Sedimentation Velocity

Sedimentation velocity experiments were made on lactoferrin in three different states: iron(III)lactoferrin in 0.1 M KCl-HEPES at pH 6.5; lactoferrin in 0.1 M KCl-HEPES at pH 6.5; and lactoferrin in 0.1 M KCl-HCl at pH 2. These results, corrected to 20° and water and extrapolated to zero protein concentration are presented in Figure 1. Non-ideal molecular electrostatic repulsion effects are quite apparent at lactoferrin concentrations below 3 g/l., but these effects could be negated by the use of an effective buffer in the experiments done at neutral pH. The lactoferrin appears to resist further conformational changes with time as a result of the pH 2 treatment since two of these experiments were repeated after 48 hr at 4° and gave essentially the same $S_{20,w}$ values (within 0.01 svedbergs) as the freshly prepared solutions. Varying the temperature and/or the ionic strength of the solvent had no effect upon the corrected sedimentation coefficient.

The decrease in $S_{20,w}$ from the neutral to the acid protein solutions indicates marked conformation change for the lactoferrin. Spectroscopic studies showed the shape change to be completely reversible and there is no loss of iron binding capability when the lactoferrin is returned to neutrality. A similar change in sedimentation coefficient of ovotransferrin when studied as a

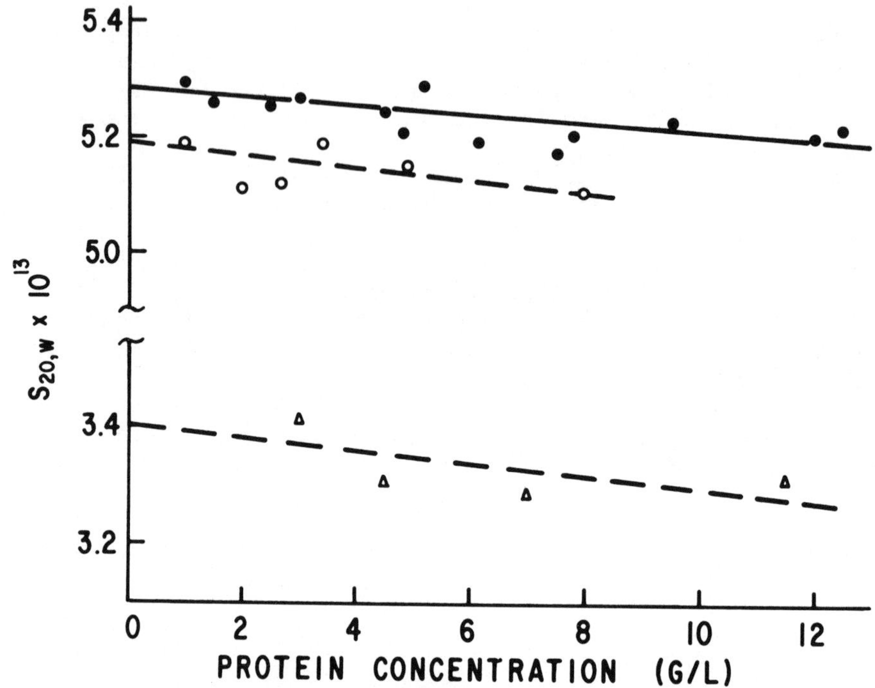

Figure 1. Sedimentation coefficients corrected for solvent and temperature plotted versus protein concentration. Top curve (———) is iron(III)lactoferrin at pH 6.5 with the data indicated by filled circles. Middle curve (— —) is lactoferrin at pH 6.5 with the data indicated by open circles. Bottom curve (- - - -) is lactoferrin at pH 2.0 with data indicated by open triangles.

function of pH has been reported (Glazer and McKenzie, 1963; Phelps and Cann, 1956).

A small but experimentally consistent difference in the $S_{20,w}^{c=o}$ between lactoferrin solutions with and without iron was observed (see Table 1). Under our experimental conditions, we were able to achieve an experimental precision of ±0.05 svedbergs. Frictional coefficients were calculated according to the equation of Svedberg and Pedersen (1940) for the lactoferrin in the three states and are reported in Table 1. These data indicate a deviation from a a spherical shape in agreement with Fuller and Briggs (1956) who

TABLE I

Physical Measurements of Lactoferrin

	Iron(III)lactoferrin pH 6.5	Lactoferrin pH 6.5	Lactoferrin pH 2
ε_m	1.35×10^5	1.09×10^5	-
$\bar{V}_{25°,app.}$	0.729 ml/g	0.723 ml/g	-
$S^{c=o}_{20,w}$	5.27 ± .05 S	5.19 ± 0.05 S	3.40 ± 0.05 S
f/f_o	1.34	1.40	2.13[a]
M_w	86,000 ± 800	86,000 ± 900	-

ε_m, molar extinction coefficient; $\bar{V}_{25,app.}$, apparent partial specific volume at 25°; $S^{c=o}_{20,w}$, sedimentation coefficient corrected to 20°, water and extrapolated to infinite dilution ($S=10^{-13}$ sec^{-1}); f/fo, frictional coefficient; M_w, weight average molecular weight.

[a]Calculated using the $\bar{V}_{25,app.}$ of lactoferrin at pH 6.5.

found iron(III)ovotransferrin to be more spherical than the metal-free protein. Rosseneu-Motreff et al. (1971) have also observed a shape difference between iron saturated and metal-free human serum transferrins. Using dielectric dispersion measurements, they found that the iron binding to transferrin resulted in a slight expansion of the molecule causing it to become more spherical as the axial ratio decreased from 2.5 to 2. The actual nature of the shape change seen with lactoferrin can not be assessed at this time, except to note that the metal does affect the protein structure.

Sedimentation Equilibrium

The molecular weight of iron(III)lactoferrin and lactoferrin in 0.1 M KCl-HEPES, pH 6.5 was determined in sedimentation equilibrium experiments. A plot of the raw apparent weight average

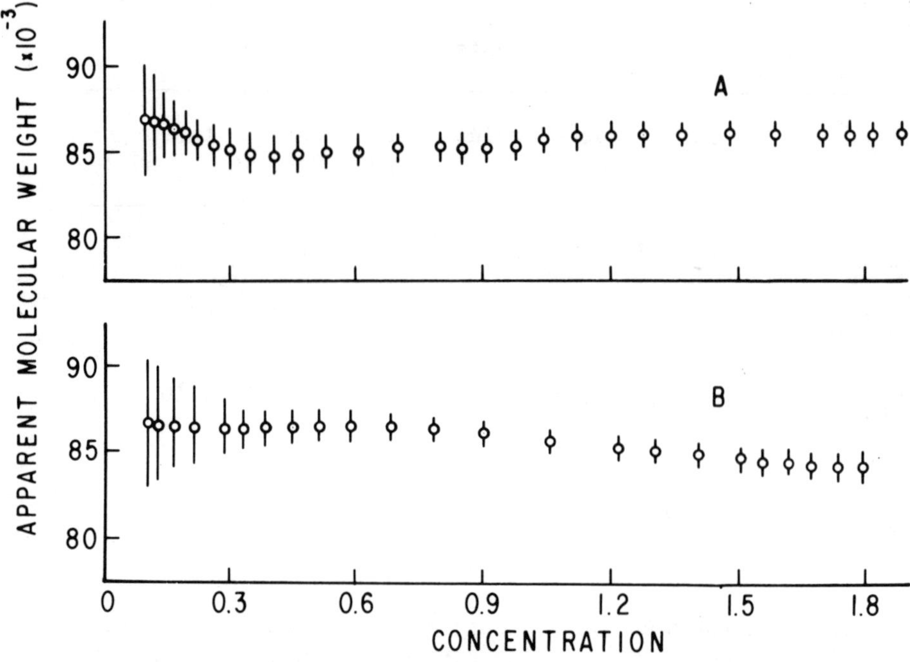

Figure 2. Sedimentation equilibrium plot of the apparent weight average molecular weight versus protein concentration expressed as fringe displacement. Curve A is lactoferrin, pH 6.5, and curve B is iron(III)lactoferrin, pH 6.5.

molecular weight data versus protein concentration in arbitrary fringe displacement units is presented in Figure 2. Each value represents the mid-point molecular weight moment at that position in the ultracentrifuge cell and the range indicated about each point has a confidence level of about 90%. Lactoferrin with and without iron(III) gave a molecular weight of 86,000. The molecular weight is in agreement with the value reported by Groves (1960) but is higher than the values obtained by Castellino et al. (1970) and Querinjean et al. (1971) who reported values of about 76,000 g. Similar flat curves were also obtained for the N and Z average molecular weights, indicating that lactoferrin exists as monomeric non-aggregating protein species at pH 6.5 either with or without Fe(III).

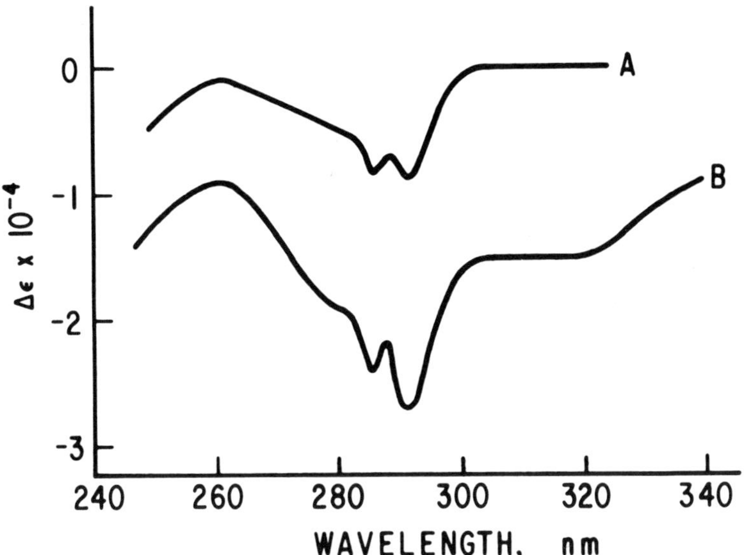

Figure 3. Difference spectra of (A) lactoferrin and (B) iron(III)-lactoferrin in acid solution. In each case the sample solution is at pH 2 and the reference at pH 6.5-7.5. Protein concentrations are 10^{-5} to 10^{-6} molar.

Absorption Spectra

Iron(III)lactoferrin and lactoferrin have an absorption maxima at 280 nm in 0.1 M KCl, pH 6.5. This maximum is shifted to 277-8 nm when the solution is made in 6 M guanidine, 8 M urea or has been titrated to pH 2 with HCl. The blue shift in the aromatic region is accompanied by a general decrease in the absorptivity in the 330-250 nm region.

Difference spectra show minima at 292 and 286 nm, characteristic of the transfer of tryptophan and tyrosine residues from a hydrophobic portion of the molecule into the aqueous phase. Acid difference spectra for the two forms are shown in Figure 3. The effects of guanidine on the spectrum of lactoferrin are immediate, but in the case of iron(III)lactoferrin spectral changes are seen to increase over a several hour period. An additional shoulder appears in the 315-300 nm region of the difference spectrum of iron(III)lactoferrin which may be due to distortion of disulfide bonds (Tan, 1971) or perturbation of tryptophan residues by the iron (Beychok, 1966), or may simply be characteristic of an iron-oxygen bonding. The maximum change as a function of guanidine concentration was achieved with a 2 M solution in the case of lactoferrin and a 6 M solution for iron(III)lactoferrin. The guanidine, urea, and acid effects on the spectrum of lactoferrin were completely reversible. Glazer and McKenzie (1963) found similar results with ovotransferrin and interpreted them as corresponding to a reversible conformational change or unfolding of the molecule.

Circular Dichroism

The 600-300 nm region of the C.D. spectrum of iron(III)lactoferrin in 0.1 M KCl, pH 6.5 showed a broad negative band centered at 455 nm and a narrower positive band or possibly an unresolved

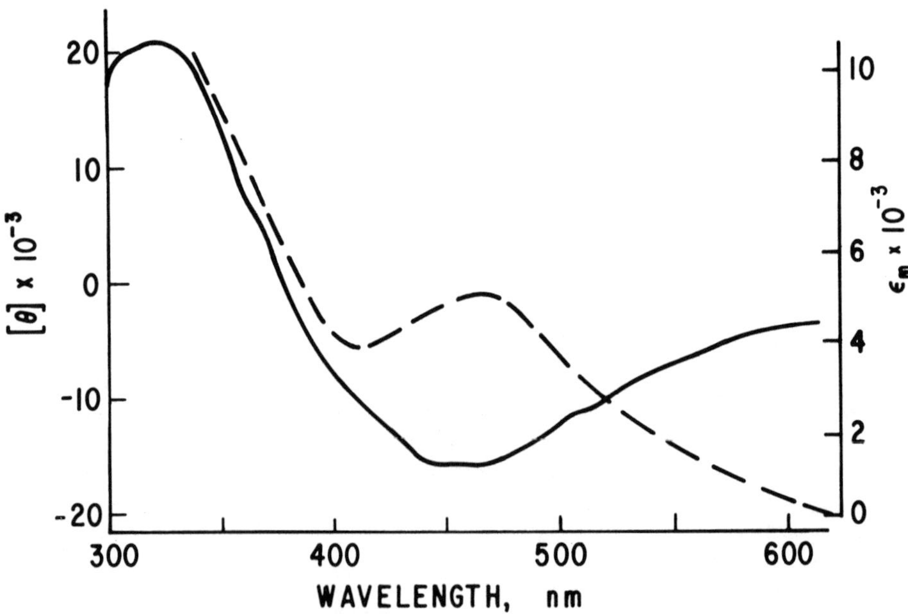

Figure 4. Visible C.D. spectrum (——) and absorption spectrum (---) of iron(III)lactoferrin in aqueous solution. The C.D. spectrum is the average of six experiments; two each in water, 0.1 M KCl, and 0.005 M HEPES all at pH 6.5-7.5.

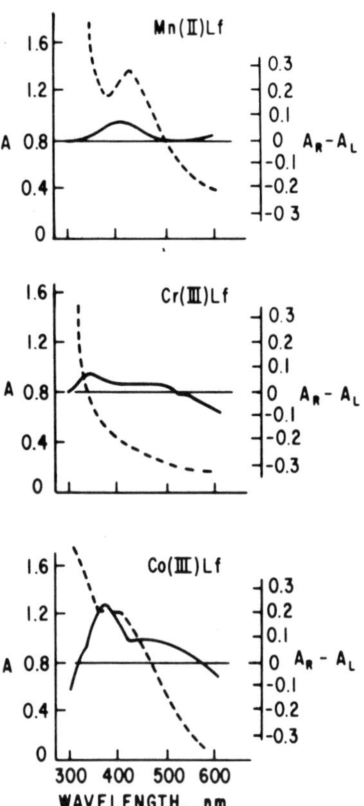

Figure 5. Visible absorption (---) and C.D. (——) spectra of manganese(II), cobalt(III), and chromium(III)lactoferrin complexes. The left axis is labeled in absorption units (A) and the right in C.D. units ($A_R - A_L$). These are relative values only.

doublet in the 330-310 nm region (Figure 4). The 455 nm band corresponds to the absorption maximum near 460 nm and is similar both in position and intensity to those reported for iron(III)ovotransferrin (Tan, 1971) and iron(III)transferrin (Nagy and Lehrer, 1972). The visible C.D. spectra of the manganese, cobalt and chromium lactoferrin derivatives with their respective absorption spectra are shown in Figure 5. The shape of the visible absorption spectra of these various metal derivatives are characteristic of some aqueous octahedral complexes of Mn(II), Co(III), and Cr(III) (Cotton and Wilkinson, 1966).

Tan (1971) has suggested that the 320 nm C.D. band in iron-(III)ovotransferrin may be due to a change in the dihedral angle of one or more of the disulfide bonds when iron is bound. Beychok (1966) has observed C.D. spectra of perturbed tryptophan in proteins at wavelengths as high as 320 nm. Though either of these explanations is possible, one should consider that if the manganese, cobalt, and chromium are bound to the protein at the same site as the iron, similar disulfide or tryptophan perturbations would be expected which were not found here.

The C.D. spectra of lactoferrin and iron(III)lactoferrin in the aromatic region of 310-250 nm closely resemble those of ovotransferrin and its iron complex as reported by Tan (1971) and Gaffield et al. (1966). Iron binding has little effect on the shape of the curve (Figure 6). Ellipticities for iron(III)lactoferrin are more positive than those for lactoferrin, particularly above 290 nm where the ellipticities for iron(III)lactoferrin are themselves positive. The aromatic spectra of the manganese, cobalt, and chromium complexes of lactoferrin are identical, within experimental error, with that of metal-free lactoferrin. Studies by Tan (1971), Tomimatsu and Vickery (1972) and Nagy and Lehrer (1972)

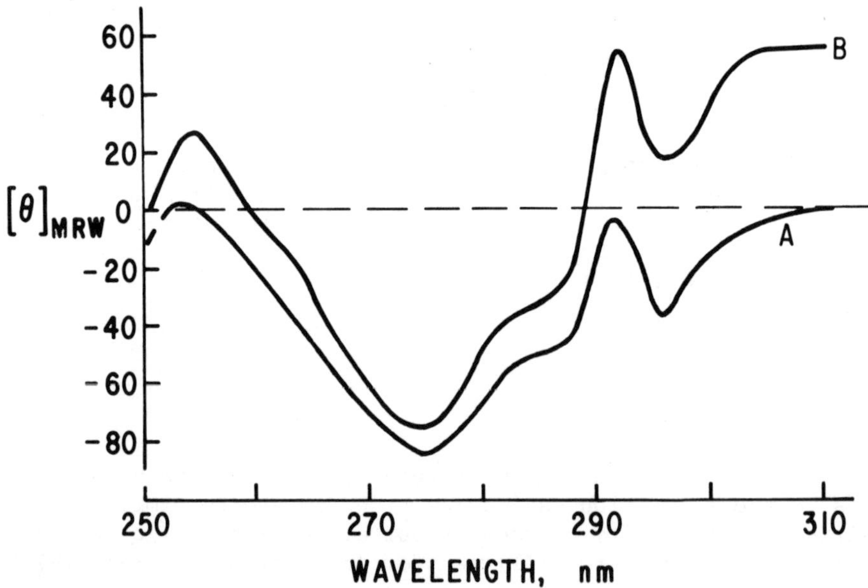

Figure 6. Near ultraviolet C.D. spectra of iron(III)lactoferrin (B) and lactoferrin (A) in neutral aqueous solution. Each curve is an average of 8 to 10 experiments with varying protein concentrations and buffer conditions.

showed that the binding of non-ferrous metals by ovotransferrin and transferrin did not affect the shape of the aromatic C.D. spectra. Analysis of the 290-250 nm portion of the C.D. spectrum in terms of specific residues is complicated by the lack of resolution, but it is a composite of the effects of tyrosine, tryptophan, phenylalanine, and disulfide linkages. Caution must be used in attributing the positive ellipticities in the 300-290 nm region to any direct iron-tryptophan interaction as they may well be due to overlap of the strong 320 nm band.

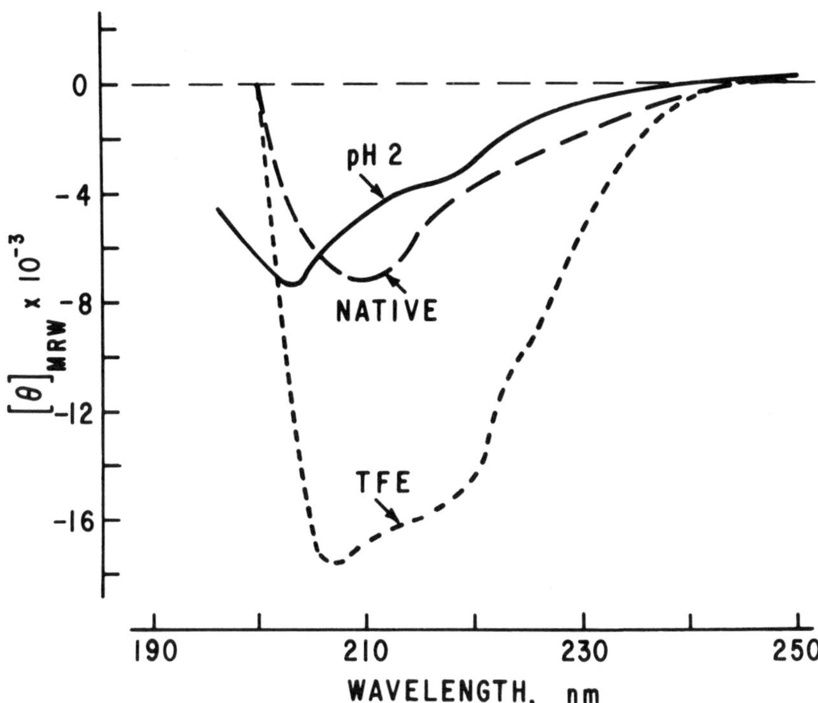

Figure 7. Far ultraviolet C.D. spectra of lactoferrin. Native (— —) is either lactoferrin or iron(III)lactoferrin in neutral aqueous solution, an average of 10 experiments each. The pH 2 spectrum (---) is for lactoferrin in an aqueous 0.1 M KCl-HCl mixture at pH 2.0, an average of 6 experiments. The TFE spectrum (---) is for lactoferrin in neutral aqueous solution diluted to give a solution which is 90% trifluoroethanol, an average of 2 experiments.

In the far ultraviolet (240-190 nm) where the C.D. spectrum is dependent primarily on protein conformation, the spectra of iron(III)lactoferrin and lactoferrin in neutral aqueous solution are indistinguishable (Figure 7). The computational method of Chen et al. (1972) using a mean residue weight of 113 gives about 15% helix and 50% unordered structure. The helical content of transferrin and ovotransferrin as calculated by this method range between 15% and 21% (Tomimatsu and Vickery, 1972). Addition of acidic methanol and trifluoroethanol to lactoferrin solutions increases the helical content to about 25% and 40%, respectively, showing that the low helicity of the native protein is not due entirely to the rigidity imparted by disulfide bonds. In solutions of 6 M guanidine, 8 M urea, or in aqueous acid solution (pH 2), the far ultraviolet C.D. spectrum is that of a completely unordered protein. These solvents also affect the near ultraviolet (aromatic) region of the C.D. spectrum by decreasing the magnitude and resolution of the bands. The effect of guanidine on the 300-290 nm portion of the C.D. spectra is shown in Figure 8. The iron(III)lactoferrin is more resistant to the effects of guanidine than is the metal-free lactoferrin.

Fluorescence

The intensity of the fluorescent emission from iron(III)lactoferrin in 0.1 M KCl-HEPES at pH 6.5 was about 50% of that for lactoferrin under the same conditions. The emission shifted to the red and became more intense at low pH or when guanidine was added to the solution. The increase in the relative fluorescence intensity for lactoferrin was linear with decreasing pH to pH 2, or when increasing the guanidine concentration to 6 M. On the other hand, the fluorescence of iron(III)lactoferrin showed no linear response when studied as a function of pH or of guanidine concentration. A greater apparent structural stability was seen with the metal-bound protein since no change in fluorescent intensity was observed until pH 4 at which point it rapidly increased to a maximum at pH 2. Similarly no change was observed until the concentration of guanidine reached 3 M; maximum change was seen at a 6 M concentration.

CONCLUSIONS

The current study on the physical and conformational properties of bovine lactoferrin has shown that this protein exists as a non-aggregating species with a molecular weight of 86,000 in 0.1 M KCl-HEPES at pH 6.5.

The evidence presented here suggests some conformation differences between lactoferrin and iron(III)lactoferrin. Sedimentation

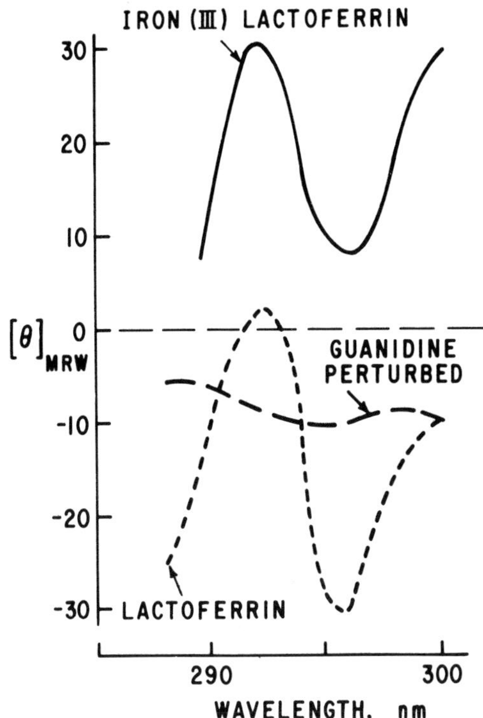

Figure 8. The effect of guanidine on the C.D. bands 300-290 nm of lactoferrin. The guanidine perturbed spectrum (— —) represents lactoferrin in 2 M guanidine or iron(III)lactoferrin in 6 M guanidine.

velocity experiments showed a small change in the $S_{20,w}^{c=o}$ and f/f_o indicating a more spherical shape for iron(III)lactoferrin. Spectroscopic and fluorescence experiments showed that the metal-bound protein was much more resistant to perturbation by acid (on approach to pH 2) or by increasing guanidine concentrations. The metal apparently does not greatly change the spectroscopically observed native conformation, but it does produce a more stable conformation.

No conclusive evidence was obtained about the role of tryptophan as being the residue contributing the nitrogen ligand for iron coordination. Other studies from this laboratory (Brown and Parry) using pH difference spectroscopy indicated four tyrosine residues may be involved in the metal-binding site.

No significant difference in secondary structure was found between iron(III)lactoferrin and lactoferrin, and the calculated helical content of 15% was similar to that reported for ovotransferrin and transferrin. The three iron binding proteins thus have remarkably similar secondary structure which was little affected by metal binding.

A much larger reversible conformational change, seen as an apparent unfolding of lactoferrin, occurred in the presence of acid, urea or guanidine. This change could be observed spectroscopically when access of solvent to chromophoric groups increased, as seen in a hypsochromic shift in the absorption maximum, a negative difference spectra, a decrease in helical structure, and a lowering in the sedimentation coefficient from 5.3 to 3.4.

The fluorescence spectrum of lactoferrin appears to be primarily due to the tryptophan residues. The increase in relative flourescence intensity which occurred upon lowering the pH, also indicated a protein conformational change. The emission red shift may indicate less tyrosine-tryptophan interaction, i.e. a more "pure" tryptophan fluorescence.

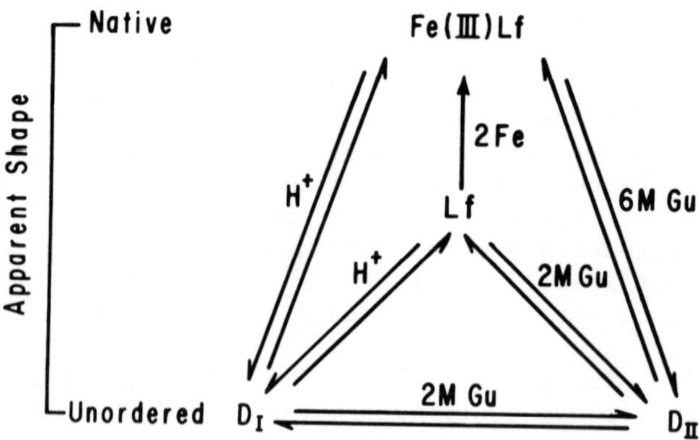

Figure 9. A summary flow sheet of bovine lactoferrin conformation changes. See text for a detailed explanation.

The shape changes observed in this study are summarized by way of a flow diagram in Figure 9. Acid addition to iron(III)lactoferrin to pH 2 causes an unfolding of the molecule and expedites iron removal shown as D_I. Similarly, 6 M guanidine unfolds the protein yielding protein form D_{II}. Our spectroscopic work showed no structural difference between D_I and D_{II}. Reduction of the disulfide bonds did not yield any evidence of further structure perturbation according to C.D. or U.V. difference spectroscopy. The metal-free protein unfolds at a lower concentration of guanidine and a small shape change was observed in the presence of bound iron as discussed above. Hence, iron binding capability of lactoferrin appears to be related to the molecular shape of the protein.

ACKNOWLEDGMENTS

We wish to express our thanks to T. T. Herskovits, Fordham University, for the use of his Cary 60-6002 and many helpful discussions, to V. G. Metzger, ERRC, for her help with the computer analysis of the data, and to Kathleen Gentilcore for her excellent technical assistance.

REFERENCES

Aasa, R.(1972). Biochem. Biophys. Res. Communs. 49:806.
Aisen, P., Aasa, R. and Redfield, A.G.(1969). J. Biol. Chem. 244:4628.
Aisen, P., Lang, G. and Woodworth, R.C.(1973). J. Biol. Chem. 248:649.
Aisen, P. and Leibman, A.(1972). Biochim. Biophys. Acta 257:3144.
Ansevin, A.T., Roark, D.E. and Yphantis, D.A.(1970). Anal. Biochem. 34:237.
Beychok, S.(1966). Science 154:1288.
Bezkorovainy, A. and Grohlich, D.(1973). Biochim. Biophys. Acta 310:365.
Brown, E. M. and Parry, R. M., Jr. Submitted to Biochemistry.
Bullen, J.J., Rogers, H.J. and Leigh, L.(1972). Brit. Med. J. 1:69.
Castellino, F.J., Fish, W.W. and Mann, K.G.(1970). J. Biol. Chem. 245:4269.
Chen, Y.-H., Yang, J.T. and Martinez, H.M.(1972). Biochemistry 11:4120.
Cotton, F.A. and Wilkinson, G.(1966). Advanced Inorgan. Chemistry, 2nd rev. ed., Interscience, New York, N. Y., chapter 29.
Elleman, T.C. and Williams, J.(1970). Biochem. J. 116:515.
Feeney, R.E. and Komatsu, S.K.(1966). Struct. Bonding 1:149.
Fuller, R.A. and Briggs, D.R.(1956). J. Am. Chem. Soc. 78:5253.
Gaffield, W., Vitello, L. and Tomimatsu, Y.(1966). Biochem. Biophys. Res. Communs. 25:35.

Glazer, A.N. and McKenzie, H.A.(1963). Biochim. Biophys. Acta 71: 109.
Greene, F.C. and Feeney, R.E.(1968). Biochemistry 7:1366.
Groves, M.L.(1960). J. Am. Chem. Soc. 82:3345.
Groves, M.L.(1965). Biochim. Biophys. Acta 100:154.
Herskovits, T.T. and Sorensen, M.(1968a). Biochemistry 7:2523.
Herskovits, T.T. and Sorensen, M.(1968b). Biochemistry 7:2533.
Johanson, B.(1960). Acta Chem. Scand. 14:510.
Luk, C.K.(1971). Biochemistry 10:2838.
Mann, K.G., Fish, W.W., Cox, A.C. and Tanford, C.(1970). Biochemistry 6:1348.
Masson, P.L. and Heremans, J.F.(1968). Eur. J. Biochem. 6:579.
Michaud, R.L.L.(1968). Ph.D. thesis, University of Vermont, Burlington, Vt.
Nagy, B. and Lehrer, S.S.(1972). Arch. Biochem. Biophys. 148:27.
Phelps, R.A. and Cann, J.R.(1956). Arch. Biochem. Biophys. 61:51.
Querinjean, R., Masson, P.L. and Heremans, J.F.(1971). Eur. J. Biochem. 20:420.
Roark, D.E. and Yphantis, D.A.(1969). Ann. N.Y. Acad. Sci. 164:245.
Rosseneu-Motreff, M.Y., Soetewey, F., Lemote, R. and Peeters, H. Biopolymers 10:1039.
Snell, F.D. and Snell, C.T.(1959). Colorimetric Methods of Analysis, Vol. 2A, 3rd ed., D. Van Nostrand Co., Princeton, N.J., p. 231.
Spartalian, K. and Oosterhuis, W.T.(1973). J. Chem. Phy. 59:617.
Svedberg, T. and Pedersen, K.O.(1940). The Ultracentrifuge, Oxford Press, London, p. 40.
Tan, A.-T.(1971). Can. J. Biochem. 49:1071.
Tomimatsu, Y. and Vickery, L.E.(1972). Biochim. Biophys. Acta 285:72.
Warner, R.C. and Weber, I.(1953). J. Am. Chem. Soc. 75:5094.
Welty, F.K., Schanbacker, F.L. and Smith, K.L.(1972). Dairy Research, Ohio Agri. Res. Dev. Center. Research summary No. 59, 23.
Yphantis, D.A.(1964). Biochemistry 3:297.

Reference to brand or firm name does not constitute endorsement by the U.S. Department of Agriculture over others of a similar nature not mentioned.

9

PHYSICOCHEMICAL STUDIES OF Ca^{++} CONTROLLED ANTIGEN ANTIBODY SYSTEMS

P. A. Liberti, H. J. Callahan and P. H. Maurer

Department of Biochemistry, Thomas Jefferson University

1020 Locust Street, Philadelphia, Pa. 19107

In this paper we will briefly discuss some basic aspects of immunoglobulin structure and function. We will then present in some detail our studies on the divalent cation controlled antigen-antibody reactions which are germane to this Symposium. And finally, we will describe some of the fundamental unresolved problems concerning the antibody molecule and show how we have used these reactions to investigate some of these problems.

By way of introduction, Figure 1 depicts some fundamentals of immune systems (1). When an animal is immunized with a foreign macromolecule called an antigen, then most times an immune response can be elicited. This is usually characterized by the appearance in the blood of protein molecules called antibodies, whose general function is to protect the organism from disease. An important class of antibody and the most predominant is called IgG, immunoglobulin G. Their molecular weights are around 140,000, they sediment with a Svedberg coefficient of about 6.5, and they are believed to have the gross conformation which is shown here, i.e. "Y" shaped. As can be seen, IgG is a very symmetrical molecule, it has a combining site (shaded area) on each lobe corresponding to the arms of the "Y". These segments of the molecule are called Fab fragments, which refers to the fraction which is antigen binding. The molecule also has a tail piece, i.e. the base of the "Y", which is called the Fc piece and named by virtue of this fraction's tendency to crystallize. A particularly advantageous quality of IgG is that pepsin cleaves off the Fc piece and two Fab fragments joined by a single disulfide bond can be isolated. This fragment is referred to as the $(Fab')_2$ fragment. Reduction of the disulfide bond yields monovalent segments or the Fab fragment itself. Thus, physical and biological properties of a whole molecule and/or its parts can be studied.

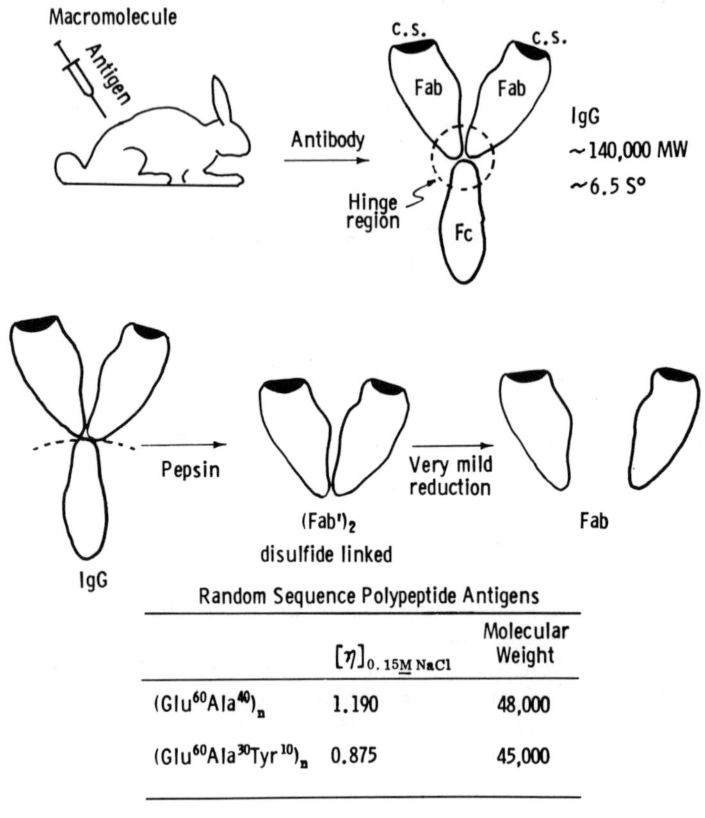

Random Sequence Polypeptide Antigens

	$[\eta]_{0.15\underline{M}\ NaCl}$	Molecular Weight
$(Glu^{60}Ala^{40})_n$	1.190	48,000
$(Glu^{60}Ala^{30}Tyr^{10})_n$	0.875	45,000

Figure 1.

In this connection one biological phenomenon associated with antibodies and whose action depends upon a particular portion of the antibody molecule, viz. the Fc piece is the complement system. Complement is composed of a series of cascading reactions which eventually leads to the activation of a number of hydrolytic enzymes of this system (2). The first of these reactions is activation of first component of complement, Cl, by its combination with Fc portion of aggregated antibodies. Because this first component has a Ca^{++} requirement the complement system can be inactivated by EDTA. As will become apparent it was this phenomenon which led to the discovery of calcium dependent antigen-antibody reactions.

Also shown in Figure 1 are some properties of the antigens we have used in this study. These antigens are synthetic random sequence polypeptides prepared via $\underline{N}$ carboxy anhydride polymerization (3). We shall refer to $(Glu^{60}Ala^{40})_n$ and $(Glu^{60}Ala^{30}Tyr^{10})_n$ as GA GAT, respectively. Both antigens have molecular weights of around 50,000 and exhibit all the polyelectrolyte properties expected for macromolecules having high glutamyl content (4).

The studies to be presented are based on the original observations and studies of Dr. Leslie Clark who discovered that some animals (especially sheep) when immunized with these antigens produce a population of antibodies which require the presence of divalent cations, particularly calcium, to affect immunospecific interaction (5,6). This phenomenon is demonstrated in Figure 2 which shows a routine quantitative test for antibody called the precipitin reaction (this is a standard immunochemical analysis) (7). The ordinate represents the amount of precipitate that is obtained as the concentration of antigen is varied along the abscissa. Such a curve is produced by adding to a series of tubes which contain an equal amount of antisera, or in other words antibody, increasing amounts of antigen. Precipitation occurs because antibody is a bivalent molecule and these antigens have as many as 20 regions to which antibody binds (4) (antigenic determinants), resulting in a cross linked network. As one exceeds the condition where antibody combining sites and antigen sites are equivalent, the cross-linking network starts to break down and the system eventually becomes soluble ("antigen excess"). The maximum amount of precipitate that one can obtain doing a precipitin reaction of this kind is indicative of how much antibody is present. These curves demonstrate that if the precipitin reaction for the GAT system is done in the presence of EDTA, the lower curve is obtained, whereas analyzing in the presence of divalent ions yields the upper curve, indicating that there is considerable loss of precipitating antibody when the reaction is done in EDTA or when calcium and other divalent ions have been removed by chelation. In some systems, precipitating antibody has been reduced by as much as 90% by EDTA (5,6). The EDTA effect can be reversed quantitatively by calcium addition.

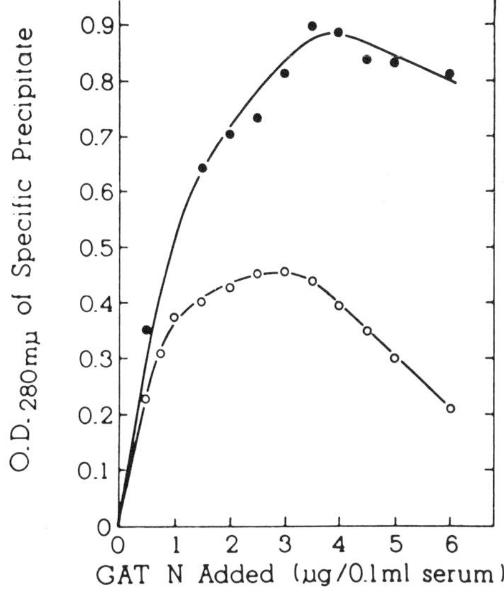

Figure 2. Precipitin reactions of undiluted sheep anti-GAT serum (sheep 5-11) in the presence (○) and absence (●) of 0.008 M EDTA.

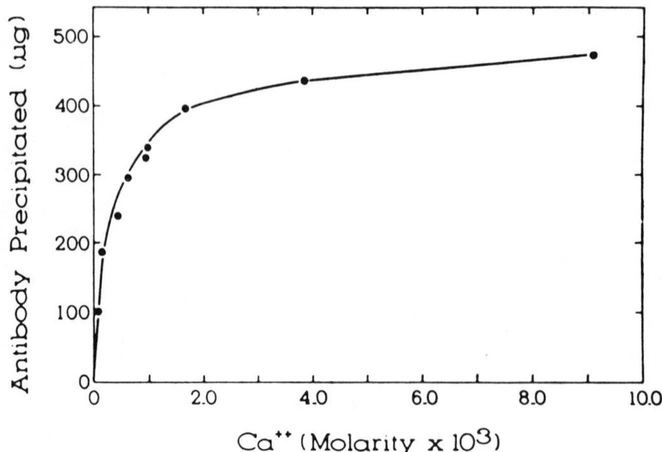

Figure 3. Effect of increasing concentrations of calcium on the specific precipitation of purified calcium-dependent antibody, (sheep 5-11).

Figure 3 shows the amount of antibody that can be precipitated under conditions of equivalence, that is when there is an equivalent amount of antigen and antibody present, as a function of calcium concentration. Note that with increased calcium the amount of antibody which precipitates, or interacts, rises and plateaus off. What is particularly interesting about this figure is that the plateau occurs at around .008 molar calcium which is roughly the physiological level of calcium found in sera, suggesting that the phenomenon we are observing is a physiologically important one.

At this point, we ask ourselves several questions. One is, does the antigen-antibody complex exist in sera and the addition of calcium cause the complex to precipitate? Or two, does calcium have its effect on the GAT antigen or antibody? Figure 4 shows the results of an experiment designed to answer these questions. This figure depicts an ultraviolet scan of mixtures of calcium dependent antibody and GAT sedimenting in an ultracentrifuge cell. In the upper portion of the figure are the results for purified antibody and GAT sedimented in the absence of calcium. As can be seen, antibody and the antigen sediment essentially independent of each other. In fact, the antibody sediments with the same sedimentation coefficient when calcium is absent as it does if the antigen were absent also, whereas in the lower half of the figure, a scan taken at the same time of sedimentation shows a great deal more of the material has moved away. From the absence of a definable plateau, there is every indication that we're dealing with the complexes, and thus it is apparent that GAT and its antibody do not recognize each

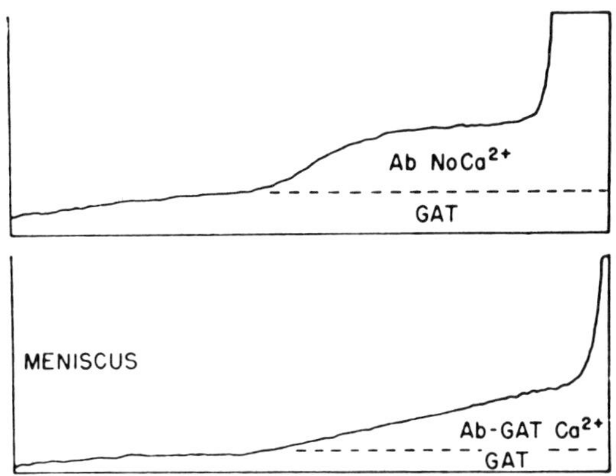

Figure 4. Scanner tracings (A_{280nm} vs. radial distance) of the sedimentation of calcium dependent antibody with GAT in the presence (lower curve) and absence (upper curve) of Ca^{2+}.

other in the absence of calcium. This result has also been substantiated using radiolabeled antigen and rabbit anti-sheep globulin; no binding occurs in the absence of calcium (4).

Table I lists the results of a study designed to determine the specificity of calcium for these reactions. In these experiments, antigen and antibody were mixed at equivalence and then either calcium or other divalent ions added and the extent of reaction as determined by precipitation analysis measured. It can be seen from the second column of the table (using percent precipitation as an indication of the amount of material that's reacted) that calcium is by far the best of the test metals; in fact, we use that as a 100% condition. The other divalent ions show a decreased ability to make this reaction take place. The data of the last column, which gives differences in ionic anhydrous radius of calcium relative to

Table I

Effect of 0.008 N cations on precipitation of Ca++ dependent antibodies

Ion	% Precipitation	Anhydrous Ionic Radius of Cation, Å	$(R_{Ca^{++}}-R_{M^{++}})$ Å
Ca^{2+}	100	0.99	0
Sr^{2+}	66	1.12	−0.13
Mn^{2+}	26	0.80	+0.19
Ba^{2+}	22	1.34	−0.35
Mg^{2+}	4	0.66	+0.33

the other divalent ions when compared to column two, shows that deviation in either direction away from the radius of calcium, results in a decreased precipitin reaction. In fact, it is possible to demonstrate a reasonably good linear relationship between the absolute difference in the anhydrous radius of calcium and that of a test metal ion, and the amount of reaction that takes place under those conditions. This data strongly suggested to us that we were dealing with a phenomenon which is peculiar to the antigen, i.e. the calcium is effecting the antigen. There were two reasons for our thinking: (a) The fact that there is a relationship to anhydrous radius indicates that we are dealing with a site-binding phenomenon which is prevalent in polyelectrolytes (8), and (b) the binding of divalent ions to dicarboxylic acids also shows the same ion functionality, that is binding decreases as anhydrous radii depart from calcium (9). Furthermore, we had good reason to suspect that the effects of calcium were on the antigens, particularly since calcium dependency had never been observed in many systems we had studied where the antigens were not highly negatively charged; in other words, systems where the antigens would not be expected to be grossly affected by calcium.

Based on these results, we studied the effects of calcium on some physical properties of GAT and GA and related antigens and their respective purified calcium dependent antibodies. These results are summarized in Table II. For GAT and GA it is apparent that calcium has a substantial effect on their solution properties, whereas for the purified antibodies the sedimentation coefficients were unchanged, as one might predict, since subtle changes could be taking place and not effect sedimentation coefficients. However, the fluorescence spectra which is more sensitive indication of change, showed no difference in the antibody spectra with and without calcium. Similarly, hydrogen exchange, which again is a very sensitive measure of local environment, showed no difference with or without calcium

Table II

Comparative Effects of Ca^{++} (0.01 M) on Physicochemical Properties of Antigens and Antibodies Involved in Calcium-Dependent Systems

Antigens (GAT, GA)	Antibodies
1. $[\eta]$, 30% decrease.	1. $s_{w,20}$ no change (6.5)
2. $s_{w,20}$ 4% increase, altered concentration dependence.	2. Fluorescence spectra unchanged.
3. ORD slight change to increased helix content.	3. Hydrogen exchange unaffected.
	4. Circular dichroism unchanged.
	5. Perturbation spectra unchanged.

for these antibodies. Similar results were obtained with circular dichroism and difference spectroscopy using various perturbants. Thus, these studies indicate that these antibodies are essentially unchanged by calcium and that calcium dependency must be a property of these antigens. Therefore, calcium must be inducing conformational changes which lead to the creation of antigenic determinants.

Although many interesting questions could have been asked about how Ca^{++} specifically effects these antigens and what role calcium plays, if any, on the antigenic determinant(s), we felt that the ability to control antigen-antibody reaction could lead to a better understanding of some fundamental properties of antibody molecules.

To begin with, one of the most obvious advantages of these calcium dependent systems is in the isolation of purified antibody against GAT or GA since all that is required is EDTA to dissociate the complex and subsequent removal of antigen. This can be done easily by passing the mixture through a DEAE sephadex column, to which the antigen is absorbed. Highly purified antibody is obtained in the void volume. To date, this is one of the most gentle means of antibody isolation. It is worth noting that these antibodies show restricted heterogeneity as determined by isoelectric focusing despite the fact that they are directed against random sequence polypeptides.

The question then became at this point, "What can you do with a system of this type?" Aside from the ease of purification, the next obvious advantage is the fact that unique control studies can be performed. For example, if one wants to measure some property of an antibody when it's combined with antigen, the question that has to be asked is, "What are the base-line corrections?" In other words, what kind of perturbations does the antigen have on the antibody when it is not reacting in an antibody-antigen type of reaction? And, of course, the kind of system that we're talking about has the ability to assess those kinds of parameters. One can take antibody, as we showed earlier, mix it with antigen in the absence of calcium, and pick up essentially all the non specific effects. Additionally, the combination of antigen-antibody reaction in a forward direction can be studied, and similarly treating a reacting system with a chelating agent such as EDTA and turning the reaction off allows the components of the reaction to be studied under dissociating conditions.

The two techniques we decided to use in these studies, at least initially, were methods designed to examine only the antibody molecule. The methods we chose were hydrogen exchange (with GAT) and spectroscopy using the calcium-dependent antibody against the GA antigen. The reason we chose the latter antigen is that it has no spectral absorption in the tyrosine-tryptophan region so that during its interaction with antibody any spectral changes observed could be attributed only to the antibody molecule. The first method which we used was hydrogen exchange. In this method antibody or any other protein is incubated with tritiated water, whereupon a certain fraction of the amide hydrogens of the protein will exchange for tritium. Subsequently, the tritiated antibody is freed from reacted tritium or unexchanged tritium and the kinetics of exchange measured as tritiums are released from the antibody molecule and replaced with hydrogens from the bulk solvent (10). As has been shown by numerous investigators, originally Lindstrom-Lang and collaborators (11) and many more since that time (12), hydrogens exchange with a rate which is commensurate with their environment; in other words, from a somewhat simplistic point of view, amide hydrogen, which is highly hydrogen bonded, exchanges very slowly and those which are involved in loose structure exchange fast. Furthermore, if a structural change occurs, the kinetics of the exchange are altered as a consequence. Hydrogen exchange is a very sensitive method of assaying conformation of a macromolecule and also subtle conformational changes.

There are some very fundamental questions that investigators who are studying antibody molecules have been interested in, particularly regarding the class of antibody molecules that we are concerned with, i.e. the IgG 7 S class of antibodies. These are essentially the following: as is obvious, for every antigen a unique antibody combining site must exist. In other words, if the antibody is going to combine with a particular antigen, then the combining site has to have a specific chemical and physical composition and configuration. Some of the questions that are important to combining site studies are: "What is the amino acid composition?" "What is the size of the antibody combining site?" and "What is the probable structure?"

Other questions of interest apply to phenomena that the antibody molecule is involved in after antigen-antibody combination such as activation of the complement system. After the initial combination of antigen-antibody, the complement system is activated and a whole cascade of reactions takes place. The function of the complement system is a story in itself. However, the important question is, "How does an antibody molecule signal the complement system to turn on?" In particular, are there conformational changes which might affect or involve the Fc segment of antibody, that is the "tail" of the antibody molecule discussed earlier? The basis for this question is that the complement system is essentially inoperative with antibody lacking Fc piece, that is the $(Fab')_2$ fragment (13).

Figure 5 shows the results of a hydrogen exchange study of the monovalent Fab fragment of the calcium dependent antibody against the GAT antigen. The lowermost curve is the exchange-out control curve of the Fab fragment in the presence of antigen and without calcium. These experiments were done with three times the equivalent concentration of antigen. The uppermost curve shows the exchange-out kinetics in the presence of Ca^{++} for the same system, i.e. with the antigen-antibody reaction taking place. The first thing to note is that the hydrogen-exchange curves diverge from each other and have a constant difference after 6-8 hours of about 20 hydrogens per molecule of Fab. This difference persists for up to 30 hrs.

The next thing which is evident from the Figure is that if a reacting system which has exchanged out for a period of three hours is dissociated by adding EDTA to stop antibody reaction, the exchange curve falls down eventually to the control curve. The same thing happens as if dissociation occurs at seven hours, or even as the insert shows, at 27 hours. From turbidity studies we have determined that the kinetics of turning this reaction on or off relative to exchange kinetics are instantaneous. Consequently, these curves are exchange kinetics uncomplicated by antigen-antibody reaction kinetics. As we shall show later, studies with divalent antibody under appropriate conditions have twice as many trapped hydrogens, which suggests that we are observing a combining site phenomenon. Now the question is, "How is the combining site involved?" There are a number of possible interpretations. One, and this is an interpretation that was given to a very similar phenomenon by Ashman and Metzger (14), is that the antibody molecules have undergone a conformational change triggered by the combining sites and some part of the molecule has become more rigid; thus, certain hydrogens are locked in position and do not exchange. An alternative

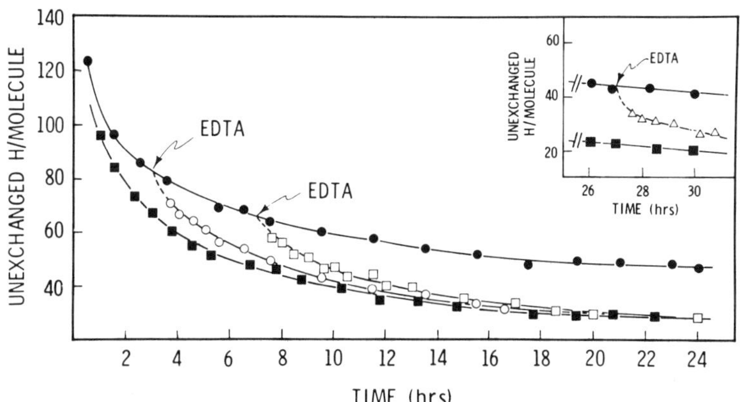

Figure 5. Exchange-out kinetics of Ca-dependent Fab in presence of 3.0 times equivalent concentration of $(Glu, Ala, Tyr)_n$. Reaction stopped with EDTA at 3 hr (○), 7 hr (□), and 27 hr (△) (see insert). Upper most curve with 0.01 M Ca^{2+} added, lower most, control.

interpretation, and one that we have chosen (15), is that the ligand or the antigen in some fashion physically blocks the hydrogens in the combining site from exchanging or freezes the combining site such that its hydrogens don't exchange. This is the type of interpretation that was given to an enzyme inhibitor reaction in the staphylococcal nuclease system of Schechter, Maravek and Anfinsen (16). If our interpretation is correct in that antigen physically keeps the combining site of the antibody fragment from exchanging, it is necessary to show that the antibody does not undergo any conformational change when it interacts with antigen. Data showing changes would support the Ashman-Metzger hypothesis. For these antibodies and most others, evidence of conformational change is very difficult to obtain. However, there is some very good recent evidence from Davies' laboratory which indicates that the X-ray diffraction patterns of a myeloma antibody with and without combined monovalent antigen (hapten) are essentially identical (17). This suggests that at least for that system filling of the combining site by antigen alone does not cause conformational change in the antibody molecule. Hence, our interpretation of ligand blocked combining site hydrogens is quite reasonable. Additionally, the value of 20 blocked amide hydrogens per combining site correlates well with Karush's calculation made some years ago that there would have to be around 20 amino acid residues required to make an antibody combining site (18). The calculations were based on the size of antigenic determinants of polysaccharide antigens (19).

Assuming our interpretation of the blocked hydrogen phenomenon is correct, then one can obtain some information about combining sites from our data. The first conclusion is that for the GAT antigen, and possibly many other antigens, the size of the combining site is such that there are 20 amide hydrogens involved, and hence there are approximately 20 amino acid residues involved in an antigen combining site.

Next, we can look at the exchange kinetics of these blocked hydrogens to obtain some information about their environment. Data for this can be obtained from the type of experiment shown in Figure 5. Consider the antibody Fab as having two kinds of exchanging hydrogens, those from the combining site which can be blocked, plus background hydrogens which cannot. Thus, if we subtract the dissociation curve (where both kinds of hydrogens are exchanging) from the reaction curve (where only background hydrogens are exchanging), the difference is the exchange curve of the blocked hydrogens. This is true providing background hydrogens are not affected by EDTA, which we know to be so, and the combining site exchanges independently of the rest of Fab. The exchange kinetics of these combining site hydrogens obtained in this fashion are shown in Table III. Neglecting the last column which is data obtained under slightly different

Table III

TIME DEPENDENCE OF EXCHANGE OF ANTIGEN-BLOCKED Fab HYDROGENS

	Reaction after Column Chromatography			Reaction on Column
Period (hrs) of Rx before Start of Dissociation	3	7	27	3
H/mol_{sd}	82	66	44	87
Time	Blocked H/molecule Unexchanged			
0	20	20	20	27
0.5	$13._5$	15	14	$16._5$
1.0	11	12	10	$10._5$
1.5	$9._8$	10	10	9
2.0	$8._5$	9	9	8
3.0	$6._8$	7	7	6
4.0	5	5	$5._5$	5
5.0	$4._5$	4		4
6.0	$3._8$	3		3
8.0	2	2		2
12.0	$0._5$	1		1

H/mol_{sd} is defined as indifferent plus blocked hydrogens at start of dissociation.

conditions, we note that the exchange-out kinetics of combining site hydrogens are the same for dissociations done at 3, 7 and 27 hrs., yet the number of indifferent or background hydrogens are different in each case. This data suggests that the combining site acts independently from the remainder of the Fab molecule. This is a very important concept and will be utilized below.

Figure 6 shows the exchange data of Table III plotted on a semi-logarithmic plot. Extrapolation of the linear portion of the curve shows that these hydrogens comprise essentially one kinetic class of about 13 amide hydrogens which exchanged with a half time of

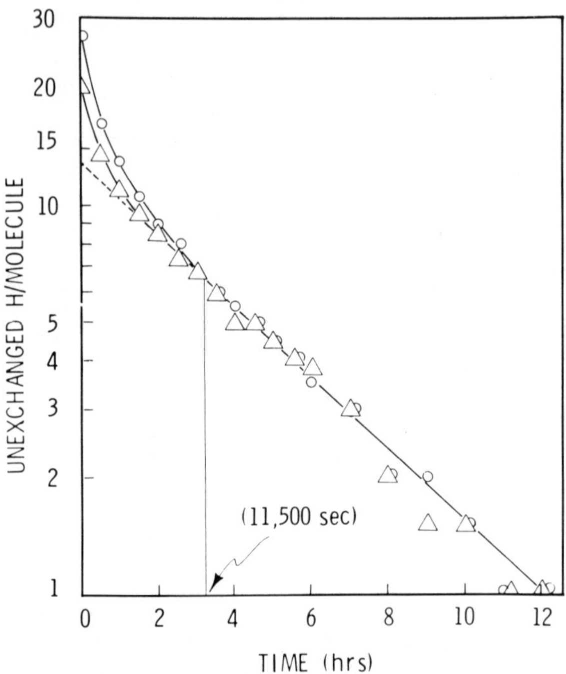

Figure 6. Exchange-out kinetics of antigen blocked Fab hydrogens, 27 hydrogens/molecule blocked (○), 20 hydrogens/molecule blocked (△).

around 11,500 seconds, and a remainder which have a half time of exchange of about 1300 sec. These exchange rates are considerably longer than the exchange rate to be expected for surface residues. They amount to about 1100 and 130 times, respectively, those rates obtained for normal surface residues of proteins. Thus, this data would indicate that the combining site of an antibody is considerably hydrogen bonded and very structured. This is not unreasonable based on the fact that we know that antibody combining sites have very exquisite specificities. It is well established that all one has to do is make a very small modification in the antigen and the antibody no longer recognizes it, or the antibody reacts much less strongly with the antigen. Thus, one could envision from our data a combining site which is very highly structured, most likely rigid, and having a very definite topology. Studies by Day and coworkers (20) on kinetic rates of antigen-antibody combination which are about 10^7 per mol per sec, and activation energies of about 4 K cal/mol, lead them to suggest that the antibody combining site is rigid. Our data, based on a completely different physical study, leads to the same conclusion. Such a conclusion would also be

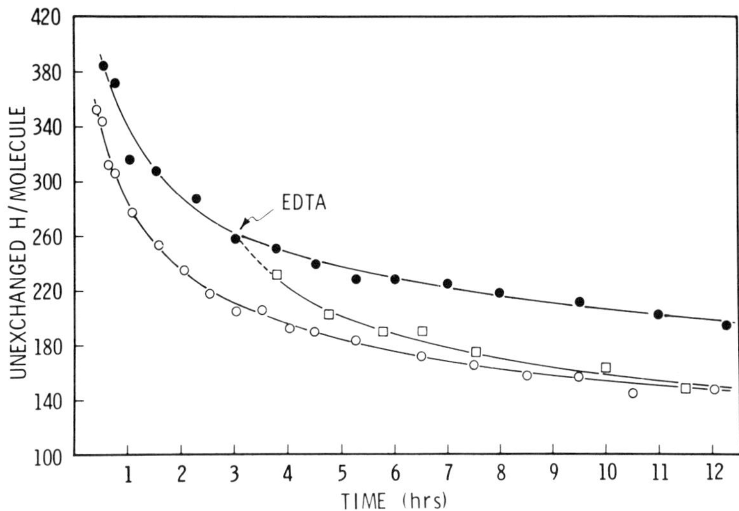

Figure 7. Exchange-out kinetics of calcium-dependent anti-(Glu,-Ala, Tyr)$_n$ at 10-times Ag excess, with 0.01 M Ca^{2+} added (●); with 0.01 M EDTA added (□); with cacodylate buffer alone (○).

consistent with the concept of combining site-Fab independence suggested from the data of Figure 6 as discussed above.

Figure 7 shows hydrogen exchange studies of the whole antibody molecule reacting with antigen. Here soluble antigen-antibody complexes formed with excess antigen were studied. The hydrogen exchange data were obtained for reacting and control antibody at 10 times antigen excess. The curves of Figure 7 show that about twice as many hydrogens are trapped for bivalent antibody as for Fab, i.e. about 50-54H/molecules. It is worth noting here that the exchange kinetics of the unreacted antibody are considerably altered at this high level of antigen as will be seen if this data is compared with subsequent control curves. Hence, the value of the controlled reactions given above became quite obvious.

It could be concluded from the data of Fig. 8 that intact antibody reacting at high levels of antigen is involved only in combining site phenomena based on the fact that twice as many hydrogens are blocked in antibody compared to Fab. Additionally, the overall shapes of the curves are similar. In this connection, it is important to note that from a physiological point of view antigen-antibody interactions at high antigen levels are incapable of triggering the

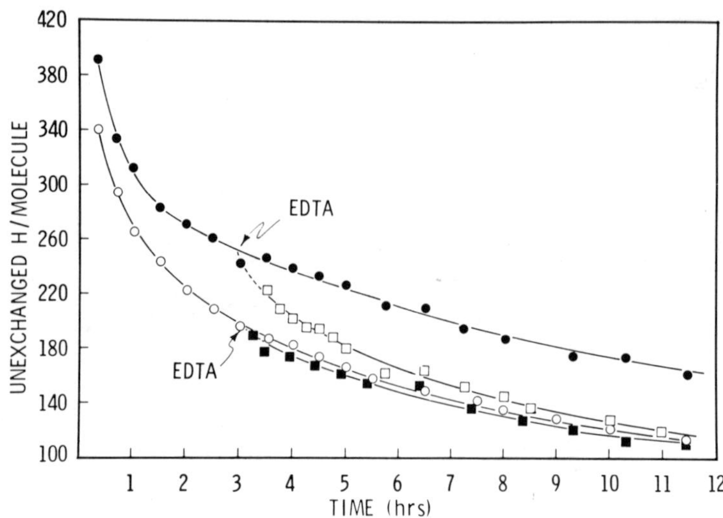

Figure 8. Exchange-out kinetics of calcium-dependent anti-(Glu,-Ala, Tyr)$_n$ at 3.4-times Ag excess in cacodylate buffer (pH 6.00) at 6°, with 0.01 M Ca^{2+} added (●); with 0.01 M EDTA added (□) and (■); with cacodylate buffer alone (○).

complement system (2,21). However, as equivalent concentrations of antigen and antibody interact, antibody can activate the complement system. Thus, we studied reactions closer to equivalence and again in the soluble complex form.

Figure 8 shows the reaction of whole antibody with antigen at three and a half times equivalence excess. It can be seen that the difference between these curves is very similar to that obtained at ten times equivalence. All we see essentially is the exchange-out kinetics of the blocked antigen combining sites and nothing more. Figure 9, however, shows something which is rather interesting. These experiments were done as close to equivalence as possible without having the soluble complexes precipitate, i.e. about 2.5 times antigen excess. In other words, it is still a soluble complex but we are dealing with complexes of antigen-antibody which are considerably large. By ultracentrifuge analysis they are approximately 20 S. The upper curve (closed circles) of Figure 9 is the exchange kinetics of antibody reacting with antigen in the presence of calcium, whereas the middle curve (open circles) is the control reaction. The two curves would indicate that there are not many hydrogens which are blocked at the combining site. However, from the dissociation curve which begins at 3 hours (squares) it can be seen that there are hydrogens which exchange out when the system is treated with EDTA. Furthermore, the number of hydrogens which are locked into the antibody upon reaction as determined from

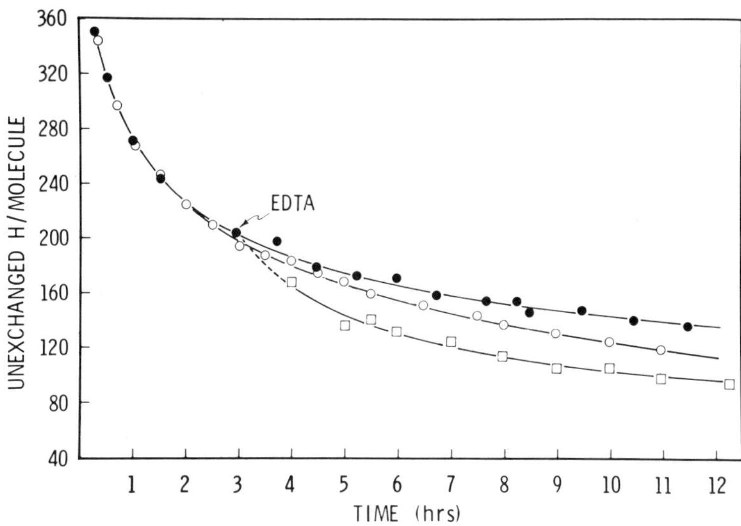

Figure 9. Exchange-out kinetics of calcium-dependent anti-(Glu,-Ala, Tyr)$_n$ at 2.6-times Ag excess, with 0.01 M Ca^{2+} added (●); with 0.01 M EDTA added (□); with cacodylate buffer alone (○).

the EDTA curve and the reaction curve is around the same as values obtained for the above systems, i.e. about 40, or about 20 per combining site. Hence, it is apparent that the whole curve of the reacting system is actually displaced downward. These curves indicate that when the antibody reacts under these conditions, there is a blocking reaction at the Fab fragments of the antibody just as is the case in isolated Fab shown earlier. However, the entire curve is displaced downward because there is a release of hydrogens from some other part of the molecule which is not concerned with the combining sites. This, we believe, indicates that the antibody molecule under these conditions undergoes some kind of a conformational change (22).

This data represents the first physico-chemical demonstration in solution of a conformation change in the antibody molecule when it interacts with antigen. The question now becomes, "Where is this conformational change taking place?" Is it taking place in the Fc piece or somewhere else? To attack this problem we studied the bivalent antibody fragment, that is the $(Fab')_2$ piece which has the "tail" of the antibody molecule cleaved off. In essence without going into all the data, we found that at high levels of antigen excess the $(Fab')_2$ fragment of the antibody has the same type of exchange kinetics as whole antibody, i.e. from 20-24 hydrogens blocked from exchange at each combining site. As equivalence is

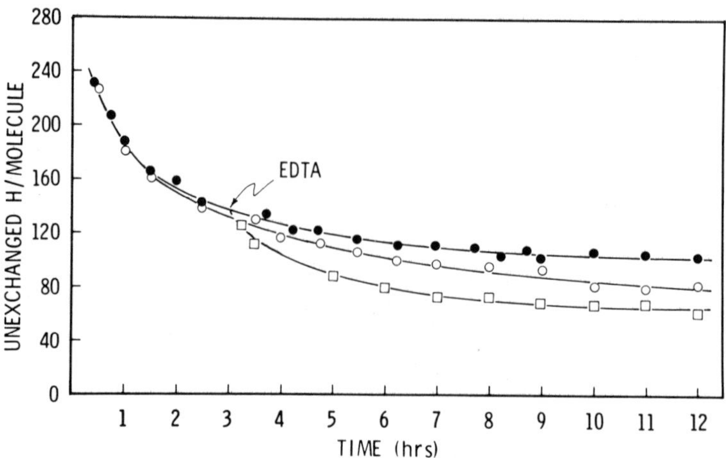

Figure 10. Exchange-out kinetics of $(Fab')_2$ fragment of calcium dependent anti-(Glu, Ala, Tyr)$_n$ at 2.4-times Ag excess, with 0.01 M Ca^{2+} added (●); with 0.01 M EDTA added (□); with cacodylate buffer alone (○).

approached, the $(Fab')_2$ undergoes the same kind of change as the whole molecule. In other words, as shown in Figure 10, the $(Fab')_2$ fragment reacting with 2.4 times excess antigen shows the same kind of exchange kinetics as whole antibody does under these conditions. These curves again indicate that some place in the molecule hydrogens have been released or there has been accelerated exchange and that something has happened to $(Fab')_2$ just as was the case with the whole antibody molecule. Since the $(Fab')_2$ fragment does not have the Fc piece of the antibody molecule, one can conclude that these changes do not occur in the Fc piece. At this point it is apparent that there is some kind of a conformational change occurring which is restricted to the region near the hinge portion where the two Fab fragments of antibody are linked. From our data we can show that the hydrogens which are involved in accelerated exchange upon antigen-antibody and antigen-$(Fab')_2$ combination are hydrogens which normally exchange very slowly in unreacted antibody. This suggests that these hydrogens are involved in some structural aspect of the antibody molecule. At present, we can only speculate as to what might be happening. Our current thinking is that the angle between the arms of the Fab segments of the antibody molecule may increase due to stresses induced via cross-linked formation.

Prior to doing the above study of hydrogen exchange, we had tried experiments with calcium dependent antibody interacting with the glutamic, alanine polypeptide, i.e. the GA antigen, which is

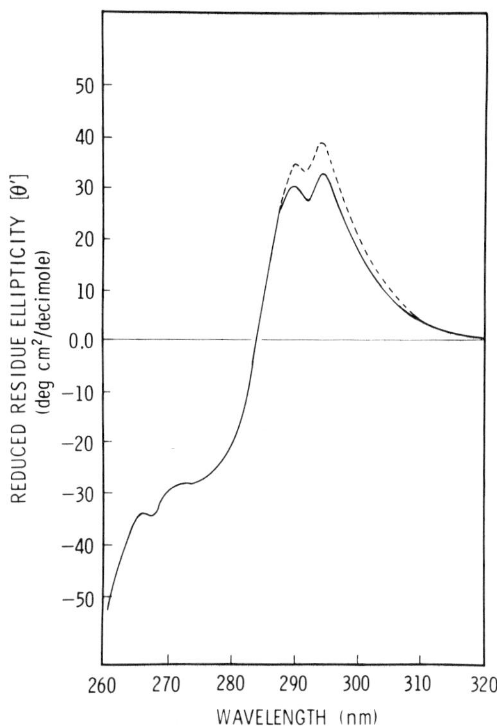

Figure 11. Near u.v. circular dichroism spectra of purified Ca Dep. sheep anti-(Glu, Ala) antibody plus (Glu, Ala) at various levels of antigen excess with and without Ca^{++}; (———) spectrum obtained with either: (a) the absence of Ca^{++}, at all levels of antigen studied, (b) 0.01 M EDTA or (c) with a six-fold equivalence excess of (Glu, Ala) and Ca^{++}. (-----) Spectrum in presence of a three-fold equivalence excess of (Glu, Ala) and Ca^{++}. Antibody concentrations were 1.3-0.18 mg/ml in 0.05 M Tris-HCl buffer, pH 7.8. The path length was 1 or 10 cm.

transparent in the tyrosine-tryptophan region of U.V. light. What we had done at that time was some very careful circular dichroism studies of these calcium dependent systems using a Cary model 61 instrument which is extremely sensitive to small dichroic changes. At very high levels of antigen excess we found essentially no change when the antibody reacts, that is in the presence of calcium as compared to the antibody mixed with the antigen without calcium.

Figure 11 shows an experiment which was done after we obtained the hydrogen exchange results where changes were apparent for the system closer to equivalence. As can be seen, there is measurable change in the circular dichroism spectra of the antibody molecule when it interacts with the GA antigen as compared to the control mixture. As the curves show, antibody and homologous antigen in the absence of Ca^{++} (or in excess EDTA) exhibit ellipticity maxima (in deg cm^2/decimole) of +33 and +30 at 295 and 290 nm, respectively (solid line).

With the addition of 0.15 ml $CaCl_2$ to the system (0.005 M final), distinct changes can be observed in the positive ellipticity bands, amounting to an increase in $[\theta']$ of ~16% (dashed line). No shifts in peak positions were evident, however. In order to insure that this difference was real and not due to instrumental variability, the peak area was rescanned after 2 cell repositionings over a period of 2 hr. Superimposable tracings were obtained in each case. The reversibility of this phenomenon was established by adding EDTA (0.005 M final), after which the ellipticity returned to that of antibody and antigen in the absence of Ca^{++} (solid line). Subsequent addition of Ca^{++} (0.001 M final) again resulted in a 16% increase in $[\theta']$ at the 2 positive peaks. Similar experiments were performed at "antigen excesses" of 6 and 12, but under these conditions no changes in $[\theta]$ were found with addition of Ca^{++} or of EDTA. It should be noted that the ellipticity obtained in the presence of Ca^{++} at 6-fold or 12-fold "antigen excess" was equivalent to that of the non-reactive system at 3-fold excess (without Ca^{++}) (solid line). This would indicate that the observed changes are not due merely to a "blocking reaction", i.e. changing the environment of a combining site chromophore by the juxtaposition of antigen, but more likely to a rearrangement of optically active centers away from the antibody active site *per se*. None of the $[\theta']$ values changed with time (up to 3 hrs), thus the final value of $[\theta']_{295}$ reported here was the same as that measured immediately after addition of Ca^{++}. Also, the ellipticity changed only at the 2 wavelengths noted, and not at the 2 other dichroic areas (~275 and 268 nm). This and the lack of changes in 6-fold excess would make it unlikely that this phenomenon is due only to scattering effects. It appears, therefore, that a specific change in ellipticity occurs during the interaction of anti (Glu, Ala) with its homologous antigen at certain ratios of antigen-antibody (23). Thus, we have been able to demonstrate by two independent methods measurable changes (presumably conformational) in the antibody molecule which depend on the size of the complex in which antibody is involved.

Further investigations into the environment of the combining site were initiated using the technique of solvent perturbation difference spectroscopy (24). The underlying phenomenon of difference spectroscopy, incorporating solvent perturbation, is described in Figure 12. Difference spectroscopy essentially means measuring the difference in absorbance between two samples of the same concen-

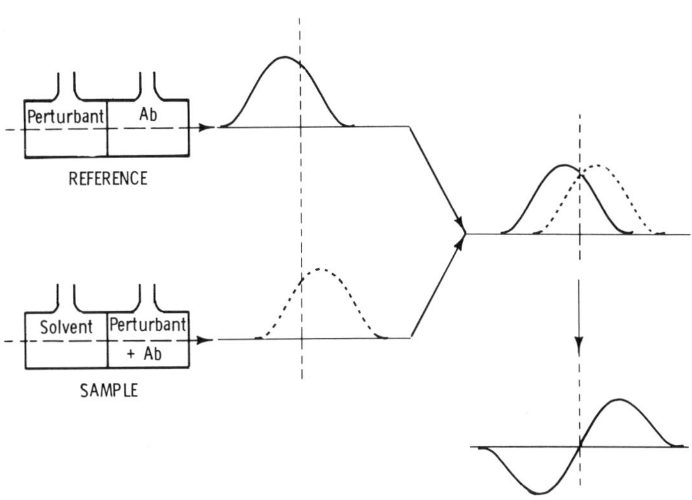

Figure 12. Tandem cell arrangement for solvent perturbation difference spectroscopy. (See text for explanation.)

tration at different wavelengths. The basis for the difference in absorbance is that protein chromophores (i.e. tryptophane and tyrosine) may absorb light (u.v.) at slightly different wavelengths (1-10 Å) depending upon their "microenvironment". In essence, there is an induced shift (either blue or red) in the protein's absorbance spectrum. If we now compare the spectrum of the original sample to that of an altered sample, we develop what is called a "difference spectrum", shown in this figure. It has been shown that almost any change in the microenvironment of chromophores will shift their spectra, but for our purposes the shift is due to the presence of small molecules (called "perturbants") which change either the refractive index and/or solvent polarity immediately surrounding the chromophores (25).

It has been possible to correlate the spectral shift for each solute ("perturbant") with the magnitude of the difference spectrum, thereby providing information about the absolute number of chromophores affected. The whole technique has been lucidly described by Herskovits (24).

In our experiments we have sought first to identify the aromatic amino acids on the surface of the anti GA antibody (26) and next

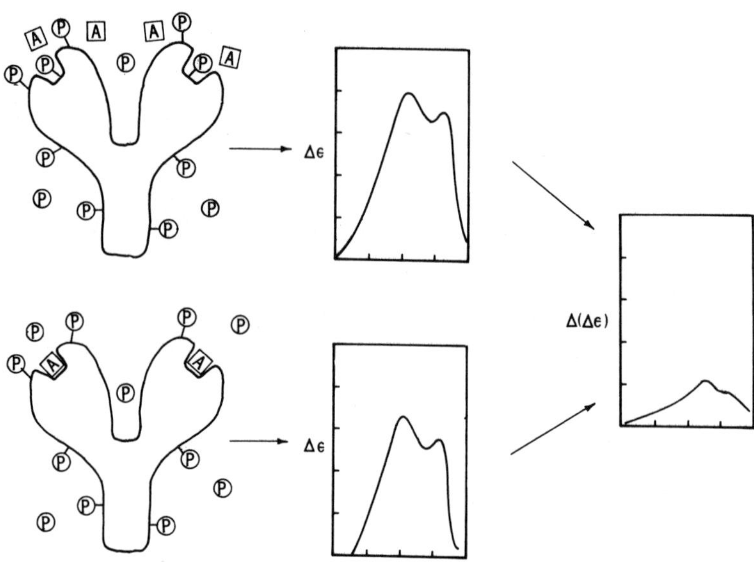

Figure 13. Schematic representation of difference spectra generated by free and antigen blocked antibody: (P) perturbant, [A] antigen.

those involved in the binding of antibody to antigen. Since the antigen (GA) has no aromatic amino acids, those affected residues must reside in the antibody. A schematic representation of such an experiment is shown in Figure 13. We see here a highly idealized concept of an antibody molecule, with 2 combining sites, as previously described. The projections represent aromatic amino acids (tyrosine and tryptophane), while the encircled p's represent perturbants and the A's, antigen molecules. In the absence of calcium (upper figure) there is no antigen-antibody interaction and the perturbant molecules may affect any "unburied" chromophore. However, when Ca^{++} ions are added, the GA interacts with antibody and prevents the perturbant molecules from affecting those chromophores in the combining site. If we examine the difference spectrum induced by perturbant in the absence of interaction (upper figure), we can identify a particular spectrum; however, during interaction the spectrum has changed, but only in amplitude. If we now subtract the reacted spectrum from the unreacted, we have a difference spectrum of the combining site chromophores, providing the combining site is independent of the remainder of the molecule. In practice, the experiment is done by first measuring a perturbation spectrum of antibody and antigen without calcium, then by adding calcium to both perturbed and non perturbed solutions and measuring again. This

Table IV

Chromophore Change upon Interaction of Sheep Anti-GA Antibody with Homologous Antigen

Perturbant	Mean Diameter (Å)	Tryptophane/Tyrosine Ratio			
		No Ca^{++}	With Ca^{++}	Difference	Residues p.c.s.*
DMSO	4.0	8.0/33.1	7.5/28.0	−0.5/−5.1	0.2/2.5
Ethylene glycol	4.3	7.4/31.3	6.4/24.8	−1.0/−6.5	0.5/3.2
Glycerol	5.2	8.0/28.8	8.0/24.1	0/−4.7	0/2.3
PEG	9.2	4.9/26.7	4.6/25.0	−0.3/−1.7	0.1/0.8

*Residues shielded per combining site, assuming bivalent antibody.

procedure subtracts any perturbations of the combining site due to antigen. At high levels of antigen excess as we used here, we know, based on our hydrogen exchange studies, that such a treatment is feasible. The next and final step is to subject this spectrum to a mathematical analysis which quantitates the number of tryptophane and tyrosine residues involved. The details of this analysis have been presented by Herskovits and Sorensen (27,28).

Table IV shows the results of perturbation studies with a number of different perturbants. The data are presented as moles of tryptophane and tyrosine available to perturbant per mole of antibody. As we see (3rd column) in the absence of antigen-antibody interaction, the smaller perturbants dimethyl sulfoxide, ethylene glycol and glycerol affect 7.5 to 8.0 tryptophanes and 29 to 33 tyrosines. The larger perturbant (polyethylene glycol) affects only 5 tryptophanes and 27 tyrosines, suggesting a partial "burial" of some residues. This is not surprising since one would expect certain amino acids to be less "available" to perturbants of varying size by virtue of their position within the protein fabric.

The fourth column lists the values after interaction (i.e. addition of Ca^{++}), and the fifth column - the difference between unreacted and reacted, that is the number of tryptophanes and tyrosines buried by antigen per molecule of antibody. The last and most significant column lists the number of tryptophane and tyrosine residues affected per combining site, assuming two sites per antibody molecule.

We now find that between 2 and 3 tyrosine residues are buried during the antigen-antibody reaction as indicated by smaller perturbants, while PEG can affect only approximately 1 residue. The meaning of this is that of the 2-3 tyrosines involved in interaction 2 are not available to PEG even before interaction, and perhaps lie in a "crevice" within the combining site.

Summary: We have characterized in some detail calcium dependent antigen-antibody reactions. Our studies, incorporating the techniques of hydrogen exchange, circular dichroism and solvent perturbation, have begun to elucidate some fundamental areas of immunochemistry and immunoglobulins. We have been able to show for the first time that conformational changes take place in antibody in solution under specific interaction conditions. Regarding our combining site studies, we can conclude that for our system the site involves about 20 amino acid residues. These residues are involved in strong hydrogen bonding and probably comprise a rigid binding surface and a surface which acts independently of the remainder of the molecule, i.e. there is no transfer of information from the site to other parts. There are aromatic residues in the anti GA combining site and our data gives some indication of the topology immediately surrounding these residues. Thus, we believe we have laid the foundation for future work which can eventually lead to an understanding of antibody specificity at the molecular level.

Acknowledgements: This research was supported by National Institute of Health Grant AI07825, National Science Foundation Grant GB 36972 and General Research Support Grant RR-5414. The authors gratefully acknowledge the secretarial assistance of Miss Mary Kemp in the preparation of this paper.

References:

1. Day, E.D. (1972) Advanced Immunochemistry, Baltimore, Md., Williams and Wilkins.
2. Muller-Eberhard, H.J. (1969) Ann. Rev. Biochem. 38, 389.
3. Katchalski, E. and Sela, M. (1958) Advan. Protein Chem. 13, 243.
4. Liberti, P.A., Maurer, P.H. and Clark, L.G. (1971) Biochemistry 10, 1632.
5. Clark, L.G., Maurer, P.H. and Liberti, P.A. (1968) Fed. Proc. Fed. Amer. Soc. Exp. Biol. 27, 259.
6. Maurer, P.H., Clark, L.G. and Liberti, P.A. (1970) J. Immunol. 105, 567.
7. Kabat, E. and Mayer, M. (1961) Experimental Immunochemistry, Springfield, Ill., C.C. Thomas.
8. Liberti, P.A. and Stivala, S.S. (1967) Arch. Biochem. Biophys. 122, 40; Strauss, U.P. (1962) Electrolytes, New York, N. Y., Pergamon Press, P. 215.
9. Jacobson, A.L. (1964) Biopolymers 2, 207.

10. Englander, S.W. (1968) Methods Enzymol. 12, 379.
11. Linderstrom-Lang, K. (1955) Chem. Soc., Spec. Publ. 2, 1.
12. Englander, S.W., Downer, N. and Teitelbaum, H. (1972) Ann. Rev. Biochem. 41, 903.
13. Reid, K.B.M. (1971) Immunology 20, 649.
14. Ashman, R.F. and Kaplan, A.P. and Metzger, H. (1971) Immunochemistry 8, 627.
15. Liberti, P.A., Stylos, W.A., Maurer, P.H. and Callahan, H.J. (1972) Biochemistry 11, 3321.
16. Schechter, A.N., Maravek, L. and Anfinsen, C.B. (1969) J. Biol. Chem. 244, 4981.
17. Davies, D., personal communication.
18. Karush, F. (1962) Advan. Immunol. 2, 1.
19. Kabat, E.A. (1961) Experimental Immunochemistry, Springfield, Ill., C.C. Thomas.
20. Day, L.A., Sturtevart, J.M. and Singer, S.J. (1963) Ann. N. Y. Acad. Sci. 103, 611-625.
21. Dixon, F.J. (1972) J. Immunol. 109, 187.
22. Liberti, P.A., Stylos, W.A. and Maurer, P.H. (1972) Biochemistry 11, 3321.
23. Callahan, H.J., Liberti, P.A., Maurer, P.H. and MacLuckie, W.R. (1973) Immunology 25, 000.
24. Herskovits, T.T. (1967) Methods Enzymol. 11, 748.
25. Donovan, J.W. (1969) in Physical Principles and Techniques of Protein Chemistry (Leach, S.L., ed.) Part A, pp. 101-170, Academic Press, New York and London.
26. Callahan, H.J., Liberti, P.A. and Maurer, P.H. (1973) Immunochem.
27. Herskovits, T.T. and Sorensen, M. (1968) Biochemistry 7, 2523.
28. Herskovits, T.T. and Sorensen, M. (1968) Biochemistry 7, 2533.

10

CALCIUM BINDING TO ELASTIN

Robert B. Rucker

Department of Nutrition
University of California
Davis, California 95616

Elastin is one of the primary structural components of arterial vessels. In its mature form, this protein is highly crosslinked and insoluble in most solvents (Partridge, 1962). Soluble fractions (α and β-elastins) of insoluble elastin may be prepared, however, by mild hydrolysis (Partridge, et al., 1955). Native soluble elastins have also been obtained after blocking the formation of crosslinks in vivo (Rucker et al., 1972, 1973; Sandberg et al., 1969; Smith et al., 1972). Native soluble elastin, often referred to as tropoelastin, is the precursor to insoluble elastin. The insoluble protein is associated with mucopolysacharide, lipid, collagen, microfibrillar and other protein in the form of elastin fibers (Ross, 1973; Ross and Bornstein, 1970).

One of the unique features of the elastic fiber is its ability to calcify (Yu and Blumenthal, 1967). Several consequences of calcium binding to elastin may have a direct bearing on the morphological changes which often occur in arterial vessels. Recent information on elastin nucleation sites for calcification (Urry, 1971) and control of arterial calcification processes (Russell and Smith, 1973) should open new avenues for future studies. The discussion and data which follow attempt to point out some of these aspects.

NUCLEATING SITES IN ELASTIN

Proposals for nucleation sites in the mineralization of elastin have previously implicated a variety of functional groups as sites for calcium binding (Schiffman et al., 1964, 1968, 1969; Eisenstein et al., 1964; Mollinari-Tosatti and Gotte 1971; Hall, 1955). Limitations exist, however, regarding these functional groups and their role in the initial binding of calcium. For example, sulfydryl groups as nucleation sites in elastin have been suggested by Schiffman et al. (1964, 1968, 1969), but it is now known that native soluble elastins contain little cysteine and other sources of sulfhydryl groups (Rucker et al., 1973; Sandberg et al., 1972; Smith et al., 1972). Elastin preparations are often contaminated by residual microfibrillar proteins, which are characterized by their high content of cysteine and cystine (Ross and Bornstein, 1969). Such proteins which contain sulfhydryl groups may facilitate mineralization, but it can be shown that their presence is not required to describe the ability of elastin to calcify.

Likewise, carboxyl groups and amino groups in elastin have also been suggested as nucleation sites (Mollinari-Tosatti and Gotte, 1971; Mollinari-Tossati et al., 1968; Hall, 1955). The arguments in this case are based on the formation of calcium chelate complexes and electrostatic associations. There are considerable data which support these arguments. Blocking or altering amino groups by treatment with fluorodinitrobenzene, nitrous acid or acetic anhydride inhibits elastin mineralization (Sobel et al., 1966; Schiffman et al., 1969). The binding of calcium ions to insoluble elastin is dependent upon pH in a manner which suggests an association involving carboxyl groups (Mollinari-Tosatti and Gotte, 1971).

Titration of the insoluble elastin usually indicates the presence of 200 μmoles of acidic groups and 100 - 150 μmoles of basic groups per gram of protein (Bendall, 1955). It should be pointed out, however, that when these groups are fully equilibrated with calcium, only 10 to 20 percent appear to be involved in binding (Mollinari-Tosatti and Gotte, 1971). Interaction between calcium and ionized groups is possibly limited because of the hydrophobic environment produced by neutral amino acid residues in the protein. Mollinari-Tosatti et al. (1971) have also suggested that reduced interaction may be due to electrostatic association between protonated amino groups and ionized carboxyl groups.

The binding of 20 to 40 μmoles of calcium per gram of elastin would allow 1.4 to 2 nucleation centers per subunit of elastin (m. w., approximately 70,000 (Smith et al., 1972)). Although it is conceivable that this represents enough bound calcium to initiate mineralization, there is evidence which indicates that when calcium is rigorously removed from elastin preparations, the protein will not initiate mineralization from calcifying solutions. Sobel et al. (1966) have demonstrated only prenucleated elastin readily initiates mineralization. Their data do not exclude, but neither do they strongly support, a role for carboxyl groups as nucleation sites, since the calcium in incubation solutions would be expected to interact with carboxyl groups. One part per million calcium, if it is present as hydroxyapatite, can represent approximately

6×10^{12} apatite nuclei per gram of protein matrix (Neuman and Neuman, 1958). Insoluble elastin prepared in the usual manner is usually nucleated. Such nucleation would probably not markedly affect the net amount of calcium bound in the absence of phosphate ion, but would promote continued mineralization if phosphate ion is present.

In order for mineralization to occur, elastin must first bind calcium. Elastin does not bind phosphate ion in the absence of calcium (Mollinari and Gotte, 1971). Excluding sulfhydryl, carboxyl, and amino groups as major nucleation sites obviously presents problems, if elastin is to act as a calcifying substrate. One other possibility has been suggested. It involves the apparent ability of elastin to bind calcium at neutral sites. Under appropriate conditions, Urry (1971) has proposed that specific acyl oxygens in elastin can act to coordinate and bind calcium. Urry's argument is based from his studies on ion-transporting antibiotics. Some antibiotics are cyclic neutral polypeptides with acyl oxygens exposed to the surrounding solvent, which can bind cations. Valinomycin is an example (Meyers and Urry, 1972). There are analogies between the amino acid sequences in valinomycin and those in elastin. In addition, the formation of metal binding centers depends on the conformational state of valinomycin. Data from several experiments are presented here in support of this relationship for elastin.

Elastin exists in both ordered and random conformations. The ordered elastin conformations are typical of α-helical polypeptides and occur when the protein is in water-alcohol mixtures (Mammi et al., 1968), solutions containing anioninc detergents (Kagan et al., 1972; Rucker et al., 1973) trifluoroethanol (Urry et al., 1971), or when the protein is in the coacervated state (Urry et al., 1969). In the studies on elastin and calcium binding described so far conditions were employed which suggest that elastins in less ordered conformations were used. When conditions are chosen so that an ordering of some of the polypeptide sequences in the protein are maintained, it can be shown that elastin acts as a calcifying substrate even in the absence of phosphate ion and at relatively high concentrations of hydrogen ion.

For our studies (Figures 1 to 5) solutions containing alcohols were used, because of the effects of alcohol on elastin conformation. The various details and conditions are outlined in the accompanying figures. The insoluble elastin preparations were obtained from the aortas of young ewes or chicks by autoclaving (Gotte et al., 1963). Solubilized elastin fractions (α-elastin) and native soluble elastin were also prepared. The elastin fractions contained only trace amounts of calcium (<100 mμ moles/g) and no measurable phosphate ion.

By increasing the concentration of the alcohols in the incubation solutions it could be shown that the amount of calcium bound to elastin was increased (Figure 1). The protein bound no more than 35 μmoles of calcium per gram in solutions containing no alcohol. When the alcohol concentrations exceeded 60% (w/v) over 200 μmoles of calcium were bound per gram of protein. In absolute methanol the binding of calcium was found to be dependent upon calcium concentrations and appeared to have two components (Figure 2). Scatchard plots of these data indicated that about 35 μmoles of calcium per gram of elastin were bound at a low concentration of calcium. At maximal saturation (15 to 20 mM calcium) 220 to 250 μmoles of calcium were bound per gram of elastin (Edsall and Wymann, 1965).

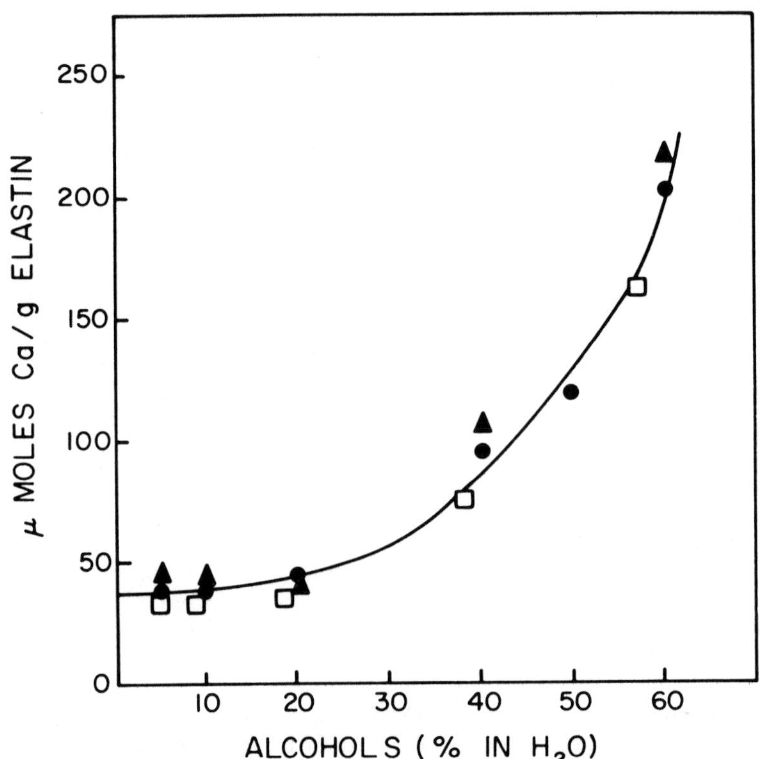

Figure 1. The relationship of alcohol concentrations and calcium binding by insoluble elastin. The incubation mixture (10 ml) contained elastin (50 mg), 25 mM $CaCl_2$, methanol (▲), ethanol (●), or n-propanol (□), and N-2-hydroxyethylpiperazine-N-2-ethanesulfonic acid (HEPES) at 0.1 M, pH 7.0. The solutions were incubated at 35° for 48 hours. The amount of calcium bound was then measured as described by Molinari-Tossatti and Gotte (1972).

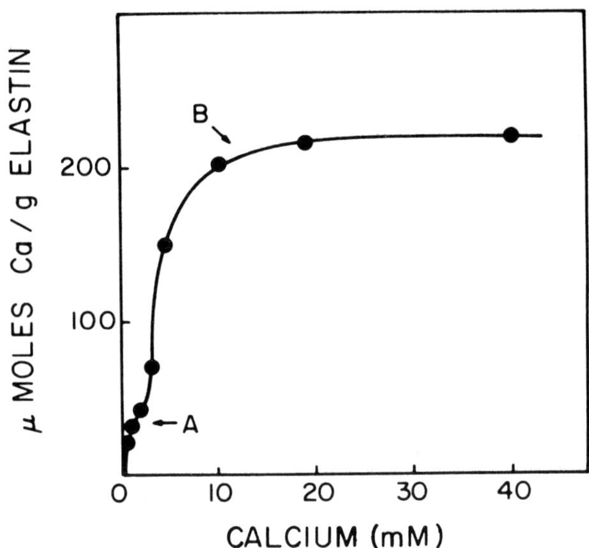

Figure 2. The binding of calcium to insoluble elastin in 100% methanol. A and B are areas which indicate the binding curve may have two components. Elastin (10 mg) and methanol (to 5 ml) containing $CaCl_2$ were incubated at 25°C for 36 hours. The amount of calcium bound was then determined.

The binding of calcium to elastin in a totally aqueous medium was always dependent upon pH (Figure 3). When a 50% buffered methanol-H_2O mixture was used, the shape of the pH vs. calcium binding curve indicated that pH had only slight effect on binding even at a relatively high concentration of hydrogen ion. Insoluble elastin in the methanol-H_2O mixture containing 5 mM $CaCl_2$ bound approximately 70 μmoles of calcium/g of protein at pH 4.0. Calcium binding increased gradually and approximately 100 μmoles of calcium were bound at pH 8.0. These observations supported the possibility of neutral binding sites. Experiments designed to exclude the possibility that enhanced calcium binding in methanol was due solely to solvent-protein partitioning effects were also performed.

Various metals differing in charge and ionic radii were added to solutions containing elastin either suspended in only the buffered solution (HEPES, 0.01 M, pH 7.0) or the solution plus methanol (50% w/v). At the end of the incubations, each sample was then resuspended and washed using the same incubation solution without the metal five times over an eight hour period. After the last washing, the elastin was collected and analysis of the individual metals in acid hydrolysates

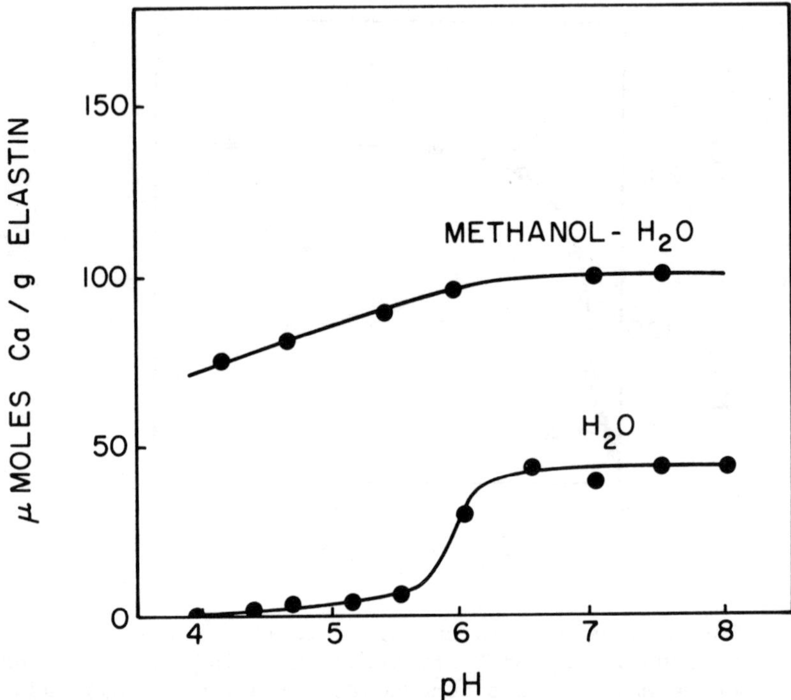

Figure 3. The effect of pH on the binding of calcium in buffered solutions with and without methanol (50% w/v). The solutions (5 ml) contained either acetate or HEPES buffers (0.01 M) plus 5.0 mM $CaCl_2$ and 10 mg of insoluble elastin.

of the protein was performed by means of atomic absorption spectrophotometry or flame emission spectrophotometry. The results are given in Figure 4. Several features of the data are unique and deserve comment.

When metal uptake by elastin is plotted versus the ionic radii of the respective metals (Pauling, 1960), it may be noted that some aspects of the plots are similar to those previously described by Spaders et al. (1970) for collagen. Collagen possesses two size-specific complexing sites with dimensions of 0.65-0.75 and 1.1-1.3 Å. For elastin, the uptake of copper and zinc was greater than the other metals tested and could be accounted for by a complexing site of the 0.65-0.75 Å dimension. The general linearity of uptake by some of the metals versus the respective ionization potentials also suggested that there may be coordination by a single type of donor (Spadaro et al., 1970). This binding site is perhaps a chelation site and differs from that proposed for calcium. The only metals which deviated markedly in this plot were sodium, magnesium, and

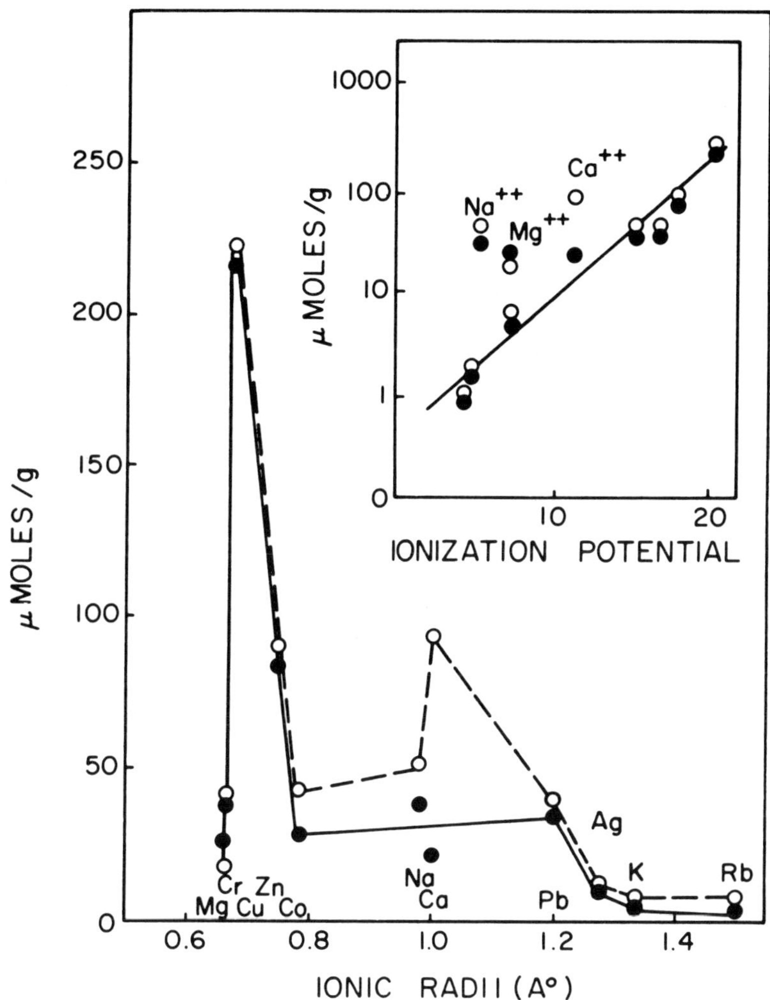

Figure 4. Binding of various metals by insoluble elastin. Incubation solutions contained elastin (50 mg), HEPES buffer (0.01 M) pH 7.0) and the various metal chlorides at 20 mM in 10 ml of solution with (O) and without (●) the presence of 50% methanol (w/v). At the end of the incubation (48 hours) the elastin was separated and washed 5 times with the incubation medium without the metal. The amount of metal bound was then determined and uptakes expressed as μmoles/g elastin plotted as indicated. Uptakes against the total ionization potentials (eV) are plotted on a semi-log scale.

calcium. Calcium, however, deviated only when methanol was also added to the incubation mixture. Methanol appeared to alter elastin so that a unique binding region specific for calcium was introduced in the protein. These data help to explain why most metals do not appear to inhibit or compete with calcium bound to elastin. If protein-solvent partitioning had a direct effect, the presence of methanol should have elicited an increase in the binding of other metals.

NEUTRAL BINDING SITES AND ELASTIN CONFORMATION

Porcine tropoelastin has been partially sequenced (Foster et al., 1973). Repeating tetra-, penta-, and hexapeptides have been observed. The tetrapeptides contain the sequence, -Gly-Gly-L-Val-L-Pro; the pentapeptides, -L-Val-L-Pro-Gly-L-Val-Gly-, and the hexapeptides, -L-Ala-L-Pro-Gly-L-Val-Gly-L-Val. These sequences are analogous to those found in some of the ion-transporting antibiotics (Urry, 1972). The presence of these sequences may contribute to the calcium binding capacity of elastin, because of their unusually high glycine content (Urry and Ohnishi, 1970). An important conformational aspect of glycine residues is that they can effect the insertion of near right angle turns in a polypeptide backbone. If glycine is followed by an amino acid residue with a bulky side chain, eg. L-Val, these near right angle turns or β-turns can become stable conformations (Geddes et al., 1968). A feature of this type of conformation is its potential ability to bind selected ions by exposed peptide oxygens in the peptide backbone. The alteration of the spatial arrangement of backbone acyl oxygens can produce unique binding sites.

As stated previously, elastin in a completely aqueous environment is in a less ordered conformation than that found in solutions containing alcohols (Mammi et al., 1971). Elastins incubated in the presence of detergents, such as sodium dodecyl sulfate, also undergo conformational change (Kagan et al., 1972; Rucker et al., 1973). Figures 5 and 6 show circular dichroism (CD) spectra for α-elastin and native soluble elastin.

For α-elastin, curve A in Figure 5 is the spectrum obtained for the protein when dissolved in H_2O. Although this spectrum contains some features analogous to those found in ordered polypeptides it is largely random in character (Greenfield and Fasman, 1969). Curve B (α-elastin in 33% methanol-H_2O) contains elements which begin to indicate some ordering of structure, ie. definition of negative elipticities centered at 204 nm and 218-222 nm. With each additional increase in the concentration of alcohol (Curve C, 66% methanol and Curve E, 99% methanol) there was a slight red shift in the trought between 204 to 208 nm, and a progressive increase in the negative elipticity centered at 220 to 222 nm. Curve D, the spectrum for

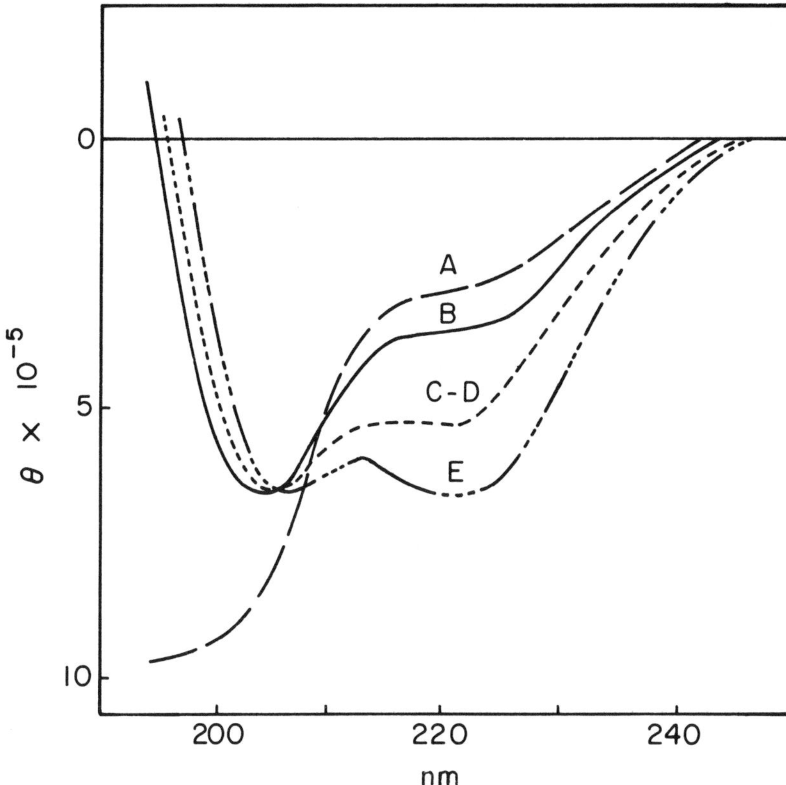

Figure 5. CD spectra of α-elastin in methanol-H_2O mixtures and 0.1% sodium dodecyl sulfate. Curve A is α-elastin in H_2O. Curve B is α-elastin in 33% methanol (w/v). Curve C is α-elastin in 66% methanol (w/v). Curve D, identical to C, is α-elastin in 0.1% sodium dodecyl sulfate. Curve E is α-elastin in 99% methanol (w/v).

α-elastin in 0.1% SDS, is almost identical to Curve C. The effects of sodium dodecyl sulfate on native soluble elastin result in similar changes (Figure 6). Although not shown, the same differences were observed when native soluble elastin is dissolved in either trifluoroethanol or methanol and then compared to that in H_2O. Only anionic detergents appear to alter the conformation of native soluble elastin. The presence of alkylmonoamines did not affect conformation. It is also important to point out that the CD spectrum for chick native soluble elastin is identical in most respects to that reported previously for coaccervated films of α-elastin (Urry et al., 1969).

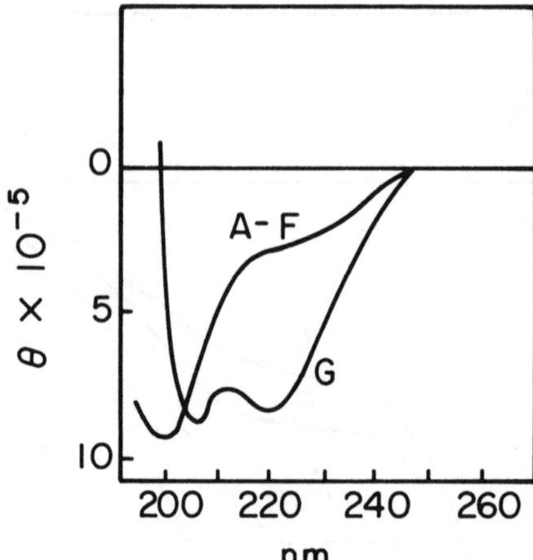

Figure 6. CD spectra of chick native soluble elastin. The spectra A through F were obtained for the protein in 0.05 M acetic acid (A), water (B), 0.1% n-butylamine (C), 0.1% n-hexylamine (D), 0.1% n-octylamine (E), or 0.1% n-decylamine (F). The spectrum obtained in 0.1% sodium dodecylsulfate (G) indicates a marked conformation change in the protein.

As the elastin molecule is extended and relaxed, it obviously undergoes conformational changes (Weis-Fogh and Anderson, 1970). These changes are possibly similar to those demonstrated in alcohol-H_2O mixtures, sodium dodecyl sulfate, and upon coacervation. Much of the elastin molecule in the ordered state should contain areas with exposed peptide acyl oxygens capable of coordinating calcium. That the coacervated elastin acts as a matrix for mineralization was recently confirmed by Starcher and Urry (1973). The elastin used in the studies by Starcher and Urry was porcine α-elastin. Normally this elastin coacervates at low pH. However, coacervation was observed to occur in a physiological pH range if the N-formyl and o-methyl ester of the protein were prepared. Since charged groups should be blocked by such treatment, it could be also assumed that the protein would act as a neutral polypeptide. In a calcifying solution containing 1.25 mM calcium and 1.25 mM phosphate ion, 650 μmoles of calcium were bound per gram of the coacervate.

In more recent studies on native soluble elastin, we have also observed increased calcium binding upon coacervation. Equilibrium dialysis of the native soluble protein at 6°C and 41°C against solution containing 1 M NaCl and calcium chloride (0.3 to 3.0 mM) indicated approximately 30 to 40 μ moles were bound per gram of elastin at 6°C, whereas over 200 μmoles per gram were at 41°C at maximal saturation. Native soluble elastin undergoes coacervation in 1 M NaCl solutions at 20-25°C (Smith et al., 1972). The chick aorta protein remains in solution below 10°C. It was assumed regions containing the binding sites were probably exposed upon coacervation. The binding properties of one of these proposed regions has been investigated.

Proton magnetic resonance studies have been utilized by Urry et al. (1973) to demonstrate calcium binding to a cyclic neutral polypeptide which contains the repeating sequence $[\text{-Gly-Gly-Val-Pro}]_3$. The peptide is analogous to valinomycin and amino acid sequences in elastin and selectively binds calcium. Chemical shifts of the peptide protons occur upon titration with Ca^{+2}, ie. a complex is formed, but there is little displacement of peptide protons. In addition, calcium elicits a change in the circular dichroism spectra of the peptide, whereas an equivalent concentration of sodium ion has no effect. The fact that this peptide interacts with calcium with little displacement of the peptide protons is strong evidence in support of neutral site binding for calcium. Other metals, eg., Cu, interact with peptides with considerable displacement in the peptide hydrogens. The cyclododecapeptide consists of a series of β-turns. Urry et al. (1973) have pointed out that similar β-turns in elastin introduces the the possibility of a β-spiral conformation for the protein. It is probably in these regions that calcium binding occurs. The proposed conformation of elastin would be amino acid sequences containing β-spiral and α-helices in series. Single chains of elastin would be arranged in parallel and held in place by intermolecular crosslinks (Figure 7). Calcium could bind when the molecule is in its extended state. Formation of amorphous calcium phosphate and other elastin complexes could then occur with charge neutralization as the driving force (Urry, 1971).

CALCIUM HOMEOSTASIS IN ARTERIAL TISSUE

The control of calcium binding to elastin and other arterial protein is a complex process. It not only involves consideration of the conformational properties of elastin, but also the control of calcium concentrations surrounding the fibers and regulation of calcium and phosphate interactions. Various nutrition status can elevate the interstitial calcium and phosphate ion concentration. Conditions, such as hypervitaminosis D (Harrison, 1933), magnesium deficiency, and calcium deficiency in the presence of excessive dietary phosphate (cf. Chen et al., 1973 and references cited) have been linked to the pathological calcification of soft tissue.

The formation of hydroxyapatite associated with calcifying matrices occurs in two phases (Weber et al., 1967; Francis et al., 1969). The first phase represents the formation of amorphous calcium phosphate. In the second phase the transition of amorphous calcium phosphate to crystalline apatite is followed by crystal growth. In this regard polyphosphates have been identified as controlling

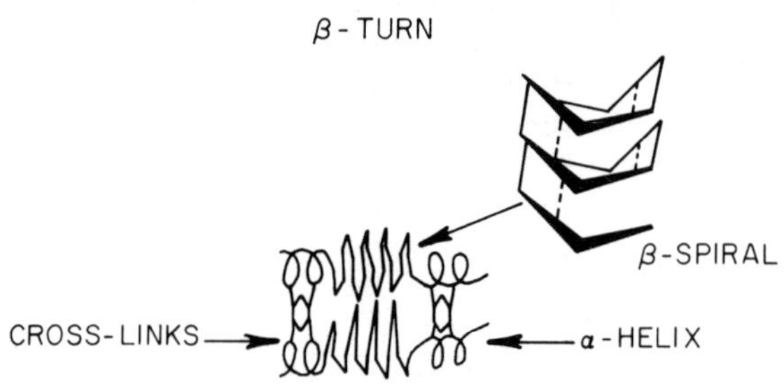

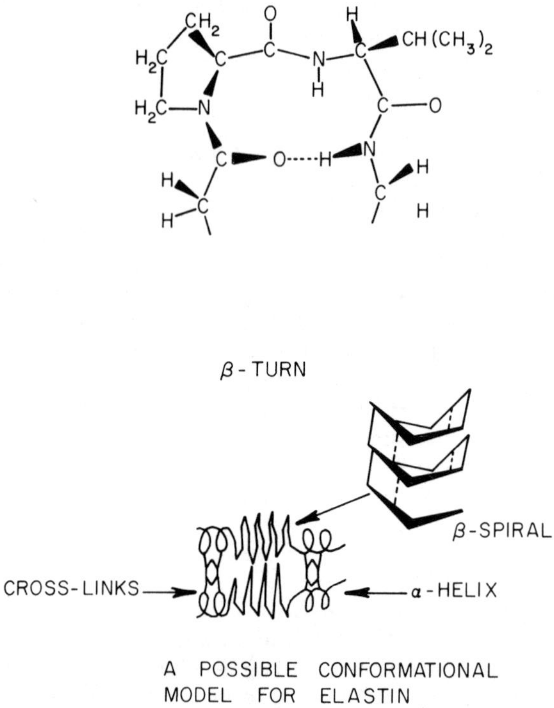

A POSSIBLE CONFORMATIONAL MODEL FOR ELASTIN

Figure 7. Represents an example of a β-turn and a conformational model for elastin. The figures are adopted from Urry (1973). The protein is viewed as having α-helical regions (crosslinking areas) and β-spiral regions (extendable and potential calcium binding areas).

agents (Fleisch and Neuman, 1961; Fleisch and Bisaz, 1962) with respect to the transition phase (crystalline apatite formation). One of these agents, pyrophosphate, and its level in arterial tissue may have a direct bearing on the mineralization of elastin fibers.

Ion concentrations of pyrophosphate ($<10\mu$ molar) inhibit the precipitation of hydroxyapaptite crystals in vitro (Francis, 1969). Pyrophosphate in plasma is present at concentrations which are inhibitory and appear in vivo to control the formation of calcium phosphate salts (Russell et al., 1971). Injection of pyrophosphate and polyphosphates in rats inhibits aortic calcification induced by excessive vitamin D (Francis et al., 1969). The mechanism appears to involve chemisorption of pyrophosphate on hydroxyapatite crystals, thus

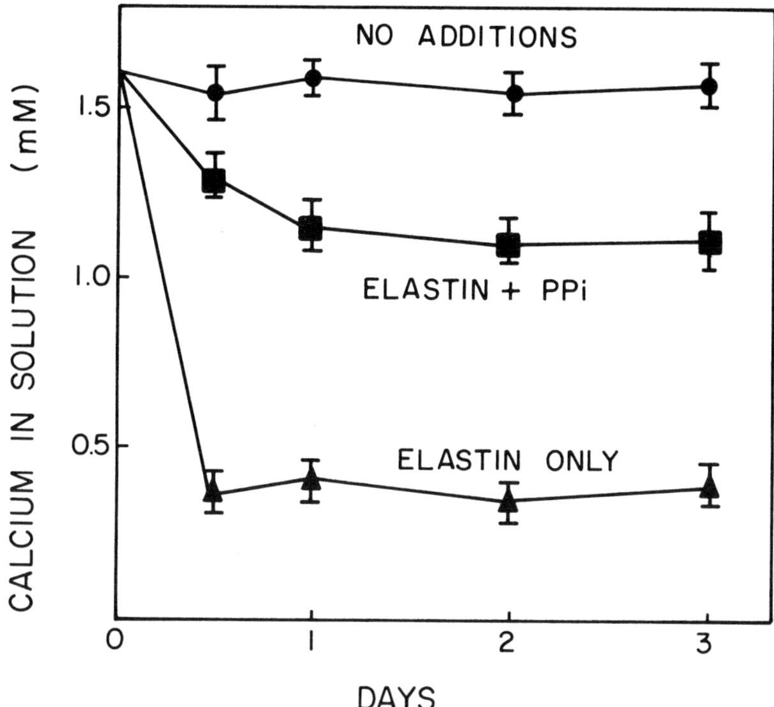

Figure 8. The relationship of pyrophosphate to calcium phosphate binding by insoluble chick aorta elastin. The calcifying solution contained 70 mM NaCl, 5 mM KCl, 1.6 mM $CaCl_2$ and 1.6 mM (NaH_2PO_4 and Na_2HPO_4 adjusted to give pH 7.6). When present disodium pyrophosphate and elastin were added to give 20 μ M and 50 mg, respectively. The calcium which remained in solution was measured at the times indicated. Each value (3 to 4 determination ± S.E.M.) expresses the millimolar concentration of calcium in solution.

preventing the continued formation of hydroxyapatite in the transition phase (Fleisch et al., 1968). Allowing that some amorphous calcium phosphate is present at the elastin fiber, the transformation and growth from amphorous to crystalline and pathological mineralization could be retarded by the presence of pyrophosphate. Disturbances in pyrophosphate metabolism may represent one of the bases for pathological mineralization. The data given in Figure 8 show inhibitory effects of pyrophosphate on elastin mineralization when chick insoluble elastin is incubated in a calcifying solution. Pyrophosphate markedly inhibited the uptake of calcium by elastin. Fleisch and Neuman (1961) have shown that the addition of pyrophosphatase, an enzyme

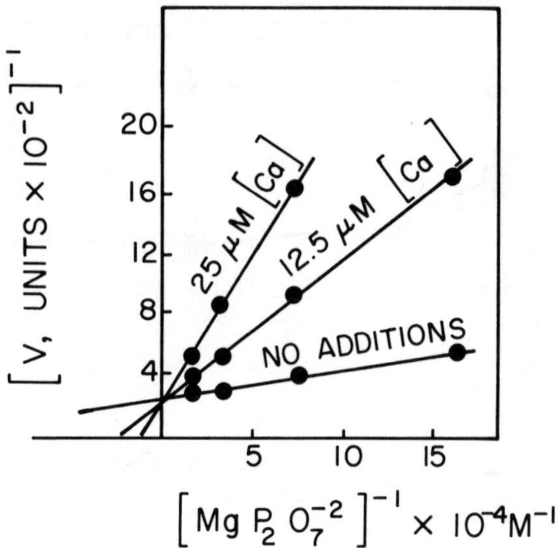

Figure 9. The effect of calcium addition on chick aorta pyrophosphatase activity. The concentrations of magnesium and pyrophosphate were adjusted to give the amounts of $MgP_2O_7^{2-}$ indicated with little or near zero concentrations of free magnesium in solution. Calcium chloride and additional disodium pyrophosphate and calcium were added to give the concentrations of 12.5 and 25 μ molar of $CaP_2O_7^{2-}$, respectively. The apparent K_m value was estimated to be 7×10^{-4} M for $MgP_2O_7^{-2}$, and the K_i value of $CaP_2O_7^{-2}$ was estimated to be 2.1×10^{-6} M. Assay conditions and other properties of the enzyme have been given by Chen et al. (1973).

which hydrolyzed pyrophosphate, can reverse inhibition allowing mineralization to proceed. From investigations in our laboratory (Chen et al., 1973) and from the work of Nordlie et al. (1968), calcium pyrophosphate is a potent inhibitor of arterial pyrophosphatase (Figure 9). This observation has provided the basis for a possible control mechanism given in Figure 10.

Other ester phosphatases and anhydrides are not actively hydrolyzed by the chick arterial enzyme at physiological pH (Chen et al., 1973). Magnesium appears to be the only cation which forms a substrate complex, ie. $[MgP_2O_7^{-2}]$. In addition, pyrophosphatase activity is closely associated with elastin and collagen fibers in arterial tissue. As a control mechanism it is envisioned that in interstitial areas where the formation of hydroxyapatite is possible, an increase in calcium would effect an elevation of the ion product

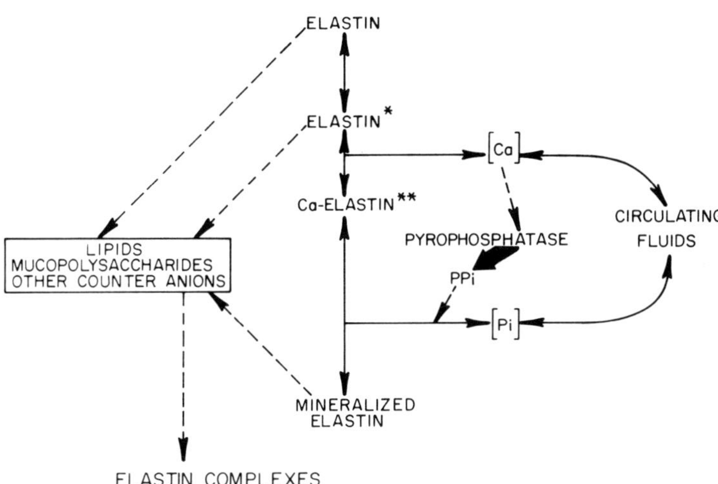

Figure 10. Possible model for the calcification of elastin. A change in conformation (Elastin⇆Elastin*) promotes the uptake of calcium. The binding of calcium may also alter the conformation of the protein (Elastin**). The binding of phosphate and formation of hydroxyapatite is controlled by pyrophosphate, the levels of which are regulated by pyrophosphatase. The calcium ion concentration is shown to provide "feed-back" regulation for pyrophosphatase. The various forms of elastin and calcified elastin may in turn form complexes with lipid, mucopolysacharides and other counter anions.

(Ca x P) necessary for apatite formation. This would be due to the elevation of pyrophosphate caused by pyrophosphatase inhibition (Chen et al., 1973).

The mineralization of elastin is not uniformly seen at all anatomical sites (Yu and Blumenthal, 1967). The aorta and larger arteries calcify more readily than smaller vessels (Blumenthal et al., 1950). Similar to collagens from different anatomical sites, elastin in larger vessels may differ in secondary structure from that found in other locations. The degree of maturation may also have an effect. Mature elastin is more highly crosslinked compared to newly formed elastin, which obviously has a direct relationship to secondary structure (Franzblau and Lent, 1968). The fact that in individuals beyond 30 to 40 years, most large vessels possess some degree of mineralization may be related to changes in secondary structure and concommitantly subtle changes in the conformation of protein. Alterations in control mechanisms related to levels of pyrophosphate and pyrophosphatase at differing anatomical sites are probably also related. The percentage of elastin as a component of the total protein is greater in large vessels compared to that in small vessels. The amount of calcifying matrix should likewise be related to the degree of mineralization.

POSSIBLE CONSEQUENCES OF ARTERIAL CALCIFICATION

Considerable information is still needed to clearly outline the sequela of events related to the regeneration and degradation of extracellular arterial components. Localized calcification of elastin may have a direct influence on these events, particularly those related to degradation processes. Several other consequences of calcium bound at elastin neutral sites have been suggested by Urry (1971). These include alterations in elasticity, attraction of cholesterol and negatively charged lipids, and facilitation of clot formation.

Elasticity could be affected, because the binding of excessive calcium to the peptide backbone of elastin should promote rigidity of the molecule. It is clear from morphological studies that elasticity of arterial vessels is reduced after mineralization (Yu and Blumenthal, 1967). Calcium binding in the absence of counter ions would facilitate the extension of adjacent β-spirals by charge repulsion. The presence of bound calcium could also alter the association of the protein with the surrounding solvent. Elasticity is an isothermal process and may be explained in part by assuming interaction between apolar residues in the protein and the polar medium (Weis-Fogh and Anderson, 1970). The association and dissociation of calcium and appropriate counter ions to elastin should alter the conformational energies for extension and relaxation. Urry et al. (1971) have shown that when α-elastin is dissolved in trifluoroethanol containing 4% trifluoroacetic acid, a CD spectra is obtained similar to that for 100% methanol as given in figure 4. The addition of calcium ion alters the spectra to that observed for α-elastin in H_2O or buffered solutions. If the calcium addition is followed by phosphate ion the resulting spectra shifts to resemble original trifluoroethanol-trifluoroacetic acid spectra. Further studies are needed along these lines. Such studies could help to clarify the extent to which and at what point elastin calcification may be regarded as a pathological condition.

Fibrous mineralized plaques are often assumed to be secondary events in the development of atherosclerotic lesions. Excessive calcium binding at localized neutral sites in elastin could play more of a primary role. In areas of active mesenchymal proliferation, lipid deposition often occurs over medial aorta undergoing mineralization (Hass et al., 1961). When quiescent, however, mesenchymal plaques over a mineralized media are resistant to lipid binding. Eisenstein et al. (1971) have demonstrated in vitro alternations in the patterns of lipid binding to calcified elastin when compared to decalcified human medial aorta elastin. If calcium and calcium phosphate bound to elastin changes the distribution and degree of binding by lipid to elastin, this may have a direct effect on elastolysis of elastic fibers. Kagan et al. (1972) have shown that the interaction of elastin with fatty acids and other lipophylic compounds enhances hydrolysis of elastin by elastase. Protein conformation seems to underlie part of the stimulating effect by polar lipids plus an ionic component, which could be attributed in vivo to calcium (Jordan et al., 1973).

Fragmentation of elastic fibers is often seen in areas of mineralization. In this regard, sulfated mucopolysacharides should also be considered. Similar to lipid, alterations in fractions which comprise the mucopolysacharides and their interrelationships with elastin may be affected by bound calcium. Lorenzen (1959) and Yu and Blumenthal (1967) have demonstrated that condroitan sulfate increases in the areas of fragmentation, particularly when mineralized, although the total amount of sulfated mucopolysacharide remains constant. Such interactions could affect the rates of degradation or regeneration of elastic fibers.

The association of elastin with collagen and microfibrillar protein also appears related to arterial calcification. The amount of collagen in arterial vessels is often reported to be inversely proportional to the amount of calcium present. The presence of collagen has been shown to markedly inhibit aorta pyrophosphatase in vitro (Chen et al., 1973). Possibly of more importance, collagen fibrils are often seen penetrating the elastic fiber (Yu and Blumenthal, 1967). This could have an effect in altering conformation changes that occur upon relaxation and stretching. The presence of collagen fibrils could also mask potential sites for calcium binding. The association of elastin with microfibrillar or other protein appears to have the opposite effect. Throughout this discussion the calcification of elastin fibers has been described solely in terms of elastin-derived neutral binding sites for calcium. The presence of sulfur-rich peptides in elastin preparations could provide additional calcium binding regions. The observations of Schiffman et al. (1964, 1966, 1969) cited earlier, ie. the suggestion of sulfydryl groups as nucleation sites, may be explained on this basis. Biochemically, elastin preparations are often ill-defined heterogeneous mixtures. The extraction procedures (Jackson and Cleary, 1962) which are used to obtain insoluble elastin most certainly do not remove completely all of the other components which comprise the elastic fiber. It is of interest that the amino acid composition of insoluble elastin preparations appears to change with aging (LaBella, et al., 1966) and in the area of atherosclerotic plaques (Kramsch et al. 1971). When cysteine is present it enhances the precipitation of mineral from calcifying solutions (Schiffman et al., 1966). Likewise, sulfydryl-metal complexes appear to enhance the mineralization of elastin preparations. After digestion of mineralized insoluble elastin by elastase the residue which remains contains calcium and a higher concentration of polar amino acids and cysteine than that from non-calcified elastin (Schiffman et al., 1969). The total amount of elastin that is digested, is not affected by the presence of calcium, but specific cleavage sites are apparently made inaccessible. When such events occur in vivo, they could also alter the normal turnover and regeneration of elastic fibers.

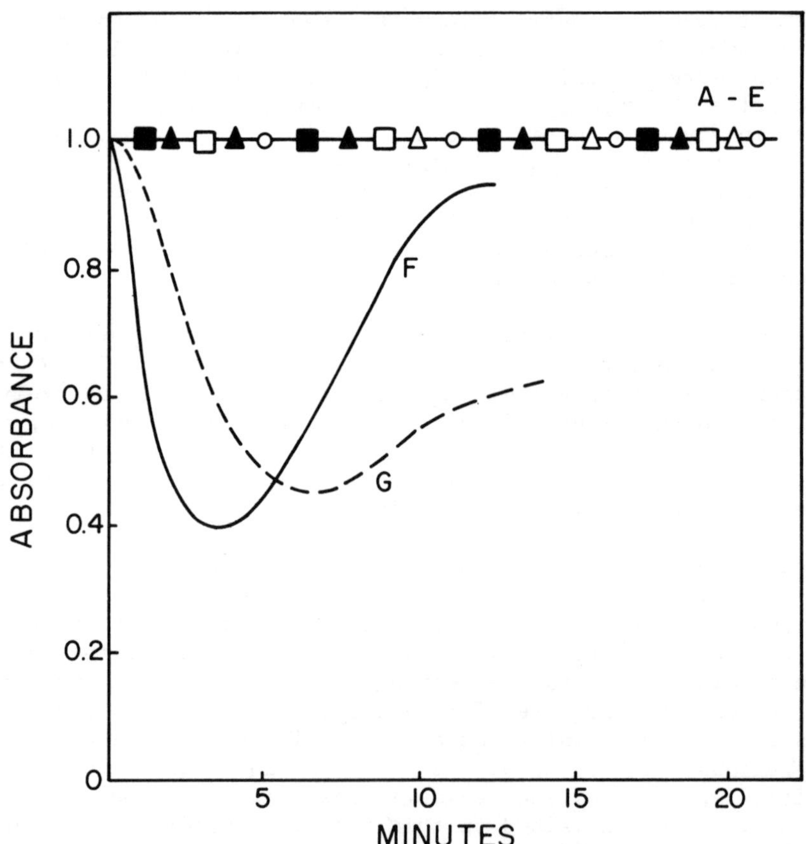

Figure 11. The effect of various elastin preparations, adenosine diphosphate, and NaCl-soluble tendon collagen on the aggregation of platelets ($8x10^8$/ml) was measured by the change in absorbance at 600 nm as described by Born and Cross (1963). Curves A through E designate the different elastin preparations. Chick native soluble elastin (■ - ■ - ■) was added at concentrations that were varied from 10 to 500 μg/ml. Chick native soluble elastin was also coacervated in a solution of

Finally the calcification of elastic fibers could serve as a loci for clot formation, which could be facilitated by a uniquely calcified matrix. The conversion of prothrombin to thrombin represents one of these steps. When the wall of a vessel is damaged, the adhesion of blood platelets to form hemostatic plugs is another possibility. We have found calcified elastin promoted coagulation of plasma in a manner similar to that for calcium (cf. Born and Cross, 1963). Whether this process involves the adhesion of platelets is still not clear (Figure 11). Using citrated plasma plus heparin or heparinized plasma, calcified elastin does not cause platelet aggregation as does adenosine diphosphate (Constantine, 1965) and collagen (Hovig, 1963). The fact that citrated plasma coagulated however, suggests that calcified elastin, as might be expected, facillitates the prothrombin to thrombin response or other similar clotting responses. The environment around fragmented and calcified elastic fibers appears ideal for the promotion of thrombosis and clots. Exposed collagen fibrils would promote platelet aggregation. Areas rich in calcium should facillitate some of the other steps involved.

When numerous steps are involved in complex reactions, it is often easy to overstate the importance of a single factor. In considering elastin and the relationship to calcium binding, the concept of neutral sites, however, resolves many previous problems. With respect to pathological considerations, the concept contains many features, which tend to unify other theories on the pathogenesis of vascular diseases. Certainly, there is nothing in the concept which lessens the importance of other factors. If charge neutralization of bound calcium affects the binding of lipids and structural components to calcified elastin, morphological changes in the elastin fiber should be a direct result. It will be necessary in future studies to distinguish between calcium bound to elastin in the absence of excessive counter ions, such as phosphate and sulfate mucopolysacharides. The properties of a positively charged lipophylic complex should be uniquely different than calcified elastin complexes studied

---

Figure 11. (Continued) 1 M NaCl and 10 mM $CaCl_2$ (30°C), centrifuged, suspended in saline, and then added (▲-▲-▲; 10 to 500 μg/ml). Rat aorta insoluble elastin (0.1 N NaOH extraction, Rucker et al., 1972) and insoluble elastin preincubated in absolute methanol containing 20 mM $CaCl_2$ were also tested (□-□-□, O-O-O, respectively). Likewise, α-elastin from the insoluble elastin was added (10-500 μg/ml). When adenosine diphosphate (10^{-6} M) was added the typical response was obtained (curve F). Collagen at 50 μg/ml elicited a similar response when added. Heparin was present in all assays at 10 units/ml.

previously, which contain calcium primarily in the form of calcium phosphate. Calcium binding to elastin in the absence of strong counter ions may serve a normal physiology function perhaps related to the extension and relaxation of the fiber. If this is not the case, a key question would be what factors inhibit calcium binding at neutral sites. In addition, considerable information is also required in order to clarify subsequent and perhaps more pathological events, such as the formation of hydroxyapatite and mineralized plaque.

In this regard, the elucidation of pyrophosphate as a controlling agent in mineralization may offer some possibilities in the management of pathological calcification. Several structural analogs of pyrophosphate, the diphosphonates, have now been used experimentally to control arterial and other soft tissue calcification. Recent publications on use of diphosphonates (Russell et al., 1973 and refs. cited) suggest consideration might be given to their eventual use as prophylaxic agents. Attempts to determine the rates of synthesis and degradation of elastin suggest there is measurable turnover of the elastin (Rucker et al., 1972; Fisher, 1971). The reversal of mineralization in arterial tissue may represent a process in which degeneration of elastic fibers is controlled and regeneration is promoted.

ACKNOWLEDGMENTS

This work was supported in part by grants from the National Institutes of Health, United States Public Health Service, the Nutrition Foundation, and the Sacramento-Yolo County Heart Association, California. The technical assistance of W. Goettlich-Riemann, K. Tom, M. Chan, D. Ford and J. Poaster is gratefully acknowledged.

REFERENCES

Bendall, J. R., (1955), The titration curves of elastin and of the derived α- and β-proteins. Biochem. J. 61: 31-32.

Blumenthal, H. T., Lansing, A. I., Gray, S. H. (1950), The interrelationship of elastic tissue and calcium in the genesis of arteriosclerosis. Am. J. Path. 26: 989-1009.

Born, G. V. R., Cross, M. J. (1963), The aggregation of blood platelets. J. Physiology 168: 178-195.

Chen, M., McCarry J., Chan, M. M., Riggins, R. S., Rucker, R. B. (1973), Chick aorta pyrophosphatase, Proc. Soc. Expl. Biol. Med. 143: 44-49.

Constantine, J. W. (1965), Temperature and the aggregation of rabbit platelets by adenosine diphosphate. Nature 205: 1075-1077.

Edsall, J. T., Wymann, J. (1965), Biophysical Chemistry. Vol. I, p. 610-620, Academic Press, New York.

Eisenstein, R., Ayer, J. P., Papajiannis, S., Haas, G. M., Hellis, H. (1964), Mineral binding by human arterial elastic tissue. Lab. Invest. 13, 1198-1204.

Eisenstein, R., Scott, R. A., Lesak, A. E. (1971), Altered lipid and calcium binding by calcified aortic tissue. Arch. Path. 92: 301-305.

Fisher, G. M., (1971), Dynamics of collagen and elastin metabolism in rat aorta. J. Applied Phys. 31: 527-530.

Fleisch, H., Neuman, W. F. (1961), Mechanisms of calcification: The role of collagen, polyphosphates, and phosphatases. Am. J. Physiol. 200: 1296-1300.

Fleisch, H., Bisaz, S. (1962) Mechanism of Calcification: Inhibitory Role of Pyrophosphate, Nature 195: 911.

Fleisch, H., Russell, R. G. G., Bisay, S., Termine, J. D., Posner, A. S., (1968), Influence of pyrophosphate on the transformation of amorphous to crystalline calcium phosphate. Calc. Tiss. Res. 2: 49-59.

Foster, J. A., Bruenger, E., Gray, W. R., Sanberg, L. B. (1973), Isolation and amino acid sequences of tropoelastin peptides. J. Biol. Chem. 248: 2876-2879.

Francis, M. D., Russell, R. G. G., Fleisch, H., (1969), Diphosphonates inhibit formation of calcium phosphate crystals in vitro and pathological calcification in vivo. Science 165: 1264-1266.

Francis, M. D. (1969), The inhibition of calcium hydroxyapatite crystal growth by polyphosphonates and polyphosphates. Calc. Tiss. Res. 3: 151-162.

Franzblau, C., Lent, R. W. (1968) Studies on the chemistry of elastin. In: Structure, Function and Evolution of Proteins. Brookhaven Symp. Biol. 21: 358-376.

Geddes, A. J., Parker, K. D., Atkins, E. D. T., Beighton, E. (1968) "Cross-β" Conformation in Proteins. J. Mol. Biol. 32: 343-358.

Gotte, L., Stern, P., Elsden, D. F., Partridge, S. M. (1963) The chemistry of connective tissue. Biochem. J. 87: 344-351.

Greenfield, N., Fasman, G. D. (1969) Computed circular dichroism spectra for the evaluation of protein conformation. Biochem. 10: 4108-4116.

Hall, D. A. (1955) The reaction between elastise and elastic tissue. Biochem. J. 59: 459-465.

Harrison, C. V. (1933) Experimental arteriosclerosis produced by cholesterol and vitamin D, J. Path. Bact. 36: 447-453.

Hass, G. M., Trueheart, R. E., Hemmens, A. (1961) Experimental athero-arteriosclerosis due to calcific medial degeneration and hypercholesterolemia. Amer. J. Path. 38: 289-323.

Hovig, T. (1963) Aggregation of rabbit blood platelets produced in vitro by "extract" of tendons. Thrombos. Diathes. Haemorrh. 9: 248-262.

Jackson, D. S., Cleary, E. G. (1962) The determination of elastin and collagen. In: Methods of Biochemical Analysis. Vol. 15: 26-73.

Jordan, R., Kagan, H. M, Hewit, N., Lewis, W., Franzblau, C., (1973) Stimulation of elastolysis by hydrophobic and ionic ligands. Fed. Proceedings 32: 614 Abs.

Kagan, H. M., Grombie, G. D., Jordan, R. E., Lewis, W., Franzblau, C., (1972) Proteolysis of elastin-ligand complexes-stimulation of elastase digestion of insoluble elastin by sodium dodecyl sulfate. Biochem. 11: 3412-3418.

Kramsch, D. M., Franzblau, C., Hollander, W. (1971) The protein and lipid composition of arterial elastin and its relationship to lipid accumulation in the atherosclerotic plaque. J. Clin. Invest. 50: 1666-1672.

LaBella, F. S. Vivian, S., Thornhill, D. (1966) Amino acid compositio of human aortic elastin as influenced by age. J. Gerontol. 21: 550-558.

Lorenzen, I. (1959) Epinephrine induced alterations in connective tissue of aortic wall in rabbits, Proc. Soc. Expl. Med. 102: 440-442.

Mammi, M., Gotte, L., Pezzin, G. (1968) Comparison of soluble and native elastin conformations by farultraviolet circular dichroism. Nature, 220: 371-372.

Mayers, D. F., Urry, D. W. (1972) Valinomycin-cation complex. Conformational energy aspects. J. Amer. Chem. Soc., 94: 77-94.

Molinari-Tosatti, M. P., Gotte, L. (1971) Some features of the binding of calcium ions to elastin. Calc. Tiss. Res. 6: 329-334.

Molinari-Tosatti, M. P., Galzignan, L., Moret, V., Gotte, L. (1968), Some features of the binding of calcium ions to elastin. Calc. Tiss. Res., 2, Suppl., 88.

Neuman, W. F., Neuman, M. W. (1958) The Chemical Dynamics of Bone Mineral. Chicago, University of Chicago Press.

Nordlie, R. C., Wright, S., Boyum, G. P., Roho, J. L. (1968) Canine Aorta Pyrophosphatase, Proc. Soc. Expl. Biol. Med. 128: 1039-1043.

Partridge, S. M., Davis, H. F. (1955) Composition of the soluble proteins derived from elastin. Biochem. J. 61: 21-30.

Partridge, S. M. (1962) Elastin Adv. Protein Chem. 17: 227-242.

Pauling, L. (1960), The Nature of the Chemical Bond, Ithaca, New York, Cornell University Press.

Ross, R., Bornstein, P. (1969) The elastic fiber. I. The separation and partial characterization of its macromolecular components. J. Cell Biol. 40: 366-376.

Ross, R., Bornstein, P. (1970) Studies of the components of the elastic fiber, In: Chemistry and Molecular Biology of the Intercellular Matrix. Edited by Balays. Academic Press, London, p. 641-656.

Ross, R. (1973) The elastic fiber. J. Histochem. Cytochem. 21: 199-208.

Rucker, R. B., Goettlich-Riemann, W., Hobe, J., Devers, K. (1972) Metabolism of chick aorta elastin. Biochem. Biophys. Acta 279: 213-216.

Rucker, R. B., Goettlich-Riemann, W. (1972) Isolation and properties of soluble elastin from copper-deficient chicks. J. Nutrition 102: 563-570.

Rucker, R. B., Goettlich-Riemann, W., Tom, K. (1973) Properties of chick tropoelastin. Biochem. Biophys. Acta, 317: 193-201.

Russell, R. G. G., Bisay, S., Donath, A., Morgan, D. B., Fleisch, H. (1971) Inorganic pyrophosphate in plasma in normal persons and patients with hypophosphatasis, osteogenesis imperfecta and other disorders of bone, J. of Clin. Invest., 50: 961-969.

Russell, R. G. G., Smith, R. (1973) Diphosphonates, J. of Bone and Joint Surgery 55: 66-86.

Sandberg, L. B., Weissman, N., Smith, D. W. (1969) The purification and partial characterization of a soluble elastin-like protein from copper-deficient porcine aorta. Biochemistry 8: 2940-2945.

Sandberg, L. B., Zeikus, R. D., Coltrain, I. M. (1971) Tropoelastin purification from copper-deficient swine: A Simplified Method. Biochem Biophys. Acta, 236: 542-545.

Sandberg, L. B., Weissman, N., Gray, W. R. (1971) Structural features of tropoelastin related to the sites of cross-links in aortic elastin. Biochemistry 10: 52-59.

Shiffman, E., Corcoran, B. A., Martin, G. R. (1966) Role of complexed heavy metals in initiating mineralization of "elastin" and the precipitation of mineral from solution. Arch. Biochem., 115, 87-94.

Shiffman, E., Lavender, D. R., Miller, E. J., Corcoran, B. A. (1969) Amino acids at the nucleating site in mineralizing tissue. Calc. Tiss. Res. 3: 125-135.

Shiffman, E., Martin, G. R, Corcoran, B. A. (1964) The role of matrix in aortic calcification. Arch. Biochem. 107: 284-291.

Smith, D. W., Brown, D. W., Carnes, W. H. (1972) Preparation and properties of salt-soluble elastin. J. Biol. Chem. 247: 2427-2433.

Sobel, A. E., Leibowitz, S., Eilberg, R. G., Lamy, F. (1966) Nucleation by elastin. Nature (Lond.) 211: 45-47.

Spadaro, J. A., Becker, R. O., Backman, C. H. (1970) Size-specific metal complexes in native collagen. Nature 225: 1134-1336.

Starcher, B., Urry, D. (1973) Elastin coacervate as a matrix for calcification. Biochem. Biophys. Res. Comm. 53: 210-216.

Urry, D. W., Starcher, B., Partridge, S. M. (1969) Coacervation of solubilized elastin affects a notable conformation change. Nature 222: 795-796.

Urry, D. W., Ohnishi, M. (1970) In: Spectroscopic approaches to biomolecular conformation, (D. W. Urry, ed.), American Medical Association Press, Chicago, Ill.

Urry, D. W., Krivacia, J. R., Haider, J. (1971) Calcium ion effects a notable change in elastin conformation by interacting at neutral sites. Biochem. Biophys. Res. Comm. 43: 6-11.

Urry, D. W. (1971) Neutral sites for calcium ion binding to elastin and collagen: A charge neutralization theory for calcification and its relationship to atherosclerosis. Proc. Nat. Acad. Sci. 68: 810-814.

Urry, D. W. (1972) A molecular theory of ion-conducting channels: a field-dependent transition between conducting and nonconducting conformations. Proc. Nat. Acad. Sci., 69: 1610-1614.

Urry, D. W., Cummingham, W. D., Osbnishi, T. (1973) A neutral polypeptide-calcium ion complex. Biochem. Biophys. Acta 292: 853-857.

Weber, J. C., Eanes, E. D., Gerdes, R. J. (1967) Electron microscope study of noncrystalline calcium phosphate. Biochem. Biophys. Acta 141: 723-724.

Weis-Fogh, T., Anderson, S. O. (1970) New molecular model for the long-range elasticity of elastin. Nature 227: 718-721.

Yu, S. Y., Blumenthal, H. T. (1967) The calcification of elastic tissue. In: The Connective Tissue. Edited by B. M. Wagner and D. E. Smith, Williams and Wilkins Company, Baltimore, p. 17-49.

11

THE COORDINATION OF CALCIUM IONS BY CARP MUSCLE CALCIUM BINDING PROTEINS A, B AND C

Carole J. COFFEE[+], Ralph A. BRADSHAW[†] and Robert H. KRETSINGER[++]
[+]Dept. of Biochemistry, Univ. of Pittsburgh, Pittsburgh, Pa., 15217, [†]Dept. of Biological Chemistry, Washington Univ. School of Medicine, St. Louis, Mo., 63110 and [++]Dept. of Biology, Univ. of Virginia, Charlottesville, Va., 22901

The white muscle of fish and amphibia contain a group of closely related, low molecular weight parvalbumins which bind two calcium ions per protein molecule. Because of this distinguishing property, they have been designated muscle calcium binding proteins (MCBP)(1,6,13,14,16). As can be seen in Table I, which summarizes their properties, they are quite acidic and

TABLE I.

PROPERTIES OF FISH AND AMPHIBIAN MCBP (PARVALBUMINS)

1. High solubility in H_2O
2. Low molecular weight (∿11,000)
3. Acidic isoelectric points (4.0-4.5)
4. Amino acid composition
   (A) 0-1 residue tyrosine, histidine, cysteine, arginine, tryptophan and proline
   (B) ∿10% Phenylalanine
   (C) ∿20% Alanine
5. Acetylated amino terminus
6. High antigenicity
7. 2 Ca^{++}/mole protein ($K_D \cong 10^{-6} - 10^{-7}$)

contain relatively high proportions of alanine and phenylalanine. In addition, they possess acetylated amino termini and are highly antigenic.

Heterogeneity in the parvalbumin family is not limited to interspecies differences. Usually between two to five closely homologous but distinct editions are found in a single fish (15). In most cases these proteins can be readily separated by electrophoresis or ion exchange chromatography. Such is the case with carp parvalbumins, in which four distinct MCBP species have been identified (15). Initially labelled components 2,3,5a and 5b (6), based on electrophoretic properties, more recently they have been identified as C, B, A_1 and A_2 to coincide with their elution from DEAE-cellulose (11).

The complete primary and three-dimensional structure of component B of carp muscle calcium-binding proteins has been determined by standard chemical and x-ray techniques (2,9). These studies have not only rigorously defined the location of virtually all the atoms of the protein but also have defined the location and ligand structure of the bound calcium atoms. Furthermore, the three-dimensional structure analysis has revealed aspects of the evolutionary mechanism by which MCBP's arose (7) and has provided a probable mechanism of action (8). This latter analysis is particularly interesting in view of the fact that no physiological role has yet been assigned these proteins. However it seems clear that their function is related to the binding of the calcium, a supposition that is strongly supported by the finding that the predicted (9) relationship to muscle troponin (TN-C) has proven to be correct.

MOLECULE DESCRIPTION

The amino acid sequence of carp MCBP-B is shown in Fig. 1. It was determined solely from the tryptic peptides, or subdigest thereof (2), using dansyl-Edmans, subtractive Edmans or hydrolysis by carboxypeptidase A and B. The acetyl group was determined by mass spectrometer identification of N-acetylalanine, which was released from the amino-terminal tryptic peptide by the action of thermolysin. The alignment of the tryptic peptides was achieved by means of the electron density map and corroborated by the amino acid sequence of the homologous hake protein which was reported (13) prior to the completion of the carp MCBP-B structure.

The three-dimensional structure of carp MCBP-B was determined by single crystal x-ray diffraction techniques (9). The completed structure has been refined by seven cycles of difference Fourier refinement with a residual $R=\Sigma(F_{obs}-F_{calc})/\Sigma F_{obs}$, for all observed

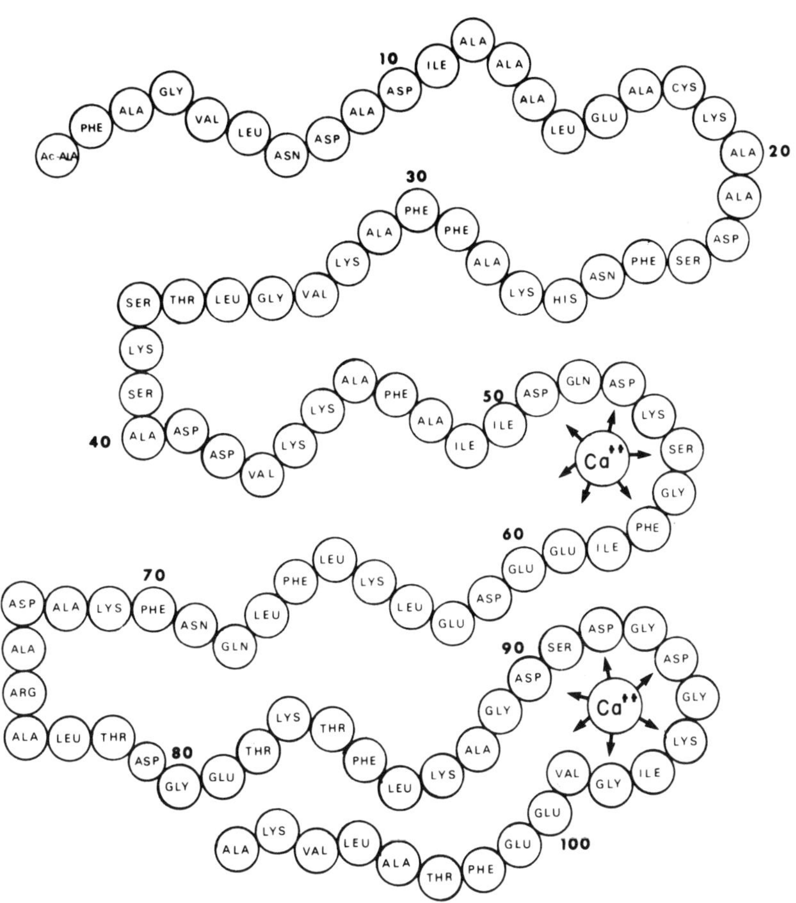

FIGURE 1.

Schematic representation of the amino acid sequence of carp muscle calcium-binding protein B. The coordination of the two atoms of calcium ion is represented schematically by the arrows.

reflections of 0.27. Bond distances and angles have been constrained by the Diamond model building program (4) to near canonical values. The average error in atomic coordinates, with the exception of the first seven residues at the amino terminus which are definitely disordered, is less than 0.25 Å.

The main chain conformation of carp MCBP-B is shown in Fig. 2. As expected, the molecule is quite globular containing little β-pleated sheet but a quite considerable amount of helical structure. There are four features of the structure that deserve emphasis. First, arginine 75 and glutamic acid 81 form a salt bridge or polar hydrogen bond that is located in the interior of the molecule. The segments of the protein involved are illustrated schematically in Fig. 3. The amino terminal segment, particularly residues 19 through 27 shield this strong dipole from the solvent. In view of the invariant nature of this pair in all of the sequences of the muscle calcium binding proteins known, it is likely that this feature of the structure plays an important role in the structure-function relationships of these proteins.

Second, there are some ten "lost" hydrogen bonds in the structure, i.e., peptide protons not involved in hydrogen bonds. These sites are particularly prevalent in critical regions of helix D and the antiparallel β-pleated sheet connecting the two calcium binding loops. This aspect of the structure, which would normally be expected to exert a destabilizing effect, may in fact provide a strong clue to the conformational changes that probably accompany the removal of the calcium atoms.

Third, the molecule contains a compact hydrocarbon core made up of about 60% of the side chains of valine, isoleucine, leucine and phenylalanine. This core accounts for about 15% of the protein volume and contains none of the protein backbone which passes beside but not through the core. As with other globular proteins this hydrophobic interior clearly is a strong stabilizing influence for the native structure of the protein.

Fourth, the two atoms of calcium are coordinated by ligands of oxygen that form two similar octahedra. As shown in Fig. 4, these can be designated as +X, +Y, +Z, -X, -Y and -Z. (The numbers in parenthesis indicate the number of the residue donating the ligand.) Clearly, the sequence of ligands follows the same pattern in each case (+X, Asp-51 and Asp-90; +Y Asp-53 and Asp-92; +Z, Ser-55 and Asp-94; -Y, Phe-57 and Lys-96 (peptide carbonyl as ligand); -X, Glu-59 and Gly-98 (H_2O); -Z, Glu-62 and Glu-101). With the exception of Phe-57 and Lys-96, the ligands are donated by side chain atoms. The most difference in the two chelate structures occurs at the -X position. In the first

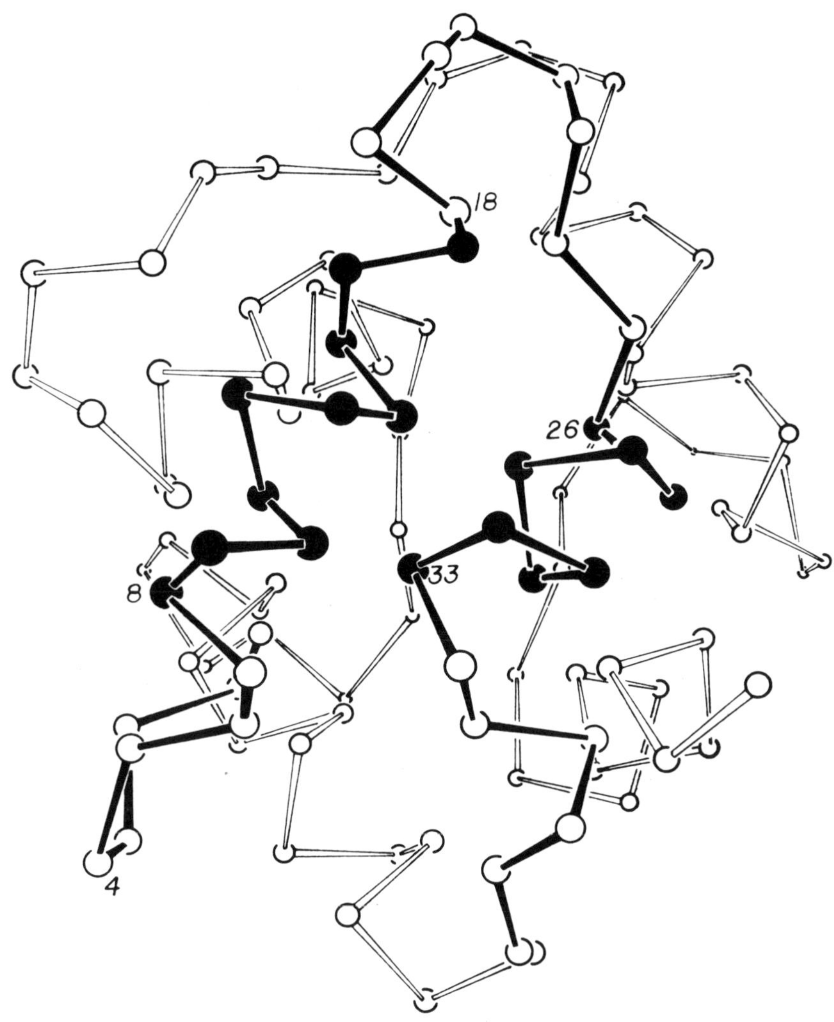

FIGURE 2.

Schematic representation of the α-carbon structure of carp muscle calcium-binding protein B. The pre-A bend, A helix, AB bend, and B helix are indicated by solid lines between α-carbons. (Copyright American Society of Biological Chemists, Inc., 1973)

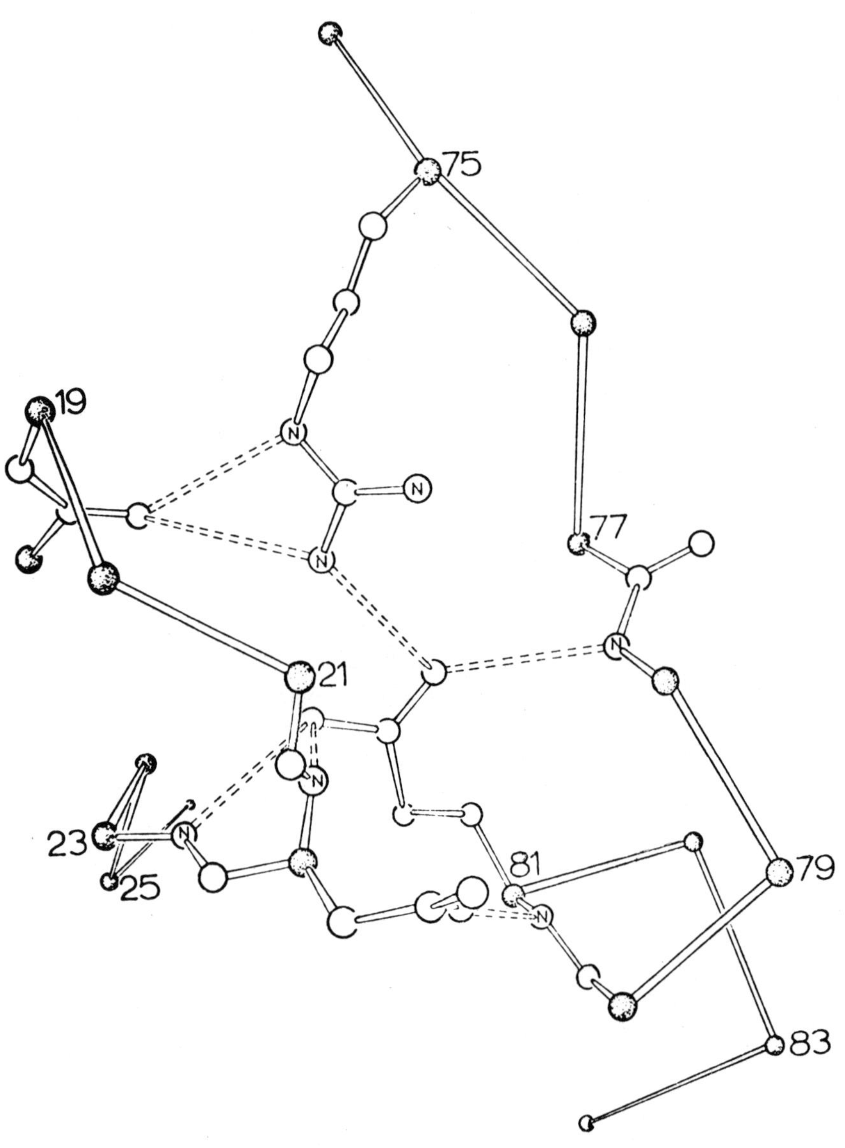

FIGURE 3.

Schematic representation of the salt linkage between arginine 75 and glutamic acid 81 (Copyright, American Society of Biological Chemists, Inc., 1973).

segment (CD loop), this position is occupied by Glu-59 while at the second site (EF loop), there is a glycine residue, the ligand being supplied by water. These observations suggest that the EF calcium can be more readily displaced, a hypothesis that has been substantiated by terbium exchange experiments (10).

Table II summarizes the relevant calcium-oxygen and oxygen-oxygen bond distances and angles in the two octahedra. The calcium-oxygen distances are listed on the diagonal with those values which are considered to be too great to participate in bonding given in parentheses. A comparison of the values in the upper and lower halves of the table underscore the similarity of the two calcium-binding loops.

Table III lists the hydrogen bonding pattern found in the calcium binding loops. The donors and acceptors as well as the distances and angles are identified. The first loop begins with a type I β bond (17) indicated on Table III by the 54→51 (93→90)

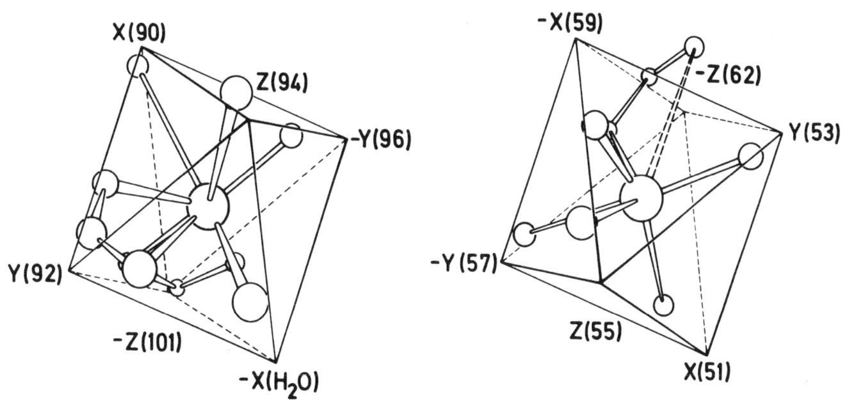

FIGURE 4.

The coordination spheres of calcium ions CD (right) and EF (left) as viewed from the surface of the molecule looking down the intra-molecular approximate two fold axis. The calcium coordinating ligands can be visualized as related to the vertices of an octa-hedron. At the CD calcium the first ligand in sequence, a carboxylate oxygen atom of aspartic acid 51, defines the +X vertex and the second, from aspartic acid 53, defines the +Y vertex. Similarly, aspartic acid 90 and aspartic acid 92 define the +X and +Y vertices of the EF calcium. Note that register of residues about the vertices is the same in both octahedra.

hydrogen bond. The peptide nitrogen atoms of residues 55, 56, 57 and 59 (94, 95, 96 and 98) provide protons to bind to oxygen atoms which are involved in calcium coordination. This intricate pattern emphasizes again the similarity in geometry of the two calcium binding octahedra.

The two calcium binding octahedra are also related with respect to their orientation to the main polypeptide chain. There are six helical segments present in the protein, spaced in relatively even manner. The two octahedra are formed from loops located between the C and D and E and F helices, respectively. As shown in Fig. 5, these helices are related in space. Helix C which leads into the first calcium binding loop is related to helix E by the intramolecular approximate two fold axis that relates the two calcium octahedra. Helix D, the helix leaving the CD loop, is likewise related to helix F.

This relationship is symbolically represented in Fig. 6 by two right hands. The extended forefingers represent the C and E helices running in the direction indicated. The calcium binding loops, CD and EF, enclose the two octahedra with the middle fingers, which run in approximate antiparallel β-pleated sheet conformation relative to each other. The two thumbs represent helices D and F as noted. The "bend" in the thumb representing helix D corresponds to the distortion present in this substructure. Finally, the four arrows indicate the one formed and the three "lost" hydrogen bonds in the conformation of the holo-protein.

The amino-terminal third of the molecule also contains two helical segments, A and B. Their orientation is also that of a right hand but with the thumb (helix B) drawn in so that the two helices run in an almost antiparallel sense. The AB loop (residues 19 through 27) joining these two helices does not coordinate calcium or presumably any other ligand. In the protein, this loop shields the salt bridge of Arg-75-Glu-81 from the solvent.

The close similarity of the two calcium binding octahedra including the related helical segments clearly suggests a contiguous gene duplication event in the evolution of this protein in which the portion of the gene coding for the CD "hand" was copied to allow a second (EF) "hand" to be added to the structure (7). However, a comparison of the amino acid sequences of these regions shows homology of only borderline significance. In fact, the AB region shows a greater amino acid sequence homology to the EF region than does the CD region. Thus, considering the general three-dimensional similarity of this region to the CD and EF regions as well, it has been

TABLE II.

The oxygen-oxygen distances, if less than 4Å, are listed in the upper right halves of the matrices. The angles, relative to the calcium ion as vertex, are listed in the lower left. The calcium-oxygen distances are on the diagonal; those in parentheses are considered too great to be designated as bonding. Coordinating groups are listed in their order of occurrence along the polypeptide chain to illustrate the homology between the CD and EF calcium binding loops. The labels X, Y and Z refer to the vertices of the octahedra in Fig. 4.

			ASP-90		ASP-92		ASP-94		LYS-96	H_2O	GLU-101	
			δ_1	δ_2	δ_1	δ_2	δ_1	δ_2	o		ϵ_1	ϵ_2
		CA	X		Y		Z		-Y	-X	-Z	
ASP-90	δ_1	X	2.87			2.5	3.1		3.1			
	δ_2			(4.5)								
ASP-92	δ_1	Y	110°		2.48		3.5	3.1		2.2	3.2	
	δ_2		58°			2.06	3.2				2.5	
ASP-94	δ_1	Z	68°		88°	84°	2.63		2.9			
	δ_2				62°			(3.4)		2.8		
LYS-96	o	-Y	75°		156°	133°	72°		2.19			2.7
H_2O		-X	155°		53°	107°	92°		114°	2.39		
GLU-101	ϵ_1	-Z	90°		83°	69°	152°		121°	103°	2.37	
	ϵ_2		100°		128°	120°	144°		72°	105°		2.44

			ASP-51		ASP-53		SER-55	PHE-57	GLU-59		GLU-62	
			δ_1	δ_2	δ_1	δ_2	γ	o	ϵ_1	ϵ_2	ϵ_1	ϵ_2
		CA	X		Y		Z	-Y	-X	-Z		
ASP-51	δ_1	X	2.13		3.6		3.7	2.9			3.3	
	δ_2			(3.8)								
ASP-53	δ_1	Y	105°		2.35		3.6		3.3		3.8	3.2
	δ_2					(4.1)						
SER-55	γ	Z	110°		100°		2.42	2.8	3.0			
PHE-57	o	-Y	82°		172°		73°	2.30	3.5		2.8	
GLU-59	ϵ_1	-X	170°		84°		74°	91°	2.60		3.0	3.1
	ϵ_2	-Z	97°		109°		133°	75°	75°	(3.6)		
GLU-62	ϵ_1				68°				63°		2.31	
	ϵ_2											(3.27)

TABLE III.

Distances (D) and angles (α) are listed for the hydrogen bonds of the calcium-binding octahedra. Alpha is the proton-donor-acceptor angle (0° for a linear bond). For donor glutamic acid 60 to acceptor glycine 95, as well as 97→58 and 99→58, the distances and angles are in parentheses to indicate that these bonds are not formed in the holo-protein.

Pept. N	Other	Pept. O	Other	D	α	Pept. N	Pept. O	Other	D	α
54 Lys		51 Asp		3.2	34°	93 Gly	90 Asp		2.5	28°
55 Ser			53 Asp δ_1	3.4	8°	94 Asp		90 Asp δ_1	3.2	18°
56 Gly			51 Asp δ_2	2.8	36°	95 Gly		90 Asp δ_2	2.5	24°
57 Phe			55 Ser γ	3.2	26°	96 Lys		94 Asp δ_1	2.6	22°
58 Ilu		97 Ilu		2.6	18°	97 Ilu	58 Ilu		(3.6	23°)
59 Glu			62 Glu ϵ_1	3.2	29°	98 Gly		101 Glu ϵ_2	3.3	32°
60 Glu		95 Gly		(4.4	38°)	99 Val	56 Gly		(4.7	29°)
61 Asp						100 Asp				
62 Glu		59 Glu		2.4	28°	101 Glu				
63 Leu						102 Phe	98 Gly		3.0	17°
	54 Lys	51 Asp		2.8	5°					
	54 Lys	48 Ala		2.4	5°					
	55 Ser		59 Glu ϵ_2	2.2	6°					

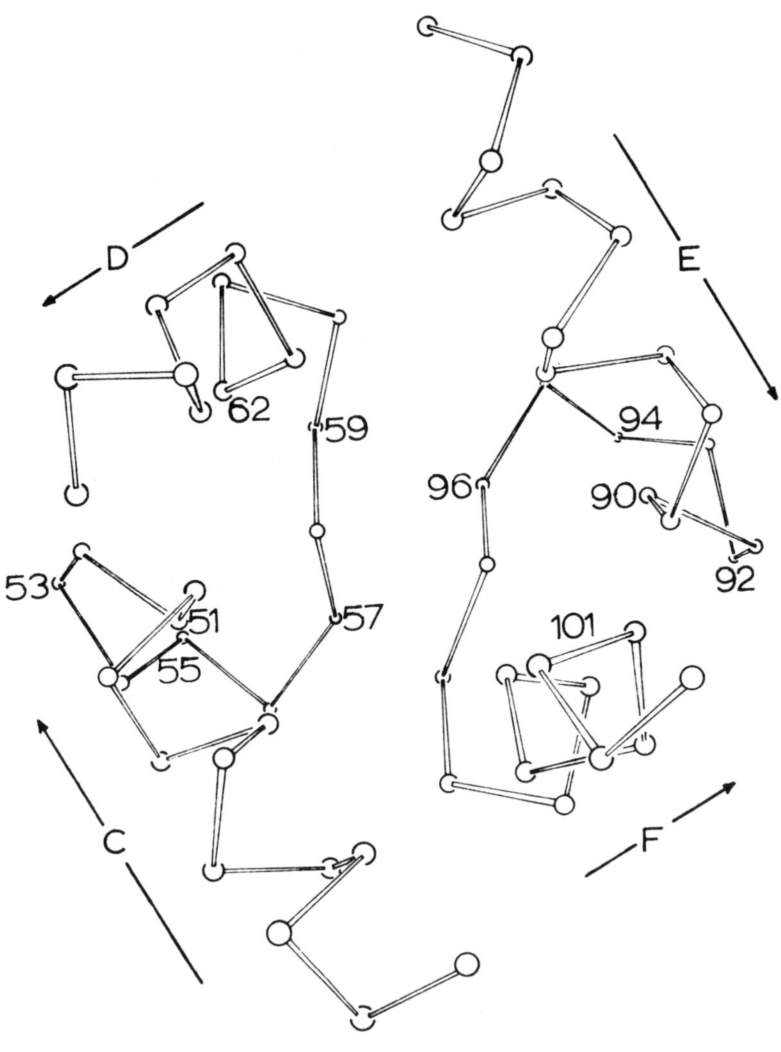

FIGURE 5.

The C helix, CD calcium-binding loop and the D helix are related to the homologous EF region by an intramolecular approximate two-fold axis. (Copyright American Society of Biological Chemists, 1973)

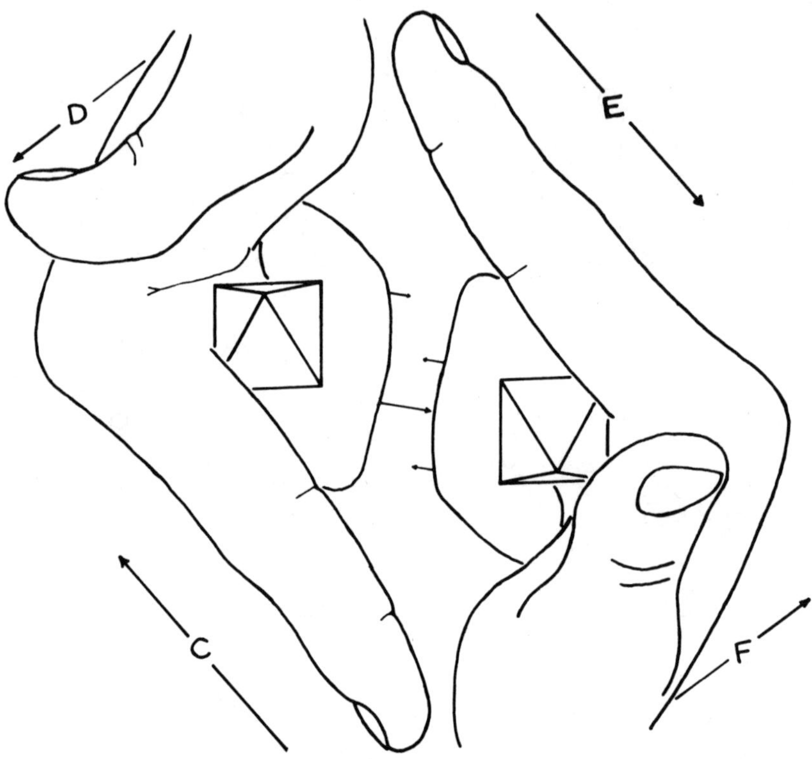

FIGURE 6.

The CD and EF regions can be easily and rather accurately visualized by a pair of right hands. The C (and E) helix is represented by the forefinger and the D (and F) helix by the thumb. The CD (and EF) loops are represented by the middle fingers holding the octahedra. The bend in the D thumb represents the bend found in the D helix.

proposed (7) that this protein actually evolved as a gene triplication. Regardless of the correctness of this supposition, it is clear that three-dimensional structure is a better indication of homology in distantly-related proteins than the primary structure.

Based upon the implication that MCBP is not an enzyme but mediates its function, as yet unknown, through the calcium ions and the structural features of the protein described, a model for the cooperative binding and release of calcium has been postulated (8). This model also draws on the supposition that gene triplication to ultimately produce two calcium ion binding sites in the present version of the protein was selected for to confer functional advantages. The proposed steps may be summarized as follows:

1. The EF calcium ion can exchange rapidly with solvent via the -X vertix of the octahedron.

2. In the absence of calcium, the coordinating carboxylate groups of residues 90 (+X), 92 (+Y), 94 (+Z) and 100 (-Z) will repel one another causing (or allowing) a definitive conformational change in the EF loop.

3. The antiparallel β-pleated sheet between the EF and CD loops can regain the three "lost" hydrogen bonds between 99→56, 97→58, and 60→95 (see Fig. 6).

4. The resulting movement of the EF loop distorts the binding conformation of the CD loop.

5. The affinity of the CD calcium is markedly lowered and it can diffuse away.

6. The four carboxylate groups coordinating the calcium ion in the CD loop become mutually repulsive without the calcium ion and move away.

7. The D helix is no longer constrained in its bent conformation and assumes a normal α-helical geometry. It regains four "lost" hydrogen bonds in this process.

8. The Arg-75 - Glu-81 salt bridge is broken. Arg-75, located at the end of the D helix, would naturally be drawn out to the solvent as a result of the "untensing" of the D helix. Glu-81 would, however, remain inside.

9. The AB loop would shift to allow the exposure of Glu-81.

10. Thus formed, the apo MCBP would have altered AB and DE loops that presumably would interact differently (negatively or positively) with the cellular components with which the function of MCBP is associated.

Although this scheme is hypothetical three lines of evidence support it: 1) calcium ions are bound cooperatively (1); 2) lanthanides easily displace the EF but not the CD calcium (10); 3) modification of Arg-75 with cyclohexanedione greatly reduces the affinity of MCBP for calcium (5). Clearly, further experimentation will be required to substantiate the details of the binding and release of the calcium and to establish the functional significance of such events.

CHARACTERIZATION OF THE MULTIPLE FORMS OF CARP MCBP

All species examined to date have been shown to have multiple forms of the muscle calcium binding proteins (15). The functional significance for these genetic variations remains as obscure as the primary physiological role of the proteins themselves. However, based on general properties and amino acid composition it appeared that a high degree of similarity would be found among the various forms of MCBP found in any given species (12). Preliminary sequence analyses described below, have borne this out. In addition, substitution of the replacement residues of carp MCBP A_1 and C in the three-dimensional model of the B protein has shown that all replacements that have been found so far can be incorporated into this structure without any significant alterations in the main chain conformation.

A representative separation of the parvalbumin fraction of the carp myogen on a column of DEAE-cellulose is shown in Fig. 7. Six distinct components, designated 0, A_1, A_2, B, C and D can be seen. Component 0 has not been confirmed to be a parvalbumin and will not be discussed further. On the hand, component D does appear to be a protein of this class. However, too little material has been obtained for sequence analysis.

Components A_1, B and C have been characterized in greater detail. Fig. 8 shows the tryptic peptide map, performed on columns of Dowex 50 X 8, of S-[^{14}C]-carboxymethyl-carp MCBP-A_1, B and C. The pattern for A_2 is entirely similar to A_1. As can be seen, the patterns of A_1 and B, in which the corresponding tryptic peptides have been designated with the same number are rather similar. Likewise, the elution profile for the tryptic peptides of carp MCBP-C is somewhat different with several distinct changes evident particularly in the latter part of the pattern. An examination of the corresponding sequences corroborates this picture. As shown in Fig. 9, MCBP-A_1 differs from the B protein at three sites, Thr→Ala 12 , Thr→Ala 28 , and Gly→Asp 73 while MCBP-C differs at 19 sites. Preliminary data suggests that MCBP-A_2 differs from MCBP-A_1 by only one Ala→Thr interchange; however it is not clear whether this occurs at position 12 or 28.

For the purposes of comparison, the amino acid sequence of the main parvalbumin of hake, as determined by Pechere _et al_. (13), is also listed. It differs from both carp MCBP-B and MCBP-C by 27 residues, although not at the same sites. As with the carp components, the replacement residues found in hake can be incorporated into the three-dimensional model of carp MCBP-B without conflict.

Two features of the carp MCBP-C sequence deserve mention. First, there is a Pro→Ala substitution at position 40. This residue marks the beginning of the C helix with Φ and Ψ angles of -37° and -50° which readily accomodate the proline residue. Second, the tyrosine residue has been tentatively assigned to position 2. This is particularly important since the first seven residues of the polypeptide chain have been located in the electron density map only with difficulty (9). The location of the side chain of Phe-2 has been particularly troublesome, with a buried and exposed site being equally probable. The assignment of tyrosine to the site in carp MCBP-C provides an opportunity to probe this site. Fig. 10 shows a spectrophotometric titration of carp MCBP-C in buffer and 6 M urea, 1 mM EGTA. Clearly, in the absence of the denaturant, the tyrosyl side chain is perturbed over one full pH unit compared to the same titration in urea. These results strongly support a buried site for the tyrosyl residue and, by implication from three-dimensional homology arguments, a similar location for Phe-2 of MCBP-A and B.

FISH MUSCLE CALCIUM-BINDING PROTEINS AND TROPONIN-C

Although the function of fish muscle calcium-binding proteins is not known, it has been suggested that it will be closely related to the binding and release of calcium (9). More, over, it has been noted (13) that MCBP shares a number of similar properties to mammalian troponin-C, the subunit of the muscle regulatory protein, troponin, which binds calcium. These include, in addition to the calcium chelating capacity, acidic isoelectric points, high antigenicity, and amino compositions rich in phenylalanine and poor in other aromatic residues.

The determination of the complete amino acid sequence of rabbit muscle troponin C by Collins, _et al_. (3) has confirmed

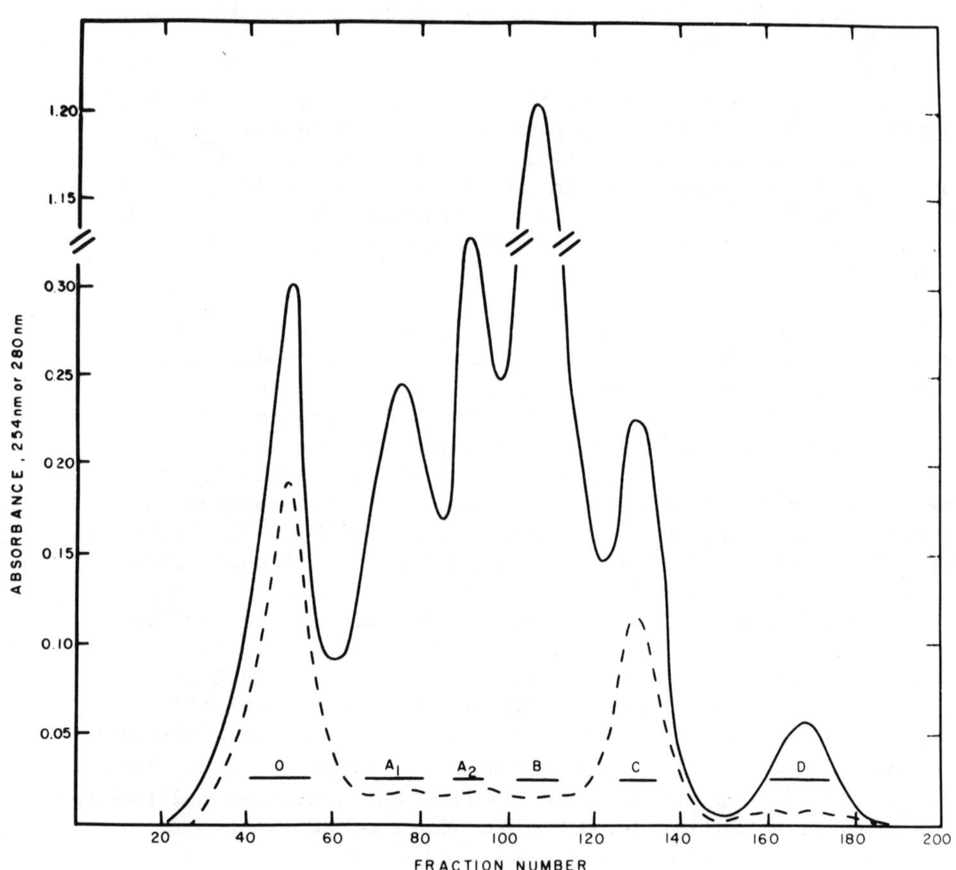

FIGURE 7.

Electron profile of the separation of the parvalbumin fraction of carp myogen on a column of DEAE-cellulose. The dashed line is the absorbance at 280 nm and the solid line the absorbance at 254 nm (9).

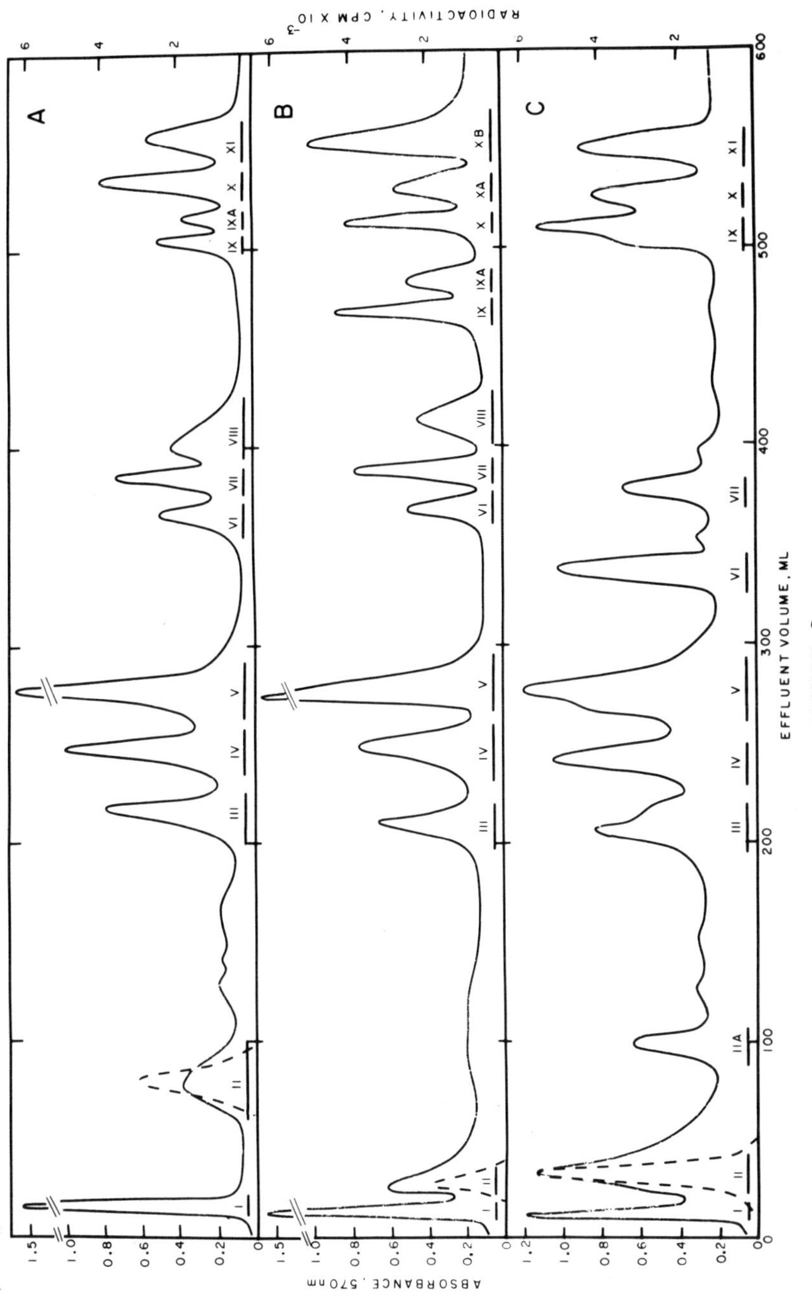

FIGURE 8.

Elution profiles of the tryptic peptides of S-[^{14}C]-carboxymethyl carp muscle calcium binding proteins, A_1, B and C on a column of Dowex 50 X 8 (2). The radioactivity is given by the dashed line and the ninhydrin absorbance at 570 nm after alkaline hydrolysis, by the solid line.

FISH MUSCLE CALCIUM-BINDING PROTEINS

```
                                                  10                                      20
Carp B:    Ac-Ala-Phe-Ala-Gly-Val-Leu-Asn-Asp-Ala-Asp-Ile-Ala-Ala-Ala-Leu-Glu-Ala-Cys-Lys-Ala-Ala-Asp-Ser-Phe-Asn-His-Lys
Carp A1:                                                  Thr
Carp C:          (Tyr-Gly)   (Ile)                       (Thr)                       (Glx)   (Glx)               (Ala)
Hake:                         Ile     Ala                 Thr             Ala                 Glu-Gly         Lys     Gly

                      30                                      40                                      50
Carp B:       Ala-Phe-Phe-Ala-Lys-Val-Gly-Leu-Thr-Ser-Lys-Ser-Ala-Asp-Asp-Val-Lys-Lys-Ala-Phe-Ala-Ile-Ile-Asp-Gln-Asp-Lys
Carp A1:      Thr
Carp C:       Ser                             Ser-Ala     Thr-Pro         Ile                     Val
Hake:         Glu         Thr     Ile         Lys-Gly             Ala     Ile         Val     Gly

                                  60                                      70                                      80
Carp B:       Ser-Gly-Phe-Ile-Glu-Glu-Asp-Glu-Leu-Lys-Leu-Phe-Leu-Gln-Asn-Phe-Lys-Ala-Asp-Ala-Arg-Ala-Leu-Thr-Asp-Gly-Glu
Carp A1:                                                                              Gly
Carp C:                                                                       Ser     Gly                     Ala
Hake:             Asp     Val                                                 Ser     Gly                     Ala

                                              90                                      100
Carp B:       Thr-Lys-Thr-Phe-Leu-Lys-Ala-Gly-Asp-Ser-Asp-Gly-Asp-Gly-Lys-Ile-Gly-Val-Asp-Glu-Phe-Thr-Ala-Leu-Val-Lys-Ala
Carp A1:
Carp C:               Ala                                                                         Ala
Hake:             Ala                                                             Glu         Ala     Met         Gly
```

FIGURE 9.

The amino acid sequences of carp muscle calcium-binding proteins A_1, B and C. Residues in parentheses have been only tentatively assigned. The amino acid sequence of hake parvalbumin was determined by Pechère *et al.* (13).

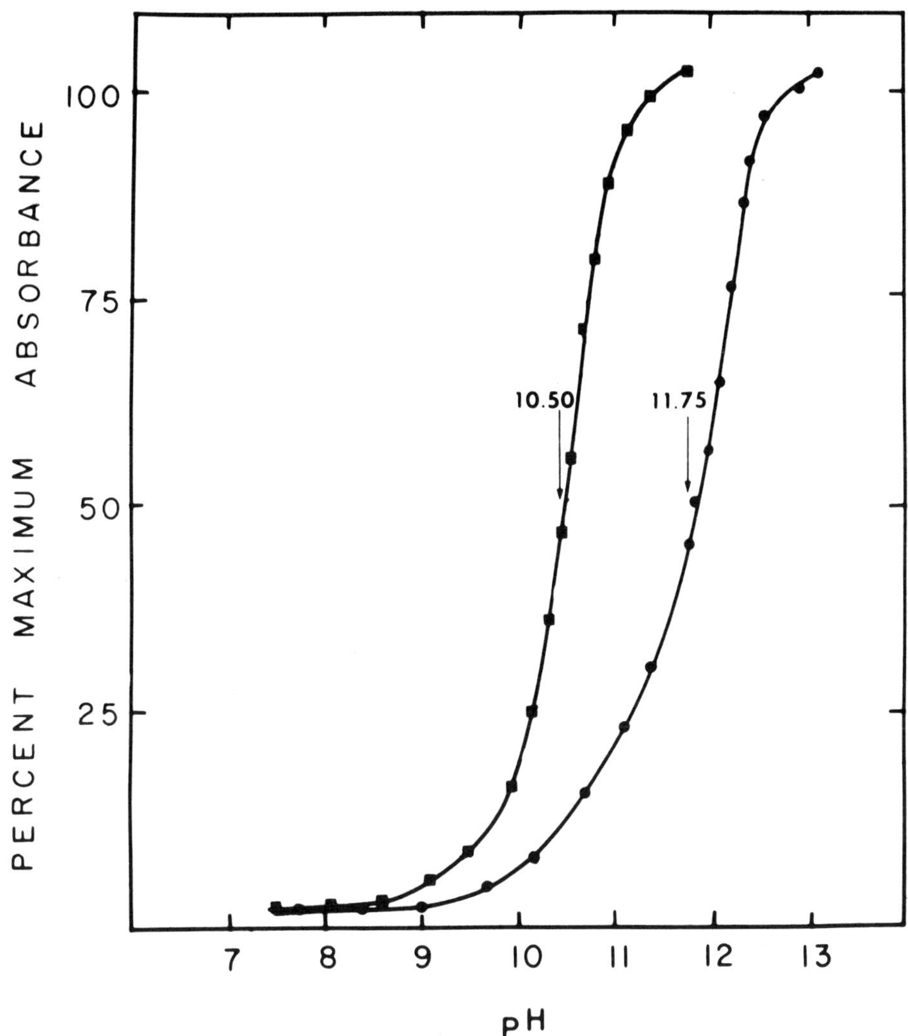

FIGURE 10.

Spectrophotometric titration of carp muscle calcium-binding protein C in aqueous buffer (■) and 6 M urea, 1 mM EGTA (●). The numbers indicate the estimated pK_a values for the single tyrosyl residue.

COMPARISON OF CARP MUSCLE CALCIUM-BINDING PROTEIN B AND RABBIT MUSCLE TROPONIN-C

```
                                            10                                      20
TNC:       Gln-Asp-Gln-Thr-Ala-Glu-Ala-Arg-Ser-Tyr-Leu-Ser-Glu-Glu-Met-Ile-Ala-Glu-Phe-Lys
MCBP-B:                                Ac-Ala-Phe-Ala-Gly-Val-Leu-Asn-Asp-Ala-Asp-Ile-Ala

                                            30                                      40
TNC:       Ala-Ala-Phe-Asp-Met-Phe-Asp-Ala-Asp-Gly-Gly-Gly-Asp-Ile-Ser-Val-Lys-Glu-Leu-Gly
MCBP-B:    Ala-Ala-Leu-Glu-Ala-Cys-Lys-Ala-Ala-Asp-Ser-Phe-Asn         His-Lys-Ala-Phe-Phe

                                            50                                      60
TNC:       Thr-Val-Met-Arg-Met-Leu-Gly-Gln-Thr-Pro-Thr-Lys-Glu-Glu-Leu-Asp-Ala-Ile-Ile-Glu
MCBP-B:    Ala-Lys-Val         Gly-Leu-Thr-Ser-Lys-Ser-Ala-Asp-Asp-Val-Lys-Lys-Ala-Phe-Ala

                                            70                                      80
TNC:       Glu-Val-Asp-Glu-Asp-Gly-Ser-Gly-Thr-Ile-Asp-Phe-Glu-Glu-Phe-Leu-Val-Met-Val-Arg
MCBP-B:    Ile-Ile-Asp-Gln-Asp-Lys-Ser-Gly-Phe-Ile-Glu-Glu-Asp-Glu-Leu-Lys-Leu-Phe-Leu-Gln

                                            90                                     100
TNC:       Gln-Met-Lys-Glu-Asp-Ala-Lys-Gly-Lys-Ser-Glu-Glu-Glu-Leu-Ala-Glu-Cys-Phe-Arg-Ile
MCBP-B:    Asn-Phe-Lys-Ala-Asp-Ala-Arg-Ala-Leu-Thr-Asp-Gly-Glu-Thr-Lys-Thr-Phe-Leu-Lys-Ala

                                           110                                     120
TNC:       Phe-Asp-Arg-Val-Ala-Asp-Gly-Tyr-Ile-Asp-Ala-Glu-Glu-Leu-Ala-Glu-Ile-Phe-Arg-Ala
MCBP-B:    Gly-Asp-Ser-Asp-Gly-Asp-Gly-Lys-Ile-Gly-Val-Asp-Glu-Phe-Thr-Ala-Leu-Val-Lys-Ala

                                           130                                     140
TNC:       Ser-Gly-Glu-His-Val-Thr-Asp-Glu-Glu-Ile-Gly-Ser-Leu-Met-Lys-Asp-Gly-Asp-Lys-Asp

                                           150                         158
TNC:       Asn-Asp-Gly-Arg-Ile-Asp-Phe-Asp-Glu-Phe-Glu-Lys-Met-Met-Glu-Gly-Val-Gln
```

FIGURE 11.

Comparison of the amino acid sequences of carp muscle calcium-binding protein B (2) and rabbit muscle troponin (TN-C) (3).

Troponin C					Carp MCBP A_1, B, C		
	9S		84E		1A	34	72A
	Y		D		FFY	G	GDG
	L	47G	A	122G	AAG	L	A
	S	Q	K	E	G	TTS	R
	E	T	G	H	WI	SSA	A
	E	P	K	V	L	K	L
	M	T	S	T	N	SST	T
	I	K	E	D	D	AAP	D
	A	E	E	E	A	D	GGA
	E	E	E	E	D	D	E
	F	L	L	I	I	WI	T
	K	D	A	G	TAT	K	K
	A	A	E	S	A	K	TTA
	A	I	C	L	A	A	F
	F	I	F	M	L	F	L
	D	E	R	K	E	A	K
	M	E	I	D	A	IIV	A
	F	V	F	G	C	I	G
	D	D	D	D	KKZ	D	D
	A	E	R	K	A	Q	S
	D	D	N	D	AAZ	D	D
	G	G	A	N	D	K	G
	G	S	D	D	S	S	D
	G	G	G	G	F	G	G
	D	T	Y	R	N	F	K
	I	I	I	I		I	I
	S	D	D	D		E	G
	V	F	A	F	HHA	E	V
	K	E	E	D	K	D	D
	E	E	E	E	TAS	E	E
	L	F	L	F	F	L	F
1Q	G	L	A	L	F	K	TTA
D	T	V	E	K	A	L	A
Q	V	M	I	M	K	F	L
T	M	V	F	M	33V	L	V
A	R	R	R	E		Q	K
E	M	Q	A	G		N	108A
A	46L	H	121S	V		F	
8R		83K		158Q		71KKS	

FIGURE 12.

Comparison of the amino acid sequences of rabbit muscle troponin (TN-C) (3) and the carp muscle calcium-binding proteins A_1, B and C. The sequences are divided into segments, representing calcium-binding loops, to illustrate the internal and external homology. Amino acids are enclosed in solid boxes if one residue type occurs three or more times in the seven segments. Dashed boxes show chemically similar or conservative replacements at sites already having three or more conserved residues.

the predictions (9,13) that FMCBP and TN-C are indeed homologous. As shown in Fig. 11, in which rabbit TN-C and carp MCBP-B are compared, 22 of the 108 residues of carp MCBP-B are identical. Only four deletions are required for this alignment although there are eight additional residues at the amino terminal and 38 additional residues at the carboxyl terminal in TN-C.

The relationship of these molecules is considerably reinforced when they are compared in the manner shown in Fig. 12. In this diagram, the sequence of TN-C is presented in four basic segments and that of carp MCBP in three. The segments in carp correspond to the AB, CD and EF "hands" described above. When TN-C is aligned this way, a striking homology, not only to the carp segments but internally as well, is immediately obvious. In each segment of the TN-C, the requisite structural requirements to build a helix-calcium-binding loop-helix is found. Thus, it is clear that TN-C will possess a similar three-dimensional structure to carp MCBP, composed of four calcium-binding "hands". It appears certain that the origin of such a molecule has been via gene tetraplication, analogous to the heavy chain component of immunoglobulins.

The striking features of the calcium binding loop, exemplified by the EF "hand" of carp MCBP-B, has been reduced by examination of all FMCBP sequences and the TN-C sequence to an easily recognizable pattern (8):

L*--L*L*--L*D-D*-D*G*-I*D*--EL*--L*L*--L*

where a single letter indicates an invariant residue and a single letter with an asterisk indicates that even though the principle of minimizing the number of mutations clearly indicates one residue type in this position, it is not invariant. It is suggested that by recognizing this pattern in the sequence of a calcium-binding protein, it will reveal the presence of a "hand" structure and suggest a fundamental informational as opposed to structural, role for calcium in the function of that protein.

ACKNOWLEDGEMENTS

This work has been supported by research grants AM 16253, AM 16728 and AM 13362 from the National Institutes of Health.

R.A.B. is a research career development awardee of the National Institutes of Health, AM 23968.

REFERENCES

1 BENZONANA, G., CAPONY, J.P., RYDEN, L. and DEMAILLE, J., Biochem. Biophys. Res. Comm. 43 (1971) 1106.
2 COFFEE, C.J. and BRADSHAW, R.A., J. Biol. Chem. 248 (1973) 3305.
3 COLLINS, J.H., POTTER, J.D., HORN, M.J., WILSHIRE, G. and JACKMAN, N, FEBS Lett., 36 (1973) 268.
4 DIAMOND, R., Acta Crysta. B21 (1966) 253.
5 GOSSELIN-REY, C., BERNARD, N. and GERDAY, C., Biochim. Biophys. Acta 303 (1973) 90.
6 KONOSU, S., HAMOIR, G. and PECHÈRE, J-F., Biochem. J. 96 (1965) 98.
7 KRETSINGER, R.H., Nature New Biology 240 (1972) 85.
8 KRETSINGER, R.H., in "Perspectives in Membrane Biology" (S. Estrada and C. Gittler, eds) Academic Press, N.Y., (1974) in press.
9 KRETSINGER, R.H. and NOCKOLDS, C.E., J. Biol. Chem. 248 (1973) 3313.
10 MOEWS, P.C. and KRETSINGER, R.H., Biochemistry (1974) in press.
11 NOCKOLDS, C.E., KRETSINGER, R.H., COFFEE, C.J. and BRADSHAW, R.A., Proc. Natl. Acad. Sci. 69 (1972) 581.
12 PECHÈRE, J-F. and CAPONY, J.P., Comp. Biochem. Physiol. 28 (1969) 1089.
13 PECHÈRE, J-F., CAPONY, J.P. and RYDEN, L., Eur. J. Biochem. 23 (1971) 421.
14 PECHERE, J.-F., CAPONY, J.P., RYDEN, L., and DEMAILLE, J., Biochem. Biophys. Res. Comm. 43 (1971) 1106.
15 PECHÈRE, J-F., DEMAILLE, J. and CAPONY, J.P., Biochem. Biophys. Acta 236 (1971) 391.
16 PECHÈRE, J-F. and FOCANT, B., Biochem. J. 96 (1965) 113.
17 VENKATCHOLAM, C.M., Biopolymers 6 (1968) 1425.

12

FERROXIDASES AND FERRIREDUCTASES: THEIR ROLE IN IRON METABOLISM

E. Frieden and S. Osaki
Department of Chemistry
Florida State University
Tallahassee, Florida 32306

INTRODUCTION

The metal most precious to man is iron, not gold. Man can survive without gold, but without iron in his diet he soon loses his capacity to carry and to utilize oxygen and his metabolism grinds to a deadly halt. This need of iron for essential biological processes is not exclusive to man, but is found throughout nature, in plants, insects, fish, birds and bacteria. Despite his desperate need for iron the human being is relatively poorly equipped to capture iron from his food. Even in iron deficiency, only a small proportion of dietary iron is absorbed. Once absorbed, iron is carefully conserved and the amount of iron lost by excretion is extremely low. In spite of this the absorption of iron frequently cannot keep pace with the organism's need for it, especially in crucial periods of life. Even in our relatively overfed society in the United States, iron deficiency may appear as a nutritional disease. It has been estimated by Dr. Louis K. Diamond (1971) of the Univ. of California School of Medicine (San Francisco) that at least 5,000,000 Americans suffer some degree of physical or mental impairment due to iron deficiency. Recent surveys in some under-developed tropical countries revealed moderate to severe anemia, responsive to iron therapy, in more than 50 percent of the women examined. Thus, if we extrapolate from our relatively well fed society to the underfed areas of the world, the number of iron deficient individuals must surely be in the hundreds of millions.

Our problems with iron are not limited to iron deficiency. In the prosperous countries, under certain dietary and medical conditions, a significant fraction of the population is faced with hemochromatosis, excess iron storage in key tissues, e.g. liver, pancreas, heart, brain. There is no known physiological route for the elimination of more than a trace of iron once it is absorbed. So we are faced with a complete spectrum of iron problems ranging from severe deficiency to dangerous excess. These are among the many reasons why iron metabolism is a subject deserving of intensive analysis.

The development of our dependence on useful iron compounds is surely one of the major trends in the evolution of higher animals (Frieden, 1973b). Iron in virtually every possible form is involved in essential biochemical processes; as the free iron ion or complexed with small molecules, but mostly as iron ions integrated into certain proteins or into complicated organic structures such as protoporphyrin to form heme which is then associated with protein. An average adult man weighing 154 pounds (70 kgs) contains about 4 grams of iron (Table I). The circulating oxygen-carrier protein, hemoglobin, accounts for 2.6 grams (65%) of this. An oxygen-binding protein in muscle, myoglobin, makes up 3% of the total. Storage forms of iron, ferritin and hemosiderin, account for one-fourth of the iron. About five percent of the iron is as yet unidentified. Less than one percent is contributed by a group of highly essential iron enzymes and the iron transport protein, transferrin (see Table I for more details of these iron compounds).

Ferrous to Ferric Cycles in Iron Metabolism

We have become accustomed to think of the interconversion of these two oxidation states of iron in relation to the role of the cytochromes in the hydrogen-electron transport system, but in considering the metabolism of iron we have just begun to appreciate the central role of the interconversion of ferrous to ferric ion. This is illustrated in Fig. 1, modified from a recent paper (Frieden, 1973a). This has been discussed in detail in the earlier paper and here we need only note that iron must go through ferrous ion to be mobilized from ferritin or to be incorporated into the most prevalent iron compounds, hemoglobin, myoglobin and ferritin. For storage as ferritin or for transport as transferrin, iron must be converted to the ferric form. Thus, at virtually every known step in iron metabolism including storage, transport, biosynthesis and degradation, ferrous to ferric cycles appear to play a most significant role in iron pathways.

Ferro-oxidation and Ferri-reduction

Let us focus on these two reactions. These two basic reactions are the oxidation of ferrous to ferric ion and the reduction

TABLE 1

IRON COMPOUNDS IN MAN (150 LBS.)

Compound	& Type	Total g.	Iron g.	% Total Iron	Function
Hemoglobin	Heme-protein	750	2.6	65	Oxygen transport in blood
Myoglobin	Heme-protein	40	.13	3	Oxygen storage in muscle
Ferritin	Fe(III)+protein	2.4	.52	13	Iron storage
Hemosiderin	Fe-protein	1.6	.48	12	Iron storage excess
Transferrin	Fe(III)-protein	10	.006	<1	Iron transport
Iron Enzymes	Heme, non-heme proteins	10	.01	<1	Oxidative, other enzymes
Unidentified	Mostly protein	25	0.2	5	Probably enzymes or useful proteins

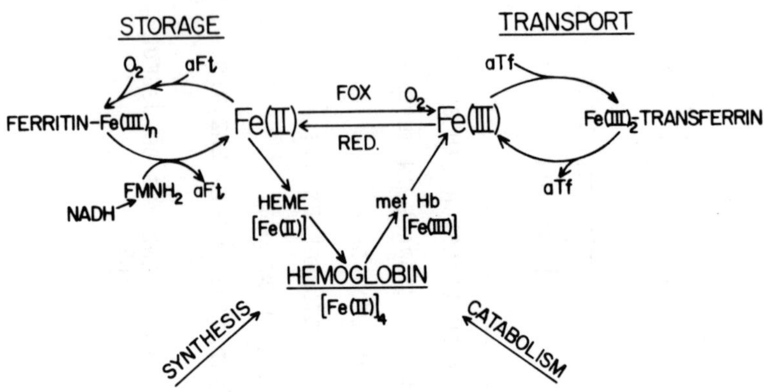

Figure 1. The ferrous to ferric cycles in iron metabolism and their importance in the storage, transport, biosynthesis, and catabolism of iron. A key influence is exerted by the ferroxidases (Fox I, etc.), copper enzymes which catalyze the oxidation of Fe(II) to Fe(III) and the ferrireductases. [Diagram modified from Frieden, 1973a].

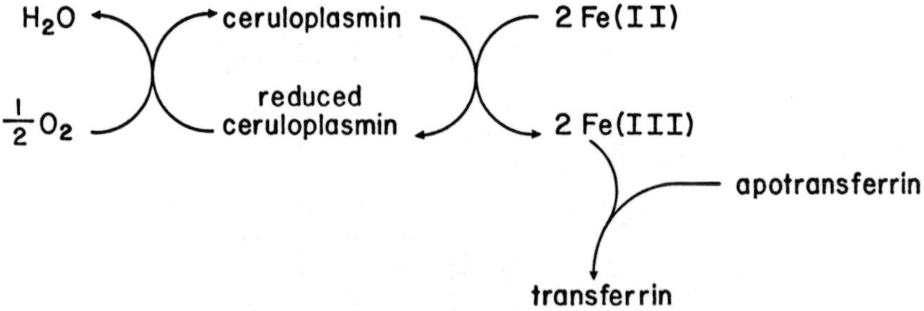

Figure 2. Reaction sequence showing how ceruloplasmin (Fox) activity enhances the rate of formation of $Fe(III)_2$-transferrin. Ferroxidase, Ferrous-O_2-oxidoreductase, E. C. 1.16.3.1.

of ferric to ferrous ion. The factors which influence these reactions most significantly are the chemical identity of the ionic iron--what type of complex or chelate? How the ionic state of iron affects its solubility in the physiological pH range? What are their relative oxidation-reduction potentials? The role of oxygen and reductants also must be considered and, finally, the question of whether these are exclusively enzymic processes?

Now let us delineate how these reactions are regarded as being essential for the transport, storage and release of iron and even biosynthetic reactions leading to iron proteins. In Table 2, we list the key reactions involved in the transport of iron as ferri-transferrin, its storage as ferritin and its release from ferritin via a ferri-reductase system. In the formation of Fe(III)-Tf, the copper protein, ceruloplasmin, which has ferroxidase activity, appears to be significantly involved. Next the production of ferritin seems to involve a ferroxidase activity of apoferritin prior to the formation of native ferritin. Finally, the release of iron from ferritin involves an NAD-FMN ferritin reductase system in which $FMNH_2$ becomes the direct reductant of the Fe(III) of ferritin resulting in its release as Fe(II).

Ferroxidase Activity

We will consider, first, ferroxidase activity. In 1966, we (Osaki *et al.*, 1966) proposed that the serum copper protein, ceruloplasmin, might be involved in iron metabolism by affecting the rate of ferri-transferrin formation. This idea was primarily stimulated by an appreciation of the ferroxidase activity of ceruloplasmin and the use of apo-transferrin for the determination of its oxidase activity as shown by the sequence in Fig. 2. How does ceruloplasmin mobilize iron? Is it due to its ferroxidase activity or is some specific responsive receptor site(s) on the iron storage cells activated by the copper protein? This kind of question is difficult to answer with any certainty. Ceruloplasmin has two important properties which are related to iron mobilization: (1) its well-documented ferroxidase activity, discovered by Curzon (1961), and confirmed and extended in a series of papers from our laboratory (Osaki *et al.*, 1966; McDermott *et al.*, 1968; Huber and Frieden, 1970a, 1970b), and (2) its almost unique ability to complex with Fe(II). The details of the ferroxidase reaction have been studied extensively by Osaki and coworkers (Osaki, 1966; Osaki and Walaas, 1967). The rate constants for the key steps in the Fox reaction are shown in Figure 3. The formation of the Fe(II)-Fox complex, the first reaction is extremely fast ($k_1 \geqq 10^7$ M^{-1} $sec.^{-1}$) and may be a limiting factor in iron mobilization from the iron storage cells in the liver. The subsequent steps are believed to reflect the control of Fe(II) oxidation and

TABLE 2

BASIC REACTIONS INVOLVED IN IRON TRANSPORT, STORAGE AND RELEASE

<u>TRANSPORT VIA TRANSFERRIN (Tf)</u>:

1) $Fe(III) + aTf + HCO_3^- \longrightarrow Fe(III)_2 - Tf - (HCO_3^-)$

2) $Fe(II) + Fox \longrightarrow [Fe(II)\text{-}Fox] \longrightarrow Fe(III) + Fox;\quad Fe(III) \xrightarrow{aTf} Fe(III)_2 - Tf$

3) $Fe(II) + aTf \longrightarrow Fe(II)\text{-}aTf \longrightarrow Fe(III)_2 - Tf$

<u>STORAGE VIA FERRITIN (Ft)</u>:

1) $Fe(II) + aFt \longrightarrow Fe(II) - Ft \longrightarrow Fe(III)_n - Ft$

<u>RELEASE FROM FERRITIN</u>:

1) XANTHINE OXIDASE SYSTEM

2) <u>FERRIREDUCTASE SYSTEM</u>:

$Fe(III)_n - Ft + FMNH_2 \longrightarrow 2Fe(II) + aFt + FMN$

$NADH + FMN + H^+ \xrightarrow[\text{REDUCTASE}]{\text{NADH-FMN}} NAD^+ + FMNH_2$

the rate of formation of Fe(III)-Tf in plasma. The slowest step (k_4 = 2.3 sec.$^{-1}$) is a conformational change in the reduced form of Fox which appears to be required prior to the reoxidation of the enzyme. Either or both of these properties could provide the impetus for the increased rate of removal of Fe(II) from liver iron storage cells and the subsequent incorporation of iron into transferrin.

In our paper in 1966, we reported a careful comparison of the rate of formation of ferri-Tf starting with Fe(II) with and without ceruloplasmin (Cp) (Osaki, *et al.*, 1966). At pH 7.4-7.5, phosphate buffer (a powerful stimulator of ferrous to ferric auto oxidation by O_2), 30°, apotransferrin = 55-100 μM, oxygen = 100 μM, a rate constant 0.5-1.1 min.$^{-1}$ was estimated without Cp and a rate constant of 12 min.$^{-1}$ for Cp. Assuming 3.3 liters of plasma, the daily Fe(III) generated would be 2-1/2 - 5 mgs. per day without Cp and over 60 mgs. per day with 2 μM Cp, the normal plasma concentration. The estimated daily requirement for iron is 35-40 mgs. per day. Thus, despite any enhancing effect of apoTf on Ferri-Tf formation from Fe(II), Cp may still be required to provide the capacity of the serum for Fe(II) oxidation (Osaki *et al.*, 1966). Crucial to these calculations is the estimate of serum free Fe(II) used. This is based on the normal urinary excretion which would arise from an estimated plasma level of 100 ug/100 ml or 20nM. The rate of formation of Fe(III)-Tf without Cp, 2-5 mgs. per day, is less than one-tenth the minimum estimate of the amount of iron that must pass through Fe(III)-Tf in one day. The Cp catalyzed system supplies about twice the daily requirement of Fe(III)-Tf. Other Fox activity in serum might also increase the capacity for Fe(II) oxidation (Planas and Frieden, 1970).

Bates *et al.*, (1973), has proposed that Fe(II) forms a weak complex with apoTf and this complex results in a faster oxidation of Fe(II) to Fe(III) with the subsequent formation of $Fe(III)_2$-Tf as:

$$2Fe(II) + ApoTf \longrightarrow Fe(II)_2\text{-Tf} \xrightarrow{1/2\ O_2} Fe(III)_2\text{-Tf} + H_2O$$

Bates *et al.* terms this effect a "ferroxidase activity" and contended that Fe(II) must be attached to apo-transferrin prior to oxidation to form ferri-transferrin at pH 7.4, unless a chelator was present, based on the fact that little color formation was observed in a mixture of Fe(II), apotransferrin, and H_2O_2, which served as an oxidant. Based on their observations, they concluded that it was doubtful that ferroxidase, Cp, has any effect on the formation of ferri-transferrin *in vivo*.

In our opinion their argument does not apply to the actual conditions in plasma because they used a non-physiological oxidant,

H_2O_2. When we used precisely the same conditions as reported by Bates *et al.*, (1973), except for the substitution of human ferroxidase, 0.20-2.3 μM, instead of 200 μM H_2O_2, Fe(II) was rapidly and stoichiometrically converted to ferri-transferrin, with the speed depending on the enzyme concentration (Osaki and Frieden, 1973). A kinetic analysis of this data excluded the possibility that ferroxidase catalyzes the oxidation of Fe(II)-transferrin. The data also suggested that the enzyme not only oxidized iron but effectively retarded its polymerization.

The fact also remains that the calculations summarized previously were made with apoTf present and in phosphate buffer and that the rate of ferri-transferrin formation without ferroxidase still was inadequate to account for the amount of iron required in the form of $Fe(III)_2$-Tf. Finally, we emphasize that these comparisons were made in the absence of Cp, which would compete successfully with apoTf for any Fe(II) (See Fig. 3).

These considerations emphasize the necessity of making a distinction between a true ferroxidase and a "ferroxidase" or iron oxidizing effect. Complexes, or chelators, preferential for Fe(III) will produce a "ferroxidase" effect but do not involve an intermediate catalytic protein which turns over and can be used over and over again. The ferroxidase activity of ceruloplasmin constitutes an authentic catalytic turnover of Fe(II) as indicated by the reactions outlined in Fig. 4.

An appropriate derivation from relatively simple assumptions (Table 3) shows that any effect on the system which tends to reduce free Fe(III) concentration will have a positive effect on the rate of Fe(II) disappearance or Fe(III)-complex formation. The assumptions made in this derivation are the steady state of the Fe(III) and the rate of formation of the ultimate Fe(III) product. This will produce an apparent iron oxidizing activity but does not prove a catalytic role, in the rigorous sense, for the complexing agent.

Ceruloplasmin (Ferroxidase) and Iron Mobilization

It is now established that ceruloplasmin is a molecular link between iron and copper metabolism, a nutritional fact known for almost half a century. This has been confirmed by numerous *in vitro* and *in vivo* experiments (Osaki *et al.*, 1971; Ragan *et al.*, 1969; Roeser *et al.*, 1970) and summarized in recent reviews (Frieden and Osaki, 1970; Frieden, 1971). Strong evidence that Cp (ferroxidase) plays a direct role in plasma iron levels comes from the recent work of the group at the Univ. of Utah. Ragan *et al.* (1969) studied the effect of injected copper and Cp into copper-deficient swine primed with adequate stores of iron. In addition

Table 3

IRON OXIDIZING EFFECT OF Fe(III) CHELATORS

$$Fe(II) + H^{+} + O_2 \underset{k_{-1}}{\overset{k_1}{\rightleftharpoons}} Fe(III) + O_2^{-}$$

$$Fe(III) + aTf \xrightarrow{k_2} Fe(III){-}Tf$$

$$\frac{dFe(III)}{dt} = k_1 \left[Fe(II)\right] (H^{+})(O_2) - k_{-1}\, Fe(III) - k_2 (FeIII)(aTf)$$

ASSUME STEADY STATE SO $\frac{dFe(III)}{dt} = 0$ $\quad H^{+}$, $_{P}O_2$ CONSTANT

$$0 = k_1' \left[Fe(II)\right] - k_{-1} \left[Fe(III)\right] (O_2^{-}) = k_2 \left[Fe(III)\right] (aTf)$$

$$v_3 = k_2 \left[Fe(III)\right] (aTf) = k_1' \left[Fe(II)\right] - k_{-1} \left[Fe(III)\right] (O_2^{-})$$

ASSUME $v_3 = \frac{-dFe(II)}{dt} = \frac{+d\left[Fe(III)Tf\right]}{dt}$

ANYTIME Fe(III) IS REDUCED BY (aTf) CHELATION, v_3 ALSO INCREASES

$$Fe(II) + Fox \xrightarrow[fast]{k_1} Fe(II){-}Fox \xrightarrow{k_2} Fe(III){-}Fox_r \xrightarrow{k_3} Fox_r \xrightarrow{k_4} \left[Fox_r'\right] \xrightarrow[O_2 \to H_2O]{k_5} Fox$$

$$Fe(III) + aTf \longrightarrow Fe(III)Tf$$

$k_1 = 10^{7} M^{-1} sec^{-1}$

$k_3 = 13\ sec^{-1}$; $k_4 = 2.3\ sec^{-1}$; $k_5 = 5.7 \times 10^{8} M^{-1} sec^{-1}$

A LIMITING FACTOR IN IRON MOBILIZATION BY... LIVER IRON STORAGE CELLS

CONTROL OF IRON OXIDATION AND RATE OF FORMATION OF Fe(III)-Tf IN PLASMA.

Figure 3. The rate constants of the various steps in the ceruloplasmin (Fox) catalyzed oxidation of Fe(II) to Fe(III). See Frieden and Osaki (1970) for details.

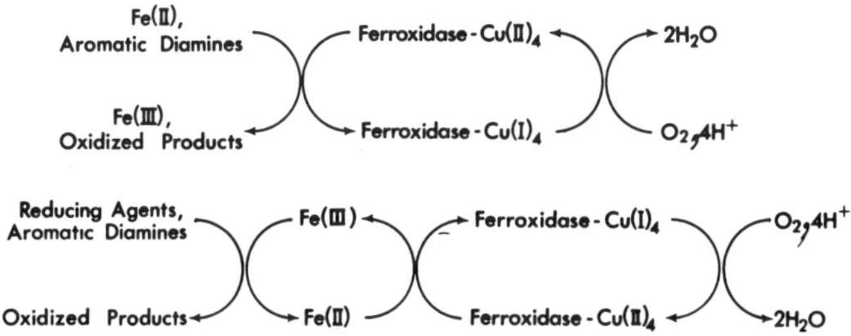

Figure 4. A summary of the various substrate groups and reactions of ferroxidase (ceruloplasmin).

to the typical copper and iron deficient milk diet, each pig was given intramuscularly a total of 2.0 g iron as iron dextrin (Pigdex) from 5-30 days of age. When a profound state of copper deficiency (80 days) was evident, the effects of injected pig Cp, $CuSO_4$ or Cu-deficient pig plasma were determined (Fig. 5). The amount of Cp injected was only enough to increase the plasma concentration to 15 μg% or 10% of the normal level. A remarkably rapid rise in plasma iron accompanies the Cp injection with a peak in 3-4 hours. The increase in plasma iron is significant after 5 min. The maximum increase in plasma iron is about twice the normal level and persists for 6 days. Neither $CuSO_4$ nor other pig plasma factors could produce this increase in iron. In fact, $CuSO_4$ actually reduced the iron levels after 2 days, presumably by stimulating RBC formation.

Roeser and coworkers (1970) have performed many additional *in vivo* experiments which further support the role of ferroxidase in iron metabolism. In copper deficient pigs, the Cp level falls to less than 1% of the normal, to usually about .5% ferroxidase activity. This deficiency of serum ferroxidase precedes the development of hypoferemia in the copper deficient pig with the accumulation of iron in liver. They demonstrated that a rise in serum ferroxidase activity precedes a rise in serum iron following copper injection. Serum iron does not increase until the serum Cp reaches about 1% of normal. However, any hypoferemia can be corrected immediately *in vivo* by the administration of Cp. intravenously injected. Fe(II) disappears rapidly from the circulation in the absence of ferroxidase activity and does not bind as readily to apotransferrin as iron injected Fe(III). In other words, in the absence of adequate ferroxidase, while there was no difference between control and copper deficient pigs in serum levels maintained when Fe(III) is injected, there was a 50% reduction in serum iron levels when Fe(II) was injected. This was interpreted as a demonstration of the direct physiological role of Cp in the control of serum Fe. Finally, they prepared asialo-Cp which was rapidly removed by the liver and shows little or no iron mobilizing activity when injected although it does have ferroxidase activity in *in vitro* tests. This supports the idea that Cp must function in the circulatory system. These comprehensive experiments by the Utah group lead to the conclusion that the defect in the release of iron in the copper deficient animal can be reversed promptly by intravenous Cp, an effect that could not be due to copper alone. In no case did any of their *in vivo* observations conflict with the hypothesis that the ferroxidase activity of Cp is directly involved in iron mobilization.

A study of iron mobilization by perfusing livers from normal dogs and normal or Cu-deficient pigs was completed during the same period in our laboratory (Osaki *et al.*, 1971). The excised

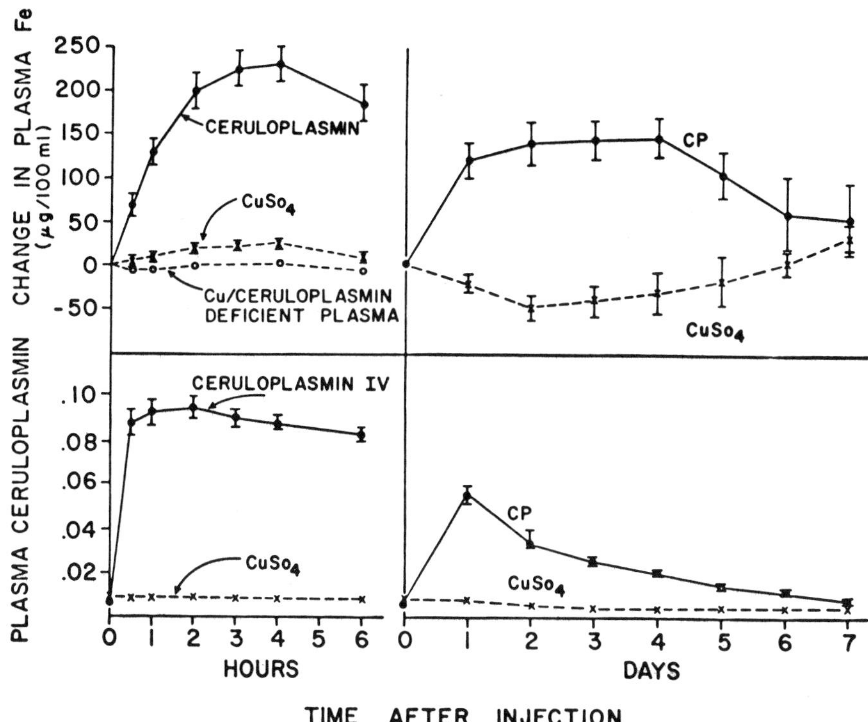

Figure 5. The effect of the intravenous administration of ceruloplasmin (e——e), copper sulfate (x-----x), or copper-deficient plasma (o-----o) on the plasma iron and plasma ceruloplasmin (expressed as PPD oxidase activity) in copper-deficient swine. Sufficient Cp or $CuSO_4$ were given to achieve 10% of the normal level (15 g/100ml plasma). Brackets indicate standard error of the mean. If deviation is within the symbol it is not shown. [From Ragan *et al.*, 1969].

liver was thoroughly flushed with an appropriate perfusion medium and the perfusate was then allowed to recycle and apotransferrin and other supplements were added. The iron appearing in the perfusate was detected in a flow cell at 460 nm as the Fe(III)-transferrin complex or at the 530 nm band of the α,α-dipyridyl complex using an aliquot of the perfusate.

A recording of the time course of the absorbance change at 460 nm is shown in Fig. 6. Each arrow indicates the time when an additional infusion was made in the reservoir. First 3 μM apo-transferrin; next 3.5 mN bicarbonate, and 800 μM ascorbate; with the last infusion of 4.9 μM Cp, an almost instantaneous rise in the 460 nm band was observed. The rate of Fe(III)-transferrin formation was estimated to be 3 μM per 100 sec. or about 220 μg iron released from the liver in 20 min. If man were to mobilize Fe at the same rate as observed in normal dog liver, up to 192 mg Fe/day could be produced or six times the estimated daily need for iron.

These experiments were extended to find out how sensitive iron efflux from perfused livers was to Cp concentration. The rate of iron efflux was determined from the slopes of the recorded responses as in Fig. 7. These were plotted as relative rates versus Cp concentration (shown in Fig. 8). The effect of ferroxidase was observed at concentrations as low as 4×10^{-9} M with a maximum effect at 2×10^{-7} M, which is 10% of the normal human serum level. This data corresponds closely to the *in vivo* results of Roeser *et al.* (1970) mentioned earlier in which only 1/10 of the normal level of Cp was required to produce maximum iron mobilization. Thus, normal serum has a ten fold excess of Cp as far as the apparent maximum requirement for iron mobilization. However, an appreciable response is noted at 1% of the normal Cp level.

The specificity of the iron mobilization response in the perfusate system was also studied. Only Cp among the compounds tested proved to have any activity in the perfused livers. No activity was shown by 30 μM apo-transferrin, HCO_3^-, 10μM $CuSO_4$, 5 mM glucose, 0.6 mM fructose, 120 μM citrate, or 36 μM bovine serum albumin ± 21 μM $CuSO_4$.

In the growing rat, Evans and Abraham (1973) showed that Hb and Cp levels in the blood parallel one another during copper deprivation and repletion (Fig. 9). Low copper leads to low blood Cp and to high liver Fe, since it can not be mobilized (Evans and Abraham, 1973). All these deviations from normal metabolism were corrected after several days feeding with 35 ppm $CuCO_3$.

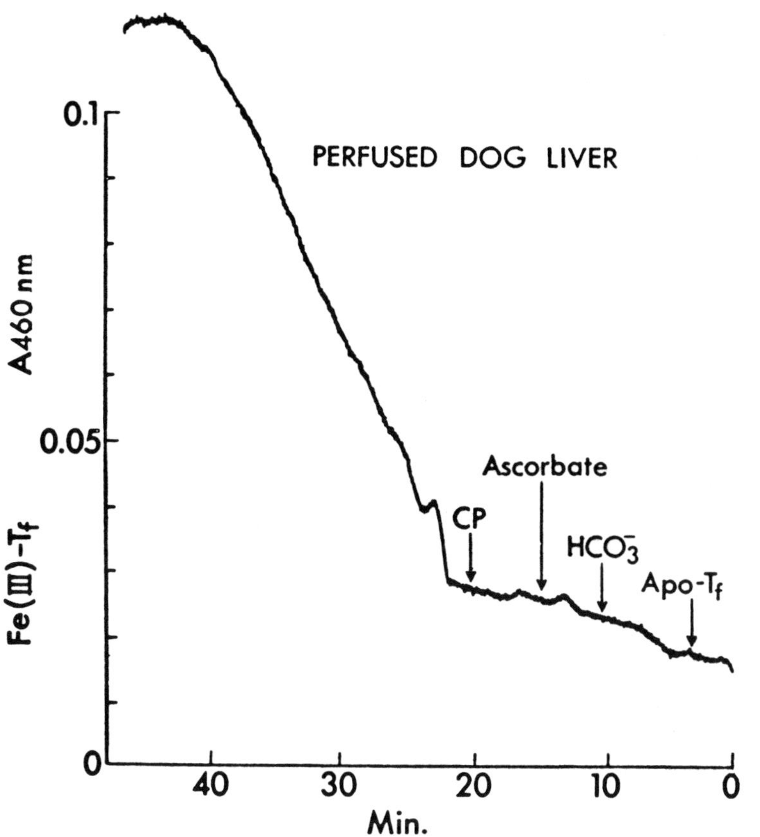

Figure 6. Mobilization of Iron by ceruloplasmin (CP) in the perfused dog liver. Time course of the absorbance change at 460 nm [Fe (III)-transferrin formation].
[Data from Osaki *et al.*, 1971].

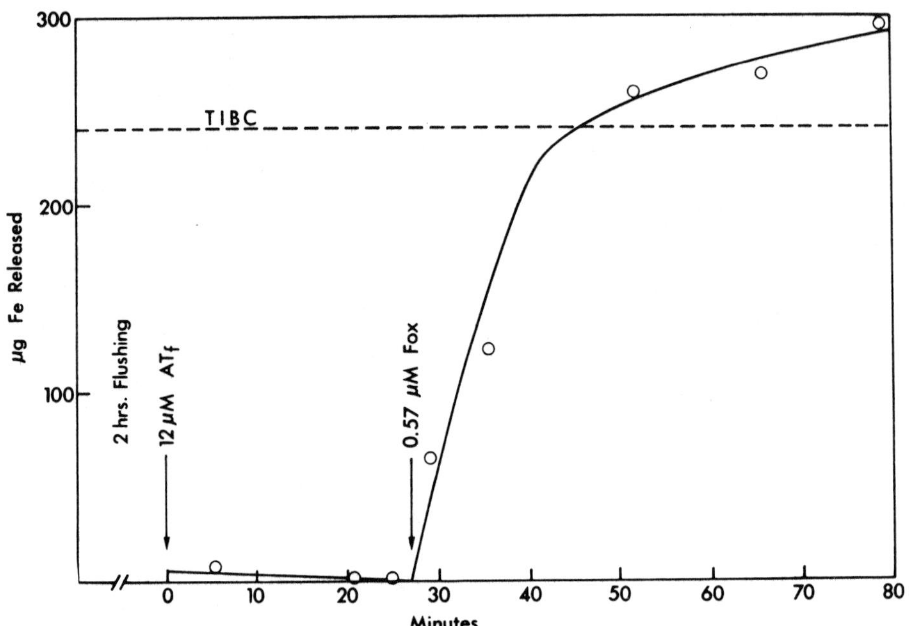

Figure 7. The effect of ferroxidase I on a perfused liver prepared from a copper-deficient pig. The arrows in the figure indicate the times when the infusions were made. The total volume of the perfusate was 350ml and its TIBC was 245µg of Fe(III) per total perfusate. The mobilized iron exceeded TIBC by 22%. ATf and Fox represent apotransferrin and ferroxidase I respectively. [Data from Osaki et al., 1971].

Incorporation of Iron into Ferritin

The uptake of iron by apoferritin requires Fe(II), and two groups (Macara et al., 1972; Bryce and Crichton, 1973) have reported that a "catalytic" effect of apoferritin on the oxidation of Fe(II) aids in the formation of Fe(III)-ferritin. The enhancement of Fe(II) oxidation by molecular O_2 occurs under conditions when several other proteins have no effect, as shown in the accompanying Fig. 10. The rate of oxidation is directly proportional to apoferritin concentration. Crichton (1973) has also recently emphasized that this particular ferroxidase activity could greatly influence intracellular iron metabolism. While this hypothesis remains to be tested, we must point out that a true ferroxidase activity has not been proved and that the role of apoferritin could resemble the "ferroxidase effect" described previously for apotransferrin.

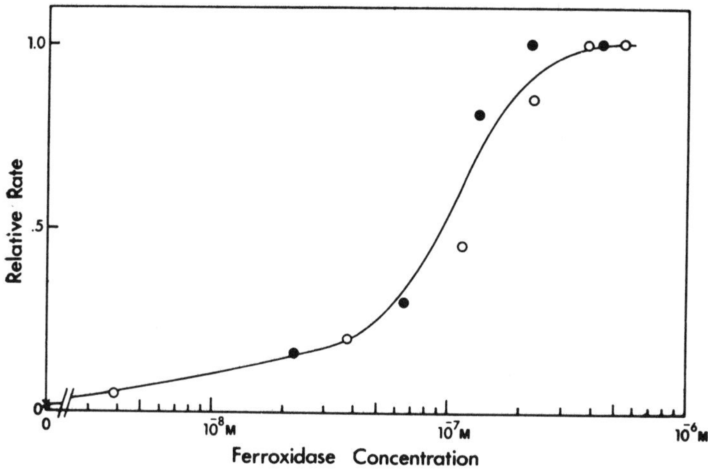

Figure 8. The effect of perfusate ferroxidase I concentration on iron mobilization from dog livers. The open and filled circles represent two independent experiments. An additional observation at 1.2μM ferroxidase I gave the same relative rate of 1.0 (data not shown). At zero enzyme concentration (X), the rate of iron inflow into the perfusion system was less than 0.15μM of Fe(III)/100 sec (a relative rate of 0.030) or about one-third of the value observed at 4μM enzyme concentration. [Data from Osaki et al., 1971].

Distribution and Role of Ferritin

The importance of intracellular ferritin in iron metabolism has been emphasized as a result of finding ferritin widely distributed in the animal and plant kingdom, from the lowest plant to the highest vertebrate, and also the fact that ferritin is found in virtually every tissue studied so far (Crichton, 1973). Liver, spleen and bone marrow are still regarded as the iron storage tissues because of their accumulation of the greatest amounts of ferritin. It is also found in serum and red blood cells and the possibility of a correlation between the serum ferritin level and the overall state of iron metabolism has been suggested by Jacobs et al., (1972). Though its role in the intestinal mucosal block has been questioned, it is still believed that an appreciable fraction of iron passes through the ferritin of the mucosal cell prior to absorption of iron (Hübers et al., 1971), (see Fig. 5).

The synthesis of ferritin has proved to be subject to sensitive regulation. Exposure to iron results in an increased rate

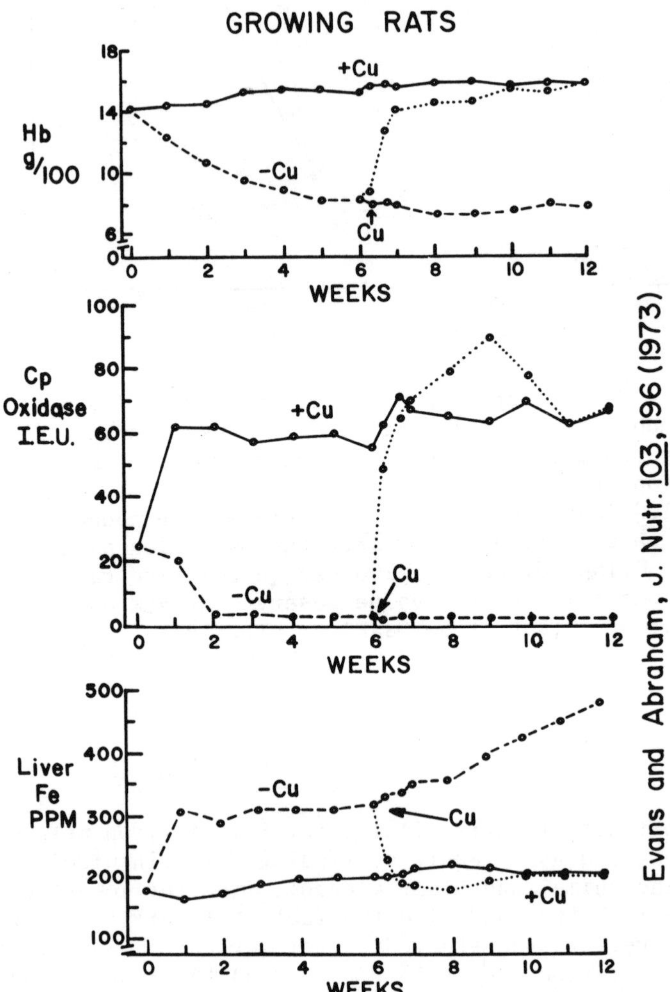

Figure 9. Changes in hemoglobin levels, plasma ceruloplasmin, and liver iron associated with dietary copper. The recovery of copper-repleted rats is indicated by an arrow. Copper fed———, copper depleted -----, and copper repleted ····· [Data from Evans and Abraham, 1973].

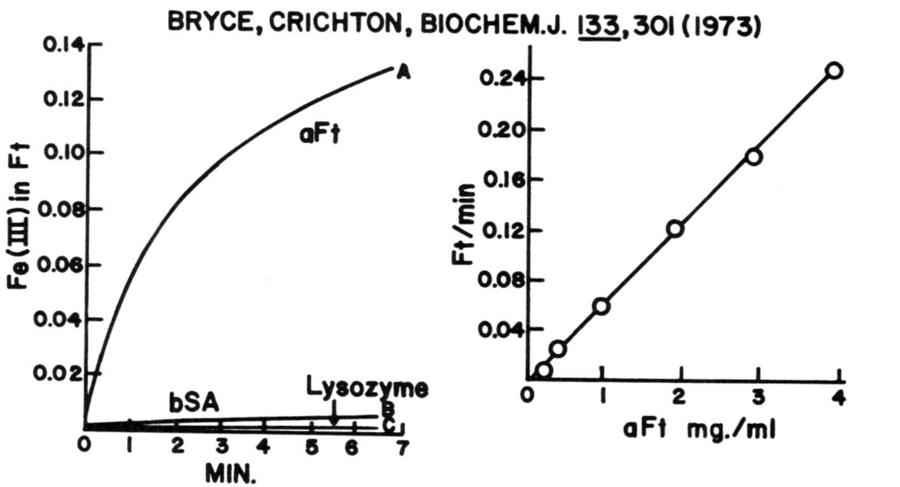

Figure 10. Effect of the addition of several proteins on the oxidation of Fe^{2+} in 100mM borate---cacodylate buffer, pH 5.50. Effect of apoferritin concentration on the initial velocity of Fe^{2+} oxidation. Apoferritin concentrations of 0.1-4mg/ml were used and the initial velocities calculated. [Data from Bryce and Crichton, 1973].

of apoferritin biosynthesis, primarily due to an effect at the translational level in tissues such as liver, reticulocytes, kidney, heart, and intestinal mucosa (Drysdale and Munro, 1966). The type of ferritin synthesized in liver cells changes with development. For example, ferritin from the livers of young rats is similar to that of adult kidney and spleen rather than to adult liver ferritin (Linder *et al.*, 1970). Two specific isoferritins have been recently identified in human and horse liver and spleen. In certain cancer cells, normal ferritin is replaced by electrophoretically distinct forms (Linder *et al.*, 1970).

Ferritin Reduction

Since iron is stored in the ferric form as ferritin and was presumed to be most stable in this form, early studies on the release of iron from ferritin were directed toward reductive mechanisms. In 1943, Michaelis *et al.* reported that iron was liberated from ferritin *in vitro* by the reducing action of sodium dithionite. Later, Tanaka (1950), found that a reducing process was necessary to release iron from ferritin. He reported that milk xanthine oxidase and a diaphorase preparation from pig heart muscle were active in the release of iron from ferritin. Tanaka suggested that the enzyme involved in the liberation of ferritin iron would be either a pyridino- or flavo- protein. The biochemical mechanism for the release of iron from tissue ferritin to plasma transferrin was investigated more extensively by Mazur and coworkers (Green and Mazur, 1957; Mazur *et al.*, 1958). They confirmed the fact that ferrous ion could be liberated from ferritin by the action of milk xanthine oxidase. Using rat liver slices *in vitro* and rabbit, guinea pig and dog liver *in vivo* , they attempted to correlate action of xanthine oxidase in these tissues and in iron release. As shown in Fig. 11, ferritin would accept electrons from reduced xanthine oxidase and thereby release its iron as ferrous ion. Correlations of iron metabolism and xanthine oxidase were observed in the early developmental period of the rat, (Mazur and Carlton, 1965). In the fetal rat, iron needs are satisfied by the maternal plasma iron and ferritin iron accumulates in the complete absence of xanthine oxidase. Xanthine oxidase appears in the liver after birth and liver ferritin iron undergoes a precipitous fall. After weaning, dietary iron provides sufficient iron for hemoglobin biosynthesis and for storage. However, these correlations, as interesting as they are, do not prove a direct molecular link between xanthine oxidase and ferritin reduction.

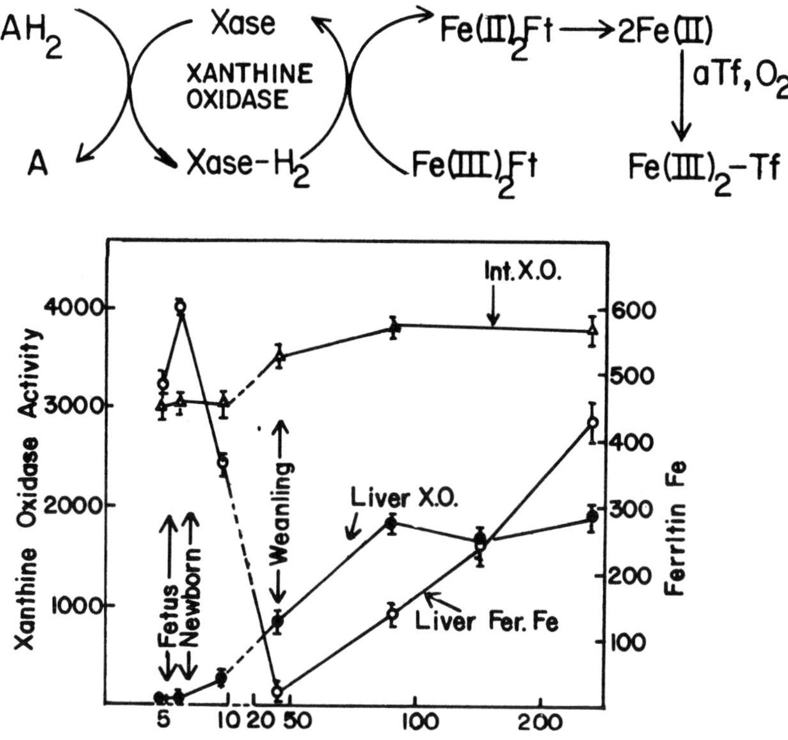

Figure 11. The mechanism of ferritin iron release proposed by Mazur and coworkers. The graph is taken from Mazur and Carlton (1965) and represents ferritin iron and xanthine oxidase activity changes in the developing rat.

In contrast, Kozma et al. (1967), reported that liver xanthine oxidase activity could be substantially reduced by allopurinol (a powerful inhibitor of xanthine oxidase) without noticeably affecting iron metabolism. More recently, Grace et al. (1970), also showed that the suppression of xanthine oxidase activity by allopurinol did not influence the mobilization of hepatic iron. We have also found that in liver homogenates, xanthine oxidase substrates did not produce significant release of iron from ferritin; neither did milk xanthine oxidase or chicken liver xanthine dehydrogenase. However, we remained convinced from observations on the mobilization of iron from perfused livers (Osaki et al., 1971), that a reductive mechanism must be available for the release of ferritin iron.

A Ferritin Reductase System

An enzyme(s) system which can reduce ferritin Fe(III) has been demonstrated in homogenates from the livers of various species of vertebrates (Osaki and Sirivech, 1971; Osaki and Frieden, 1973), cow, pig, dog, rat, chicken, tadpole, and frog. This sytem catalyzes the complete reduction of ferritin bound Fe(III) to free Fe(II) in the presence of NADH and FMN, FAD or riboflavin, at oxygen concentrations less than 3 μM. The isolated enzyme is NADH-specific--NADPH, hypoxanthine, succinate, ethanol, or glucose are not utilized as electron donors. This activity is not inhibited by 1 mM allopurinol and does not use hypoxanthine as an electron donor, indicating that the activity is not associated with liver xanthine oxidase. This enzyme system has been purified more than 350 fold from a beef liver microsomal fraction. The pH optimum for ferritin reduction with the beef liver enzyme is 7.2-7.8. The enzyme has been also differentiated from microsomal NADH-cytochrome b_5-oxidoreductase by several criteria. Another FMN dissociable system was examined and it was found that bacterial luciferase (Gibson *et al.*, 1966) can reduce ferritin in a manner similar to the microsomal enzyme. The ferritin reducing system thus requires an NADH-FMN oxidoreductase in a sequence of reactions as follows:

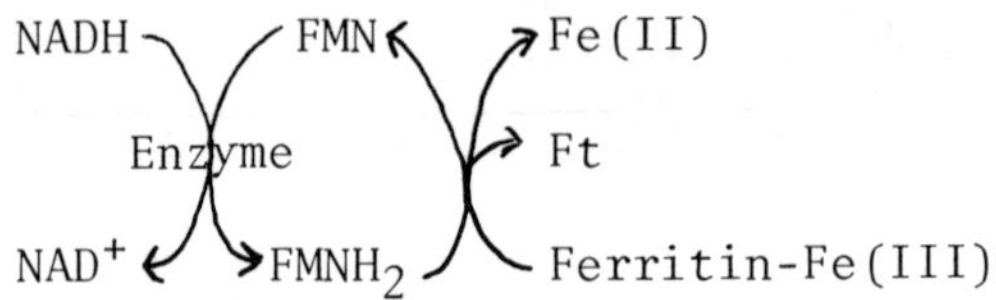

The first reaction requires the microsomal enzyme described, but the second step can be studied independently as a separate reaction. We have made a comparison of the reaction of horse spleen ferritin with riboflavin derivatives and with typical tissue reducing agents and found that only the riboflavins can reduce ferritin-Fe(III) at a rate and to an extent that is likely to be significant physiologically (Table 4) (Sirivech *et al.*, 1974). Ferritin Fe(III) was rapidly and quantitatively reduced and liberated as Fe(II) by $FMNH_2$, $FADH_2$ and reduced riboflavin. Dithionite also released Fe(II) from ferritin but at less than one percent of the rate with $FMNH_2$. Cysteine, glutathione and ascorbate gave a similar slower rate and yielded less than 20% of the total iron from ferritin. The reduction of ferritin-Fe(III) by the three riboflavin compounds and dithionite gave complex second order kinetics with overlapping fast and slow reactions. The fast reaction appeared to be due to a reduction of Fe(III) of a lower degree of polymerization, equilibrated to ferritin iron. The amount of ferric ion initially reduced was small, in the range of less than 0.3 percent of the total iron. Addition of FMN to the ferritin-dithionite system was found to enhance the reduction, apparently

TABLE 4

RATES AND RATE CONSTANTS FOR THE REDUCTION OF FERRITIN BY VARIOUS REDUCING AGENTS

The values for reduced flavins and sodium dithionite were obtained at 15° using a stopped flow apparatus. The experiments were performed at 30° using ascorbate, cysteine and glutathione in a Cary 15 spectrophotometer. The ferritin concentration was 4.74 mM as Fe(III) and the pH was 7.4 in 25 mM Tris buffer.

Compounds	Concentration	Rate	k
	μM	$\Delta A_{530}sec^{-1}$ x 10^{-3}	$M^{-1}sec^{-1}$
$FMNH_2$	50	7.1	10.8
$FADH_2$	50	2.5	3.1
Reduced Riboflavin	50	7.6	13.0
Dithionite	2000	2.4	0.052
Ascorbate	100	0.16	---
Cysteine	100	0.24	---
Glutathione	100	0.23	---

due to the reduction of FMN by dithionite to form $FMNH_2$ which then reduces ferritin Fe(III). The data does not contradict previous data on other reduction mechanisms reported by Mazur *et al.*, (1958) and the recent data of Duggan and Streeter (1973), but does explain why these non-specific mechanisms are not likely to prevail in tissue (Sirivech *et al.*, 1974).

A comparison of the thermodynamic parameters of $FMNH_2$-ferritin and dithionite-ferritin complex formation showed that $FMNH_2$ required a lower activation energy and a negative entropy change, whereas dithionite required 50 percent more activation energy and showed a positive entropy change in ferritin reduction. The effectiveness of $FMNH_2$ in ferritin Fe(III) reduction may be due to a specific binding of the riboflavin moiety to the protein portion of the ferritin molecule.

Ferritin Iron Reduction or Oxidation?

The question may also be raised as to what determines whether iron is incorporated by apoferritin or released from ferritin? If apoferritin enhances the oxidation of Fe(II) under certain conditions, why should it release Fe(II) under other conditions? While this question has not been directly explored experimentally, we can point to two factors which might determine the direction a Fe(II) ion may take. First, the degree of completeness or saturation of the ferritin molecule may affect the path of Fe(II). If the ferritin molecule is relatively saturated, the sites which bind Fe(II) may be occupied or shielded so that the flow of Fe(II) is outward from the inner iron core. Second, the concentration of oxygen and $FMNH_2$ could play a vital role. We have emphasized that the ferritin reductase system is highly sensitive to oxygen--showing strong inhibition when the oxygen concentration is above 3 μM. This may be due to a competition with Ferritin-Fe(III) by oxygen for $FMNH_2$ necessary for reduction. Thus, low oxygen concentration would favor the reduction and release of Fe(II) from ferritin, while higher oxygen (>3 μM) would slow the reduction of ferritin-Fe(III) and promote the oxidation of Fe(II) to Fe(III) and its assimilation into the ferritin macromolecule. Finally, we should note that Mazur *et al.* (1965) have observed that hypoxia stimulates the *in vivo* release of iron from storage cells.

Mobilization of Serum Iron and Ferroxidase in Birds

If the ferroxidase and ferrireductase systems play a vital role in iron metabolism, these activities and their metabolic sources and products, e.g. transferrin, ferritin, etc., may vary greatly under metabolic impetus. The possible importance of these systems in normal development has been emphasized by Frieden (1971, 1973a,b), Crichton (1973) and others. Though these correlations are at an early stage, we can point to several evidences for these effects.

In vertebrates the functional relationships between serum iron and ferroxidase activity, iron binding proteins and, ultimately, hemoglobin biosynthesis are being established. A recent study (Fig. 12) of the dramatic increase (5-10x) in serum iron ferroxidase activity and other blood parameters in normal, copper-deficient and estrogenized roosters was reported by Planas and Frieden (1973). Serum ferroxidase activity, iron, hemoglobin, and hematocrit values were greatly reduced when roosters were maintained on a copper and iron deficient diet for 17 days. Their anemic condition was exacerbated after 40 days, correlating with the total disappearance of ferroxidase activity. Dietary supplementation with copper, with or without iron, produced the largest increase in ferroxidase activity. The injection of copper salts also produced an increase in ferroxidase activity and, later, an increase in serum iron. The administration of estrogen, either as estradiol or diethylstilbestrol (DES), also induced the appearance of ferroxidase in both normal and copper or iron deficient animals. The maximum ferroxidase activity enhancement was obtained on the second day and the serum iron increase on the third day. There was no change in transferrin; apparently a more metabolically labile iron-binding protein, phosvitin, is produced to serve as a supplementary iron carrier. DES increased the ferroxidase activity almost to normal levels in copper and iron deficient roosters and those fed high doses of zinc and silver.

Planas (1973) has extended these observations and found a 5-10x increase in serum iron and copper and up to a 20x increase in ferroxidase serum activity in laying hens after estrogen treatment. Planas thus has confirmed the metabolic link between serum iron, copper and ferroxidase but his work also emphasizes the uniqueness of the special iron-binding protein, phosvitin, which is produced in the liver of several different vertebrates, e.g. birds and amphibians, in response to estrogen stimulation.

Changes in Iron and Copper Metabolism During Amphibian Metamorphosis

Remarkable changes in metalloprotein metabolism must occur during the metamorphosis of the amphibian tadpole (Frieden and Just, 1970), thus providing a useful model system for testing many functional relations of iron and copper proteins. There is a complete switch in the hemoglobin chains--the three tadpole hemoglobins have no chain in common with the 3-4 frog hemoglobins. Extensive iron reutilization must be involved since all new hemoglobin synthesis is believed to proceed *de novo*. The total resynthesis of

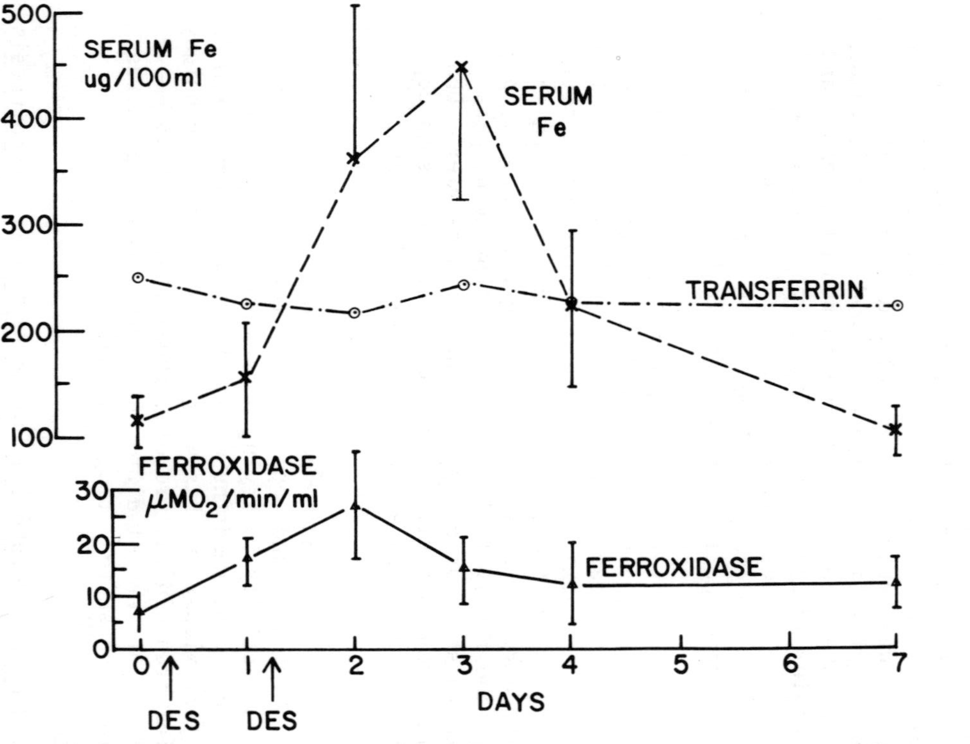

Figure 12. Effect of DES on serum ferroxidase, iron, and transferrin in the rooster. Each point represents average of 12 roosters with standard deviations indicated. DES, 2.0mg/kg, was injected at 0 and 1st day immediately after bleeding. In this experiment ferroxidase was determined using oxygen electrode. An identical pattern was obtained with experimental groups of 15 and 19 roosters using DES and 17β-estradiol at 2.0 and 1.0mg/kg, respectively, using spectrophotometric-transferrin method for ferroxidase measurement. [Data from Planas and Frieden, 1973].

hemoglobins coincides with the beginning of the metamorphic climax (stage XXI) and is preceded by several fold increases in ferroxidase activity and transferrin levels.

We have also studied changes in liver iron and ferritin reductase activity during metamorphosis (Osaki and Frieden, 1973), since this enzyme should play a role in the mobilization of iron for heme biosynthesis. Liver homogenates from bullfrog tadpoles were active in releasing Fe(II) from ferritin in the presence of NADH and FMN. This activity (per gram of tissue) declined by two-thirds as the young tadpole (stage V) approached metamorphic climax and increased five fold at the height of the metamorphic climax (stage XXIII). A rapid increase in liver iron at the beginning of climax (stages XX-XXI) is followed by the sharpest increase in ferritin reducing activity and then another sharp drop in liver iron due to iron mobilization. These phenomena coincided with or preceded the switch to adult hemoglobin.

Since these metalloprotein systems seem to be related in the evolutionary history of the iron and copper metalloproteins (Frieden, 1973b), we are seeking further evidence for the sequence-- liver iron → ferrireductase → ferroxidase → transferrin → hemoglobin.

Conclusion: Pathways of Iron Metabolism in Typical Mammalian Cells

With this background on the ferroxidases and the ferrireductases we will conclude with a description of the normal path of iron metabolism in typical prototype cells in man and related animals. Let us first consider a typical liver iron storage cell as diagrammed in Fig. 13. Transferrin delivers iron from one of its two sites releasing it to a specific receptor in the cell membrane. This process of iron exchange is possibly reductive, perhaps similar to the ferritin reductase system; it could also involve the release of an anion, usually bicarbonate ion, which is essential to iron binding. Ultimately, the iron is converted to Fe(II) which is required for interaction with apoferritin prior to oxidation and polymerization into ferritin. This iron can then be released on demand via the ferritin reductase system. Cytoplasmic levels of oxygen and $FMNH_2$ may play a major role in the regulation of ferritin reducing activity. Fe(II) can be used for any intracellular iron requirements such as iron enzymes or other iron proteins, or the Fe(II) can be mobilized with the aid of ferroxidase (ceruloplasmin) into the blood stream with its ultimate destination $Fe(III)_2$-transferrin.

The potential route of an iron atom is shown for a typical erythroid cell (Fig. 14) whose biosynthetic efforts are mainly directed towards the production of hemoglobin. Incoming iron, after conversion to Fe(II), may be used directly for heme production via the ferrochelatase system. The ferritin iron pool is still available with the appropriate ferritin reductase system to expedite the release of iron as Fe(II) as needed for hemoglobin or other iron utilizing systems.

The spleen cell diagrammed in Fig. 15 is known to represent a graveyard for old erythrocytes; thus iron is readily accepted but it is only sluggishly exported. It must have active iron incorporating systems possibly similar to those in the iron storage or red cells, but its iron releasing system is relatively inactive. In fact, the ferrireductase system described earlier cannot be detected in spleen cells. Fig. 15 includes the possibility of iron mobilization from spleen utilizing alternative routes which might involve low or high molecular iron transporting agents (X). As a matter of fact, an animal can survive without a spleen, but an enlarged spleen causes severe anemia, suggesting that mobilization of iron from spleen must be retarded.

The mucosal cells appear to have both transferrin-like and ferritin-like iron trapping mechanisms to satisfy its special function in iron absorption. The possible role of low molecular weight chelates of ferric ion, Fe(III)X in Fig. 15, e.g. ferric fructose, ferric citrate, have been emphasized by Saltman (1973). Endogenous complexers of Fe(II) could also be involved as suggested by Frieden and Osaki (1973). The presence of a ferroxidase activity in intestinal mucosal cells has also been suggested by Manis (1973). Mucosal cells may also have a special need for an active ferrireductase system to implement iron export. Potent activity of the NADH-FMN ferrireductase system has been observed in homogenates of rat duodenum.

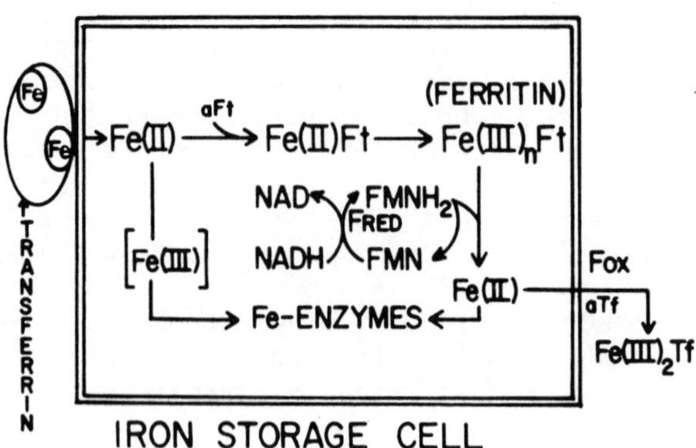

Figure 13. The possible metabolic paths of iron in iron storage cells, e.g. the reticulo-endothelial cells of the liver. See text for more details.

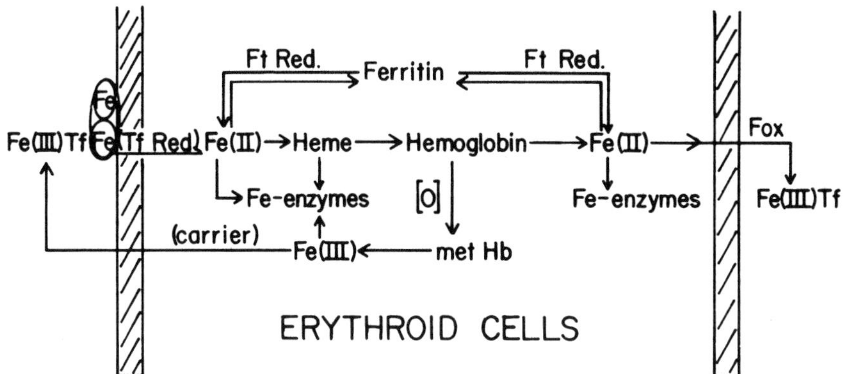

Figure 14. Iron metabolism in erythroid cells.

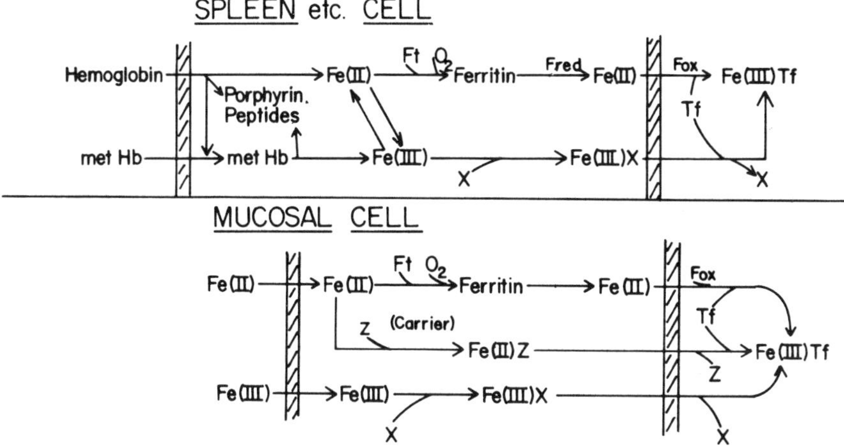

Figure 15. Iron metabolism in spleen and intestinal mucosal cells.

Summary

In this paper we have pointed out the significance of the ferrous to ferric interconversions in the iron metabolism of the higher animal. We have focused upon the processes of ferroxidation and ferrireduction, which seem to be highly essential for iron transport, storage and release and the numerous biosynthetic reactions leading to essential iron compounds, both micro- and macro-molecular. The history of the recognition of ferroxidase activity and the role of ceruloplasmin in mobilizing iron and in catalyzing the oxidation of ferrous to ferric ion is summarized. Quantitative estimations were presented illustrating how this reaction may be limiting in the rate of formation of ferri-transferrin. The *in vivo* and *in vitro* evidence relating ceruloplasmin to the mobilization of iron in the copper deficient pig and rat and in perfused livers was reviewed. A ferroxidase mechanism also seems to be involved in the incorporation of iron into ferritin. The distribution and the role of ferritin in iron metabolism also was emphasized and a new mechanism for ferritin reduction proposed and documented. This system involves an NAD, FMN reductase followed by a highly specific reduction of ferritin iron by $FMNH_2$. It was confirmed that only riboflavin derivatives can reduce ferritin iron at a rate and to an extent that is likely to be significant physiologically. The controlling factors which determine whether ferritin iron is reduced or oxidized were attributed mainly to the degree of completeness or saturation of the ferritin molecule and the concentrations of oxygen and $FMNH_2$. Examples of how ferroxidase and ferrireductase systems play a vital role in the iron metabolism of other vertebrate systems were outlined. The mobilization of serum iron, copper and ferroxidase in birds was discussed. Changes in iron and copper metabolism during amphibian metamorphosis also were described. We concluded with a description of the normal path of iron metabolism in typical prototype cells in man, including liver iron storage cells, erythroid cells, spleen cells, and intestinal mucosal cells. These considerations emphasize the importance of the ferroxidase and ferrireductase systems in those cells which accentuate iron metabolism.

ACKNOWLEDGEMENTS

This work was supported by NIH Grants HL-08344 from the National Heart and Lung Institute and HD-01236 from the Institute of Human Health and Development. This is paper No. 47, in a series on the biochemistry of the copper and iron metalloproteins.

REFERENCES

Bates, G.W., Workman, E.F., and Schlabach, M.R. (1973). Biophys. Res. Comm., 50, 84.

Bryce, C.F.A., and Crichton, R.R. (1973). Biochem. J., 133, 301.

Crichton, R.R. (1973). FEBS Letters, in press.

Curzon, G. (1961). Biochem. J., 79, 656.

Diamond, L.K. (1971). In Extent and Meanings of Iron Deficiency, published by the U.S. Nat. Acad. Sci., Washington, D. C. p. 19

Drysdale, J.W., and Munro, H.N.J. (1966). J. Biol. Chem., 241, 3630.

Duggan, D.E., and Streeter, K.B. (1973). Arch. of Biochem. & Biophys., 156, 66.

Evans, J.L., and Abraham, P.A. (1973). J. Nutr., 103, 196.

Frieden, E. (1973a). Nutr. Rev., 31, 41.

Frieden, E. (1973b). In TEMA II Symposium, in press.

Frieden, E. (1971). Adv. in Chem. Series, Bioinorg. Chem., Am. Chem. Soc., 292.

Frieden, E., and Just, J.J. (1970). Biochem. Actions of Hormones, (ed., G. Litwack), Academic Press, N.Y., pp. 1-52.

Frieden, E., and Osaki, S. (1970). In Heavy Metals and Cells, 2nd Rochester Conference on Toxicity, J. Wiley & Sons, p. 39.

Gibson, A.H., Hastings, J.W., Weber, G., Duane, W., and Massa, J. (1966). Flavins and Flavoproteins, (ed. E. Slater), p. 341.

Grace, N.D., Greenwald, M.A., and Greenberg, M.S. (1970). Gastroenterology, 59, 103.

Green, S., and Mazur, A. (1957). J. Biol. Chem., 227, 653.

Huber, C.T., and Frieden, E. (1970). J. Biol. Chem., 245, 3973.

Hübers, H., Hübers, E., Forth, W., and Rummel, W. (1971). Life Sciences, 10, 1141.

Jacobs, A., Millar, F., Worwood, M., Beamish, M.R., and Wardrop, C.A. (1972). Brit. Med. J., 862, 206.

Kozma, C., Salvador, R.A., and Elion, G.B. (1967). Lancet, 2, 1040.

Linder, M., Munro, H.N., and Morris, H.P. (1970). Cancer Res., 30, 2231.

Macara, T.G., Hoy, T.G., and Harrison, P.M. (1972). Biochem. J., 126, 151.

Mazur, A., and Carlton, A. (1965). Blood, 26, 317.

Mazur, A., Green, S., Saha, A., and Carlton, A. (1958). J. Clin. Invest., 37, 1809.

Mazur, A., Baez, S., and Shorr, E. (1955). J. Biol. Chem., 213, 147.

McDermott, J.A., Huber, C.T., Osaki, S., and Frieden, E. (1968). Biochem. Biophys. Acta, 151, 541.

Michaelis, L., Coryell, C.D., and Granick, S. (1943). J. Biol. Chem., 148, 463.

Osaki, S., and Frieden, E. (1973). Abstracts, Conference on Iron Storage and Transport Proteins, London, England.

Osaki, S., and Sirivech, S. (1973). Fed. Proc., 30, abstr. 1394.

Osaki, S., Johnson, D., and Frieden, E. (1971). J. Biol. Chem., 246, 3018.

Osaki, S. (1966). J. Biol. Chem., 241, 5053.

Osaki, S., and Walaas, O. (1967). J. Biol. Chem., 242, 2643.

Osaki, S., Johnson, D., and Frieden, E. (1966). J. Biol. Chem., 241, 2746.

Planas, J. (1973). Private communication.

Planas, J., and Frieden, E. (1973). Am. J. Physiol., 225, 423.

Ragan, H.A., Nacht, S., Lee, G.R., Bishop, C.R., and Cartwright, G.E. (1969). Am. J. Physiol., 217, 1320.

Roeser, H.P., Lee, G.R., Nacht, S., and Cartwright, G.E. (1970). J. Clin. Invest., 49, 2408.

Saltman, P. (1973). TEMA II Symposium, in press.

Sirivech, S., Frieden, E., and Osaki, S. (1974). In press.

Tanaka, S. (1950). J. Biochem. (Japan), 37, 129.

Topham, R., and Frieden, E. (1970). J. Biol. Chem., 245, 6698.

13

COPPER AND AMINE OXIDASES IN CONNECTIVE TISSUE METABOLISM

E. D. Harris* and B. L. O'Dell

Department of Agricultural Chemistry

University of Missouri, Columbia 65201

That copper is an essential biocatalyst has been recognized for nearly a half century but its critical role in connective tissue metabolism has come into focus only in recent years. It is now clear that copper plays a role in the crosslinking of the two important connective tissue proteins, elastin and collagen. The gross pathology that results from this copper deficiency is most dramatic when it occurs in large arteries and results in spontaneous and massive hemorrhage. Because of this unique pathology and the importance of a sound vasculature to human health, our research has been focused on the aorta. The effect of copper deficiency on the integrity of collagen and elastin and the activity of the amine oxidases in this organ has been investigated. This paper summarizes earlier observations and presents recent results relating to a copper-dependent amine oxidase which plays a key role in the crosslinking process.

Pathology of Connective Tissues Due to Copper Deficiency and Lathyrism. Although the earlier literature describing copper deficiency referred to skeletal pathology, such as, spontaneous fractures and osteoporosis (See review by Asling and Hurley, 1963), renewed interest in the role of copper in connective tissue metabolism began with the nearly simultaneous observation of aortic rupture in pigs (Shields et al., 1962) and chicks (O'Dell et al., 1961). The literature related to the pathology in arteries has been extensively reviewed by Carnes (1968). Sustained interest in the observed pathology resulted from the

*Present address: Department of Biochemistry and Biophysics, Texas A&M University, College Station, 77843.

recently acquired knowledge relating to the chemistry of collagen and elastin and particularly from discovery of the crosslinking compounds in elastin, such as, the desmosines (Partridge et al., 1963) and lysinonorleucine (Franzblau et al, 1965) and in collagen (See review by Gallop et al., 1972).

Aortas from chicks fed a copper deficient diet contain less elastin and a higher proportion of soluble collagen than controls (O'Dell et al., 1966a). Data showing the effect of copper on aortic elastin in chicks of various ages are presented in Table 1. Furthermore, the elastin isolated from deficient aortas contains less desmosine and isodemosine and more lysine than that from controls (O'Dell et al., 1966b). Similar observations have been made on chick aortic elastin (Starcher et al., 1964), on pig aortic collagen (Weissman et al., 1963), and on chick bone collagen (Rucker et al., 1969).

TABLE 1

Effect of Dietary Copper Deficiency and Age on the Elastin Content of Chick Aorta

Age	Elastin Content -Cu*	Elastin Content +Cu
Weeks	% of Dry Weight	
1	22.0	45.2
2	19.2	50.6
3	13.7	43.9
4	13.5	48.9

Data from D. W. Bird's PhD thesis, University of Missouri, Columbia, 1966. *Deficient diet based on non-fat milk solids; supplemented with 200 ppm of zinc to accentuate the deficiency.

The total collagen content of chick aorta is not affected by copper deficiency but there is less intermolecular crosslinking as indicated by a higher proportion of solubility in cold saline and acetic acid (Chou et al., 1968). Not only is intermolecular crosslinking depressed but intramolecular bonding is similarly affected (Chou et al., 1969). This is illustrated in Figure 1 which shows lower proportion of β and higher proportion α chains in the salt soluble collagen from copper deficient chick tendon.

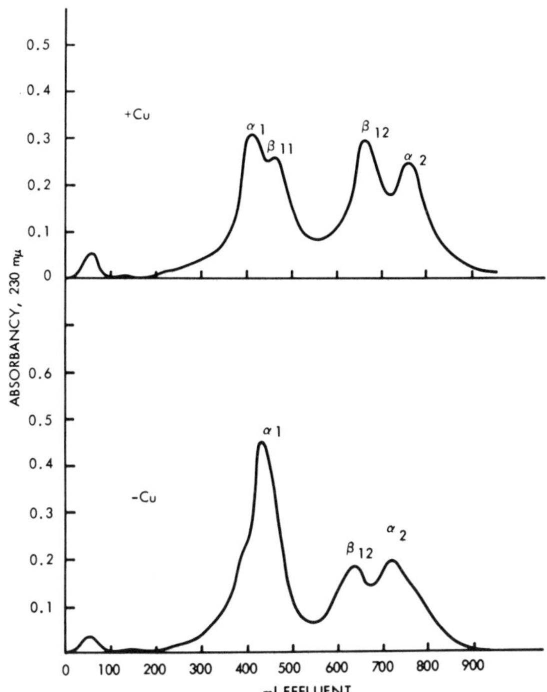

Figure 1. Chromatographic separation of the alpha and beta chains of denatured 1 M NaCl-soluble collagen on carboxymethyl cellulose. The lower pattern, obtained from copper-deficient chick tendon, is compared to that of copper supplemented controls above.

Lathyrogens, such as, β-aminopropionitrile (BAPN) and penicillamine produce an arterial pathology analogous to that of copper deficiency. In turkey poults the histopathology and changes in the chemical composition of aortas due to lathyrism and copper deficient were hardly distinguishable (Savage et al., 1966). However, copper supplementation does not alleviate the pathology induced by BAPN. BAPN also inhibits the formation of desmosine and isodesmosine in elastin of aortas grown in culture(Miller et al., 1965; O'Dell et al., 1966b). These results suggest that the pathology of copper deficiency and lathyrism results from the same biochemical lesion, failure of crosslink formation, i.e., BAPN inhibits a copper dependent enzyme which is critical in the formation of the crosslinking compounds in collagen and elastin.

Maturation of Collagen and Elastin. It appears that copper has no specific effect on synthesis of the peptide chains of elastin and collagen. The decreased concentration of elastin in aortas of deficient animals is due to the failure of the protein to crosslink in the normal manner and thus form highly insoluble, mature elastin. Elastin is a hydrophobic and highly crosslinked protein which is insoluble in all solvents that do not break peptide bonds. It is usually determined as the residue that is not solubilized by hot 0.1 NaOH or by warm formic acid. An elastin precursor protein that is not fully crosslinked would likely be extracted by the solvents used to remove non-elastin proteins. Thus, the lower concentration of elastin in copper deficient aortas reflects lack of crosslinking rather than lack of polypeptide synthesis.

A soluble elastin-like protein, presumably an elastin precursor, has been isolated from aortic tissue of copper deficient pigs (Sandberg et al., 1971; Smith et al., 1972) and chicks (Roensch et al., 1972). This protein, which has been termed tropoelastin, contains no desmosine but contains more lysine than normal elastin. Lysine residues are the precursors of the desmosines and the crosslink is formed from lysine while it is in peptide linkage (Partridge et al., 1966). Four lysine residues are involved in the synthesis of a desmosine molecule, three of which are oxidatively deaminated to give α-aminoadipic acid-δ-semialdehyde (allysine) residues. This oxidation is catalyzed by a specific deaminase, a peptidyl lysine oxidase which will be discussed below.

Normally tropoelastin does not accumulate in isolable quantity in tissues and a soluble precursor of elastin has not been identified except in copper-deficient and lathyrogenic (Sykes and Partridge, 1972) animals. Soluble elastin can be extracted from chick aortas with 0.02Mformic acid, precipitated with 40% ammonium sulfate and then solubilized with a mixture of water:propanol:butanol (2:3:5). At this stage of purification the alcoholic extract of copper deficient aortas contained 6.7 mg of protein per gram of fresh tissue compared to 0.3 mg for control tissue (Roensch and O'Dell, 1973). The amino acid composition of the protein after electrophoretic purification is compared with mature chick elastin in Table 2.

As shown normal elastin is a hydrophobic protein characterized by its high content of alanine, valine, glycine, proline and low content hydroxyproline. Its lysine content is low and it contains desmosine & isodesmosine. Elastin also contains lysine, derived crosslink intermediates, such as lysinonorleucine, and the aldol condensates including merodesmosine, which were not determined here. The composition of tropoelastin compares closely

TABLE 2

Amino Acid Composition of Chick Aorta Elastin and Tropoelastin

Residues/1000 Total Residues

	Mature Elastin	Tropoelastin
Hydroxyproline	3.7	0.0
Aspartic Acid	2.5	6.7
Threonine	11.8	12.9
Serine	3.2	8.0
Glutamic Acid	11.8	14.8
Glycine	361.0	370.0
Alanine	191.0	168.4
Proline	136.0	132.0
Valine	165.0	143.0
Methionine	0.1	1.9
Isoleucine	13.6	21.4
Leucine	61.5	42.9
Tyrosine	7.8	9.9
Phenylalanine	16.1	16.5
Lysine	3.5	37.0
Histidine	0.0	1.6
Arginine	5.2	9.6
Desmosine	1.0	0.0
Isodesmosine	0.7	0.0

Tropoelastin purified after alcoholic extraction by acrylamide gel electrophoresis followed by extraction from pooled gel sections. Mature elastin isolated from aortas of normal broiler age chickens by removal of other proteins with hot 0.1 N NaOH.

----

with that of mature elastin except for its much higher lysine content and lack of desmosines. Its somewhat higher content of polar amino acids suggests either slight contamination or the presence of a peptide unit which is cleaved off before final crosslinking. The concentrations of the intermediates in desmosine biosynthesis have not been determined in copper deficient elastin. Whether or not this soluble elastin-like protein is the true precursor of elastin is not clear because no one has succeeded in effecting its *in vitro* conversion to mature elastin. However, its amino acid composition and physical properties strongly suggest that it is either the precursor or a closely related molecule.

Although desmosine and isodesmosine do not occur in collagen, other lysine derived compounds serve as crosslinks (Gallop et al., 1972). Among these compounds are lysinonorleucine, the "aldols" including the hydroxylysine derivatives, merodesmosine, aldol-histidine and histidino-hydroxymerodesmosine (Gallop et al., 1972; Tanzer et al., 1973). All involve initial formation of an allysine residue and require a catalyst to oxidatively deaminate the epsilon-carbon of a lysine residue. It is not clear whether or not there is enzyme specificity for oxidation of lysine residues in collagen and elastin.

Aortic Amine Oxidases. Aortic tissue contains at least three types of amine oxidases that act on monoamines. No diamine oxidase could be detected in chick aorta (Bird et al., 1966). Benzylamine serves as a substrate for two of the enzymes, one of which is soluble in salt buffers (Rucker and O'Dell, 1971). The other benzylamine oxidase can be solubilized by the use of a detergent such as Triton X-100 and is presumably membrane bound (Harris and O'Dell, 1972). These enzymes will be referred to as *soluble* and *insoluble benzylamine oxidase*, respectively. The third amine oxidase catalyzes the oxidation of peptidyl lysine and will be referred to as *lysyl oxidase*, using the nomenclature of Pinnell and Martin (1968) who first purified the enzyme from embryonic chick cartilage. Lysyl oxidase catalyzes the formation of allysine residues from peptidyl lysines. The enzyme has been purified from cartilage by Siegel and Martin (1970a) and from bone by Siegel et al.,(1970b). The properties of a similar enzyme purified from chicken aortic tissue are described below (Harris and O'Dell, 1973). The relative proportions of the soluble and insoluble benzylamine oxidase activities in chick and bovine aortas are shown in Table 3. Of the total homogenate activity, approximately 8% was soluble in chick and 14% in bovine aorta; the remainder was solubilized by detergent. The characteristics of thebenzylamine oxidases in beef aorta will be described below. Lysyl oxidase in aorta is largely insoluble in salt buffers but it can be extracted by 4 M urea solutions. Its extractability from nearly mature chick and bovine aortic tissue is shown in Table 4. Bovine aorta has a lower concentration of lysyl oxidase activity than chick aorta and none of the activity was soluble in salt buffer.

TABLE 3

Benzylamine Oxidase Activities in Chick and Bovine Aorta Fractions

Fraction	Benzaldehyde Formed Chick	Bovine
	moles/hr/g X 10^7	
Homogenate (13,000 x g)	2.58 ± .84(100%)	16.7 ± 4.5(100%)
Buffer Soluble (1)	0.20 ± .41(7.8%)	2.3 ± 0.9(13.8%)
Buffer Insoluble (2)	2.46 ± .58(95.5%)	12.2 ± 5.0(73.2%)

1. Mean of 5 determinations. Tissue homogenized in 0.05 M phospate, pH 7.8, and centrifuged at 105,000 X g to obtain the soluble (1) fraction. The resulting sediment was suspended in buffer (2) before assay.

TABLE 4

Solubility of Lysyl Oxidase Activity in Chick and Bovine Aortas

Fraction	Chick ^3H_2O Released/4 hr Assay	Bovine
	CPM/g aorta	
Salt Buffer[1] Soluble (65,000 X g)	855	0
4 M Urea[2] Soluble (65,000 X g)	13,100	2,580
4 M Urea Insoluble	0	170

1. 0.1 M phosphate, 0.15 M NaCl, pH 7.7.
2. 0.01 M phosphate, 0.015 M NaCl, 4 M urea, pH 8.4.

Copper Deficiency and Amine Oxidase Activity. Soon after it became apparent that copper is involved in the formation of allysine and the elastin crosslinks, a search was made for a copper-dependent amine oxidase. Because of the ease of the determination, benzylamine was used as substrate. Benzylamine oxidase activities in connective tissue such as aorta, cartilage and tendon were markedly lower in copper deficient animals but the levels in muscle, liver and brain were not affected (Table 5). At least part of the benzylamine oxidase of connective tissue is copper dependent and its activity can be restored by dialysis against cupric ion *in vitro* (Bird et al., 1966). Furthermore, there is an inverse relationship between the enzyme activity and the proportion of soluble collagen in aorta and tendon (See Figure 2).

TABLE 5

Total Benzylamine Oxidase Activity in Homogenates of Copper Deficient and Control Chick Tissues[1]

Tissue	Dietary Status -Cu	+Cu
	moles of benzaldehyde/hr/mg N X 10^8	
Aorta	0.30	1.30
Tendon	0.20	0.75
Skin	1.50	3.60
Cartilage	0.65	3.00
Heart	1.40	6.80
Kidney	0.65	3.00
Skeletal Muscle	0.20	0.20
Liver	1.30	1.10
Brain	1.85	1.65

These data from Bird et al. (1966) and Chou et al. (1968) represent the means of 5-10 observations each. The differences between treatments are highly significant ($P < 0.005$) except for muscle, liver and brain.

Although the benzylamine oxidase activity was depressed in connective tissues of copper deficient chicks and plasma of pigs (Blaschko et al., 1965) and sheep (Mills et al., 1966), this enzyme is not sufficiently sensitive to BAPN inhibition to account for it being the key enzyme inhibited in lathyrism. This led to a search for an amine oxidase which is highly sensitive to

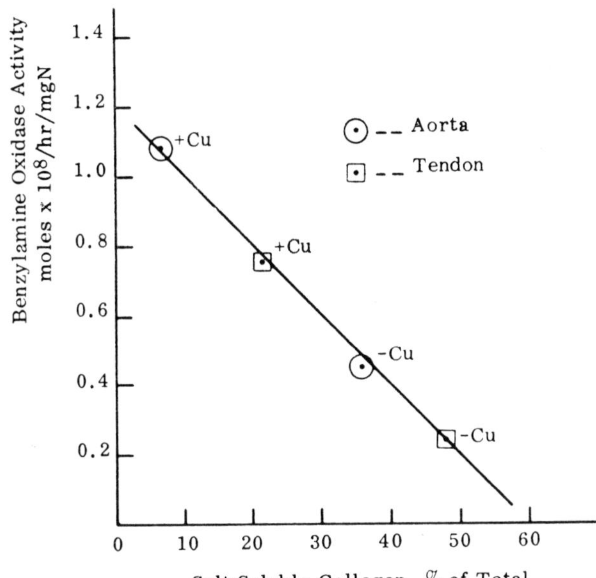

Figure 2. Inverse relationship between benzylamine oxidase activity and the proportions of salt soluble collagen in chick aorta and tendon. Adapted from Chou et al., (1968).

BAPN and specific for the oxidation of peptidyl lysine. The assay for lysyl oxidase (Pinnell and Martin, 1968) resulted. The effect of copper deficiency on the lysyl oxidase activity of chick aorta is shown in Table 6. The enzyme was barely detectable in deficient aortas, the mean activity amounting to less than 1% of that in aortas from copper supplemented controls. From these experiments it is clear that dietary copper has a marked effect on the activities of both benzylamine oxidase and lysyl oxidase activities in aortic tissue of chicks and the results suggest that both enzymes are copper dependent.

<u>Properties of the Aortic Benzylamine Oxidases.</u>

(a) Soluble benzylamine oxidase. In view of the nutritional evidence that the benzylamine oxidase activity in aortic tissue

TABLE 6

Effect of Dietary Copper Deficiency on the Lysyl Oxidase Activity of Chick Aorta

Group	^3H_2O Released by 4 M Urea Extracts Dietary Status -Cu	+Cu
	CPM mg protein in extract	
I	9 (14)[1]	3438 (16)
II	4 (16)	3508 (16)
III	9 (16)	3153 (18)
IV	44 (16)	2866 (19)
	16 ± 9	3241 ± 147

1. Number of aortas pooled.

-----

is copper dependent, efforts were made to purify and characterize the enzyme(s) involved. Bovine aorta was chosen because it was readily available in quantity. Although only approximately 15% of the total activity is soluble in phosphate buffer, this fraction was purified first and compared to bovine plasma amine oxidase (Rucker and O'Dell, 1971). Plasma amine oxidase is clearly a copper metalloenzyme (Yasunobu and Yamada, 1961) and the properties of the soluble aorta benzylamine oxidase are highly similar to those of the plasma enzyme. The soluble aorta benzylamine oxidase is inhibited by treatment with diethyldithiocarbamate and cuprizone and dialysis against cupric ion partially restores activity. Both enzymes are inhibited by carbonyl reagents such as hydroxylamine and iproniazid. The chief difference between the plasma and aorta enzyme relates to the fact that the aorta enzyme catalyzes the oxidation of the lysine residue in lysine-vasopressin while the plasma enzyme does not. Neither the physiological function of these enzymes nor the origin of the plasma enzyme is known but the similarities of the enzymes suggest that the plasma enzyme arises from connective tissues.

(b) Insoluble benzylamine oxidase. As pointed out above the major proportion of the benzylamine oxidase activity in chick and beef aorta is not soluble in salt buffers, but is solubilized

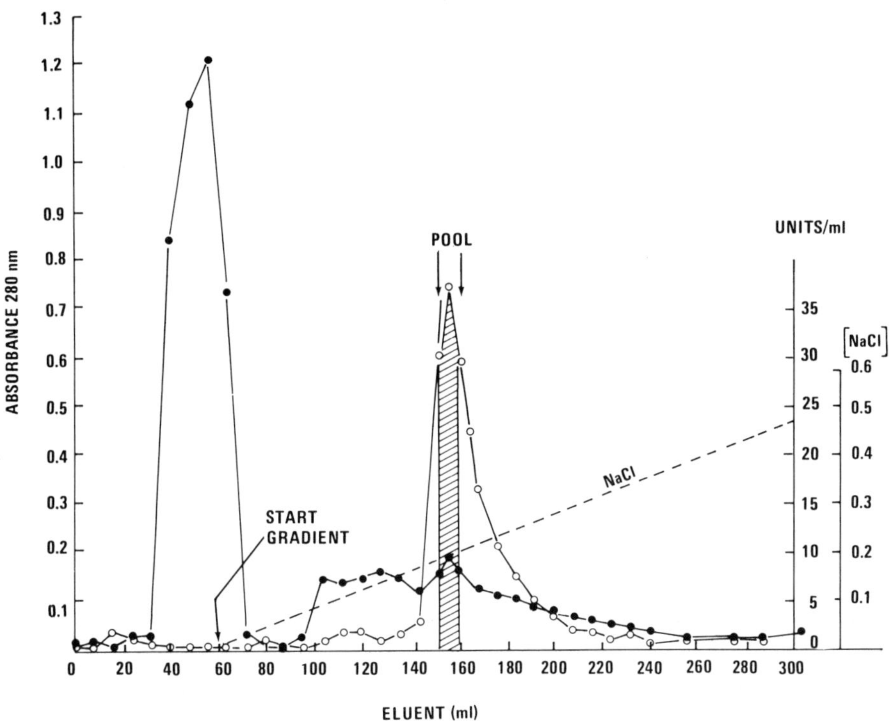

Figure 3: Chromatography of benzylamine oxidase from bovine aorta on a DEAE cellulose. The Triton X-100 extract was dialysed against a solution containing 5 mM Tris-HCl, pH 7.4, 7mM β-mercaptoethanol, 1 mM benzylamine in 0.1% v/v Triton X-100 and applied to a 2.5 cm x 50 cm column containing diethylaminoethyl (DEAE) cellulose equilibrated with the same buffer. Absorbance at 280 nm (———), enzyme activity (----).

by Triton X-100. The protein concentration must be rather dilute (2 mg/ml) in order to effect solution with 0.1% Triton (Harris and O'Dell, 1972). As shown in Figure 3 the insoluble or membrane bound enzyme is adsorbed on DEAE cellulose and eluted as a single peak by a linear NaCl gradient. DEAE chromatography purified the enzyme approximately 4-fold, but the copper content of the protein was unchanged, indicating a lack of copper dependence. This insoluble enzyme fraction catalyzes the oxidation of N-methylbenzylamine but not of spermine. In contrast, the soluble enzyme catalyzes the oxidation of spermine but not of secondary amines (Harris and O'Dell, 1972). The membrane bound or insoluble enzyme is also more strongly inhibited by pargyline than the soluble enzyme. These observations suggest that the membrane bound enzyme is not a copper containing enzyme and that it is related to the mitochondrial monoamine oxidases.

Isolation and Characterization of Aortic Lysyl Oxidase. Lysyl oxidase catalyzes the release of tritiated water from aortic protein labeled inculture with lysine-6^{3}H (Pinnell and Martin, 1968). This assay has been used in our laboratory to monitor the purification of the enzyme from chick aorta (Harris and O'Dell, 1973). Lysyl oxidase was extracted from broiler age chick aorta with 4 M urea and dialyzed against a solution of 0.12 M NaCl in 0.016 M potassium phosphate, pH 7.6. The activity was adsorbed by a DEAE-cellulose at pH 7.6; washing the column with 0.5 M NaCl buffer eluted only a small fraction of the activity along with considerable impurity. As shown in Figure 4, the major activity was eluted with a concave, NaCl gradient run in 6 M urea buffer. Three active fractions were obtained with most of the activity appearing in the first fraction which had twice the specific activity of the other two fractions. Electrophoretic analysis on sodium dodecyl sulfate (SDS) acrylamide gels showed the first peak fraction to be of highest purity (Figure 5). Further characterization and analyses were performed on this fraction. It is of interest that aorta contains multiple enzymes in contrast to cartilage which has only one (Narayanan and Martin, personal communication). This may relate to the fact that aorta synthesizes both collagen and elastin in contrast to bone cartilage which contains only collagen.

The copper content of the lysyl oxidase preparation averaged 22 ng atoms per mg of protein and the copper to protein ratio of the DEAE purified lysyl oxidase was 10-fold higher than before purification. This increase corresponded to the observed increase in specific activity. The purified enzyme does not catalyze the oxidation of benzylamine.

To establish the essentiality of copper in lysyl oxidase a procedure which reversibly removes copper from the protein structure is required. Lowering the pH of the lysyl oxidase preparation from 7.6 to 5.1 by dropwise addition of 10% acetic acid caused the enzyme to precipitate from solution. A precipitate also formed when the enzyme was dialysed against pH 5.1 buffer.

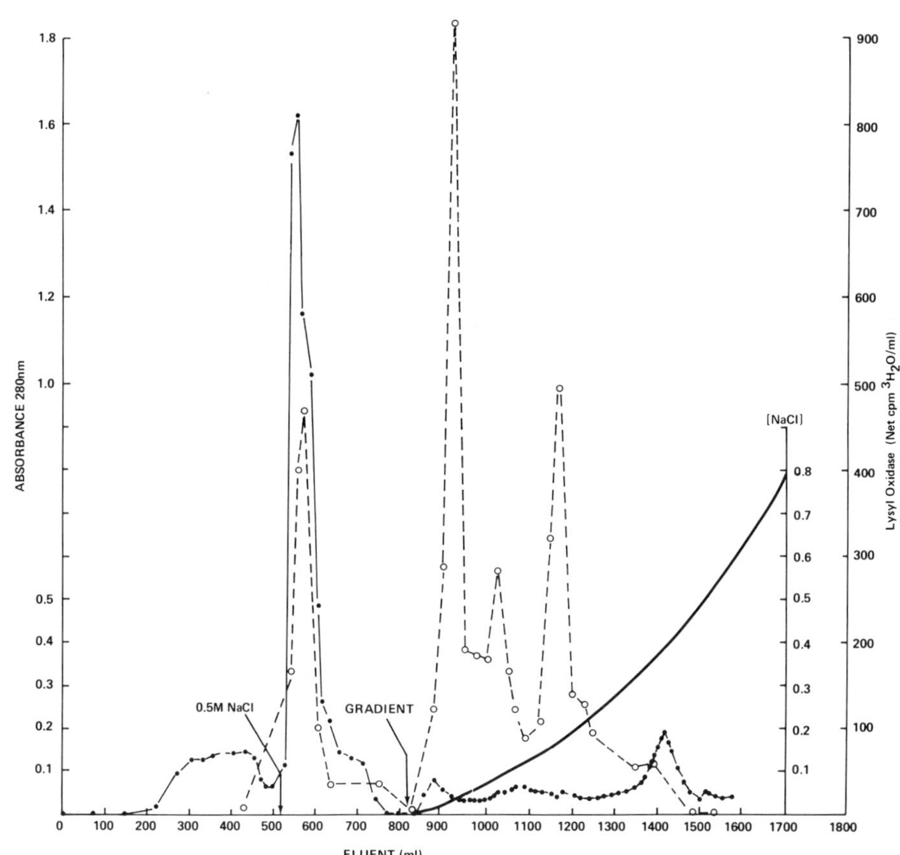

Figure 4: Purification of chick aorta lysyl oxidase. Dialysed extracts of aorta in 0.12 M NaCl, 0.016M potassium phosphate buffer, pH 7.6, were applied to a 2.5 x 50 cm column of DEAE cellulose. The column was washed with 100 ml of 0.5M NaCl in 0.016M phosphate and then eluted with NaCl (0 - 0.8M) in 6 M urea, pH 8.3. Absorbance at 280 nm (——), enzyme activity (----).

As shown in Table 7 this precipitate, when taken up in pH 7.6 buffer, showed approximately 50% of the original activity (Exp. 1). In an attempt to restore activity, the pH 5.1 precipitate was suspended in 6 M urea buffer and dialysed stepwise, first against a solution containing metal salts in 10^{-5} M β-mercaptoacetic acid, and then against assay buffer. Of the metal ions tested, only copper restored full enzymatic activity (Exp. 2). These findings

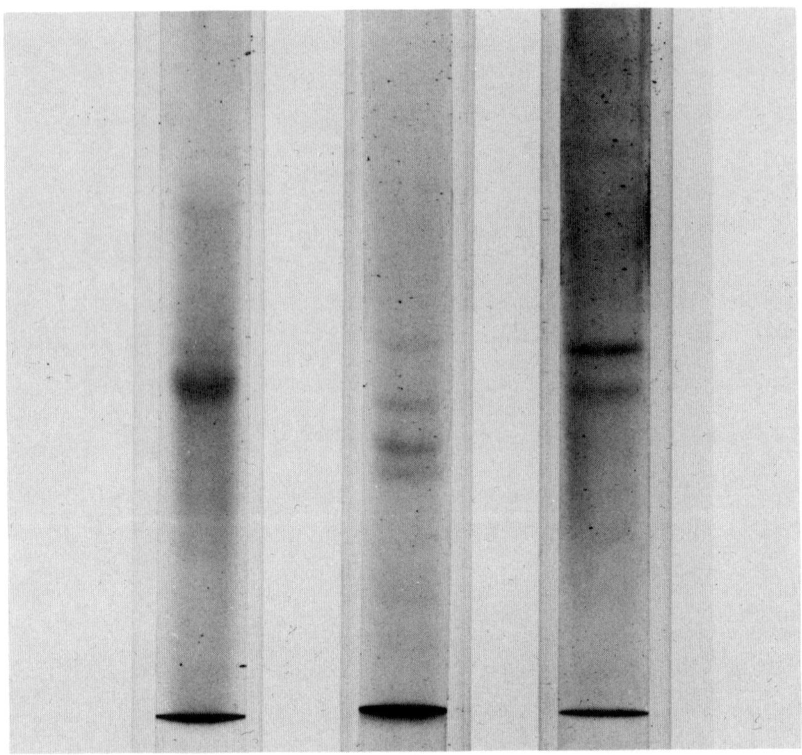

Figure 5: SDS acrylamide electrophoretograms of purified lysyl oxidase. The 3 gels shown represent the material in the 3 respective gradient elution peaks from DEAE cellulose. (Same order as in Figure 4).

support the concept that copper is specific for lysyl oxidase function, but it is not known whether or not copper is present at the active site of the enzyme.

SUMMARY

Copper deficiency in some species results in gross pathology of connective tissues due to failure of crosslink formation. Aortic tissue from copper deficient chicks and pigs accumulate a soluble elastin-like protein which appears to be an uncrosslinked precursor of elastin. There is also higher proportion of soluble collagen than in normal aortic tissue.

TABLE 7

Activity of Acid-Treated Enzyme Before and After Addition of Metal Ions

Enzyme Treatment	Cation in Dialysis Buffer	Lysyl Oxidase Activity	
		CPM/mg Protein	%
Exp. 1[1]			
None	---	5,225	100
pH 5.1 ppt	none	2,666	51
pH 5.1 ppt	Cu^{2+}	4,164	80
Exp. 2[2]			
None	---	15,561	100
pH 5.1 ppt	none	8,872	57
pH 5.1 ppt	Zn^{2+}	6,690	43
pH 5.1 ppt	Fe^{2+}	8,854	57
pH 5.1 ppt	Cu^{2+}	18,229	117

[1] $CuSO_4$, 10^{-6}M.

[2] $ZnSO_4$, $FeSO_4$, $CuSO_4$, 10^{-5}M.

Both bovine and chick aortic tissue contain at least three different amine oxidase(s), two of which catalyze the oxidative deamination of benzylamine and one of which acts on peptidyl lysine. One of benzylamine oxidases is soluble in salt buffers and the other is solubilized only by use of a detergent. Both the soluble benzylamine oxidase and the lysyl oxidase activities in aorta are depressed by dietary copper deficiency. Lysyl oxidase is also highly sensitive to BAPN and is believed to be the key initiating enzyme in collagen and elastin crosslinking.

Chick aortic tissue contains two or more chromatographically separable lysyl oxidases. One has been highly purified and shown to contain copper which is specifically essential for catalytic activity. The fact that copper remains with the protein during its isolation and that it can be reversibly removed suggests that lysyl oxidase is a copper metalloprotein.

ACKNOWLEDGEMENTS

A contribution from the Missouri Agricultural Experiment Station, Journal Series No. 6866. Supported in part by the Public Health Service Grant HE 11614.

REFERENCES

Asling, C. W. and Hurley, L. S. (1963). Clinical Orthopaedics 23:213.

Bird, D. W., Savage, J. E. and O'Dell, B. L. (1966). Proc. Soc. Exptl. Biol. Med. 123:250.

Blaschko, H. and Buffoni, F., Weissman, N., Carnes, W. H. and Coulson, W. F. (1965). Biochem. J. 96:4C.

Carnes, W. H. (1968). International Review of Connective Tissue Research, edited by D. A. Hall. New York: Academic Press, Vol. 4.

Chou, W. S., Savage, J. E. and O'Dell, B. L. (1968). Proc. Soc. Exp. Biol. Med. 128:948.

Chou, W. S., Savage, J. E. and O'Dell, B. L. (1969). J. Biol. Chem. 244:5785.

Franzblau, C., Sinex, F., Faris, B. and Lampidis, R. (1965). Biochem. Biophys. Res. Commun. 21:575.

Gallop, P. M., Blumenfeld, O. O. and Seifter, S. (1972). Ann. Rev. Biochem. 41:617.

Harris, E. D. and O'Dell, B. L. (1972). Biochem. Biophys. Res. Commun. 48:1173.

Harris, E. D. and O'Dell, B. L. (1973). Manuscript in preparation for Biochem. Biophys. Acta.

Miller, E. J., Martin, G. R., Mecca, C. E., and Piez, K. A. (1965). J. Biol. Chem. 240:3623.

Mills, C. F., Dalgarno, A. C. and Williams, R. B. (1966). Biochem. Biophys. Res. Commun. 24:537.

O'Dell, B. L., Hardwick, B. C., Reynolds, G. and Savage, J. E. (1961). Proc. Soc. Exptl. Biol. Med. 108:402.

O'Dell, B. L., Bird, D. W., Ruggles, D. L. and Savage, J. E. (1966a). J. Nutrition 88:9.

O'Dell, B. L., Elsden, D. F., Thomas, J., Partridge, S. M., Smith, R. H. and Palmer, R. (1966b). Nature 209:401.

Partridge, S. M., Elsden, D. F. and Thomas, J. (1963). Nature 197:1297.

Partridge, S. M., Elsden, D. F., Thomas, J., Dorfman, A., Telser, A. and Ho, P. L. (1966). Nature 209:399.

Pinnell, S. R. and Martin, G. R. (1968). Proc. Natl. Acad. Sci. U.S. 61:708.

Roensch, L. F., Savage, J. E. and O'Dell, B. L. (1972). Proc. Fed. Am. Soc. Exp. Biol. 31:480.

Roensch, L. F. and O'Dell, B. L. (1973). Manuscript in preparation from Roensch, PhD Thesis, University of Missouri, 1973.

Rucker, R. B., Parker, H. E., and Rogler, J. C. (1969). J. Nutrition 98:57.

Rucker, R. B. and O'Dell, B. L. (1971). Biochem. Biophys. Acta 235:32.

Sandberg, L. B., Weissman, N. and Smith, D. W. (1969). Biochem. 8:2940.

Savage, J. E., Bird, D. W., Reynolds, G. and O'Dell, B. L. (1966). J. Nutrition 88:15.

Shields, G. S., Coulson, W. F., Kimball, D. A., Carnes, W. H., Cartwright, G. E. and Wintrobe, M. M. (1962). Am. J. Pathol. 41:603.

Siegel, R. C . and Martin, G. R. (1970a). J. Biol. Chem. 245: 1653.

Siegel, R. C., Pinnell, S. R. and Martin, G. R. (1970b). Biochemistry 9:4486.

Smith, D. W., Brown, D. W. and Carnes, W. H. (1972). J. Biol. Chem. 247:2427.

Starcher, B., Hill, C. H. and Matrone, G. (1964). J. Nutrition 82:318.

Sykes, B. C. and Partridge, S. M. (1972). Biochem. J. 130: 1171.

Tanzer, M. L., Housley, T., Berube, L., Fairweather, R., Franzblau, C. and Gallop, P. M. (1973). J. Biol. Chem. 248:393.

Weissman, N., Shields, G. S. and Carnes, W. H. (1963). J. Biol. Chem. 238:3115.

Yasunobu, K. T. and Yamada, H. (1961) in Snell, E. E. Intern Symp. on Chemical and Biological Aspects of Pyridoxal Catalysis. Rome. Academic Press, New York 1962, p. 453.

14

COPPER- AND ZINC-BINDING COMPONENTS IN RAT INTESTINE[1]

Gary W. Evans[2] and Carole J. Hahn[3]

USDA, ARS, Human Nutrition Laboratory
Grand Forks, North Dakota and
Department of Biochemistry, University of North Dakota
Grand Forks, North Dakota

Protein-metal interactions encompass a wide variety of metabolic processes. Several of the papers presented in this symposium demonstrate that metals react with enzymes or proteins and thereby promote catalytic activity and/or maintain functional conformation. Protein-metal interactions are also involved in metal storage, i.e. ferritin, and metal transport in the blood, i.e. transferrin. In addition, there is an increasing awareness that proteins and peptides are involved in metal transport across biological membranes (1-3).

Protein-metal interactions are apparently involved in the absorption of trace elements from the gastrointestinal tract. Early investigations demonstrated that copper, cadmium and zinc antagonize the absorption of each other (4-8) and these observations implied that chemically similar elements may compete for binding sites on specific metal-transport proteins. In 1969, Starcher (9) identified a single metal-binding protein in chick duodenum and demonstrated that the protein would bind copper as well as cadmium and zinc. Starcher's observation provided the first biochemical evidence implicating a protein in the regulation of trace element absorption.

[1] Supported in part by USDA Cooperative Agreement 12-14-100-11, 178 (61), Amend. 1.

[2] Research Chemist, USDA, ARS, Human Nutrition Laboratory.

[3] Research Associate, Dept. of Biochemistry, University of North Dakota Medical School.

The mechanisms which regulate the intestinal absorption of copper and zinc remain obscure in spite of the fact that these elements are well established as essential nutrients in man and other animals. Assuming that ligand-metal interactions are involved in the absorption of copper and zinc, we have examined metal-binding components in the intestine in an attempt to identify some of the factors involved in the absorption of these essential elements. This paper describes the identification of copper- and zinc-binding components in rat intestine following oral administration of either copper-64 or zinc-65.

METHODS

Male Sprague-Dawley rats age 70-90 days were used as a source of intestinal tissue. The animals had been housed in stainless steel cages and had been fed a diet of Purina Laboratory Chow[4] and tap water. To produce zinc-deficiency, rats were fed a zinc-deficient diet (10) and resin-deionized water for 13 days. All animals were fasted for 18 hours prior to being used in experiments.

Radioactive zinc solutions were prepared by diluting 1.0 μCi carrier-free zinc-65 (International Chemical and Nuclear Corp., Irvine, California) in 1.0 ml distilled water which contained 0.065 μg Zn^{2+} in the form of $ZnSO_4$. Copper-64 (International Chemical and Nuclear Corp.) in the form of $Cu(NO_3)_2$ was diluted to a concentration of 0.5 mCi/ml which was equivalent to 44 μg Cu^{2+}/ml. The 1.0-ml isotope solutions were administered separately by gastric tube and one hour later, the animals were decapitated. Blood was collected in a heparinized beaker and a 15-cm segment of the small intestine, beginning at the pylorus, was removed.

The small intestine was prepared for analysis by thoroughly rinsing the lumen with cold 0.85% NaCl. Thereafter, the mucosal cells were scraped from the serosal tissue by using a glass slide. The mucosal scrapings were then suspended in 10 ml of 0.85% NaCl and the suspension was homogenized in a Potter-Elvehjem homogenizer equipped with a teflon pestle. The homogenate was centrifuged at 105,000 x g for one hour in a refrigerated centrifuge (Beckman, Model L2-65B) after which the supernatant was removed and freeze-dried. The lumen contents were centrifuged at 4,000 x g for 15 minutes after which the supernatant was removed and freeze-dried.

[4]Mention of a trademark or proprietary product does not constitute a guarantee or warranty of the product by the U.S. Department of Agriculture, and does not imply its approval to the exclusion of other products that may also be suitable.

The freeze-dried lumen contents and intestinal supernatant were dissolved separately in 1.0 ml distilled water and applied to 0.9- x 60-cm columns packed with Bio-Gel P-10 (Bio Rad Laboratories, Richmond, California) which had been equilibrated with 0.025 M KH_2PO_4, pH 7.3, and 0.05 M KCl. The same buffer was used to elute the samples. Fractions were collected in a refrigerated fraction collector and monitored for radioactivity in a gamma-well counter (Nuclear-Chicago, Model 4233).

The elution characteristics of the Bio-Gel P-10 columns used in these experiments were determined with Blue Dextran 2000 (Elution volume, Ve, = 11 ml), cytochrome C (Ve = 22 ml) and a mixture of amino acids (Ve = 34 ml).

The radioactive isotopes were extracted from the 105,000 x g pellet by suspending the pellet in 10 ml of a 0.3% solution of sodium dodecyl sulfate (SDS), pH 8.2. The suspension was stirred for two hours after which the contents were centrifuged at 105,000 x g for 60 minutes. The supernatant from the centrifuged SDS extract was then freeze-dried, dissolved in 2.0 ml of 0.3% SDS and applied to a 2.5- x 40-cm column packed with Sepharose 4B (Pharmacia Fine Chemicals, Piscataway, New Jersey) which had been equilibrated with a solution of 0.3% SDS, pH 8.2. The same solution was used to elute the sample and fractions were collected in a refrigerated fraction collector and monitored for radioactivity. The 280 nm absorbance of the sample eluted from the Sepharose 4B column was monitored with a Uvicord II (LKB Instruments, Rockville, Maryland).

The elution of copper-64 and zinc-65 from plasma was analyzed by applying 2.0 ml plasma to a 2.5- x 40-cm column packed with Sephadex G-75 (Pharmacia Fine Chemicals). The gel had been equilibrated with 0.025 M KH_2PO_4, pH 7.4, and 0.05 M KCl and the same buffer was used to elute the samples.

Ion-exchange chromatography was carried out on 0.9- x 15-cm columns packed with either Dowex-50W strongly acid cation-exchange resin in hydrogen form or Dowex-1 strongly basic anion-exchange resin in chloride form. Water and various concentrations of NaCl were used to elute the samples from the resins.

Thin layer chromatography was carried out on silica-gel impregnated glass paper (Gelman Type ITLC - SA) in two different solvent systems. One solvent consisted of butanol:acetic acid: water (4:1:1). The other solvent contained phenol:water (75:25) and the chromatogram was developed in an ammonia-rich atmosphere. Following chromatogram development, amino acids were detected with ninhydrin.

RESULTS AND DISCUSSION

Identification of Copper- and Zinc-Binding Components. The results obtained from gel-filtration analysis of the intestinal lumen and mucosal supernatant fraction are illustrated in Figure 1. In the samples from the intestinal lumen, both copper-64 and zinc-65 were eluted in a major peak and a minor peak. The major portion of copper-64 was eluted in a peak which corresponded with

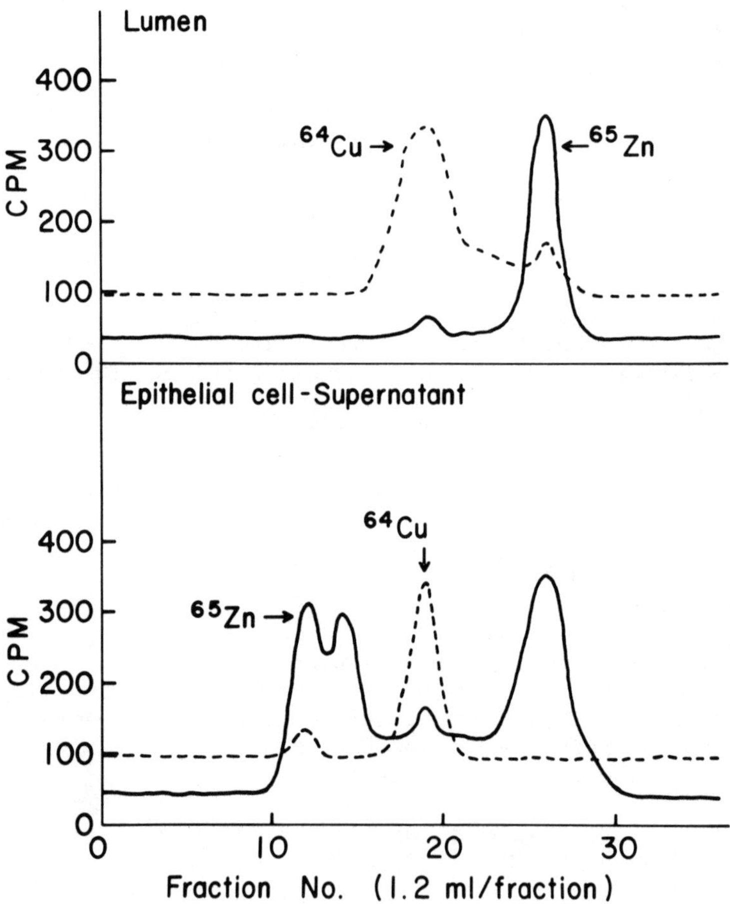

Figure 1. The elution profile of orally administered copper-64 and zinc-65 from the intestinal lumen and mucosal supernatant of normal rats.

the elution volume of proteins with a molecular weight of approximately 10,000 daltons. The second and much smaller fraction of copper-64 in the lumen was eluted at the total volume of the column. Approximately 95% of the zinc-65 recovered from the lumen was eluted in a peak at the total volume of the column while the remainder of the isotope was eluted in a small peak corresponding with the elution volume of the major copper-64 fraction.

In the mucosal supernatant, copper-64 appeared in two fractions; a small peak was eluted at the void volume of the column and a second peak, comprising approximately 90% of the isotope recovered in the supernatant, was eluted at the same volume as that of the major copper-64 peak from the lumen. Zinc-65 from the mucosal supernatant was eluted in four distinct fractions; approximately 50% of the isotope was recovered in two peaks (peaks 1-2) which eluted with high molecular weight material at or near the void volume of the column, a third fraction of zinc-65 (peak 3) was eluted at the same volume as that of the major copper-64 peak and the remainder of the isotope, approximately 40% of the total recovered, was eluted at the void volume of the column.

The results illustrated in Figure 1 and discussed above were observed in the intestines removed from rats which had been fed a stock diet. We have also analyzed copper- and zinc-binding components in the mucosal supernatant of rats fed diets deficient in either copper or zinc. In rats which had been fed a copper-deficient diet for two weeks and subsequently stomach-tubed with copper-64, we observed results identical to those discussed above. However, in rats which had been fed a zinc-deficient diet for two weeks and subsequently stomach-tubed with zinc-65, the binding of zinc-65 in the mucosal supernatant differed from that of rats fed a stock diet or given zinc supplements (Figure 2). In the mucosal supernatant of zinc-deficient rats, the fraction of zinc-65 associated with peak 3 in the mucosa of normal or zinc-supplemented rats was absent. This observation suggests that the fraction of zinc-65 associated with peak 3 in the mucosal supernatant is not involved in the transport of zinc to the blood since the carcass absorption of zinc-65 is markedly elevated in zinc-deficient rats (10).

Throughout our investigations we have observed that approximately 50% of either copper-64 or zinc-65 is sedimented during centrifugation at 105,000 x g. We attempted to fractionate copper- and zinc-binding components in the pellet by extraction and subsequent analysis by gel-filtration chromatography. Extraction of the pellet with water, ethanol, butanol or acetone was totally unsuccessful but extraction with 0.3% sodium dodecyl sulfate (SDS) resulted in a recovery of approximately 25% of the total amount of either copper-64 or zinc-65. When the SDS soluble

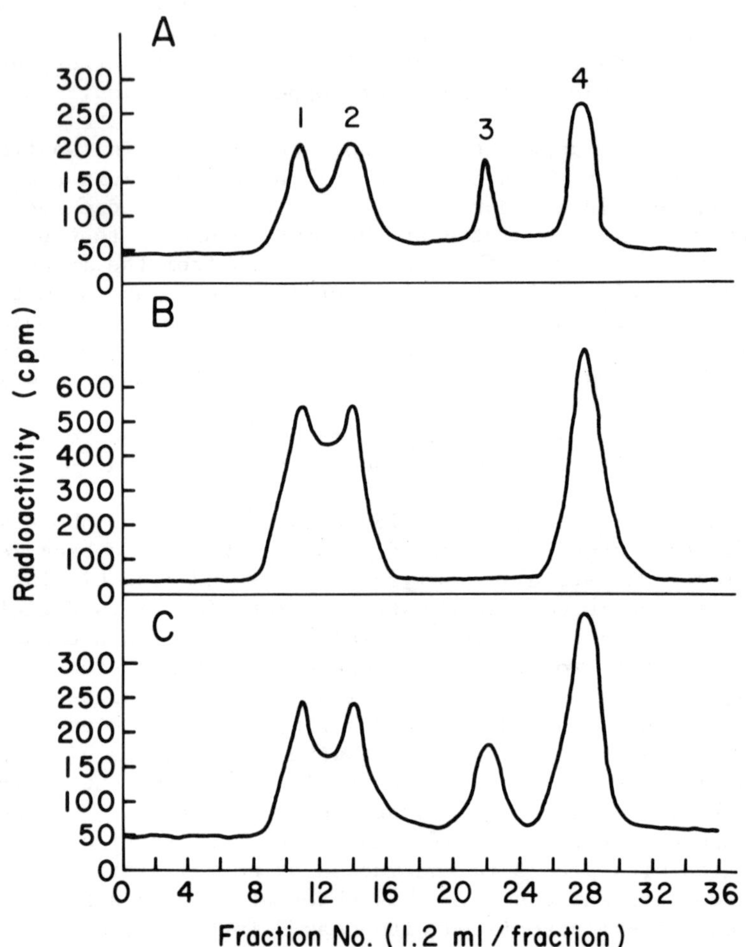

Figure 2. The elution profile of orally administered zinc-65 from the intestinal mucosal supernatant of zinc-deficient and zinc-supplemented rats. A. Rats fed a stock diet. B. Rats fed a zinc-deficient diet for 13 days. C. Rats fed a zinc-deficient diet for 13 days and supplemented with zinc by intraperitoneal injections of 200 µg Zn^{2+} in the form of $ZnSO_4$ 24 hours before the administration of the isotope.

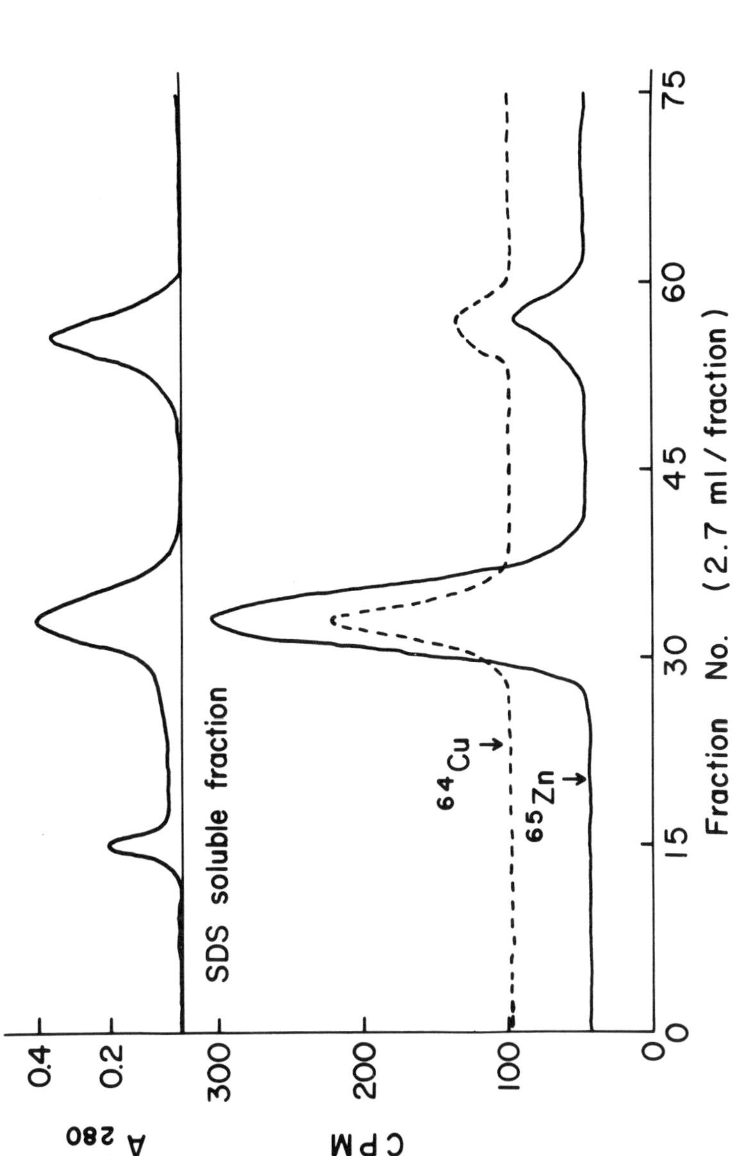

Figure 3. The elution profile of SDS - soluble copper-64 and zinc-65 from the intestinal pellet of normal rats.

fraction from the mucosal pellet was analyzed by gel-filtration, both copper-64 and zinc-65 were eluted in identical positions in two separate fractions (Figure 3). The major portion of both isotopes from the pellet was eluted approximately 40 ml behind the void volume and probably represents copper-64 and zinc-65 associated with material having a molecular weight well above one million daltons. The second fraction of copper-64 and zinc-65 from the pellet was eluted at the total volume of the column.

In plasma, both copper-64 and zinc-65 were eluted in a fraction which had an elution volume identical to that of albumin (Figure 4). In addition, a small fraction of zinc-65 in the plasma was eluted at the total volume of the column.

Characterization of Copper- and Zinc-Binding Components in Rat Intestine. Characterization of the various copper- and zinc-binding factors in rat intestine ranges from incomplete to nearly complete. To date, we have directed our efforts primarily toward elucidating the nature of the ligand associated with that fraction of intestinal zinc-65 which elutes at the total volume of a Bio-Gel P-10 column (peak 4 in Figure 1 and Figure 2). Although the following discussion deals mainly with the chemical and physical characteristics of the low molecular weight zinc-binding ligand, a brief description of some of the other copper- and zinc-binding factors is included where information is available.

As described in the results above, the major portion of copper-64 in the lumen and mucosal supernatant as well as a small fraction of zinc-65 were associated with a fraction which eluted at a volume similar to that of proteins with a molecular weight of approximately 10,000 daltons. Previous investigations have demonstrated that both orally administered copper-64 and zinc-65 are associated with a protein of approximately 10,000 daltons in the intestinal mucosal supernatant of chickens (9,11) and rats (12,13). Although neither the copper-binding protein nor the zinc-binding protein from intestine have been fully characterized, Evans et al. (14) have presented evidence suggesting that the 10,000 dalton metal-binding protein in mammalian intestine is similar to metallothionein. Metallothionein is characterized by a high sulfhydryl content, an absorption maximum near 250 nm, the presence of trace elements including copper, cadmium and zinc, and a molecular weight of approximately 10,000 daltons (15-17). The metal-binding protein from rat intestine fulfills these criteria (13) and therefore can be classified as metallothionein. However, far more research is required to completely delineate the nature of both the copper-binding metallothionein and the zinc-binding metallothionein since recent observations (18) suggest that there may be a variety of metallothioneins, each differing slightly in amino acid content and metal-binding characteristics.

The only information we have obtained regarding the SDS soluble, copper- and zinc-binding components in the mucosal pellet (Figure 3) is that the metals are associated with a protein in the high molecular weight fraction. When material from this fraction was treated with the proteolytic enzyme pronase and rechromatographed, the characteristic 280 nm absorption peak was completely absent and none of the isotope was recovered at the elution volume of the fraction from which the material was originally obtained.

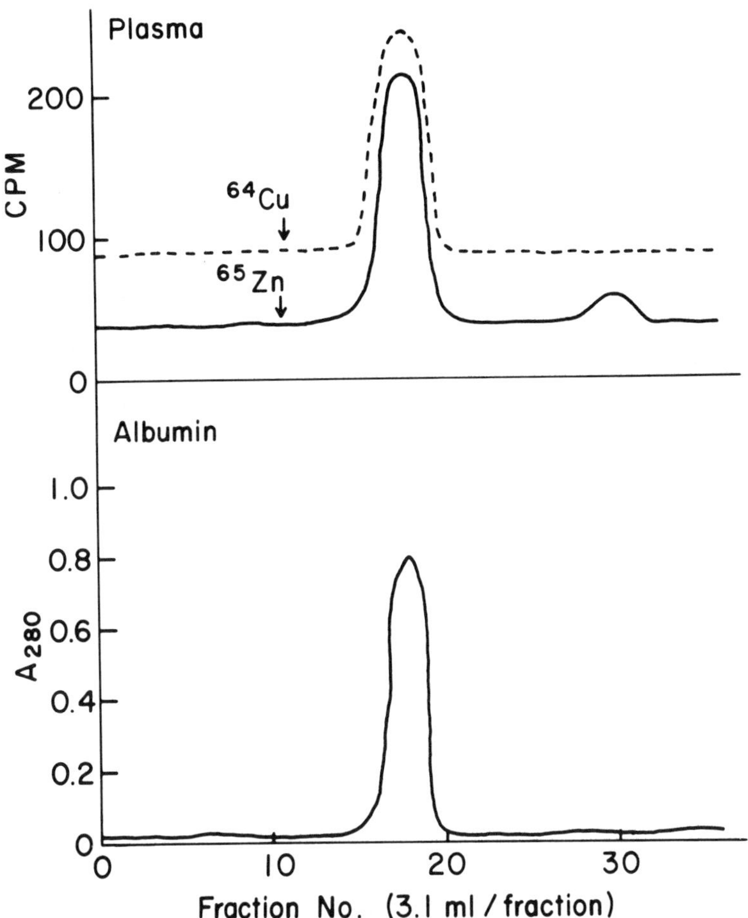

Figure 4. The elution profile of plasma copper-64 and zinc-65 compared with the elution profile of albumin.

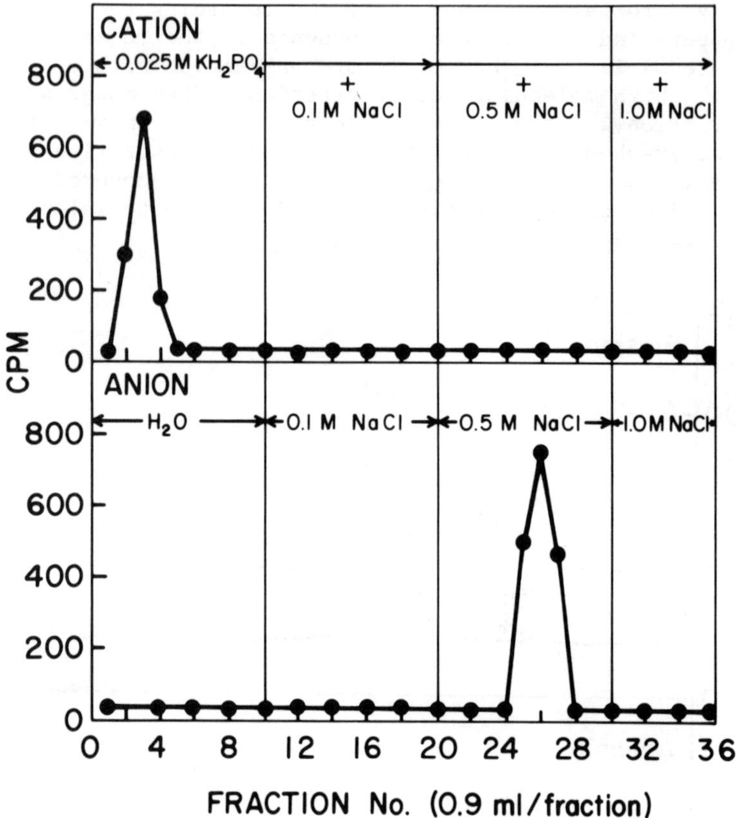

Figure 5. Ion exchange chromatography of the low-molecular-weight zinc-binding ligand from rat lumen and intestinal mucosa.

The results illustrated in Figure 1 and Figure 2 indicate that a large fraction of zinc-65 from the intestinal lumen and mucosal supernatant is eluted at the total volume of the column (peak 4). Initially, we suspected that these peaks represented unbound ionic zinc. However, when a solution of labeled $ZnCl_2$ was applied to the column, less than one percent of the radioactivity was recovered. This observation indicated that uncomplexed zinc-65 is adsorbed on Bio-Gel P-10 and suggested that the intestinal zinc-65 eluted at the total volume of the column may be associated with a specific ligand. Moreover, the results obtained from analysis by thin layer chromatography demonstrated that the low molecular weight fraction of intestinal zinc-65 differs markedly from zinc salts and simple zinc-amino acid complexes (19).

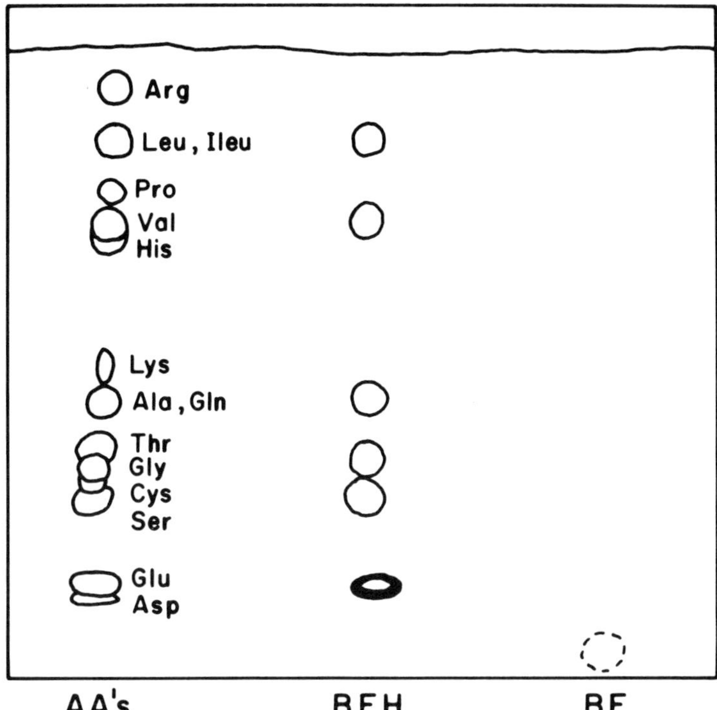

Figure 6. Thin-layer chromatography of the zinc-binding ligand from rat intestine. AA's:amino acid standards. BFH:zinc-binding factor hydrolyzed. BF:zinc-binding factor unhydrolyzed.

The ionic properties of the low-molecular weight zinc-65 complex from the intestine are depicted in Figure 5. When the sample was applied to a buffered cation resin and eluted with the same buffer, the zinc-65 was recovered readily in a single peak. In contrast, 0.5M NaCl was required to elute the zinc-65 complex from an anion resin. The anionic properties of the zinc-65 complex are probably due to at least one carboxylate group in the molecule since infrared analysis of the complex shows the presence of a carbonyl peak at 1630 cm^{-1} which is shifted to 1725 cm^{-1} by the addition of acid (unpublished observation).

Amino acid analysis of a highly purified sample of the zinc-binding ligand from intestine demonstrates that the compound is a peptide that contains at least six different amino acid residues (Figure 6). In two different solvent systems, thin-layer chromatography of a hydrolyzed sample suggested the presence of the following amino acids: leucine or isoleucine; valine; alanine; glycine, serine, threonine or possibly cysteine; and glutamic acid or aspartic acid.

The experiments described in this manuscript demonstrate that orally administered copper and zinc become associated with a variety of metal-binding ligands and macromolecules in the intestine. Although the nature of the experiments precludes any attempt to explain the mechanism of copper or zinc absorption, the results certainly suggest that the two elements may be transported from the lumen to the blood by two separate metal-binding components. In the intestinal lumen, copper is complexed with a protein similar to metallothionein whereas zinc is complexed with a small peptide. Both of these metal-binding ligands could conceivably be involved in transporting copper and zinc across the intestinal epithelium but more experimentation will be required to prove this unequivocally.

ACKNOWLEDGMENTS

The authors wish to thank Dr. J.G. Brushmiller, Department of Chemistry, University of North Dakota, for his technical advice and Mary Wolenetz and Carrie I. Grace for their technical assistance.

REFERENCES

1. Helbock, H.J. and Saltman, P., Biochim. Biophys. Acta, 135, 979 (1967).

2. Ratledge, C. and Marshall, B.J., Biochim. Biophys. Acta, 279, 58 (1972).

3. Pressman, B.C., Federation Proc., 32, 1698 (1973).

4. Magee, A.C. and Matrone, G., J. Nutr., 72, 233 (1960).

5. Hill, C.H., Payne, W.L. and Barber, C.W., J. Nutr., 80, 227 (1963).

6. Van Campen, D.R., J. Nutr., 88, 125 (1966).

7. Van Campen, D.R. and Scaife, P.U., J. Nutr. 91, 473 (1967).

8. Van Campen, D.R., J. Nutr. 97, 104 (1969).

9. Starcher, B.C., J. Nutr., 97, 321 (1969).

10. Evans, G.W., Grace, C.I. and Hahn, C., Proc. Soc. Exp. Biol. Med., 143, 723 (1973).

11. Suso, F.A. and Edwards, H.M., Jr., Proc. Soc. Exp. Biol. Med., 137, 306 (1971).

12. Van Campen, D.R. and Kowalski, T.J., Proc. Soc. Exp. Biol. Med., 136, 294 (1971).

13. Evans, G.W. The biological regulation of copper homeostasis in the rat. In "World Review of Nutrition and Dietetics", Ed., G. Bourne, S. Karger, Basel, Switzerland, Vol. 17, pp. 225-249, 1973.

14. Evans, G.W., Majors, P.F. and Cornatzer, W.E., Biochem. Biophys. Res. Commun., 40, 1142 (1970).

15. Kagi, J.H.R. and Vallee, B.L., J. Biol. Chem., 235, 3460 (1960).

16. Kagi, J.H.R. and Vallee, B.L., J. Biol. Chem., 236, 2435 (1961).

17. Pulido, P., Kagi, J.H.R. and Vallee, B.L., Biochemistry, 5, 1768 (1966).

18. Nordberg, G.F., Nordberg, M., Piscator, M. and Vesterber, O., Biochem. J., 126, 491 (1972).

19. Hahn, C., Grace, C.I. and Evans, G.W., Federation Proc., 32, 895 (1973).

15

METAL-ALBUMIN-AMINO ACID INTERACTIONS: CHEMICAL AND PHYSIOLOGICAL INTERRELATIONSHIPS

R.I. Henkin

Section on Neuroendocrinology

National Heart and Lung Institute, Bethesda, Maryland

Following the introduction and passage of transition metal ions such as copper, zinc, cadmium and nickel into appropriate portions of the gastrointestinal tract of man and other animals absorption across the gut mucosa occurs leading to the introduction of these metals into the blood. At this point, due to the functional characteristics of these metals, they appear bound to various ligands and transported to various tissues.

The purpose of this paper will be to examine the various transport forms of these transition metals in blood particularly emphasizing copper, zinc and cadmium. In addition mechanisms which control urinary excretion of these metals and which relate to the manner by which the liganded metals are presented to the kidney will be discussed in relationship to the transport forms of these metals. Finally physiological examples in man and in animals will be used to demonstrate the application of these principles.

Ligands of copper and zinc in serum

Copper

Normally copper and zinc appear in serum or plasma in two general forms, bound to macromolecular ligands or to micromolecular ligands (Figure 1). For copper, approximately 90% of this metal circulates bound to the large molecular weight glycoprotein ceruloplasmin. This glycoprotein contains approximately 6% carbohydrate, has a molecular weight of about 70,000 and contains 7 copper atoms per molecule. Copper is incorporated into ceruloplasmin in liver

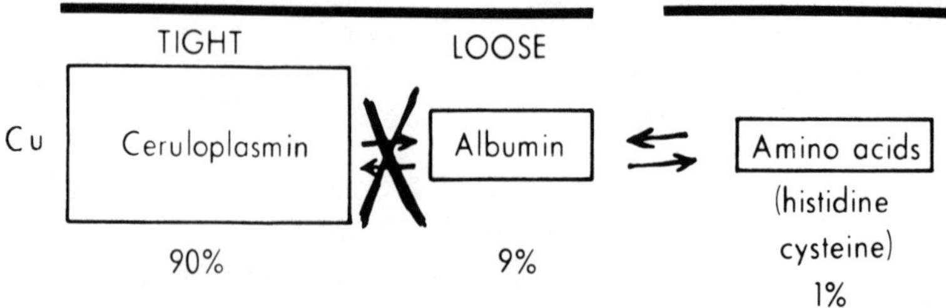

Figure 1. Major circulating ligands of copper in serum or plasma have been divided, for convenience, into macromolecular and micromolecular ligands. The major macromolecular ligand, ceruloplasmin, binds approximately 90% of the circulating copper ligands. The other macromolecular copper ligand, albumin, binds approximately 9% of the circulating copper and this copper is in equilibrium with copper bound to micromolecular ligands, mainly the amino acids histidine and cysteine, which represent about 1% of the circulating copper.

and this copper is not freely exchangeable with other potential copper ligands in serum due to its high association constant. Ceruloplasmin does not incorporate exogenous copper *in vitro* nor does it readily lose copper during dialysis or purification. Apoceruloplasmin does not appear to be normally present in serum. Removal of copper from ceruloplasmin also appears to occur only in the liver. This occurs following an approximate 20% reduction of the sialic acid from the carbohydrate-chains of the ceruloplasmin (Van den Hamer, et al., 1970) at which time ceruloplasmin is "retired" to the liver presumably to be degraded and metabolized.

Of the remaining 10% of circulating copper, approximately 9% is bound to another macromolecular species, albumin (Neumann and Sass-Kortsak, 1967; Harris and Sass-Kortsak, 1967; Sarkar and Kruck, 1966; Sarkar and Wigfield, 1968) and probably to a specific locus or binding site; i.e., to the amino acid histidine closest to the N-terminal residue of the albumin. Animals, such as the dog, which do not possess histidine in this position on their albumin exhibit differences from man in copper binding. However, in man, the copper in this position may reflect some aspect of the copper absorbed from the gastrointestinal tract since the copper in albumin is in equilibrium with other copper ligands in serum.

The remaining 1% of circulating copper is bound to micromolecular species; i.e., to amino acids (Sarkar and Kruck, 1966; Sarkar and Wigfield, 1968), and, in man, under normal conditions, primarily to the amino acids histidine and cysteine. It is primarily these micromolecular liganded species, histidine-copper and cysteine-copper and the macromolecular liganded albumin-copper that comprise the major exchangeable or "loosely bound" copper complexes in blood.

Zinc

Zinc exhibits a generally similar pattern of binding although distribution among the various ligands differs from that of copper (Figure 2). For zinc approximately 32% of the circulating metal is

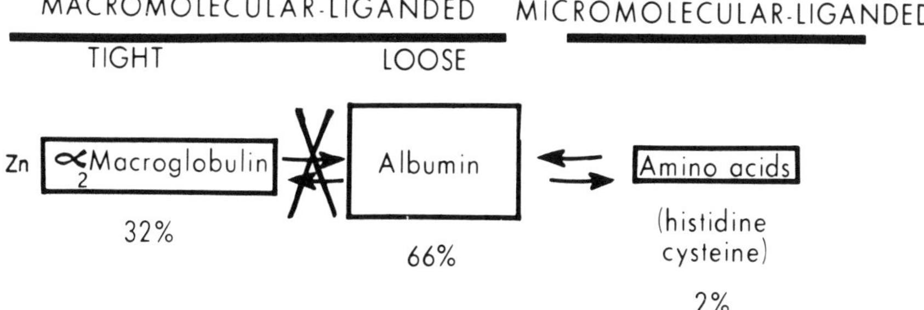

Figure 2. Major circulating ligands of zinc in serum or plasma have been divided, for convenience, into macromolecular and micromolecular ligands. The largest, in size, macromolecular ligand, an α_2-macroglobulin, binds approximately 32% of the circulating zinc but this zinc is not in equilibrium with other circulating zinc ligands. The major macromolecular zinc ligand, albumin, binds approximately 66% of the circulating zinc and this zinc is in equilibrium with zinc bound to micromolecular ligands, mainly the amino acid histidine, which represents about 2% of the circulating zinc.

bound to a macromolecular species, an α_2-macroglobulin (Parisi, et al., 1970). This protein is also a glycoprotein, containing approximately 6% carbohydrate, with a molecular weight of approximately 800,000. As with copper, zinc is presumably incorporated into this protein only in the liver and is not freely exchangeable with other potential zinc ligands in serum. α_2-macroglobulin does not incorporate exogenous zinc, *in vitro*, nor does it readily lose zinc during dialysis or purification. Apoα_2-macroglobulin does not appear to be normally present in serum. Removal of zinc from this α_2-macroglobulin appears to occur only in the liver, the degradation and metabolism of the protein occurring in the liver presumably following asialiation. Of the remaining 68% of the circulating zinc, approximately 66% is bound to another macromolecular species, albumin (Parisi, et al, 1970: Vikbladh, 1951; Wolff, 1956) and, as for copper, to its histidine moieties (Gurd and Goodman, 1952; Giroux and Henkin, 1972). It is the zinc-albumin complex which equilibrates rapidly with orally ingested or injected radio-isotopic zinc (Suso and Edwards, 1971). Since albumin has 16 histidine moieties it is theoretically possible to bind 16 zinc atoms to it. However in normal serum albumin is undersaturated with zinc. The remaining 2% of the circulating zinc is bound to micromolecular species; i.e., to amino acids, and in man, under normal condition, primarily to the amino acids histidine and cysteine. It is these micromolecular liganded species and the macromolecular liganded albumin-zinc complex that comprise the major exchangeable or "loosely bound" zinc complexes in blood.

In order to demonstrate these formulations a series of studies were carried out *in vitro* and *in vivo*. *In vitro* by competition experiments a formation constant for the 1:1 zinc-human serum albumin complex was calculated and by potentiometric titrations, formation constants for several amino-acid-zinc complexes were calculated (Giroux and Henkin, 1972). These results indicated that 2% of the "loosely bound" zinc in normal human plasma was bound to amino acids and that the only major amino acid ligands in plasma were cysteine and histidine. The distribution of loosely bound metal between macro- and micromolecular ligands in *in vitro* competition experiments was utilized to show that the avidity of the macromolecular ligands of loosely bound copper, zinc, and cadmium in normal human serum was equivalent to that of human serum albumin for these metals. These studies, using a technique which does not involve fractionation of serum proteins, showed that albumin is possibly the only macromolecular ligand of loosely bound copper, zinc and cadmium in human serum. Further, histidine, a physiologically important serum ligand of copper and zinc was shown to strip more copper than zinc from their corresponding complexes with albumin. The results of these studies showed that the cadmium-albumin complex was the stablest of the three metal-protein complexes with respect to dissociation by histidine.

In vivo the changes in the levels of zinc and copper which occur in blood and urine following the administration of L-histidine were studied. In these studies in both man and animals L-histidine administration was uniformly associated with the loss of zinc and copper from the body as reflected in an increased excretion of these metals in the urine and, for zinc, a decreased concentration in blood.

MATERIALS AND METHODS

In Vitro Studies. (Giroux and Henkin, 1972, 1972a). Potentiometric titrations of solutions of amino acids and zinc sulfate were carried out using Albert's method (Albert, 1950; Albert, 1952); aqueous 0.15 M NaCl was used as solvent. Amino acid pKa (NH_3^+) values were determined for glycine, glutamine and serine; they agreed precisely with standard reference values. Standard references values (Eichhorn, 1970) were used for the other amino acids.

Competition for zinc between albumin and amino acids was analyzed by gel chromatography on 1-cm diameter columns of Sephadex G-25 fine, 16.4-16.6 ml bed volume equilibrated with a buffer of 0.15 M NaCl, 1 mM sodium diethylbarbiturate, supplemented with different amino acids and adjusted to pH 7.4. Human serum albumin, adjusted to pH 7.4, was diluted to a solution of 4.5 g albumin and approximately 400 μg zinc per 100 ml of column buffer; a 2-ml aliquot was applied to the column, followed by additional column buffer. The 6th through the 9th ml of effluent was collected in a single 4-ml sample and analyzed for protein (Lowry et al., 1951) and zinc (Meret and Henkin, 1971). The ratio of zinc to protein in the effluent sample was compared to the ratio observed when zinc-albumin complex was prepared and chromatographed using amino acid-free buffer.

Five samples of human serum were used; each was a pooled sample of sera obtained from four or more fasting normal individuals taken at 9 a.m. using techniques previously described (Lifschitz and Henkin, 1971). The albumin concentration of each pooled sample was determined by the HABA technique (Porter and Waters, 1966).

Competition between protein and amino acid for loosely bound copper, zinc and cadmium was analyzed by gel chromatography (Giroux and Henkin, 1972). Loosely bound metal pools in serum were labelled and the albumin-metal complexes were prepared by addition of ^{64}Cu, ^{65}Zn, or ^{109}Cd to protein solutions. Less than 0.1 μg/ml copper and zinc were added to the serum samples. Total copper and zinc in normal serum is about 1 μg/ml for each metal (Christian, 1969; Schroeder and Nason, 1971; Meret and Henkin, 1971). Approximately 0.01 equivalent of cadmium was added per albumin equivalent to the protein solutions. Normal human serum has a very low and highly variable cadmium content, in the microgram per liter range (Kubota et al., 1968; Boyett and Sullivan, 1970). Sera and albumin solutions were adjusted to pH 7.4 using solutions of HCl or NaOH. Solutions of histidine and glycine were prepared in 5 mM sodium diethylbarbiturate, 0.15 M NaCl and adjusted to pH 7.4. Equal volumes of amino acid and albumin solution or serum were mixed to study copper and zinc distribution. Two volumes of amino acid solution were mixed with one volume of albumin solution or serum when cadmium distribution was studied. Aliquots of 0.2 ml of protein-metal-amino acid mixed solutions were chromatographed on columns of Sephadex G-25 fine. The columns, of approximately 1.7 ml bed volume, were formed in Pasteur pipets and equilibrated and eluted with a buffer of 5 mM sodium diethylbarbiturate, 0.15 M NaCl, pH 7.4. Five fractions were eluted. The first fraction, 1 ml, contained most of the macromolecular components of the mixture applied to the column. Three fractions of 0.1 ml each were collected in the trough between protein-bound and amino acid-complexed radioisotope peaks. After the fourth fraction the column was washed with 0.7 ml of an amino acid, or an amino acid-EDTA solution, followed by additional equilibration buffer. The effluent, 3-4 ml, was collected as the fifth fraction. In this way the metal complexed with amino acid or adsorbed to the Sephadex gel was eluted and the column prepared for the next chromatogram. Radioactivity was determined in a Packard AutoGamma Spectrophotometer. Recovery of radioactivity applied to the columns was quantitative for each isotope. An arbitrary division, almost always between the third and fourth fractions eluted from a column, was used to quantitate macromolecular-bound vs. radioisotope not bound by macromolecules. Radioisotope was removed from albumin and serum proteins when dilutions of protein solutions in amino acid-free buffer were chromatographed on Sephadex columns. The efficacy of removal of radioisotope from macromolecules by amino acids was calculated relative to such chromatograms developed in the absence of added amino acid.

In Vivo Studies

Human studies (Henkin et al, 1972): Six patients with clinically documented scleroderma and six normal volunteers served as subjects. Each subject was hospitalized at the Clinical Center, National Institutes of Health on an air conditioned ward, ate a constant diet and consumed a fixed amount of distilled water. The patients were studied under control conditions and following the oral administration of the amino acid L-histidine given in divided dosages of 4.1 g, 8.2 g, 16.4 g, 32.8 g, 49.2 g and 65.6 g usually for periods of 2-3 days each. The normal volunteers were studied under control conditions and following the oral administration of the amino acid L-histidine given in divided dosages of 4.1 g, 8.2 g, 16.4 g and 32.8 g usually for periods of 2-3 days each. Blood was collected fasting in the early a.m. from each subject during the administration of each dosage of drug and processed as previously noted (Lifschitz and Henkin, 1971). Twenty-four hour collections of urine were obtained daily in plastic containers from each subject during the study. After collection the total urine volume was measured in a plastic graduated cylinder, an aliquot was placed into a plastic container and it was stored at -20°C until assayed.

Blood and urine were analyzed as noted previously (Meret and Henkin, 1971). Means ± 1 SEM were calculated for blood and urine from patients and normal volunteers under each drug condition.

Animal studies:† Twenty-one male Sprague Dawley rats, 270 ± 20 g in weight were individually housed in stainless steel metabolic cages. The animals were kept in an air conditioned room in which temperature (23°C), humidity and light (12 hours dark 5:00 p.m. to 5:00 a.m., 12 hours light) were controlled. Rats were arbitrarily divided into 3 groups of 7 rats. One group (C) received a control diet [containing $ZnCO_3$, 58 ppm (McConnell and Henkin, 1974)] *ad libitum* throughout the study. A second group (H) received the control diet in an amount taken in by the control rats to which L-histidine monohydrochloride was added such that each rat received a dosage of L-histidine of 1 g/kg for the first two weeks of the study, 5 g/kg for the next 8 weeks and 7.5 g/kg for the last 3 weeks. A third group (HZ) received the control diet in an amount taken in by the control rats to which both L-histidine monohydrochloride in an amount similar to that given to the H group, and zinc sulfate, 8 g/kg, was added. Urine was collected over a 24 hour period daily free of feces from each rat and analyzed for total zinc and copper as noted previously (Meret and Henkin, 1971). Results under each drug condition were reported as μg metal/100 gm body weight/24 hours.

RESULTS

In Vitro Studies

Zinc-Albumin Complex

The value for the formation constant at physiological pH of the zinc-albumin complex was obtained from data shown in Figure 3. By combination of Equilibria 2 and 3 (disregarding ionic charge)

$$\mathrm{Zn} + \mathrm{A} \rightleftharpoons \mathrm{Zn}\cdot\mathrm{A} \qquad K_1 \qquad (1)$$

$$\mathrm{Zn} + 2\mathrm{A} \rightleftharpoons \mathrm{Zn}\cdot\mathrm{A}_2 \qquad \beta_2 \qquad (2)$$

$$\mathrm{Zn} + \mathrm{HSA} \rightleftharpoons \mathrm{Zn}\cdot\mathrm{HSA} \qquad K_{HSA} \qquad (3)$$

an equilibrium constant may be formulated,

† Carried out with the assistance of Dr. Mei-Lie Swenberg.

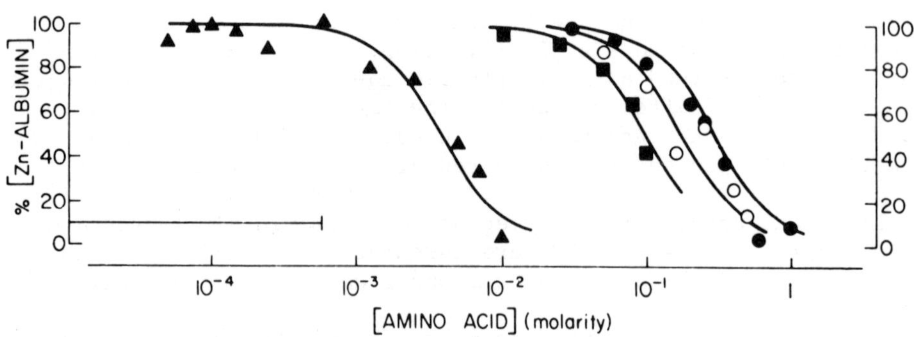

Figure 3. Percentage of albumin-bound zinc in column effluent in competition experiments, relative to amino acid-free controls, when histidine (Δ), asparagine (■), threonine (O), and glycine (●), were added to column buffer and albumin solution. Curves were drawn using formation constants at pH 7.4 derived from data in Tables I and II and Albert, 1950 and Hallman et al., 1971. The horizontal bar indicates the normal concentration range of serum amino acids.

$$\frac{[Zn \cdot (A_2)]}{[Zn \cdot HSA]} \frac{[HSA]}{[A]^2} = K \qquad (4)$$

$$K = \frac{\beta_2}{K_{HSA}} \qquad (5)$$

where A is amino acids and HSA is human serum albumin.

The equilibrium value K reflects the avidity with which any particular amino acid competes with albumin for zinc ion. Log-log plots of Eqn 4 are shown in Figure 4 for the amino acids histidine, asparagine, threonine and glycine. The insert of Figure 4 is the log-log plot of Eqn 5 employed to determine the value of 7.0 for log K_{HSA}. The proportion of Zn · A complex is small when large amounts of amino acid-complexed zinc are present in these experi-

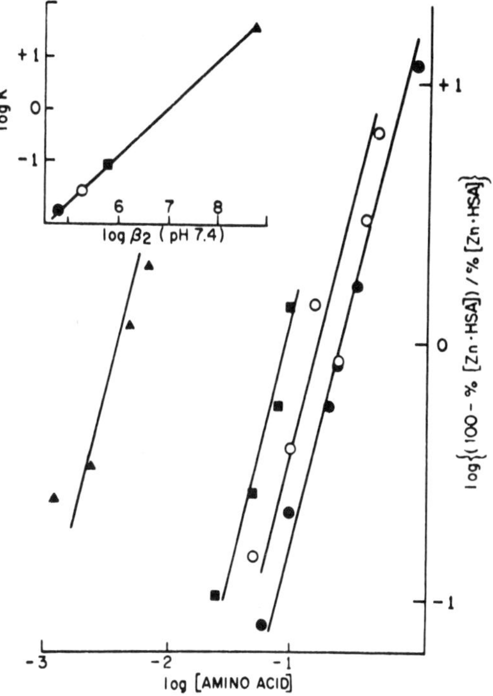

Figure 4. Ratios of amino acid-to albumin-complexed zinc as a function of amino acid concentration (Eqn 4). The percentage of zinc in the Zn · $(A)_2$ complex is 100--% [Zn · HSA]. Least squares best-fit lines of slope = 2 are indicated. K was determined for each amino acid from the value of the intercept where [Zn · HSA] = [Zn · $(A)_2$]. In the insert K values are plotted as a function of β_2 (Eqn 5). K_{HSA} was evaluated at the point where log K = o. Log β_2 values at pH 7.4 were calculated from data in Table I for histidine, threonine and glycine and from Albert's data for asparagine (Albert, 1950). Symbols are the same as in Figure 3.

ments, so use of Eqn 1 is unnecessary in this determination of K_{HSA}.

Zinc-Amino Acid Complexes

The formation constants of the first step (K_1) and overall combined reaction (β_2) for several amino acid-zinc complexes are presented in Table I. Values for histidine, threonine, glutamine, serine and glycine agree with values obtained by others (Albert, 1950; Albert, 1952; Hallman et al., 1971; Maley and Mellor, 1950;

TABLE I

Concentrations of Amino Acids, Albumin and Zinc in Normal Human Serum and Formation Constants of Zinc-Amino Acid Complexes

Only formation constants determined in the present work are listed; the literature values are presented graphically in Figure 1. Neutral acidic and basic amino acids are listed in order of increasing pK_a-(NH_3+)** except for cysteine and histidine. Concentrations of amino acids were taken from the literature; cysteine and cystine (Brigham et al., 1960), all others (Hamilton, 1968).

Serum constituent	Serum concentration (μM)	Zinc complex formation constants	
		log K_1	log β_2
Cysteine	33		
Histidine	74		12.28§
Cystine	42	4.64*	8.62
Asparagine	44		
Threonine	129	4.72	8.68
Tyrosine	52		
Glutamine	568	4.39	8.14
Phenylalanine	53	4.50	8.36
Serine	115	4.62	8.48
Methionine	23		
Citrulline	29	4.47	8.22
Tryptophan	48		
Leucine	111		
Valine	214		
Isoleucine	63		
Alanine	336		
Glycine	237	5.25	9.57
α-Aminoisobutyric acid	20		
Proline	184		
Aspartic acid	8		
Glutamic acid	58		
Ornithine	60		
Arginine	75		
Lysine	153		
Albumin	652	(4.5 g per 100 ml)	
"Loosely bound" zinc	10	(2/3 of 98 μg per 100 ml)	

*Values for the monoprotonated dianion, thus K_1' and β_2''; determined by titrations of 0.5 mM cystine and zinc.

** Eichhorn, 1970.

§Estimated from $\log \beta_2 = 2 \log K_1$ --0.66.

Li and Manning, 1955; Perkins, 1952; Monk, 1951; Bjerrum, 1950), while values for phenylalanine and citrulline have not been previously reported. From these constants, "apparent" formation constants were calculated for pH 7.4. For each amino acid this apparent value allows use of the total concentration of amino acid and depends on the pK_a of the zinc coordinating groups (O'Sullivan, 1969).

A plot of log β_2 (pH 7.4) as a function of $pK_a(NH_3+)$ for most of the serum amino acids is shown in Figure 5. Aside from cysteine and histidine, which are exceptional among the α amino acids in not coordinating zinc through the α amino and α carboxyl groups (Albert, 1952), the most avid zinc ligands at pH 7.4 tend to be those with the lowest pK_a (NH_3+), as Albert (1950) has pointed out. Theoretically, among the neutral amino acids, cystine, asparagine, threonine, tyrosine, glutamine, phenylalanine and serine might be physiologically important zinc ligands in serum and, of course, cysteine, histidine and various mixed complexes of these might be also, depending upon their concentrations in serum. Alanine, glycine and valine must be considered too, because of their relatively high concentrations in serum. Formation constants at pH 7.4 indicate that the acidic amino acids bind zinc well, but their concentrations in serum are low while the basic amino acids complex zinc poorly.

Zinc Distribution Among Albumin and Serum Amino Acids

Equilibria 1-3 may be generalized. All zinc complexes may be analytically expressed as members of a set of linear equations, provided the concentration of each zinc-ligand complex is a small fraction of free ligand concentration, a requirement met in the present work:

$$\frac{[\text{Zn complex}]}{[\,\Pi \text{ ligands}]} - [Zn^{2+}] \cdot \text{formation constant} = 0 \qquad (6)$$

Formation constants used in application of Eqn 5 to calculate distribution of loosely bound zinc among albumin and various amino acid complexes at pH 7.4 are shown in Table II. Amino acids which complex zinc poorly or are present in low concentration are unlikely to complex zinc in serum; such unlikely complexes were not considered. Where possible, formation constants were derived from data of Hallman et al., 1971 since they reported the important mixed complex constants. A computer (Kronos time sharing system, library program $SIMEX_I$) solved the simultaneous linear equations for distributions of zinc among 20 likely serum zinc complexes for the following conditions: normal concentrations of serum constituents (Table I); 2/3 normal albumin concentrations; 5-fold increased cysteine, histidine and cystine concentration. Table II lists the resultant distributions.

Increasing concentrations of the metal ligands histidine and

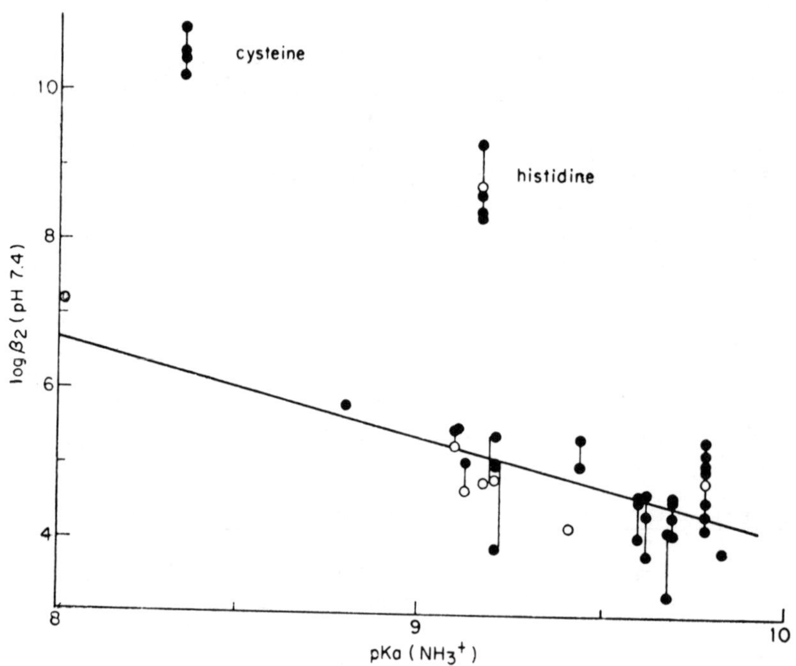

Figure 5. Log β_2 values at pH 7.4 of neutral serum amino acids as a function of amino acid pK_a (NH_3+); i.e., the values are entered left to right in the same order as the neutral amino acids are listed in Table 1. Values at pH 7.4 were calculated from literature β_2 values using the pK_a values of the original reports. Values determined in the present work are shown as O, literature values as O. The range of values reported for an amino acid is indicated by a vertical line; e.g. four different log β_2 values are shown for the zinc-albumin complex, eight for the zinc-glycine complex.

TABLE II

Complex Formation Constants at pH 7.4 and Loosely Bound Zinc Distribution Among Serum Ligands

Formation constants were calculated from data in Hallman, et al., 1971 except as noted; values in parentheses from the present work. These formation constants were used to calculate zinc distribution in human serum as described in the text. Complexes are listed in order of their calculated relative abundance in normal serum; however, if a complex accounts for less than 0.1% of the loosely bound zinc its value is not entered in the table.

Formation constant at pH 7.4 log value	Complex	Loosely bound zinc distribution in human serum (%)				
		Normal*	2/3 albumin	5-fold increase in		
				cysteine	histidine	Cystine
(6.98)	[Zn·albumin]	98.0	97.1	74.2	94.7	97.9
10.82	[Zn·(cysteine)$_2$]$^{2-}$	1.1	1.7	21.6	1.1	1.1
10.12	[Zn·histidine·cysteine]$^-$	0.5	0.8	1.9	2.5	0.5
9.77	[Zn·Hcysteine-cysteine]$^-$	0.1	0.2	1.9		0.1
4.79	[Zn·histidine]$^+$		0.1		0.4	
9.09	[Zn·H·histidine·cysteine]			0.2	0.2	
8.62 (8.72)	[Zn∘(histidine)^2]				0.9	
8.65	[Zn·H·histidine·cystine] Zn^{2+}				0.1	0.1
8.12	[Zn·histidine·cystine]$^-$					
2.82 (2.65)	[Zn·glutamine]$^+$					
(3.93)	[Zn·Hcystine]$^+$					
8.05	[Zn·H^2∘histidine·cystine]$^+$					
2.92 (2.87)	[Zn·glycine]$^-$					
3.10 (3.01)	[Zn·threonine]$^+$					
3.02 (2.80)	[Zn·serine]$^+$					
2.47	[Zn·alanine]$^+$					
3.22**,§	[Zn·asparagine]$^+$					
2.51	[Zn·valine]$^+$					
3.08§,†	[Zn·tyrosine]$^+$					
(7.20)	[Zn·(Hcystine)$_2$					

*Table I; **(Albert, 1950); §Footnote § , Table 1; †(Albert, 1952).

glycine decreased the concentrations of protein-bound copper, zinc, and cadmium in solutions of serum albumin and in serum. This is illustrated for ^{65}Zn and histidine (Figure 6), ^{65}Zn and glycine (Figure 7), ^{64}Cu and histidine (Figure 8), and for ^{109}Cd and histidine (Figure 9). In each case, the isotope distribution curves for albumin and for serum, with their respective standard deviations, overlapped.

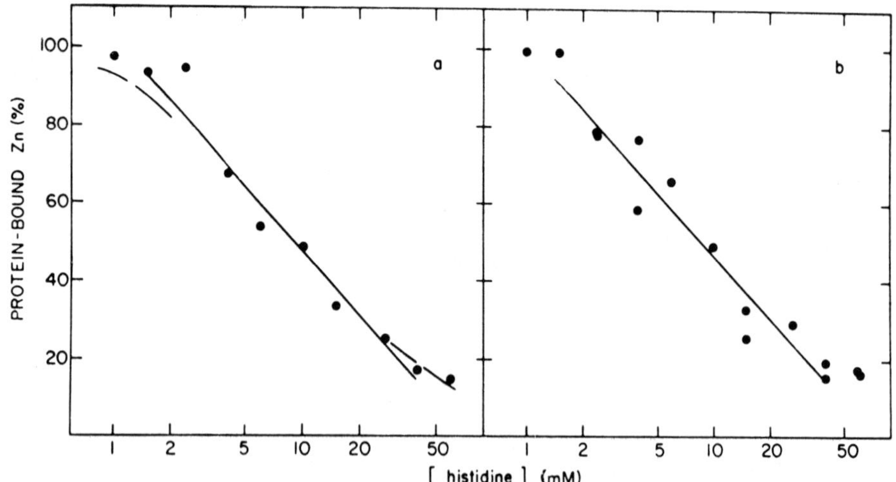

Figure 6. ^{65}Zn complexed by albumin (a) and serum macromolecules (b) as a function of added histidine. Controls having no added amino acid were used to establish a 100% protein-bound value. Other data were plotted as proportions of this value. Each point derives from one chromatogram as described in the text. The curves approach asymptotes of 100% protein-bound metal and 0% protein-bound metal at low and high amino acid concentrations, respectively. Straight lines were fitted to data in the central and approximately linear portion of each curve. The regressions were not statistically significantly different. The solid line is the least squares regression. The dashed line in (a) is a calculated curve and is explained in the text.

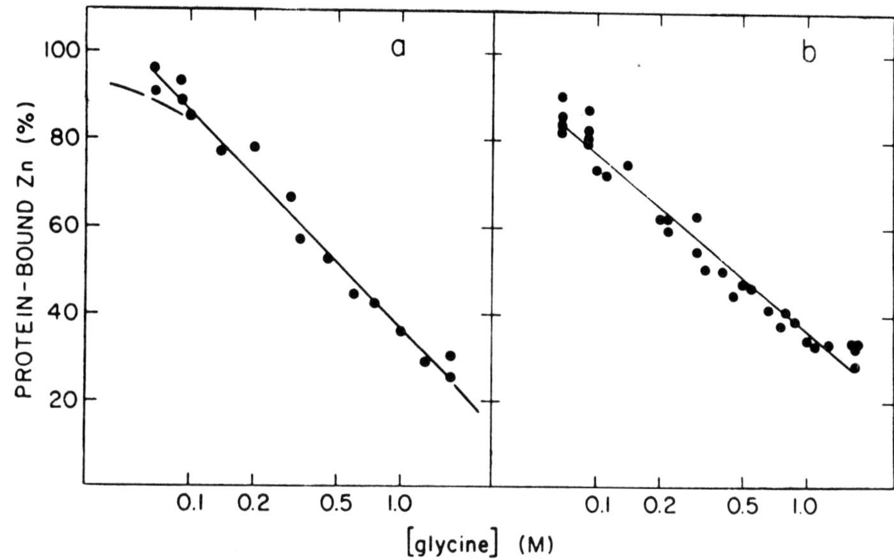

Figure 7. ^{65}Zn complexed by albumin (a) and serum macromolecules (b) as a function of added glycine. Further explanation as in Figure 6.

The amino acid concentration necessary to remove one-half of the isotope which is protein-bound in serum in the absence of added amino acid was seen to be approximately 4, 9 and 27 mM histidine for ^{64}Cu, ^{65}Zn, and ^{109}Cd, respectively, and 0.5 M glycine for ^{65}Zn. Previously the relative effectiveness with which several amino acids compete with human serum albumin for zinc was shown to depend upon their relative avidity for zinc under the experimental conditions (Giroux and Henkin, 1972).

In Vivo Studies

Histidine Administration in Man: Table III illustrates the changes observed following administration of graded series of doses of histidine to patients with scleroderma and to normal volunteers. During the administration of 8.2 g of the amino acid there was a significant increase ($p < 0.01$) in the urinary excretion of zinc in both the patients and the volunteers without any significant drop in the serum concentration of this metal. Although urinary copper excretion increased in both groups the change was not significant. During the administration of 16.4 g of L-histidine, urinary

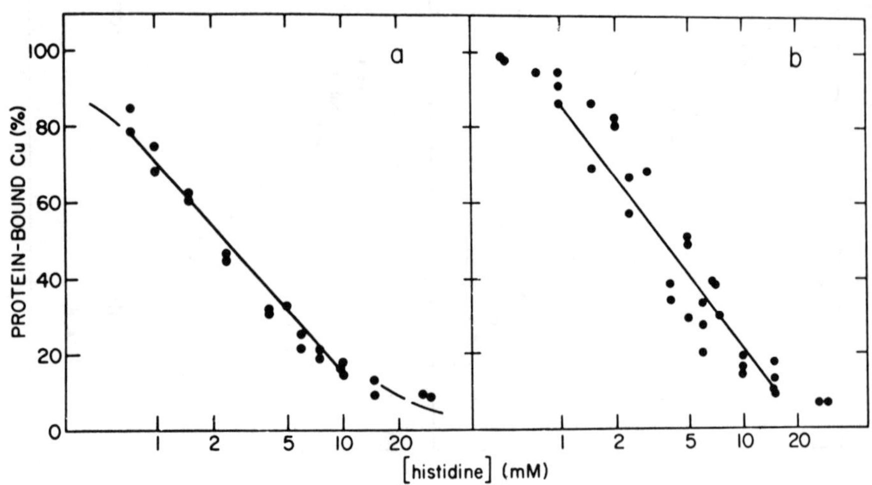

Figure 8. ^{64}Cu complexed by albumin (a) and serum macromolecules (b) as a function of added histidine. Further explanation as in Figure 6.

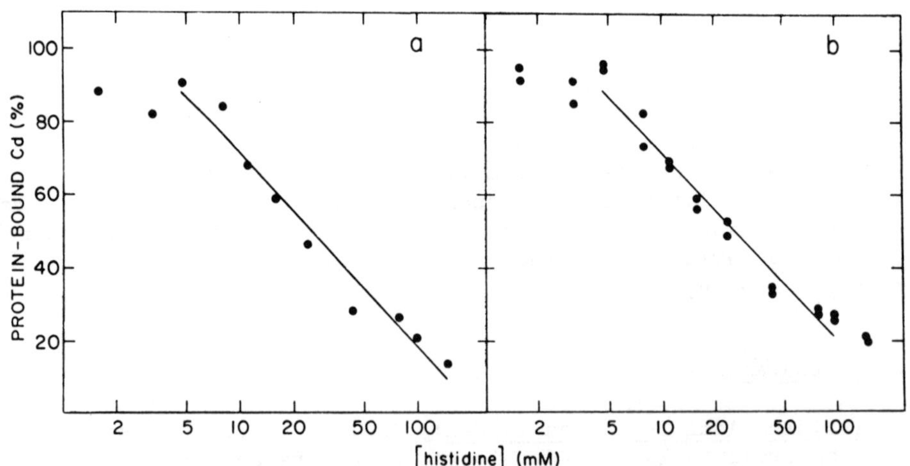

Figure 9. ^{109}Cd complexed by albumin (a) and serum macromolecules (b) as a function of added histidine. Further explanation as in Figure 6.

TABLE III

Changes in Serum and Urine Zinc and Copper Following L-histidine in Patients with Scleroderma and in Normal Volunteers

		Patients (6)	Normal Volunteers (6)
BASELINE			
Serum - μg/dl	Zn	81 ± 6[+]	90 ± 3
	Cu	121 ± 10	98 ± 8
Urine - μg/24 hr	Zn	422 ± 28	508 ± 60
	Cu	34 ± 3	34 ± 4
HISTIDINE 8.2 g			
Serum - μg/dl	Zn	81 ± 8	81 ± 1[$]
	Cu	118 ± 12	98 ± 9
Urine - μg/24 hr	Zn	793 ± 100*	977 ± 119*
	Cu	47 ± 9	39 ± 2
HISTIDINE 16.4 g			
Serum - μg/dl	Zn	70 ± 3	78 ± 2*
	Cu	117 ± 9	100 ± 10
Urine - μg/24 hr	Zn	903 ± 115*	1303 ± 108*
	Cu	40 ± 4	54 ± 4*
HISTIDINE 32.8 g			
Serum - μg/dl	Zn	69 ± 5	66 ± 2*
	Cu	115 ± 12	95 ± 9
Urine - μg/24 hr	Zn	2584 ± 420*	2979 ± 311*
	Cu	53 ± 7¢	66 ± 6*
HISTIDINE 49.2 g			
Serum - μg/dl	Zn	55 ± 6*	
	Cu	113 ± 11	
Urine - μg/24 hr	Zn	5938 ±1416*	
	Cu	112 ± 21*	
HISTIDINE 65.6 g			
Serum - μg/dl	Zn	48 ± 3*	
	Cu	109 ± 10	
Urine - μg/24 hr	Zn	7876 ±1873*	
	Cu	217 ± 21*	

[+]M ± 1 SEM *P < 0.01 with respect to baseline condition (two tailed t test) [$]P < 0.02 with respect to baseline condition (two tailed t test) ¢P < 0.05 with respect to baseline condition (two tailed t test). Number in parenthesis indicates subject number.

zinc excretion increased still further in both groups to levels greater than twice baseline and serum zinc concentration fell significantly ($p < 0.01$) in the volunteers. In this latter group given this dose of amino acid, urinary copper exretion increased significantly over baseline levels. During administration of 52.8 g and higher doses of the amino acid a similar pattern of increases in urinary zinc and copper and decreases in serum copper occurred in both groups. In the patients at the highest dose given mean urinary zinc excretion reached levels approximately 20 times greater than baseline whereas urinary copper excretion increased by a factor of 7.

<u>Histidine Administration in Rat</u>: Figures 10 and 11 illustrate the changes observed in urinary zinc and copper excretion among rats fed control diet and those fed either L-histidine or L-histidine and zinc. Given a dose of L-histidine of 1 g/kg, there were no significant differences in urinary zinc or copper excretion among the groups

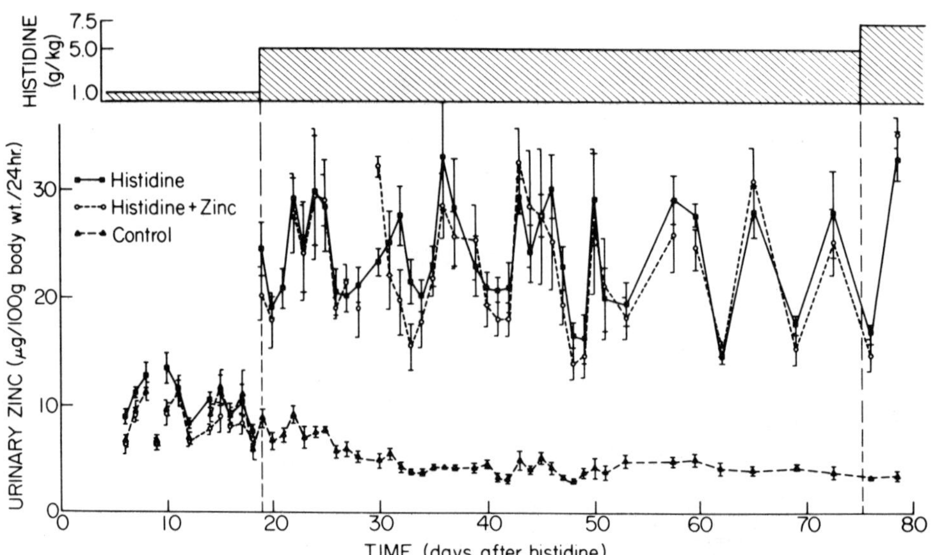

Figure 10. Urinary zinc excretion in three groups of 7 rats each; control (control diet throughout), histidine (control diet with 1.0, 5.0 and 7.5 g/kg histidine added) and histidine + zinc (control diet with 1.0, 5.0 and 7.5 g/kg histidine and with 8 g/kg zinc sulfate added). Histidine and histidine + zinc groups were pair-fed with the control group. Mean ± 1 SEM for each group are plotted at each time point.

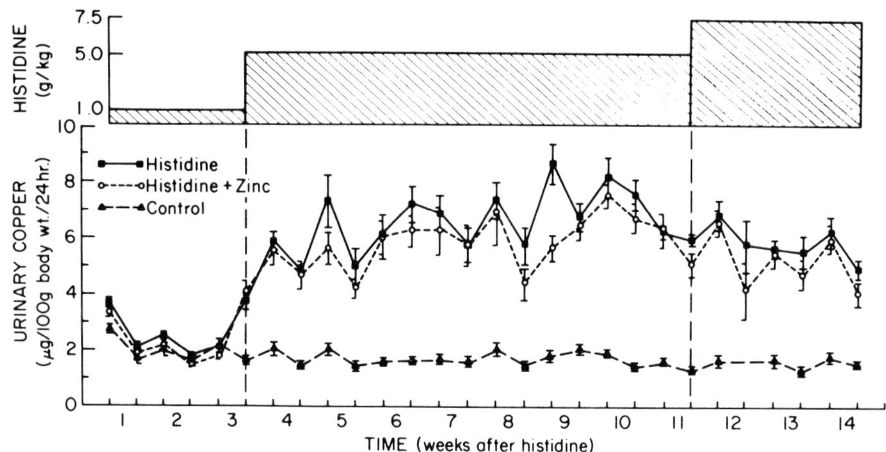

Figure 11. Urinary copper excretion in three groups of 7 rats each; control (control diet throughout), histidine (control diet with 1.0, 5.0 and 7.5 g/kg histidine added) and histidine + zinc (control diet with 1.0, 5.0 and 7.5 g/kg histidine with 8 g/kg zinc sulfate added). Histidine and histidine + zinc groups were pair-fed with the control group. Mean ± 1 SEM for each group are plotted at each time point.

studied. However, on both 5 g/kg and 7.5 g/kg there were significant ($p < 0.01$) increases in both zinc and copper excretion in the rats fed L-histidine alone or L-histidine with zinc with respect to the rats fed the control diets alone. For zinc and copper the increase was the same being somewhat greater than a factor of 3. There was little change in the excretion of either urinary copper or zinc when the dose was increased from 5 to 7.5 g/kg.

DISCUSSION

In Vitro Studies

Stability Constants: Under physiological conditions considerably less than one zinc equivalent is bound to albumin in plasma. Hypothesis of a single albumin zinc-binding site (Eqn 3) is adequate for describing distribution of loosely bound zinc. The value of log K_{HSA} in Table II is a formation constant at pH 7.4 for the 1:1 zinc-albumin complex determined in amino acid competition experiments in which the ratio of zinc to albumin at no time exceeded 0.1 on a molar basis. Whether histidyl residues (Gurd and Goodman, 1952; Tanford, 1952), the sulfhydryl group (Klotz et al., 1952), carboxylate groups (Tanford, 1952; Perkins, 1964), or the amino terminal portion of albumin (including a histidyl residue), which is the locus of the strongest copper-binding site (Peters, 1960; Breslow, 1964; Peters and Blumenstock, 1967; Tsangaris, et al., 1969), are involved in complexing the first zinc equivalent cannot be determined from a stability constant.

Competition for Zinc Among Albumin and Amino Acids: Amino acids which most avidly complex zinc at pH 7.4 are most effective in stripping zinc from the zinc-albumin complex. In systems of zinc, albumin and amino acids the amount of non-protein bound zinc may be calculated using Eqn 6 and formation constants from Tables I and II. Prasad and Oberleas (1970) reported that 660 μM histidine added to dialyzed serum rendered 7% of an added amount of ^{65}Zn ultrafilterable (total zinc:albumin = 2.0); values for 660 μM threonine, glycine and asparagine were 4, 1.5 and 1%, respectively. However, at less than 1 equivalent of zinc per albumin equivalent, non albumin bound zinc values in these simple systems are 3.5% in 660 μM histidine and less than 0.1% in 660 μM threonine, glycine or asparagine (Figure 1). The values found by Prasad and Oberleas (1970) presumably are a reflection of their own observation that the second zinc equivalent is bound to albumin less firmly than the first.

The macromolecular serum ligand(s) of loosely bound copper, zinc and cadmium has (have) the same affinity for these metals as does serum albumin, within the experimental uncertainties of these studies. In order to reject the hypothesis that serum albumin is the macromolecular ligand in the exchangeable pools of these metals in serum a specific difference between the curves presented in each figure would be required: less metal cannot be bound by serum proteins than is bound by albumin alone. The isotope distribution curve for serum, supposedly containing other proteins besides albumin which complex the metal in question, must lie, at every point, above the isotope distribution curve for serum albumin. In none of the figures is this the case. Thus, the data provide no compelling evidence for the presence in serum of proteins which both bind exogenous copper, zinc, or cadmium more strongly than does albumin, and bind a significant portion of metal in the loosely bound pool.

Since 2/3 of total human serum zinc is normally loosely bound (Vikbladh, 1951; Parisi and Vallee, 1970; Wolff 1956) the amount of stable and radioisotopic zinc added to serum in these experiments did not markedly expand the exchangeable pool. Five to nine percent of total human serum copper is normally loosely bound (Wintrobe et al., 1953; Gubler et al., 1953; Cartwright and Wintrobe, 1964); this is a ratio of 0.002 copper equivalent per albumin equivalent. This pool was more than doubled by addition of stable and radioisotopic copper in these experiments. The total amount of cadmium added was approximately 0.01 equivalent per albumin, perhaps one hundred times more cadmium than is normally present in serum (Kubota et al., 1968; Imbrus et al., 1963). There may exist, in serum, proteins which complex small amounts of loosely bound cadmium; their contribution to cadmium distribution might not be noticed in such a cadmium loaded system.

Neumann and Sass-Kortsak (1967) have shown that human serum albumin and dialyzed serum have equal avidity for added copper, from a copper/albumin ratio of near zero to five, and that addition of physiological concentrations of serum amino acids to serum albumin solution containing copper increases the amount of ultrafilterable copper to the level found in native serum. Similar experiments were carried out with zinc by Prasad and Oberleas (1970). Our experiments support and extend those observations. Jammers and Catsch (1967) have studied competition for ^{65}Zn between human serum and EDTA and another chelating agent by a chromatographic technique similar to the one we used.

The association constant of the copper-human serum albumin complex has been reported to be 10^{16} (Lau and Sarkar, 1971) and that of the zinc-human serum albumin complex to be 10^{7} (Giroux and Henkin, 1972). The value of the cadmium complex association constant is unknown, although it may be similar to the value of the zinc complex (Tanford, 1952). The data presented here can be used to calculate association constants. However, the sephadex matrix adsorbs these radioisotopes so that in our experimental system competition for metal is not simply between protein and amino acid solutes. This does not invalidate side-by-side comparison of metal binding by serum and serum albumin, but in the absence of additional data it is not known to what extent observed values of the association constants must be modified to correct for this factor. If literature values (Hallman et al., 1971) for histidine and glycine complexes of zinc and copper are assumed and if the metal-protein association constant is treated as the adjusted parameter, values of $10^{6.3}$, $10^{6.1}$, and $10^{9.1}$ generate precise fits of calculated curves to the isotope distribution data from serum albumin in Figures 6(a) (zinc), 7(a) (zinc), and 8(a) (copper), respectively.

These fitted curves are indicated in Figures 6(a)–8(a) by dashed lines; in the central portion of each curve dashed and solid lines are coincident. The association constants, although biased, do have some utility: they can be used to estimate characteristics of a second (hypothetical) macromolecular metal ligand in serum which would raise the isotope distribution curve for serum so far above the curve for albumin that the hypothesis of equivalency would be rejected and the presence of this second metal-complexing protein would be detected.

Let us assume, for example, the value of $10^{6.3}$ for the association constant (K_{HSA}) of the zinc–albumin complex [Figure 6(a)]. Further, assume the presence in serum of protein P, which competes with albumin and with histidine for loosely bound zinc, and whose total zinc-binding capacity is equal to the size of the loosely bound zinc pool, i.e., 10 μM in normal human serum. Isotope distribution curves which we have calculated indicate that if the association constant (K_p) of zinc–protein P complex is approximately 10 K_{HSA}, the curve for serum containing protein P would at no point lie greater than 5, 20 and 50 percentage points above the serum albumin curve, respectively, if the value of K_p were set equal to 10,100 and 1000 x K_{HSA}, respectively. (The absolute difference is greatest in the central portion of the curves.) The first-mentioned situation would not force one to postulate the presence of protein P, because the experimental data are insufficiently precise, but the presence in serum of protein P with K_p approximately 100 K_{HSA} most likely would be noticed. If one assumes that protein P is unable to bind more than 10% of the loosely bound zinc pool in serum the isotope distribution curve for serum obviously could never lie more than 10 percentage points above the serum albumin curve and the presence of protein P probably would go unrecognized. These considerations indicate limits within which one may accept the hypothesis that albumin is the only macromolecular ligand of loosely bound zinc in normal human serum. Figure 6 shows that in the case of competition with histidine for ^{65}Zn the isotope distribution curves for serum and serum albumin are superimposable.

In spite of stronger binding of copper than zinc by albumin, a given concentration of histidine in these experiments strips a greater percentage of loosely bound copper than zinc or cadmium from serum macromolecules. This observation may have physiological relevance, since histidine, cystine, and cysteine have been calculated to be preferred amino acids for forming copper and zinc complexes in serum, especially cysteine and mixed histidine–cysteine, histidine–cystine complexes (Giroux and Henkin, 1972; Hallman, et al., 1971), and may be involved in membrane transport processes (Harris and Sass-Kortsak, 1967).

Loosely bound zinc and copper in serum have been considered the transport form of these metals (Vikbladh, 1951; Bearn and Kunkel,

1954; Prasad and Oberleas, 1970; Wolff, 1956; Neumann and Sass-Kortsak, 1967). Our studies show that serum albumin is highly representative of the macromolecular serum component of the exchangeable copper, zinc, and cadmium pools.

In Vivo Studies

In order to demonstrate the physiological role of the amino acids in the regulation of zinc and copper transport and metabolism, L-histidine was administered to both man and animals with the resultant increase of urinary zinc and copper, decrease in serum zinc concentration (shown only for man) and the presumed loss of total body zinc. In balance studies with L-histidine performed at the National Institutes of Health some patients lost as much as 1% of their total body zinc daily on maximal doses of this amino acid. This loss was primarily via the urine with little contribution from the stool. The urinary loss began as soon as L-histidine was administered and stopped as soon as it was terminated. In studies carried out over periods as long as two weeks with a fixed dose of 32.4 g L-histidine there was no significant diminution in the excretion of urinary zinc or copper. Although serum zinc was lowered significantly by L-histidine administration serum copper was not. This may be explained by the relatively large amounts of albumin-bound zinc available and the relatively small amounts of albumin-bound copper. Since 90% of circulatory copper is tightly bound to ceruloplasmin and relatively unavailable for exchange with histidine very large losses of total body copper might have to occur in order to observe any significant drop in serum copper.

Physiological Implications of Competition for Zinc: Zinc Distribution in Human Plasma

The concentration of zinc in normal human serum is 15 μM. α_2-macroglobins firmly bind approximately 1/3 of the total (Parisi and Vallee, 1970). Other zinc-binding proteins and enzymes are present in serum but account for a small amount of zinc (Wolff, 1956; Prasad and Oberleas, 1970; Prasad and Oberleas, 1968; Himmelhoch et al., 1966; Boyett and Sullivan, 1970; Steinbach et al, 1966; Vessel and Bearn, 1957), while albumin is the major, and probably the only, serum macromolecule which binds exchangeable zinc and loses zinc when serum is treated with zinc-chelating agents (Vikbladh, 1951; Wolff, 1956; Parisi and Vallee, 1970; Suso and Edwards, 1971; Fritze and Gietz, 1968; Boyett and Sullivan, 1971; Giroux and Henkin, 1972). Table II indicates that in normal serum, albumin, in competition with amino acids, binds 98% of the loosely bound zinc. The concentration of zinc distributed among all amino acid complexes is 0.2 μM. (This corresponds to about 1 μg of zinc per 100 ml of serum). Only cysteine and histidine among the amino acids complex zinc under normal conditions and the amount of ionic zinc is insignificant. Variations in serum concentrations of these amino acids alter zinc distribution, while even a 5-fold increase in serum concentration of cystine, the third best zinc ligand among the amino acids, does not change the normal zinc distribution (Table II). In normal serum less than 1% of total cysteine is complexed with zinc. A consequence of such marked undersaturation of both albumin and all amino acid ligands is that loosely bound serum zinc distribution is independent of zinc concentration over a wide range. Indeed, were ligand saturation approached use of Eqn 6 would be inappropriate and the more sophisticated approach of Hallman et al., (1971) then would be necessary.

Employing zinc: albumin ratios of 0.33-2.5, Prasad and Oberleas (1970) found that at the lowest ratios ultrafilterable zinc in normal human serum amounted to about 2% of total zinc, a value in agreement with the calculated value for amino acid-bound zinc in Table II.

Ultrafilterable Zinc and Copper in Plasma and Urinary Excretion

Ultrafilterable zinc and copper are operational terms and include, at least, two potential sources of the metals; e.g., zinc and copper bound to amino acids and zinc and copper which are relatively easily removed from albumin under the specific conditions of the method used. In this sense estimates of ultrafilterable zinc and copper in normal plasma may vary somewhat depending on differences in the manner by which the specific studies have been carried out. It is thus obvious that comparison of values between normal and pathological samples must be made using similar techniques. In one series of systematic studies (Henkin, R.I., in Mertz, W. and Cornatzer, W.E. Newer Trace Metals in Nutrition) estimates of ultrafilterable or diffusible zinc in the plasma of 24 normal volunteers was found to be 19 ± 2 μg/100 ml (Mean ± 1 SEM), ultrafilterable or diffusible copper, 11 ± 1 μg/100 ml. Other studies, using different techniques, reported lower values for ultrafilterable zinc.

By extrapolation from the *in vitro* studies the unhindered passage of zinc-amino-acid complexes in normal plasma across the renal glomerulus would result in a calculated filtered load of 2 g of zinc in a normal 24 hr glomerular filtrate of 183 liters (Smith, 1956). Observations of several investigators including our laboratory indicate that under normal conditions less than 0.5 mg of zinc is excreted in urine daily and this apparently bound to either amino acids or porphyrins. Thus, the major part of the normal filtered load of amino-acid-complexed zinc must be reabsorbed by the kidney. Some studies have suggested that there is renal tubular secretion of zinc (Rubini et al., 1961).

Shifts in plasma zinc such that more or less zinc is complexed with micromolecular ligands may upset the normal zinc transport and storage processes. Any increase in micromolecular liganded species crossing the renal glomerulus could lead to excessive urinary loss of zinc. The albumin-amino acid distribution of zinc may be abnormal in many physiological and pathological states including histidinemia (LaDu, 1966), acute viral hepatitis (Felig et al., 1970; Walshe, 1953; Henkin and Smith, 1972) cirrhosis (Boyett and Sullivan, 1970; Walshe, 1953; Vallee et al., 1956; Vallee et al., 1957) pregnancy (Soupart, 1959; Henkin et al., 1971; Henkin, 1971) and during administration of various drugs, such as 6-azauridine (Slavik et al., 1973a).

During the first two years of life in man there are significant alterations in plasma concentrations of total and diffusible zinc and copper (Henkin et al., 1973). In acromegaly, elevated plasma growth hormone concentration is associated with increased renal excretion of both copper and zinc (Henkin, 1974) as well as aminoaciduria; these changes all revert to or toward normal following reduction of growth hormone subsequent to treatment of the acromegaly (Henkin, 1974). On the other hand isolated growth hormone deficiency is associated with decreased levels of urinary zinc and copper excretion which increase significantly following administration of exogenous growth hormone (Henkin, 1974). A similar sequence of events occurs in Cushing's syndrome where elevated levels of circulating carbohydrate-active steroids are associated with excessive excretion of urinary zinc and copper, aminoaciduria and lower than normal levels of serum zinc (Henkin, 1974a); correction of this abnormality is also associated with a return of these changes to or toward normal (Henkin, 1974a). Analogous to isolated growth hormone deficiency patients with Addison's disease or panhypopituitarism and adrenalectomized or hypohysectomized cats exhibit lower than normal levels of urinary zinc and copper (Henkin, 1974a). Bile zinc and copper levels in the adrenalectomized or hypophysectomized cats are significantly higher than normal (Henkin, 1974a). Replacement with appropriate adrenal corticosteroids in man or in cat returns each of these abnormalities to or toward normal (Henkin, 1974a).

The chemical basis for each of these changes is not fully understood. For hepatitis there is a significant increase in diffusible or ultrafilterable zinc during the acute phase of the disease (Henkin and Smith, 1972) during which time there is a significant decrease in total serum zinc, an increase in urinary zinc excretion and an increase in urinary histidine. As the disease wanes diffusible zinc decreases to normal, serum zinc increases, and urinary zinc excretion decreases (Henkin and Smith, 1972) as does the histidinuria. During 6-azauridine administration urinary zinc increases, serum zinc decreases and significant increases in serum histidine occur (Slavik et al., 1973a); in addition homocysteine appears in the serum (Slavik et al., 1973). With the discontinuation of this drug all of these changes revert to normal.

Some similar changes appear to occur in acromegaly and untreated Cushing's syndrome (Henkin, 1974; Henkin, 1974a). Each of these results is consistent with the concept that zinc and to a lesser degree copper, (since zincuria always precedes any cupruria) is stripped from its binding sites on albumin with the subsequent renal excretion of the amino-acid metal complexes. In this sense regulation of body stores of zinc, copper and perhaps cadmium as well, is dependent upon the interrelationship between the binding of these metals to albumin and the circulating levels of histidine. Since albumin bound metal cannot normally cross the renal glomerulus a convenient form of body conservation based upon molecular size may be indicated. Other ligands, perhaps including porphyrins and other factors not yet treated in model studies, may also influence the distribution of these metals in plasma and their excretion in urine. Nevertheless the albumin-amino acid-metal interrelationship represents a major factor in the manner by which metals are transported in blood, excreted in urine and conserved by the body.

REFERENCES

Albert, A. (1950). Quantitative studies of the avidity of naturally occurring substances for trace metals. 1. Amino-acids having only two ionizing groups. Biochem. J. 47: 531-538.

Albert, A. (1952). Quantitative studies of the avidity of naturally occurring substances for trace metals. 2. Amino-acids having three ionizing groups. Biochem. J. 50: 690-697.

Bearn, A.G. and Kunkel, H.G. (1954). Localization of Cu^{64} in serum fractions following oral administration: an alteration in Wilson's disease. Proc. Soc. Exptl. Biol. Med. 85: 44-48.

Bjerrum, J. (1950). On the tendency of the metal ions toward complex formation. Chem. Rev. 46: 381-401.

Boyett, J.D. and Sullivan, J.F. (1970). Distribution of protein-bound zinc in normal and cirrhotic serum. Metabolism 19: 148-157.

Boyett, J.D. and Sullivan, J.F. (1971). The effect of oral zinc sulfate on protein-bound zinc in serum. Ala. J. Med. Sci. 8: 124-131.

Breslow, E. (1964). Comparison of cupric ion-binding sites in myoglobin derivatives and serum albumin. J. Biol. Chem. 239: 3252-3259.

Brigham, M.P. Stein, W.H. and Moore, S. (1960). The concentrations of cysteine and cystine in human blood plasma. J. Clin. Invest. 39:1633-1638.

Cartwright, G.E. and Wintrobe, M.M. (1964). Copper metabolism in normal subjects. Am. J. Clin. Nutr. 14: 224-232.

Christian, G.D. (1969). Medicine, trace elements, and atomic absorption spectroscopy. Anal. Chem. 41: 24A-40A.

Eichhorn, M.M. (1970). Data on the naturally-occurring amino acids. In H.A. Sober, Handbook of Biochemistry, 2nd ed., Chemical Rubber Co., Cleveland, Ohio, pps. B3-B49.

Felig, P., Brown, W.V., Levine, R.A. and Klatskin, G. (1970). Glucose homeostasis in viral hepatitis. New Engl. J. Med. 283: 1436-1440.

Fritze, K. and Gietz, R.J. (1968). Contamination problems in trace analysis or protein bound metals. J. Radioanal. Chem. 1: 265-268.

Giroux, E.L. and Henkin, R.I. (1972). Competition for zinc among serum albumin and amino acids. Biochim. Biophys. Acta 273: 64-72.

Giroux, E.L. and Henkin, R.I. (1972a). Macromolecular ligands of exchangeable copper, zinc and cadmium in human serum. Bioinorg. Chem. 2: 125-133, 1972.

Gubler, C.J., Lahey, M.E., Cartwright, G.E. and Wintrobe, M.M. (1953). Studies on copper metabolism. IX. The transportation of copper in blood. J. Clin. Invest. 32: 405-414.

Gurd, F.R.N. and Goodman, D.S. (1952). Preparation and properties of serum and plasma proteins. XXXII. The interaction of human serum albumin with zinc ions. J. Amer. Chem. Soc. 74: 670-675.

Hallman, P.S., Perrin, D.D. and Watt, A.E. (1971). The computed distribution of copper (II) and zinc (II) ions among seventeen amino acids present in human blood plasma. Biochem. J. 121: 549-555.

Hamilton, P.B. (1968). Free amino acids α in blood plasma of newborn infants and adults. In H. A. Sober, Handbook of Biochemistry, 1st ed., Chemical Rubber Co., Cleveland, Ohio, pp. B55.

Harris, D.I.M. and Sass-Kortsak (1967). The influence of amino acids on copper intake by rat liver slices. J. Clin. Invest. 46: 659-667.

Henkin, R.I. (1971). Newer aspects of copper and zinc metabolism. In Mertz, W. and Cornatzer, W.E. Newer Trace Metals and Nutrition, Marcel Dekker, N.Y., pps. 255-311.

Henkin, R.I., Marshall, J.R. and Meret, S. (1971). Maternal-fetal metabolism of copper and zinc at term. Am. J. Obstet. Gynecol. 110: 131-134.

Henkin, R.I., Keiser, H.R. and Bronzert, D. (1972). Histidine dependent zinc loss, hypogeusia, anorexia and hyposmia. J. Clin. Invest. 51: 44a.

Henkin, R.I. and Smith, F.R. (1972). Zinc and copper metabolism in acute viral hepatitis. Amer. J. Med. Sci. 264: 401-409.

Henkin, R.I., Schulman, J.D., Schulman, C.B. and Bronzert, D.A. (1973). Changes in total, nondiffusible, and diffusible plasma zinc and copper during infancy. J. Ped. 82: 831-837.

Henkin, R.I. (1974). Growth hormone dependent changes in zinc and copper metabolism in man. In Hoekstra, W.G., Suttie, J.W., Ganther, H., and Mertz, W. Trace Elements in Man and Animals II, University Press, Baltimore Md., Chapter 92 (In press).

Henkin, R.I. (1974a). On the role of adrenocorticosteroids in the control of zinc and copper metabolism. In Hoekstra, W.G., Suttie, J.W., Ganther, H. and Mertz, W. Trace Elements in Man and Animals II, University Press, Baltimore, Md., Chapter 91 (In press).

Himmelhoch, S.R., Sober, H.A., Vallee, B.L., Peterson, E.A. and Fuwa, K. (1966). Spectrographic and chromatographic resolution of metalloproteins in human serum. Biochemistry 5: 2523-2530.

Imbrus, H.R., Cholak, J., Miller, L.H. and Sterling, T. (1963). Boron, cadmium and nickel in blood and urine. Arch. Environ. Health 6: 286-295.

Jammers, W. and Catsch, A. (1967). Isotopischer austausch von zink zwischen proteinen und polyamino-polycarboylsäuren. Naturwissenschaften 54: 588.

Klotz, I.M., Urquhart, J.M. and Kiess, H.A. (1952). Interactions of metal ions with the sulfhydryl group of serum albumin. J. Am. Chem. Soc. 74: 5537-5538.

Kubota, J., Lazar, A. and Losee, F.L. (1968). Copper, zinc, cadmium and lead in human blood from 19 locations in the United States. Arch. Environ. Health 16: 788-793.

LaDu, B.N. (1966). Histidinemia. In Stanbury, J.B., Wyngaarden, J.B. and Fredrickson, D.S., The Metabolic Basis of Inherited Disease, 2nd Ed., McGraw-Hill, N.Y., pps. 366-375.

Lau, S.J. and Sarkar, B. (1971). Ternary coordination complex between human serum albumin, copper (II) and L-histidine. J. Biol. Chem. 246: 5938-5943.

Li, N.C. and Manning, R.A. (1955). Some metal complexes of sulfur-containing amino acids. J. Am. Chem. Soc. 77: 5225-5228.

Lifschitz, M.D. and Henkin, R.I. (1971). Circadian variation in copper and zinc in man. J. Appl. Physiol. 31: 88-92.

Lowry, O.H., Rosebrough, N.J., Farr, A.L. and Randall, R.J. (1951). Protein measurement with the folin phenol reagent. J. Biol. Chem. 193: 265-275.

Maley, L.E. and Mellor, D.P. (1950). Stability of some metal complexes of histidine. Nature 165: 453.

McConnell, S.D. and Henkin, R.I. (1974). Altered preference for sodium chloride, anorexia and changes in plasma and urinary metabolism in rats fed a zinc deficient diet. J. Nutrition (in press).

Meret, S. and Henkin, R.I. (1971). Simultaneous direct estimation by atomic absorption spectrophotometry of copper and zinc in serum, urine, and cerebrospinal fluid. Clin. Chem. 17: 369-373.

Monk, C.D. (1951). Electrolytes in solutions of amino acids. Part IV- Dissociation constants of metal complexes of glycine, alanine and glycylglycine from pH titrations. Trans. Farnday Soc. 47: 297-302.

Neumann, P.Z. and Sass-Kortsak, A. (1967). The state of copper in human serum: Evidence for an amino acid-bound fraction. J. Clin. Invest. 46: 646-658.

O'Sullivan, W.J. (1969). Stability Constants of Metal Complexes. In Dawson, R.M.C., Elliott, D.C., Elliott, W.H. and Jones, K.M. Data for Biochem. Res. 2nd Ed., Oxford University Press, N.Y. p. 423.

Parisi, A.F. and Vallee, B.L. (1970). Isolation of a zinc α_2-macroglobulin from human serum. Biochemistry 9: 2421-2426.

Perkins, D.J. (1952). A study of the amino-acid complexes formed by metals of Group II of the periodic classification. Biochem. J. 51: 487-490.

Perkins, D.J. (1964). Zn^{2+} binding to poly L-glutamic acid and human serum albumin. Biochim. Biophys. Acta. 86: 635-663.

Peters, J., Jr. (1960). Interaction of one mole of copper with the alpha amino group of bovine serum albumin. Biochim. Biophys. Acta. 39: 546-547.

Peters, T., Jr., and Blumenstock, F.A. (1967). Copper-binding properties of bovine serum albumin and its amino-terminal peptide fragment. J. Biol. Chem. 242: 1574-1578.

Porter, E.J. and Waters, W.J. (1966). A rapid micromethod for measuring the reserve albumin binding capacity in serum from newborn infants with hyperbilirubinemia. J. Lab. Clin. Med. 67: 660-668.

Prasad, A.S. and Oberleas, D. (1968). Zinc in human serum: evidence for an amino acid-bound fraction. J. Lab. Clin. Med. 72: 1006.

Prasad, A.S. and Oberleas, D. (1970). Binding of zinc to amino acids and serum proteins *in vitro*. J. Lab. Clin. Med. 76: 416-425.

Rubini, M.E., Montalvo, G., Lockhart, C.P. and Johnson, C.R. (1961). Metabolism of zinc-65. Amer. J. Physiol. 200: 1345-1348.

Sarkar, B., and Kruck, T.P.A. (1966). Copper-amino acid complexes in human serum. In Peisach, J., Aisen, P. and Blumberg, W.E., The Biochemistry of Copper, Academic Press, N.Y., p. 183-196.

Sarkar, B. and Wigfield, Y. (1968). Evidence for albumin-Cu (II)-amino acid ternary complex. Can. J. Biochem. 46: 601-607.

Schroeder, H.A. and Nason, A.P. (1971). Trace-element analysis in clinical chemistry. Clin. Chem. 17: 461-474.

Slavik, M., Danilson, D., Keiser, H.R. and Henkin, R.I. (1973a). Alterations in metabolism of copper and zinc after administration of 6-azauridine triacetate. Biochem. Pharm. 22: 2349-2352.

Slavik, M., Lovenberg, W., and Keiser, H.R. (1973). Changes in serum and urine amino acids in patients with progressive systemic sclerosis treated with 6-azauridine triacetate. Biochem. Pharm. 22: 1295-1300.

Smith, H.W. (1956). Principles of Renal Physiology, Oxford University Press, N.Y., p. 30.

Soupart, P. (1959). L'aminoacidurie de la grossesse. Ann. Soc. R. Sci. Med. Nat. Brux. 12: 105-182.

Steinbach, M., Audran, R., Reuge, C. and Blatrix, C. (1966). Etude d'une d-globuline contenant du zinc: la proteine. In H. Peeters Protides of the Biological Fluids, Elsevier, Amsterdam, pps. 185-188.

Suso, F.A. and Edwards, H.M., Jr. (1971). Binding capacity of intestinal mucosa and blood plasma for zinc. Proc. Soc. Exp. Biol. Med. 137: 306-309.

Tanford, C. (1952). The effect of pH on the combination of serum albumin with metals. J. Am. Chem. Soc. 74: 211-215.

Tsangaris, J.M., Chang, J.W. and Martin, R.B. (1969). Cupric and nickel ion interactions with proteins as studied by circular dichroism. Arch. Biochem. Biophys. 130: 53-58.

Vallee, B.L., Wacker, W.E.C., Bartholomay, A.F. and Robin, E.D. (1956). Zinc metabolism in hepatic dysfunction. I. Serum zinc concentrations in Laënnec's cirrhosis and their validation by sequential analysis. New Eng. J. Med. 255: 403-408.

Vallee, B.L., Wacker, W.E.C., Bartholomay, A.F. and Hock, F.L. (1957). Zinc metabolism in hepatic dysfunction. II. Correlation of metabolic patterns with biochemical findings. New Eng. J. Med. 257: 1055-1065.

van den Hamer, C.J.A., Muvall, A.G., Scheinberg, I.H., Hickman, J. and Ashwell, G. (1970). Physical and chemical studies on ceruloplasmin IX. The role of galactosyl residues in the clearance of ceruloplasmin from the circulation. J. Biol. Chem. 245: 4397-4402.

Vessel, E.S. and Bearn, A.G. (1957). Localization of lactic acid dehydrogenose activity in serum fractions. Proc. Soc. Exp. Biol. Med. 94: 96-99.

Vikbladh, I. (1951). Studies on zinc in blood. Scan. J. Clin. Lab. Invest. 3, Suppl. 2: 1-74.

Walshe, J.M. (1953). Disturbances of aminoacid metabolism following liver injury. Quart. J. Med. 22: 483-505.

Wintrobe, M.M., Cartwright, G.E. and Gubler, C.J. (1953). Studies of the function and metabolism of copper. J. Nutr. 50: 395-419.

Wolff, H.P. (1956). Unterschungen zur pathophysiologie des zinkstoffwechsels. Klin. Wochenschr. 34: 409-418.

16

THE EFFECT OF ZINC DEPRIVATION ON THE BRAIN[1]

G.J. Fosmire[2], Y.Y. Al-Ubaidi[3], E. Halas[4], and H.H. Sandstead[5]

USDA, ARS, Human Nutrition Laboratory
Grand Forks, North Dakota and
Department of Psychology, University of North Dakota
Grand Forks, North Dakota

More than a century ago, zinc was shown to be an essential trace element for Aspergillus niger by Raulin (1869). It was not, however, until 1934 that zinc was shown to be essential in a mammalian species when Todd, Elvehjem and Hart (1934) produced zinc deficiency in rats. The manifestations of zinc deficiency have been reviewed by Underwood (1971); they include growth retardation, anorexia, parakeratosis of the skin and esophagus, abnormal bone maturation, and unsteady gait. In severe deficiency, sterility occurs, while in less severe deficiency, the young produced are small and may exhibit deformities of many tissues.

---

[1]Supported in part by USDA Cooperative Agreement 12-14-100-11, 178 (61), Amend. 1.

[2]Research Associate, Department of Biochemistry, University of North Dakota.

[3]Biochemist, 8350 South 86 Avenue, Justice, Illinois.

[4]Professor, Department of Psychology, University of North Dakota.

[5]Director, USDA, ARS, Human Nutrition Laboratory.

It is apparent that zinc deficiency has it most profound effects on rapidly proliferating tissues (Swenerton and Hurley, 1968; Hurley and Swenerton, 1966). The brain is such a tissue in very young animals. In the rat, brain maturation is largely completed by the 20th day postnatally. DNA biosynthesis has nearly ceased (Enesco and Leblond, 1962) and the brain has about 70% of its final weight (Kishimoto et al., 1965). Much of the remainder of the increase in size occurs by hypertrophy of existing cells (Winick, 1970). Since the brain of the rat grows so rapidly during the initial 20 days of postnatal life, a deficiency in zinc during this period might be expected to have adverse effects on brain maturation.

Brain maturation occurs as a precisely timed progression of biochemical, morphological, and functional events. Each developmental event is to some extent dependent on the preceding occurrences. In an altricial animal, such as the rat, macroneurons differentiate largely in utero, while the bulk of microneuronal development occurs during the suckling period, and most of the neuroglial maturation occurs during the early post-weaning period (Altman, 1971). It may be expected, therefore, that the effects of zinc deficiency or of any other nutritional insult on the brain will be dependent upon the time in the maturational cycle at which the deficiency is produced.

Studies of zinc deficiency must be controlled for the effects of starvation. This is because a conspicuous effect of zinc deprivation is anorexia. The effects of undernutrition and protein deficiency on brain development have been studied during both prenatal and postnatal development. When dams were given a protein deficient diet, the total DNA content of the fetal brains was shown to be reduced (Zamenhof, et al., 1968; Zeman and Stanbrough, 1969). Protein deprivation has also been found to decrease the number of large, multipolar neurons in the spinal cord and brain as well as to reduce the number of cells in the surface layers of the cerebral cortex and Purkinje and granule cells in the cerebellum (Shrader and Zeman, 1969).

Depriving the suckling rat of sufficient nutriture by placing large litters (18-20 pups) on dams results in decreased brain weight, along with reduced total protein, and DNA and RNA content (Winick and Noble, 1966, Guthrie and Brown, 1968). Subsequent to rehabilitation after weaning, such animals do not demonstrate normal growth. When undernourished from 5 to 20 days of postnatal life, a significant reduction in the phospholipid, cholesterol, and cerebroside content also occurs (Culley and Mertz, 1965; Dobbing and Widdowson, 1965). These effects appear to be poorly reversible. This is especially true for the cerebellum (Culley and Lineberger, 1968).

Undernutrition after weaning appears to have much less impact on the brain. A reduction in brain size occurs which appears to be due to a reduction in the size of individual cells (Winick and Noble, 1966). This effect is apparently reversible (Dickerson and Walmsley, 1967). On the other hand, cholesterol levels have been found to be permanently depressed if malnutrition is continued from birth until the 35th day of age (Guthrie and Brown, 1968). Thus certain elements of the brain appear vulnerable to malnutrition after weaning. The mature brain appears to be resistant to undernutrition (Dobbing, 1968).

As was indicated earlier, zinc has a biochemical role in the growth of the brain. Hurley and Shrader (1972) have shown that zinc is essential for neural development of the fetal rat. By feeding a soy protein (hence high in phytate) diet containing less than 1 mg Zn/kg diet they produced gross malformations of the brain in 47% of the fetuses compared with no developmental abnormalities in pups of control dams fed the same diet supplemented with 100 mg Zn/kg diet.

The response of the brain of the neonatal rat to zinc deprivation has been incompletely documented. It has been shown that growth of pups suckled by zinc deficient dams is severely retarded by the 9th day of life. Brains from 12-day-old zinc deficient pups showed decreased amounts of lipid in cerebellum compared with brains from pups nursed by pair fed or ad libitum fed dams. The zinc deficient pups also displayed a decreased incorporation of thymidine into brain DNA and an impaired incorporation of inorganic sulfate into the trichloroacetic acid percipitable fraction of the brain (Sandstead et al., 1972).

The present studies on the effect of zinc deficiency on the brain were stimulated by the finding of abnormal behavior in zinc deficient animals. Adult zinc deficient animals display apathy, lethargy, and decreased physical activity (Apgar, 1968a, 1968b) along with diminished sexual activity (Whitenach, et al., 1970). Abnormal behavior occurs at parturition. Delivery is difficult, excessive bleeding may occur, and the dams often do not consume the after-births or care for the pups and do not prepare a nest site (Apgar, 1968a, 1968b). Acutely zinc deficient adult rats are reported to have impaired learning of a one-way conditioned avoidance, and an eight blind water maze. Reduced activity/ emotionality as determined by the open field test has also been observed (Caldwell, et al., 1970). These studies, however, were done on animals that were zinc deficient at the time of testing. Thus the results of the tests were confounded by effects of inanition and illness. Experiments comparing the behavior of the offspring of dams mildly zinc deficient during pregnancy have also been reported. Caldwell and Oberleas (1969) found that the pups of zinc adequate dams were superior in the Lashley III water maze test and in a platform avoidance conditioning test compared with pups from the zinc deficient dams. These data were obtained after the pups had been fed a zinc adequate diet for 3 weeks. The findings seem to indicate that residual damage was present in the pups of the zinc deprived dams.

METHODS

The studies reported in this paper are concerned with the effects of inadequate levels of zinc on the development of the rat brain during the suckling period. Time pregnant dams of the Sprague-Dawley strain were housed in plexiglass cages and fed ad libitum until delivery. Soon after parturition, they were placed in clean plexiglass cages with equivalent numbers of pups. All animals then were fed a biotin enriched, sprayed egg white diet (Luecke et al., 1968), which contained less than 1 mg Zn/kg of diet. One group of dams was given deionized water and fed the diet ad libitum. Dams pair fed with the individuals in the deficient group were given the amount of food consumed by the deficient dam on the previous day and water containing 100 mg Zn/liter. A third group of dams was fed the diet ad libitum and given the zinc enriched water.

At intervals the entire litter of each animal in the group was killed by decapitation. The brains were rapidly removed and chilled. In most experiments only the forebrain (brain anterior to the cerebellum) was used. The tissues were pooled, weighed as quickly as possible, and homogenized in 0.25 M sucrose containing 25 mM KCl, 5 mM $MgCl_2$ and 50 mM Tris·HCl, pH 7.5. Aliquots were taken for analysis of DNA, RNA, and protein. DNA and RNA were determined using the fluorescence technique of Karsten and Wollenburger (1972). DNA was also determined using the Burton modification of the diphenylamine procedure (Burton, 1956). Protein was measured by Hartree's modification (1972) of the Lowry procedure (Lowry et al., 1951).

The remainder of the homogenate was centrifuged at 4000 rpm and 4°. Portions of the supernatant were taken for polysomal profiles and assay of RNAse activity. RNAse was measured using modifications of the technique of Gagnon and de Lamirande (1973). Alkaline and total RNAse were determined at pH 7.8 in Tris buffer while acid RNAse was measured at pH 5.8 in citrate phosphate buffer (Ambellan, 1972). The portion of the supernatant used for the polysomal profile was made 0.5% in deoxycholate, recentrifuged, and the resultant supernatant layered over a 15.5

to 35.5% (w/v) sucrose gradient in Hoaglands medium (5 mM Tris, 25 mM KCl, 5 mM $MgCl_2$, pH 7.5). After centrifugation for 160 min., 284,000 x g max, 4°, the gradients were removed at a constant rate from the top and monitored at 260 nm.

The sediment from the first centrifugation was resuspended in homogenizing medium and layered over 2.0 M sucrose containing 50 mM Tris, 25 mM KCl, 5 mM $MgCl_2$, pH 7.5. Centrifugation at 50,000 x g max for 1 hour yielded the nuclei. The nuclei were resuspended in 0.25 M sucrose, 1 mM $MgCl_2$ and used for the assay of RNA polymerase activity. The high magnesium, low salt assay mixture described by Dutton and Mahler (1968) was used. Aliquots were withdrawn at 0, 5, 10, and 15 minutes; precipitated with trichloroacetic acid; and collected on Whatman GF/C glass fiber papers for determination of the incorporation of ^{3}H-uridine triphosphate into RNA.

Behavioral studies which have been previously reported (Lokken et al., 1973) were done on rats subsequent to nutritional rehabilitation. Acquisition of an elevated Tolman-Honzig maze was used as the index of learning ability. Pups were raised as described previously, and after 21 days rehabilitated by feeding a zinc adequate diet ad libitum for 23 days. They were then tested using the Tolman-Honzig elevated maze following maze habituation for 3 days. Prior to each day's test, the animals were deprived of food for 23 hours. Following completion of the maze, the animals were fed for 45 minutes. All were then returned to their cages.

RESULTS

One of the manifestations of zinc deficiency in the young animal is growth failure. Figure 1 shows the growth of pups exposed to the three experimental conditions for 16 days. Growth proceeded more slowly in the litters nursed by zinc deficient and pair fed dams. Of note is the similarity in the growth of these two groups. The forebrain weights of the animals were similarly reduced in the zinc deficient and pair fed pups (Figure 2). At 6 days, the forebrain weights appear very similar, but by the 9th day, the pups suckled by ad libitum fed, zinc adequate dams displayed greater forebrain weights. This continued through the 21-day suckling period. Comparison of the effects of zinc deficiency on body weight, with its effect on forebrain shows an obvious sparing effect on the brain.

The analysis of the forebrains for DNA content revealed that the pups suckled by the ad libitum fed, zinc adequate dams contained more total DNA (Figure 3). The zinc deficient pups

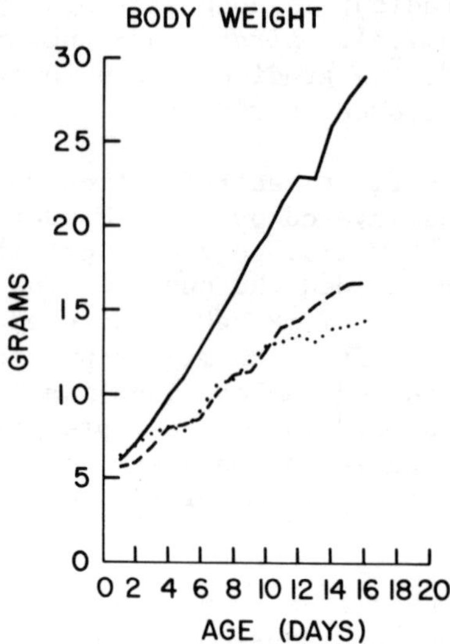

Figure 1. Growth of rat pups suckled by dams fed a zinc deficient diet (....), pair fed, given adequate zinc (---) or fed ad libitum, given adequate zinc (-). These are average weights of eight pups in each litter.

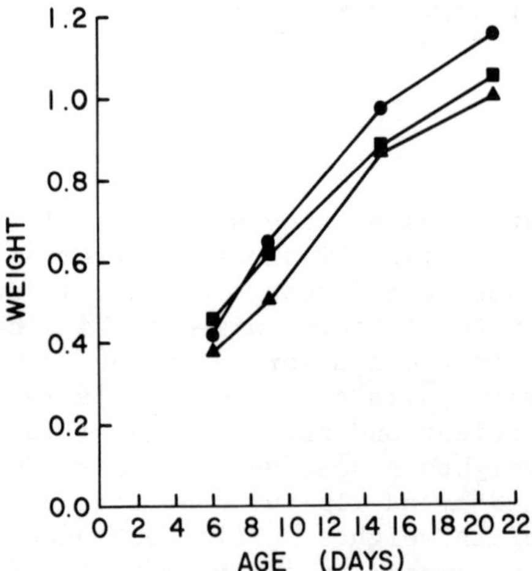

Figure 2. Forebrain (anterior to the cerebellum) weights in grams as a function of age in pups suckled by zinc deficient (▲-▲), pair fed (■-■) or ad libitum fed (●-●) dams. (Average of eight pups in each test group at each time period.)

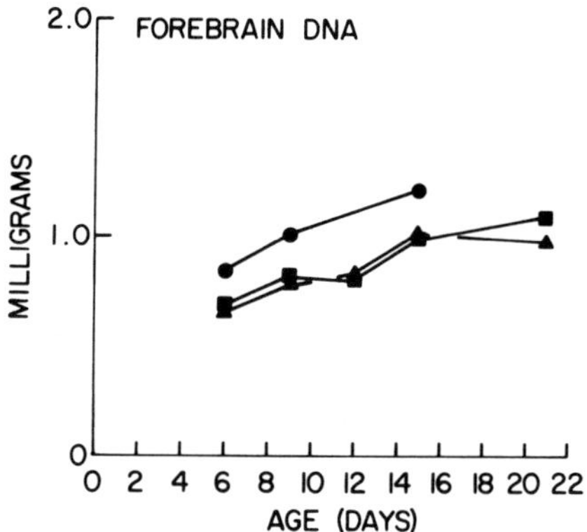

Figure 3. DNA content of forebrain as a function of age in rat pups suckled by zinc deficient (▲-▲), pair fed (■-■), or ad libitum fed (●-●) dams. The values are derived from eight pooled forebrains at each time period for each test group.

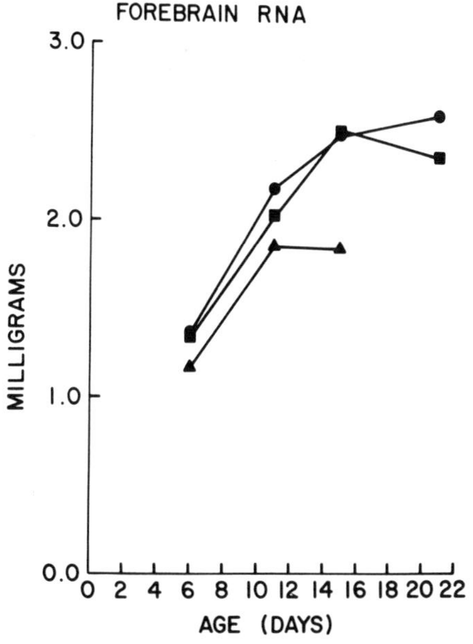

Figure 4. RNA content of forebrain as a function of age in rat pups suckled by zinc deficient (▲-▲), pair fed (■-■), or ad libitum fed (●-●) dams. Values represent an average content of four individual forebrains at each time period.

and pups from pair fed dams contained less DNA, and were very similar to each other. Analysis of the forebrain for RNA showed that the pups from zinc adequate pair fed or ad libitum fed dams had higher levels of total RNA than did the zinc deficient pups (Figure 4).

The total protein and protein/DNA were reduced in the forebrains of pups nursed by zinc deficient dams compared with the forebrains of pups suckled by pair fed dams. Both of these were reduced compared with values from pups of the zinc adequate ad libitum fed dams (Figure 5). The ratio of protein to DNA content as a function of age is given in Table 1.

TABLE 1

	mg Protein/mg DNA		
Age days	Ad Lib	Pr Fed	Zn Def.
6		52.6	43.0
11	76.7	66.8	61.9
15	72.5	75.0	62.0
21	86.5	65.0	63.0

It can be seen that the ratio of protein to DNA is smaller for

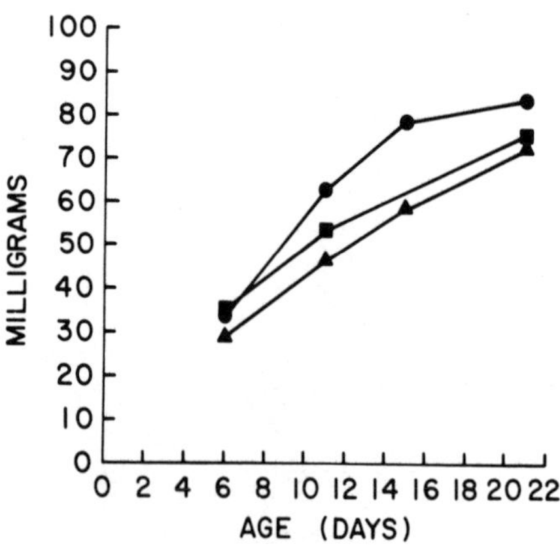

Figure 5. Total protein in forebrain as a function of age in rat pups suckled by zinc deficient (▲-▲), pair fed (■-■), or ad libitum fed (●-●) dams. Average values were obtained for four individual forebrains in each test group at each time period.

the zinc deficient animals throughout the 21-day suckling period when compared with the two control groups. Thus the smaller amounts of total protein seen in the zinc deficient animals is a reflection of a decrease in cell size which was greater than that which occurred in the brains of the starved pups at 6, 11, and 15 days. Because the amount of protein per cell was reduced, presumably reflecting some interference with normal protein biosynthesis, polysomal profiles were examined. The profiles displayed in Figure 6 were obtained using homogenates of whole brain from 5-day old pups. The profiles from zinc deficient animals displayed an increase in height of the 80s peak (monosomes) with a coincident decrease in the heavier polysomes. The pups from pair fed dams yielded profiles intermediate between the deficient and ad libitum controls in the distribution of

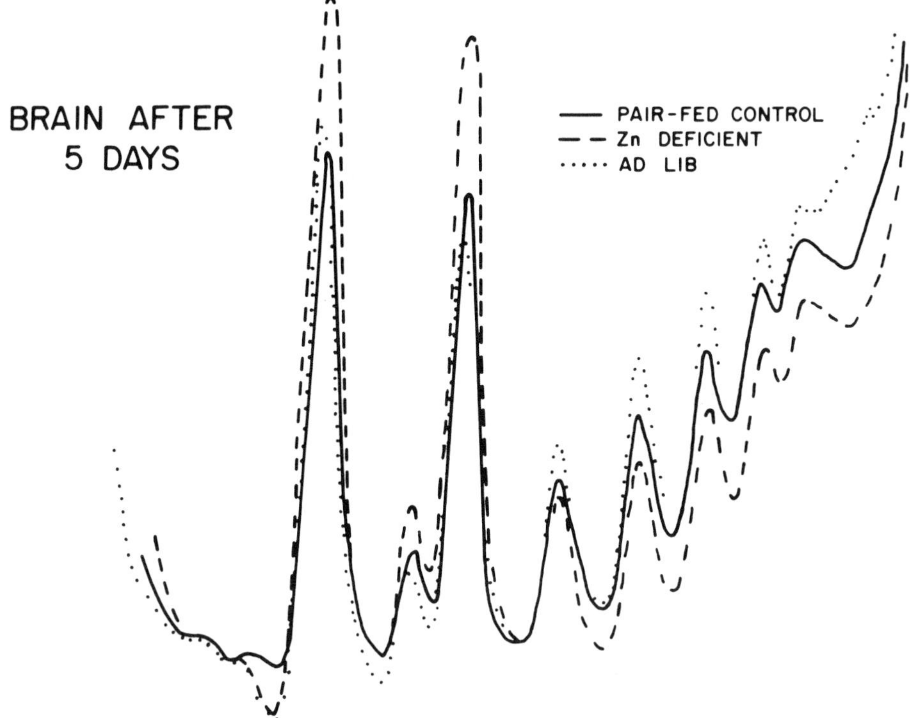

Figure 6. Sedimentation profiles of rat brain polysomes on a linear sucrose gradient (15.5 - 35.5% w/v). Each profile was isolated from the deoxycholate treated (0.5% final concentration) post mitochondrial supernatant derived from the whole brain of four pups. The absorbance at 260 nm was monitored automatically in a Gilford Model 2400 Spectrophotometer.

material between monosomes and heavier polysomes. Thus, it appears that the protein synthetic apparatus of the cell was adversely affected in brains of the zinc deficient pups to a greater extent than was produced by the undernutrition subsequent to pair feeding of the nursing dams. Similar observations made at other times during the first 13 days postnatally are summarized in Table 2.

TABLE 2

Age (Days)	Treatment	Monosome	Polysome
3	Zn Def	25.9	74.1
	Pr Fed	25.4	74.5
5	Zn Def	39.4	60.9
	Pr Fed	27.2	72.7
	Ad Lib	26.1	73.8
5	Zn Def	42.7	57.3
	Pr Fed	24.9	75.1
7	Zn Def	46.3	53.7
	Pr Fed	31.8	68.2
	Ad Lib	17.9	82.1
13	Zn Def	44.6	55.4
	Pr Fed	29.8	70.2

RNA polymerase was assayed because of its role in RNA synthesis and indirect effect on protein biosynthesis and because of previous evidence that it may be impaired in zinc deficiency (Terhune and Sandstead, 1972). Assays were done for the RNA polymerase which catalyses the synthesis of ribosomal RNA. The results are given in Table 3.

TABLE 3

RNA Polymerase Activity (CPM x 10^{-3}/mg DNA)

Age (Days)	Source	Zn Def	Pr Fed	Ad Lib
6	Forebrain	4.5	5.1	6.8
9	Forebrain	4.1	8.0	-
16	Whole Brain	52.5	59.2	60.6

By the 6th day, the incorporation of ^{3}H-UTP into RNA at 5 minutes as a function of RNA polymerase was lowest in the forebrain of the zinc deficient pups, intermediate in the pair fed controls and highest in the forebrain of the ad libitum fed controls. At 9 days, the zinc deficient group demonstrated about half as much incorporation as the pair fed controls. When the whole brain was used, more incorporation was observed presumably because of the inclusion of the cerebellum. This is shown by the values obtained at 16 days. Again the zinc deficient animals displayed less incorporation of ^{3}H-UTP while the incorporation of the pair fed and ad libitum fed controls was similar.

A possible explanation for the abnormalities in the polysomal structure is an increase in RNAse activity. Therefore the active and total alkaline RNAse activity and the acid RNAse activity were determined. These assays were done on forebrain. The results are described in Table 4.

TABLE 4

Treatment	Age (Days)		
	Active RNAse (ng/mg DNA		
	6	11	15
Zn Def	4.9	4.1	4.1
Pr Fed	5.2	3.9	4.6
Ad Lib	5.1	4.0	4.6
	Total RNAse (ng/mg DNA)		
Zn Def.	25.2	18.4	37.2
Pr Fed	27.5	19.3	88.7
Ad Lib	26.0	21.8	79.5
	Acid RNAse (Δ Abs/mg DNA)		
Zn Def	5.1	8.6	7.9
Pr Fed	5.6	9.2	9.2
Ad Lib	6.1	9.1	9.6

It appears that there was no consistant difference in RNAse activity (active, total, or acid) between the test groups. If there was any trend at all, perhaps the activity of the total alkaline RNAse and the acid RNAse did not increase with age in the zinc deficient animals to the same extent as in the controls.

The maze acquisition (average total errors) of the nutritionally rehabilitated (from day 21) young adult rat was determined (Figure 7). The running time on the maze is Figure 8. The running times for the three groups were similar throughout the testing period. The motivation of the three groups appears to have been similar. The slightly increased time for the zinc deficient group can be accounted for by their larger number of errors. The number of total errors made by animals which had been zinc deficient during the suckling period was greater than for either control group ($P < .001$). Thus is appears that the animals deprived of adequate zinc during the suckling period were disadvantaged with regard to acquiring the maze. Nutritional rehabilitation for 24 days did not reverse the defect. On the other hand, nutritional rehabilitation of the pair fed controls was followed by normal acquisition of the Tolman-Honzig maze.

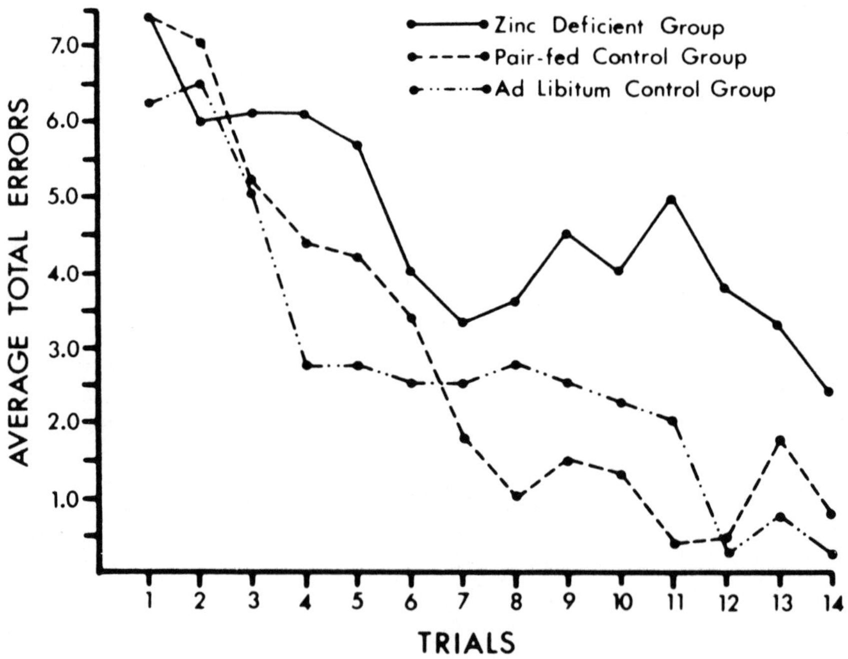

Figure 7. Average total errors made by the three test groups over 14 trials.

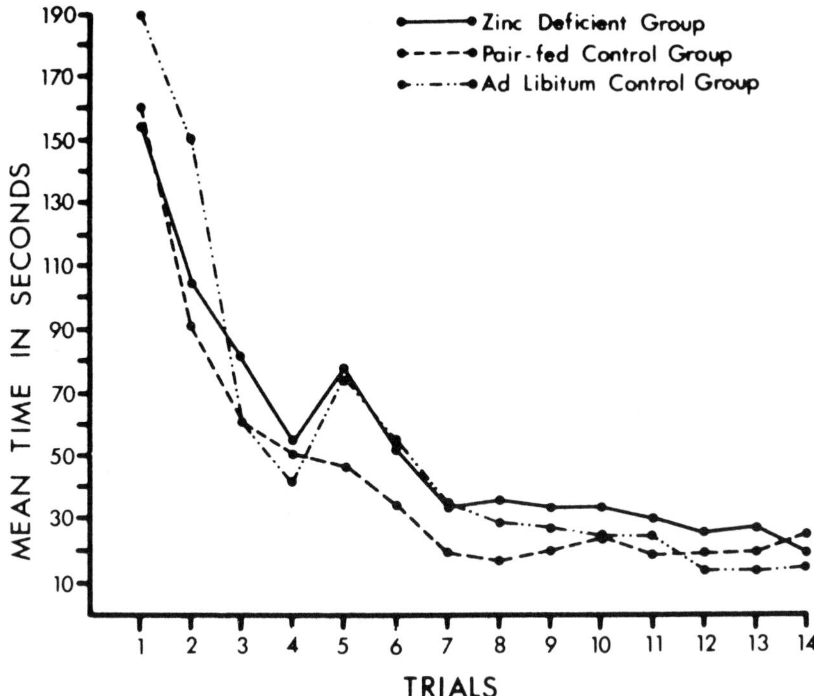

Figure 8. Mean running time in seconds for the three test groups over 14 trials.

DISCUSSION

The biochemical and behavioral findings indicate that the brain is sensitive to zinc deprivation during its critical neonatal period of growth. The biochemical differences observed were in some instances subtle while in others, gross. Undernourishment of the pups caused by pair feeding of the nursing dams had an impact which in some instances was the same as that produced by zinc deficiency while in others it was less severe or was not apparent (i.e., the data were similar to those of the ad libitum controls). Thus DNA levels in the brain appear to be similarly depressed in pups from zinc deficient and pair fed controls. Protein biosynthesis, on the other hand, seems to be more severely affected by zinc deficiency than by undernourishment sufficient to result in a similar degree of growth retardation. The total amount of protein as well as the protein/DNA ratio were reduced in the zinc deficient animals indicating that size of the brain cells was reduced in zinc deficiency. Consistent with an impairment of protein synthesis were the polysomal profiles. The zinc deficient animals showed a decrease in polysomes which was greater than that which occurred in the undernourished group. RNA polymerase was also shown severely depressed in the zinc deficient group. These observations support the concept that zinc is essential for normal brain development and function.

REFERENCES

Altman, J. (1971). Nutritional deprivation and neural development. in Sterman, M.B., McGinty, D.J., and Adinolfi, A.M. (Ed.), Brain Development and Behavior, Academic Press, New York, pp. 365-367.

Ambellan, E. (1972). Assays for acid RNAse activity in crude or partly purified cell fractions. Anal. Biochem. 50: 453.

Apgar, J. (1968a). Comparison of the effects of copper, manganese, and zinc deficiencies on parturition in the rat. Amer. J. Physiol. 215: 1478.

Apgar, J. (1968b). Effect of zinc deficiency on parturition in the rat. Amer. J. Physiol 215: 160.

Burton, K. (1956). A study of the conditions and mechanism of the diphenylamine reaction for the colorimetric estimation of deoxyribonucleic acid. Biochem. J. 62: 315.

Caldwell, D.F. and Oberleas, D. (1969). Effects of protein and zinc nutriture on behavior in the rat. Pan Amer. Health Organ. Sci. Publ. 185, 2.

Caldwell, D.F., Oberleas, D., Clancy, J.J., and Prasad, A.S. (1970). Behavioral impairment in adult rats following acute zinc deficiency. Proc. Soc. Exp. Biol. Med. 133: 1417.

Culley, W.J. and Lineberger (1968). Effect of undernutrition on the size and composition of the rat brain. J. Nutr. 96: 375.

Culley, W.J. and Mertz, E.T. (1965). Effect of restricted food intake on growth and composition of preweanling rat brain. Proc. Soc. Exp. Biol. Med. 118: 233.

Dickerson, J.W.T. and Walmsley, A.L. (1967). The effect of undernutrition and subsequent rehabilitation on the growth and composition of the central nervous system of the rat. Brain 90: 897.

Dobbing, J. (1968). Effects of experimental undernutrition on development of the nervous system, in Scrimshaw, N.S. and Gordon, J.E. (Eds) Malnutrition, Learning, and Behavior, Massachusetts Institute of Technology, Cambridge, Mass.

Dobbing, J. and Widdowson, E.M. (1965). The effect of undernutrition and subsequent rehabilitation on myelination of rat brain as measured by its composition. Brain 88: 357.

Dutton, G.R. and H.R. Mahler (1968). *In vitro* RNA synthesis by intact rat brain nuclei. J. Neurochem. 15: 765.

Enesco, M. and Leblond, C.P. (1962). Increase in cell number as a factor in the growth of the organs of the young male rat. J. of Embryol. Exp. Morph. 10: 530.

Gagnon, C. and de Lamirande, G. (1973). A rapid and simple method for the purification of rat liver RNAse inhibitor. Biochem. Biophys. Res. Comm. 51: 580.

Guthrie, H.A. and Brown, M.L. (1968). Effects of severe undernutrition in early life on growth, brain size and composition in adult rats. J. Nutr. 94: 419.

Hartree, E.F. (1972). Determination of protein: A modification of the Lowry method that gives a linear photometric response. Anal. Biochem. 48: 422.

Hurley, L.S. and Shrader, R.E. (1972). Congenital malformations of the nervous system in zinc-deficient rats, in Pfeiffer, C.C. (Ed.) Neurobiology of the Trace Metals Zinc And Copper, International Review of Neurobiology, Supplement 1, Academic Press, New York.

Hurley, L.S. and Swenerton, H. (1966). Congenital malformations resulting from zinc deficiency in rats. Proc. Soc. Exp. Biol. Med. 123: 692.

Karsten, U. and Wollenberger, A. (1972). Determination of DNA and RNA in homogenized cells and tissues by surface fluorometry. Anal. Biochem. 46: 135.

Kishimoto, Y., Davies, W.E., and Radin, N.S. (1965). Developing rat brain: Changes in cholesterol, galactolipids, and the individual fatty acids of gangliosides and glycerophosphatides. J. Lipid Res. 6: 532.

Lokken, P.M., Halas, E.S., and Sandstead, H.H. (1973). Influence of zinc deficiency on behavior. Proc. Soc. Exp. Biol. and Med. 144: 680.

Lowry, O.H., Rosenbrough, N.J., Favr, A.L., and Randall, R.J. (1951). Protein measurement with the folin phenol reagent. J. Biol. Chem. 193: 265.

Luecke, R.W., Olman, M.E., and Baltzer, B.V. (1968). Zinc deficiency in the rat: Effect on serum and intestinal alkaline phosphatase activities. J. Nutr. 94: 344.

Raulin, J. (1869). Etudes cliniques sur la vegetation. Ann. Sci. Nat. Botan. Bio. Vegetale 11: 93.

Sandstead, H.H., Gillespie, D.D., and Brady, R.N. (1972). Zinc deficiency: Effect on brain of the suckling rat. Pediat. Res. 6: 119.

Shrader, R.E. and Zeman, F.J. (1969). Effect of maternal protein deprivation on morphological and enzymatic development of neonatal rat tissue. J. Nutr. 99: 401.

Swenerton, H. and Hurley, L.S. (1968). Severe zinc deficiency in male and female rats. J. Nutr. 95: 8.

Terhune, M.W. and Sandstead, H.H. (1972). Decreased RNA polymerase activity in mammalian zinc deficiency. Science 177: 68.

Todd, W.R., Elvehjem, C.A., and Hart, E.B. (1934). Zinc in the nutrition of the rat. Amer. J. Physiol. 107: 146.

Underwood, E.J. (1971). Trace Elements in Human and Animal Nutrition, 3rd Edition, Academic Press, New York, pp. 222-236.

Whitenach, D.I., Luecke, R.W., and Whitehair, C.K. (1970). Pathology of the testes in zinc depleted and repleted mature and immature rats. Fed. Proc. 29: 297.

Winick, M. (1970). Nutrition and nerve cell growth. Fed. Proc. 29: 1510.

Winick, M. and Noble, A. (1966). Cellular response in rats during malnutrition at various ages. J. Nutr. 89: 300.

Zamenhof, S., Van Martheus, E., and Margolis, F.L. (1968). DNA (cell number) and protein in neonatal brain: Alteration by maternal dietary protein restriction. Science 160: 322.

Zeman, F.J. and Stanbrough, E.C. (1969). Effect of maternal protein deficiency on cellular development in the fetal rat. J. Nutr. 99: 274.

17

BIOCHEMICAL AND ELECTRON MICROSCOPIC STUDIES OF RAT SKIN DURING ZINC DEFICIENCY

J. M. Hsu, K. M. Kim, W. L. Anthony

Veterans Administration Hospital;

Department of Pathology, University of Maryland

School of Medicine; Department of Biochemical

and Biophysical Sciences, The Johns Hopkins University;

Baltimore, Maryland

I. INTRODUCTION

Zinc has long been recognized as essential to normal growth and development in animals and in man. Its role in regulating metabolic events at the molecular level is becoming evident, but it is not yet possible to correlate biochemical function to the pathology of deficiency. Studies by a number of investigators have indicated that feeding animals with a diet low in zinc causes an abnormal condition on skin in addition to growth retardation. These include the rats (Todd et al., 1934), mice (Day and Skidmore, 1947), swine (Kernkam and Ferrin, 1953), cattle (Miller and Miller, 1962), lambs (Ott et al., 1964) goats (Miller et al., 1964), dogs (Robertson and Burns, 1963), rabbits (Graham and Telle, 1967), squirrel monkeys (Macapinlac et al., 1967), hamsters (Boquist and Lernmark, 1969), guinea pigs (McBean et al., 1972). In man (Prasad et al., 1963) the clinical aspects of zinc deficiency are remarkably similar to those described above for animals. They consisted of growth retardation and rough skin with hyperpigmentation. These observations indicate the importance of zinc in skin metabolism.

Another significant aspect of zinc in relation to the skin is the observation of enhanced wound healing by zinc therapy.

These have been demonstrated in patients with pilonidal sinuses (Pories et al., 1967), venous leg ulcers (Husain 1969), sickel-cell leg ulcers (Serjeant et al., 1970) and burns (Larsen et al., 1970).

The present paper reviews published work from this laboratory on the effects of zinc deficiency on amino acid incorporation, collagen synthesis and nucleic acid metabolism in rat skin. For comparison, the other selected tissues from the zinc-deficient and zinc-supplemented rats were examined. In addition, some current unpublished work is described.

The basic procedures of diet composition, rat origin and environmental control have been followed in all the work summarized in this review.

Diets. The zinc-deficient diet had the following percentage composition: sucrose, 65.97; dried egg albumin, 15.0; salt-free casein hydrolysate, 3.0; corn oil, 10.0; salt mixture, 5.74 (Hsu et al., 1969); and vitamin supplement, 0.29 (Hsu et al., 1969). The zinc content of this diet was approximately 2 ppm. The zinc-supplemented diet is the same zinc-deficient diet supplemented with 65 ppm of zinc as zinc carbonate.

Experimental animals. Male weanling rats of the Sprague-Dawley strain, obtained from a commercial laboratory were used in all experiments. The animals were randomly separated into three groups and placed in individual stainless steel cages in a temperature-controlled room. The first group, zinc-deficient rats, was fed the zinc-deficient diet; the second group was fed the zinc-supplemented diet; and the third group, used as pair-fed controls to the zinc-deficient rats, received the same diet as rats in the second group. Distilled, deionized water was freely available.

Measurement of radioactivity. All measurements of radioactivity were made in a Tricarb liquid scintillation spectrometer. Diotol, composed of 4.6 g 2,5-diphenyloxazale (PPO), 0.091 g p-bis-2-(5-phenyloxazole)-benzene (POPOP), 73 g naphthalene, 210 ml methanol, 350 ml dioxane and 350 ml toluene was used as scintillation fluid. All samples were corrected for decay and for quenching by the internal standard techniques of Herberg (1963).

II. GROWTH AND COMPOSITION OF THE SKIN

After sixteen days feeding, the average body weight gains and standard deviations were 13±3g for the zinc-deficient rats; 35±7g for the zinc supplemented pair-fed rats and 70±3g for supplemented ad libitum-fed rats. The growth inhibition results partly from reduced food intake and partly from an impaired utilization of the food (Hsu et al., 1969, Somers and Underwood, 1969) and from an increased oxidation of amino acids (Theuer and Hoekstra 1966). In addition, one quarter of the deficient rats showed loss of hair and scabby epidermal lesions.

The effects of zinc deficiency upon chemical composition of rat skin are presented in Table I. Water content in the skin of the deficient rats was significantly higher than that of the zinc supplemented pair-fed controls. The contents of lipid and nitrogen appear to be unaffected by zinc deficiency. The amount of total hydroxyproline was increased in the zinc-deficient rats as compared to control animals.

TABLE I

EFFECT OF ZINC DEFICIENCY UPON CHEMICAL COMPOSITION OF RAT SKIN

	Water %	Lipid %	Nitrogen mMoles/ g dry skin	Hydroxyproline μ Moles/ g dry skin
Zinc-supplemented (pair-fed)	61.9±1.4[a,c]	19.4±1.8	4.64±0.44	236±11[b]
Zinc-deficient	66.2±1.1	15.8±2.5	5.52±0.48	304±12

a Mean ± SD. Six rats in each group.
b Difference between Zn-supplemented and Zn-deficient rats is statistically significant ($p<0.01$)
c Difference between Zn-supplemented and Zn-deficient rats is statistically significant ($p<0.05$)

The concentrations of zinc and other minerals were determined in dried skin samples. A significant decrease of zinc concentration in the skin of the zinc deficient rats as compared to the zinc-supplemented ad libitum-fed rats was observed (Table II). No significant difference was found between the zinc-deficient and controls in relation to the concentrations of the copper, manganese, sodium and potassium at observation time.

III. INCORPORATION OF ^{35}S FROM CYSTINE-^{35}S AND METHIONINE-^{35}S INTO SKIN PROTEINS

Previously it was reported that after cystine-^{35}S injection, the zinc-deficient rats excreted two to three times more urinary

TABLE II

EFFECT OF Zn-DEFICIENCY ON THE CONTENT OF SKIN - Zn, Cu, Mn, Na and K

Type of diet	Zn	Cu	Mn	Na	K
	μg/g dry wt.			mg/g dry wt.	
Zn-supplemented (ad lib)	41.28 ± 3.25[a,b]	7.76 ± 1.34	7.19 ± 2.36	2.89 ± 0.61	3.48 ± 0.77
Zn-supplemented (pair-fed)	36.21 ± 9.02	7.83 ± 3.39	7.74 ± 1.81	3.01 ± 0.77	3.85 ± 1.04
Zn-deficient	28.13 ± 5.11	12.52 ± 6.71	10.17 ± 3.18	3.33 ± 1.32	2.95 ± 0.66

a Mean ± SD. Six rats in each group were on experimental diet for 15 days

b Difference between Zn-supplemented (ad lib) and Zn-deficient rats is statistically significant ($p<0.05$).

total ^{35}S and inorganic sulfate-^{35}S than the zinc-supplemented controls (Hsu and Anthony, 1970; Hsu et al., 1970). Although the mechanism by which zinc deficiency increases the urinary excretion of sulfate is not clear, it is conceivable that cystine metabolism in the zinc-deficient rats is altered. Studies were therefore carried out to determine the effect of zinc deficiency on cystine incorporation into tissue proteins. Each rat was injected intramuscularly with cystine-^{35}S and sacrificed at various time intervals thereafter. The results in Table III indicate that at 4-hours the protein specific activity present in the skin of the zinc-deficient rats was one-seventh that of ad libitum-fed zinc-supplemented rats, and one-fifth that of pair-fed controls. Although more cystine-^{35}S was incorporated into skin protein of the zinc-deficient rats at 8-hours than that at 4-hours, the amounts were still only 35% of the normal value. Similar trends were found between the zinc-supplemented and the zinc-deficient rats at the end of 16 and 24 hours of cystine-^{35}S incorporation into protein.

The results of the effect of zinc deficiency on the incorporation of cystine-^{35}S into other tissue proteins are given in Table IV. No significant differences were observed in the protein-^{35}S activity of the liver, kidney, testes and muscle between the two groups. However, more cystine-^{35}S was incorporated into pancreas protein of rats fed zinc-deficient diet than those fed the zinc-supplemented diet.

The results summarized in Table V indicate that skin protein in the zinc-repleted rats had the same radioactivity as the zinc-supplemented rats. In the pancreas, similar findings were noted between the two groups. Thus, the defects in incorporation observed in the zinc-deficient rats seemed to be reversible.

Recently several investigators have attempted to relate zinc to protein synthesis. Macapinlac et al. (1968) found that there was no difference in the rate of incorporation of leucine-U-^{14}C in the testes protein between the zinc-supplemented and the zinc-deficient rats. Likewise, O'Neal et al. (1970) were unable to find any significant changes in the relative specific activities of the brain protein isolated from the zinc-deficient and pair-fed zinc-supplemented rats. It occurred to us that studies of protein metabolism in zinc deficiency might be carried out most profitably in the skin of the rat, since an inadequate intake of zinc causes abnormalities of the dermal system in all the mammalian species that have been examined. Data in the present investigation clearly demonstrate that the zinc-deficient rats show markedly diminished capacity to incorporate injected cystine-^{35}S into skin protein. To our knowledge this is the first biochemical evidence in regards to

TABLE III

INCORPORATION OF L-CYSTINE-^{35}S INTO SKIN PROTEIN

Type of Diet	Days of feeding	No. of rats	Hours after injection	Specific activity DPM/mg protein
Zn-supplemented (ad lib)	14	7	4	1302 ± 585[a,b]
Zn-supplemented (pair-fed)	14	5	4	881 ± 319[b]
Zn-deficient	14	6	4	174 ± 66
Zn-supplemented (ad lib)	16	5	8	1444 ± 136[b]
Zn-supplemented (pair-fed)	16	6	8	1337 ± 108[b]
Zn-deficient	16	5	8	506 ± 118
Zn-supplemented (ad lib)	16	6	16	1080 ± 227[b]
Zn-deficient	16	6	16	352 ± 83
Zn-supplemented (ad lib)	15	5	24	1267 ± 112[b]
Zn-supplemented (pair-fed)	15	5	24	1117 ± 295[c]
Zn-deficient	15	5	24	306 ± 155

a Mean ± SD.

b Difference between Zn-supplemented (ad lib) and Zn-deficient rats is statistically significant ($p < 0.01$).

c Difference between Zn-supplemented (pair-fed and Zn-deficient rats is statistically significant ($p<0.01$).

TABLE IV
24-HOURS INCORPORATION OF L-CYSTINE-^{35}S
INTO OTHER TISSUE PROTEIN

Type of Diet	Specific Activity DPM/mg. Protein				
	Pancreas	Liver	Kidney	Testes	Muscle
Zn-supplemented (ad lib)	341 ± 35[a,b]	213 ± 39	397 ± 48	301 ± 38	120 ± 32[b]
Zn-supplemented (pair-fed)	458 ± 82[a]	216 ± 28	382 ± 9	304 ± 24	129 ± 44[b]
Zn-deficient	587 ± 64	201 ± 46	382 ± 61	257 ± 61	84 ± 28

a Mean ± SD. Six rats in each group were on experimental diet for 15 days.
b Difference between Zn-supplemented (ad lib and pair fed) and Zn-deficient rats is statistically significant ($p < 0.05$).

TABLE V
EFFECT OF Zn REPLETION ON INCORPORATION OF L-CYSTINE-^{35}S INTO TISSUE PROTEINS

Type of diet	Body wt. before repletion	Body wt. 3-days after repletion	Specific Activity DPM/mg. Protein		
			Skin	Pancreas	Liver
	g	g			
Zn-supplemented (pair-fed)	104 ± 4[a,c]	110 ± 5	343 ± 53	155 ± 35	99 ± 6
Zn-deficient repleted[b]	83 ± 12	92 ± 14	309 ± 86	180 ± 32	107 ± 9

a Mean ± SD. Six rats in each group were killed 24-hours after L-cystine-^{35}S injection. (2 μCi/100 g. body wt.)

b Each rat received an intraperitoneal injection of 400 μg Zn daily on last 3-days of 18-day experiment.

c Difference between Zn-supplemented (pair-fed) and Zn-repleted rats is statistically significant ($p<0.05$).

the functional impairment of skin induced by zinc deficiency. Since the major portion of the epidermal layer of skin is made up of albuminoid proteins, keratins which have a high content of cystine (Carruthers 1962), it would be of interest to determine the relationship between zinc deficiency and the biosynthesis of keratins.

Since ^{35}S of cystine-^{35}S incorporation of skin protein in the zinc-deficient rats is markedly decreased (Hsu and Anthony 1972), and since methionine is readily converted to cystine, it was of interest to determine the effect of zinc deficiency on the metabolic fate of methionine-^{35}S. As shown in Table VI the uptake of ^{35}S in the skin-hair of the zinc-deficient rats was considerably lower than that of the control rats. Zinc deficiency also drastically reduced the amount of L-methionine-^{35}S and DL-methionine-2-^{14}C incorporation into skin proteins (Table VII).

Unlike cystine, methionine is incorporated into both epidermal and dermal proteins. Although data shown in Table VII reveal that the specific activities of skin proteins were less in zinc-deficient rats than in zinc-supplemented rats, no information is available in regard to the nature of the radioactivity in skin protein. It is believed that either labeled methionine itself, or its derivative, cystine or both are incorporated into skin proteins.

IV. INCORPORATION OF ^{14}C-LABELED AMINO ACID

To determine the effect of zinc deficiency on the incorporation of amino acids other than those containing sulfur, various labeled amino acids were tested.

Figure I shows that the incorporation of all five non-essential amino acids into skin proteins was significantly reduced in the zinc-deficient rats as compared with the zinc-supplemented pair-fed rats. The most striking difference was the 70% decrease in the specific activity of skin protein in the zinc-deficient rats after glycine-2-^{14}C injection. The incorporation of L-proline-U-^{14}C, DL-alanine-1-^{14}C and L-cystine-3-^{14}C into skin proteins was reduced to less than 45% of the normal value. Zinc-deficient rats also incorporated 30% less of L-glutamic acid-U-^{14}C into skin protein than zinc-supplemented controls.

Of the essential amino acids studied, the incorporation of L-tryptophan-2,3-^{3}H, L-histidine-2-^{14}C, L-lysine-U-^{14}C, and L-methionine-2-^{14}C into skin protein of zinc deficient rats was reduced 35, 50, 34 and 40% respectively (Figure 2).

TABLE VI

UPTAKE OF ^{35}S BY SKIN AND HAIR AFTER L-METHIONINE-^{35} INJECTION[a]

Hours after L-methionine-35S injection	Days of feeding	% of injected ^{35}S: Zn-supplemented ad lib	Zn-supplemented pair-fed	Zn-deficient
24	18	16.28 ± 1.03[b,c]	16.61 ± 2.60[d]	8.45 ± 1.78
48	15	20.47 ± 1.50[c]	16.29 ± 2.26[d]	7.98 ± 1.80
96	16	18.14 ± 3.19[c]	19.39 ± 2.76[d]	8.12 ± 3.20
144	18	24.94 ± 2.59[c]	19.29 ± 4.63[d]	6.01 ± 1.86

a Six rats were in each group. Each received im injection of L-methionine-^{35} (5 μCi/100 g body wt).

b Mean ± SD

c Difference between Zn supplemented (ad lib) and Zn-deficient rats is statistically significant ($p < 0.01$).

d Difference between Zn-supplemented (pair-fed) and Zn-deficient rats is statistically significant ($p < 0.01$).

TABLE VII

INCORPORATION OF LABELED METHIONINE INTO SKIN PROTEIN

Type of diet	Number of rats used	Days of feeding	Isotope injected	Hours after injection	Specific activity dpm/mg protein	Specific activity dpm/mg wet tissue
Zn-supplemented (pair-fed)	6	16	L-Methionine-^{35}S	4	640 ± 91[a,b]	92 ± 29[b]
Zn-deficient	5	16	L-Methionine-^{35}S	4	239 ± 81	30 ± 15
Zn-supplemented (pair-fed)	6	15	L-Methionine-^{35}S	24	654 ± 170[b]	80 ± 25[b]
Zn-deficient	6	15	L-Methionine-^{35}S	24	220 ± 84	32 ± 11
Zn-supplemented (pair-fed)	6	16	DL-Methionine-2-^{14}C	24	100 ± 16[b]	18 ± 1.4[b]
Zn-deficient	5	16	DL-Methionine-2-^{14}C	24	58 ± 19	10 ± 2.8

a Mean ± SD

b Difference between Zn-supplemented (pair-fed) and Zn-deficient rats is statistically significant ($p<0.01$).

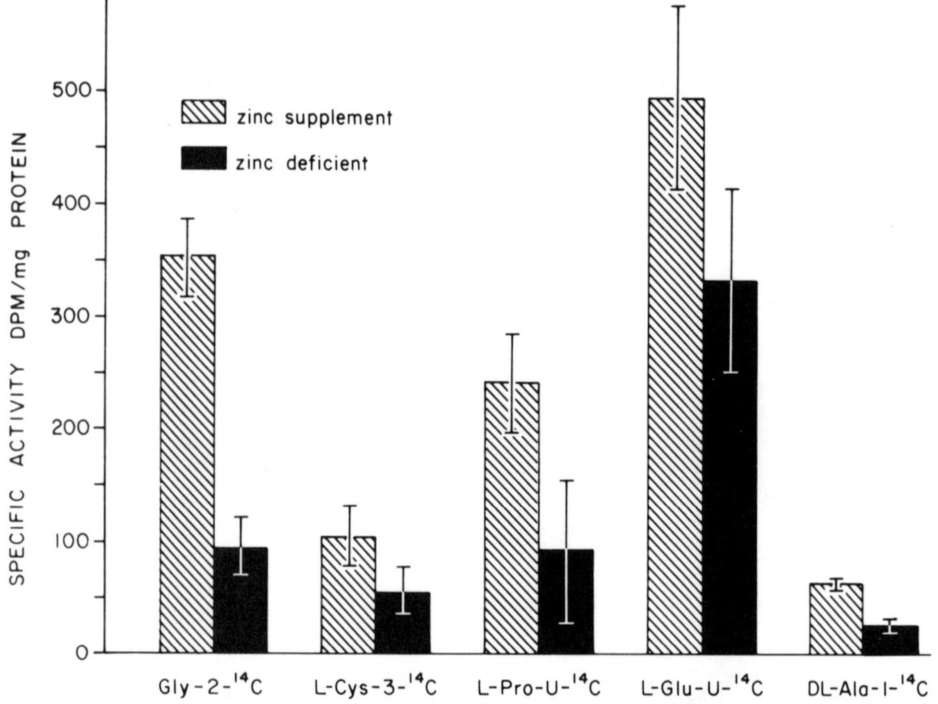

Figure 1. Effects of zinc deficiency on 24-hours incorporation of glycine-2-^{14}C, L-cystine-3-^{14}C, L-proline-U-^{14}C, L-glutamic acid-U-^{14}C, and DL-alanine-1-^{14}C into skin protein. The heights of the solid bars and the cross bars represent the mean specific activity of the six supplemented and six Zn-deficient rats. Vertical lines denote standard deviations.

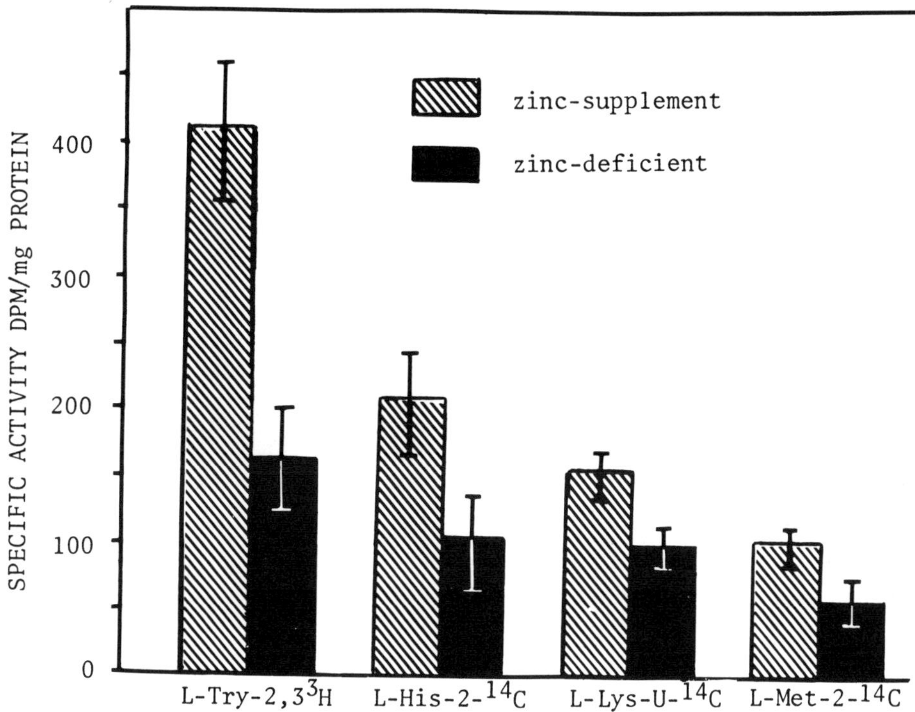

Figure 2. Effects of zinc deficiency on 24-hours incorporation of four essential amino acids into skin protein. The heights of the solid bars and the cross bars represent the mean specific activity of six zinc-supplemented and six zinc-deficient rats. Vertical lines denote standard deviations.

TABLE VIII

24-HRS. INCORPORATION OF ^{14}-C-ARGININE INTO SKIN PROTEIN

Type of Diet	Number of Rats Used	Isotope Injection	Specific Activity DPM/mg Proteins
Zn-supplemented (pair-fed)	6	L-arginine-guanido-^{14}C	418 ± 22[a,b]
Zn-deficient	6	L-arginine-guanido-^{14}C	220 ± 102
Zn-supplemented (pair-fed)	8	L-arginine-U-^{14}C	262 ± 15[b]
Zn-deficient	7	L-arginine-U-^{14}C	63 ± 12

a Mean ± SD
b Difference between Zn-supplemented (pair-fed) and Zn-deficient rats is statistically significant ($p<0.01$).

Table VIII indicates that a significant reduction of radioactivity in proteins were observed in the skin of the Zn-deficient rats after injection of both L-arginine-guanido-^{14}C and L-arginine-U-^{14}C. As shown by Swick and Handa (1956), L-arginine-guanido-^{14}C is not subject to isotope reutilization in the liver because of the large amount of arginase, which hydrolyzes the label to urea. If this were found to be true in skin, it would suggest that the decreased incorporation of amino acids into skin protein of the zinc-deficient rats is not due to amino acid reutilization. The defects in arginine incorporation into skin protein can be restored to normal as indicated in Table IX.

TABLE IX

EFFECT OF ZINC REPLETION ON 24-HOURS INCORPORATION OF L-ARGININE U-^{14}C INTO SKIN PROTEIN

Type of Diet	Number of Rats Used	Specific Activity DPM/mg Proteins
Zn-supplemented (pair fed)	6	279 ± 74[b]
Zn-deficient and repleted[a]	7	311 ± 35

a Zinc repleted rats received subcutaneously 400 µg of zinc in the form of $ZnCO_3$ daily for last three days of an 18-day experiment.
b Mean ± SD

The above findings suggest that a common defect may exist in the integument of the zinc-deficient rats which causes a general reduction of amino acid utilization. However, this may represent an indirect effect and not a specific role of zinc in the incorporation of amino acids into skin protein. That is zinc-deficiency may induce more detrimental effects to the skin cells than to other tissues because they have no direct means in receiving the nutrient supply.

V. BIOSYNTHESIS OF SKIN COLLAGENS

Incorporation of ^{14}C-labeled Amino Acids into Skin Collagen

The findings of reduced incorporation of glycine and proline, two major precursors of collagen, into skin protein of the zinc-deficient rats indicate that zinc may be involved in collagen synthesis. To test this possibility, insoluble and soluble fractions of skin collagens were isolated and their specific activities were determined 24-hours after an injection of labeled amino acid. Table X shows that the radioactivity recovered as insoluble fraction was markedly reduced in the skin of zinc-deficient rats regardless of the origin of isotope used. The ratios were approximately 5.0, 2.5, 1.5 and 1.7 between the zinc-supplemented pair-fed rats and the zinc-deficient rats after injection of glycine-2-^{14}C, L-proline-U-^{14}C, L-glutamic acid-U-^{14}C and L-histidine-2-^{14}C, respectively. Table X also shows that zinc-deficient rats incorporated significantly less labeled glycine and proline into soluble fractions of skin collagens. The above finding is of interest because it points out that the biological role of zinc is specifically involved in one type of protein, namely skin collagen.

Effect of Insulin and Growth Hormone

Zinc is closely associated with insulin, the characteristic dark-staining granules of β-cells of the islets of Langenhans (Logothetopoulos et al. 1964). The data of Quarterman et al. (1966) strongly suggest that zinc is required for normal insulin secretion. The endocrine abnormalities of the zinc-deficient patients have been described as those of hypopituitarism (Sandstead et al. 1967). To understand that the decrease of amino acid incorporation into skin collagen of zinc-deficient rats is not related to the lack of hormonal secreation, the synthesis of skin collagen was studied in zinc-deficient rats after the injection of insulin or growth hormone.

The results from Table XI indicate that the treatment of insulin appears to have a greater effect than that of growth hormone in glycine-2-^{14}C Incorporation into skin collagens

TABLE X

24-HRS. INCORPORATION OF ^{14}C-AMINO ACID INTO SKIN COLLAGEN

Type of Diet	^{14}C-Amino Acid Injected	No. of Rats	Skin Collagens DPM/100 μg hydroxyproline	
			Insoluble	Soluble
Zn-supplemented (pair-fed)	Glycine-2-^{14}C	6	267 ± 28[a,b]	993 ± 79[b]
Zn-deficient	Glycine-2-^{14}C	6	50 ± 18	412 ± 141
Zn-supplemented (pair-fed)	L-Glutamic-U-^{14}C	6	450 ± 169[c]	1630 ± 536
Zn-deficient	L-Glutamic-U-^{14}C	5	274 ± 39	1450 ± 302
Zn-supplemented (pair-fed)	L-Proline-U-^{14}C	7	290 ± 39[b]	879 ± 135[b]
Zn-deficient	L-Proline-U-^{14}C	6	139 ± 32	446 ± 222
Zn-supplemented (pair-fed)	L-Histidine-2-^{14}C	5	78 16[b]	- - -
Zn-deficient	L-Histidine-2-^{14}C	5	44 11	- - -

a Mean ± SD

b Difference between Zn-supplemented (pair-fed) and Zn-deficient rats is statistically significant ($p<0.01$).

c Difference between Zn-supplemented (pair-fed) and Zn-deficient rats is statistically significant ($p<0.05$).

regardless of their zinc status. There was a significant increase in the formation of labeled collagens by insulin-injected zinc-deficient and zinc-supplemented rats as compared to their respective animals without hormone treatment. Table XI also shows that the increase of specific activity in insoluble skin collagen by insulin was greater in zinc-deficient rats than in zinc supplemented controls.

TABLE XI

ZINC DEFICIENCY AND HORMONE EFFECT ON 4-HOUR INCORPORATION OF GLYCINE-2-^{14}C INTO SKIN COLLAGEN

Type of diet	Hormone treatment	Skin Collagens DPM/100 μg hydroxyproline	
		Insoluble	Soluble
Zn-supplemented (pair-fed)	None	136 ± 18[c,d]	1043 ± 99[d]
Zn-deficient	None	49 ± 30	582 ± 303
Zn-supplemented (pair-fed)	Growth hormone[a]	205 ± 74[d]	1356 ± 375[d]
Zn-deficient	Growth hormone	75 ± 23	500 ± 104
Zn-supplenented (pair-fed)	Insulin[b]	236 ± 73[e]	1848 ± 787[e]
Zn-deficient	Insulin	155 ± 56	1025 ± 261

a Each received 0.1 mg of NIH GH B_{14} subcutaneously daily for last 6-days of 14-day experiment.

b Each received 0.5U of Iletin (insulin injection, Lilly and Co., Indianapolis, Ind.)

c Mean ± SD

d Difference between Zn-supplemented (pair-fed) and Zn-deficient is statistically significant ($p<0.01$)

e Difference between Zn-supplemented (pair-fed) and Zn-deficient is statistically significant ($p<0.05$)

Since zinc deficiency in a wide range of experimental animals and in man produces severe growth retardation, the syndrome has been suggested to be associated with the function of the pituitary, an organ which has a high content of zinc (Millar et al. 1961). However, significant weight gains were not observed in zinc-deficient rats treated with bovine growth hormone (Macapinlac, et al. 1966). Similarly, this hormone did not change quantitatively the glycine incorporation into skin collagen (Table XI).

The effect of zinc on insulin metabolism is far from clear. On the basis of a study of four pairs of zinc-deficient and control rats, Quarterman et al. (1966) have stated that the zinc-deficient animals had decreased concentration of plasma insulin. Boquist and Lernmark (1969) observed that intravenous and intraperitoneal glucose tolerance were impaired in zinc-deficient hamsters with the normal value of serum immunoreactive insulin. The fact that the injection of insulin significantly increased the formation of skin collagen in both zinc-deficient rats and zinc-supplemented rats (Table XI) indicates that insulin has its own action on protein biosynthesis regardless of the status of zinc. This is in agreement with the knowledge of the effect of insulin in the promotion of amino acid uptake and in the synthesis of protein (Wool et al. 1965).

Collagen Fractions

Additional experiments were carried out to determine the effect of zinc deficiency on various fractions of skin collagens. The results indicate (Table XII) that zinc-deficient rats had a significant increase of hydroxyproline content in insoluble fraction with a concomitant decrease of hydroxyproline in 0.5 M neutral salt fraction. Only a trace amount of hydroxyproline was found in the acid soluble fraction and there was no quantitative difference between zinc-deficient and zinc-supplemented rats.

The results presented in Table XIII reveal that zinc-deficient rats incorporated significantly less glycine-2-^{14}C into 0.5 M neutral soluble, acid soluble and insoluble fractions of skin collagen than zinc supplemented rats. These findings confirm and extend the earlier observations (Table X) indicating an impairment of collagen synthesis in zinc-deficient rats.

Crosslinking

All of the above studies strongly link zinc to the biosynthesis of rat skin collagen. Further studies are found to elucidate the influence of zinc deficiency on crosslinking of

TABLE XII

HYDROXYPROLINE CONTENTS IN VARIOUS FRACTIONS OF RAT SKIN

Type of diet	0.15 M NaCl Extract	0.5 M NaCl Extract	Acid Soluble pH 3.6	Insoluble Residue
		μ Moles/g fresh skin		
Zn-supplemented (ad lib)	0.455 ± 0.155[a]	18.84 ± 3.83[b]	0.14 ± 0.03	75.4 ± 7.01[d]
Zn-supplemented (pair-fed)	0.443 ± 0.114	19.63 ± 4.22[c]	0.26 ± 0.14	85.5 ± 10.62[e]
Zn-deficient	0.433 ± 0.134	14.95 ± 2.13	0.38 ± 0.21	134.1 ± 16.96

a Mean ± SD. Six rats in each group.

b Difference between Zn-supplemented (ad lib) and Zn-deficient rats is statistically significant ($p<0.05$)

c Difference between Zn-supplemented (pair-fed) and Zn-deficient rats is statistically significant ($p<0.05$)

d Difference between Zn-supplemented (ad lib) and Zn-deficient rats is statistically significant ($p<0.01$)

e Difference between Zn-supplemented (pair-fed) and Zn-deficient rats is statistically significant ($p<0.01$)

TABLE XIII

24-HOUR INCORPORATION OF GLYCINE-2-^{14}C INTO SKIN COLLAGEN

Type of Diet	0.15 M NaCl Extract	0.5 M NaCl Extract	0.5 M Acetic pH 3.6	Insoluble Residue
	DPM/μ Mole Hydroxyproline			
Zn-supplemented (ad lib)	19973 ± 6971[a]	4017 ± 406[b]	1371 ± 315[d]	5834 ± 1952[d]
Zn-supplemented (pair-fed)	26324 ± 8136	4268 ± 523	1022 ± 206[e]	6281 ± 207[e]
Zn-deficient	27436 ± 7197	2986 ± 383	776 ± 242	1278 ± 603

a Mean ± SD. Six rats in each group.
b Difference between Zn-supplemented (ad lib) and Zn-deficient rats is statistically significant ($p<0.05$)
c Difference between Zn-supplemented (pair-fed) and Zn-deficient rats is statistically significant ($p<0.05$)
d Difference between Zn-supplemented (ad lib) and Zn-deficient rats is statistically significant ($p<0.01$)
e Difference between Zn-supplemented (pair-fed) and Zn-deficient rats is statistically significant ($p<0.01$)

skin collagen. The α_1 and α_2 chains of salt-soluble fraction were isolated by carboxymethyl cellulose chromatography (Morris and McClain 1972). The data shown in Table XIV reveal a reduction in the incorporation of glycine-2-^{14}C into the α_1 and α_2 chains. The specific activity in the α_1 and α_2 chains from the zinc-deficient rats was 49 and 30% less, respectively than that from the zinc-supplemented animals.

TABLE XIV

INCORPORATION OF GLYCINE-2-^{14}C INTO α_1 AND α_2 CHAINS OF SALT-SOLUBLE FRACTION OF SKIN COLLAGEN

Type of Diet	α1	α2
	DPM/mg Collagen	
Zn-supplemented (pair-fed)	4254 ± 124[a,b]	3785 ± 80[b]
Zn-deficient	2243 ± 98	2641 ± 141

a Mean ± SD. Six rats in each group.
b Difference between Zn-supplemented (pair-fed) and Zn-deficient rats is statistically significant ($p<0.01$)

The amino acid composition of the α_1 and α_2 chains of neutral soluble skin collagen from zinc-supplemented and zinc-deficient animals is shown in Table XV. No marked variations in α-chain amino acid composition are apparent in either experimental group, suggesting that zinc deficiency had no effect on the primary-secondary structure of the collagen molecule.

The results of this investigation suggest that zinc may play a fundamental role in the collagen crosslinking process. Certainly the increased content of the β-components and increased aldehyde content of the salt soluble collagen pool indicate that the formation of covalent intramolecular crosslinks is enhanced in the skin collagen from zinc-deficient rats. The reason for this increase in intra-molecular crosslinking is not clear from the results of the present study (Figures 3 and 4).

The results of this study, however, do not rule out the possibility that zinc deficiency is adversely affecting other factors which could alter collagen synthesis, such as an inhibition of the proly and lysyl hydroxylases of the glycosylation process which would prevent release of newly synthesized collagen from the fibroblast (Traub et al. 1971).

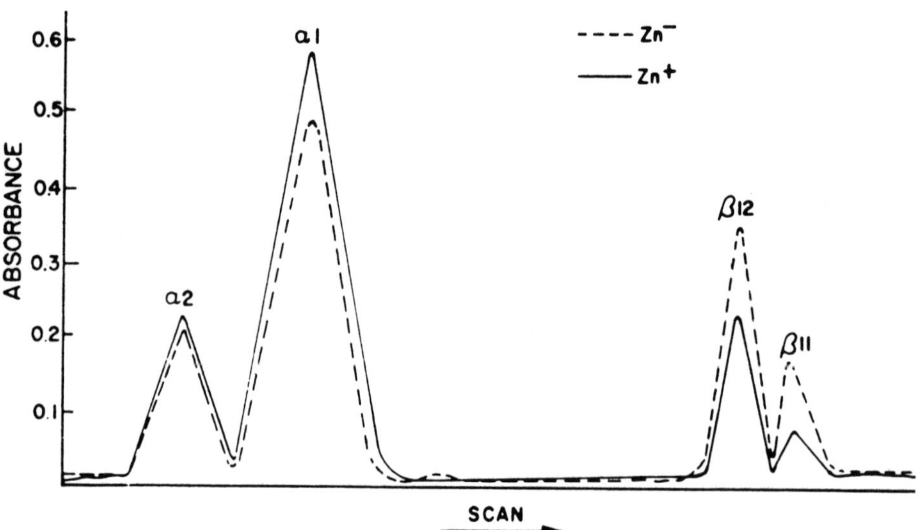

Figure 3. Densitometer tracings of disc gel patterns obtained from salt-soluble collagen.

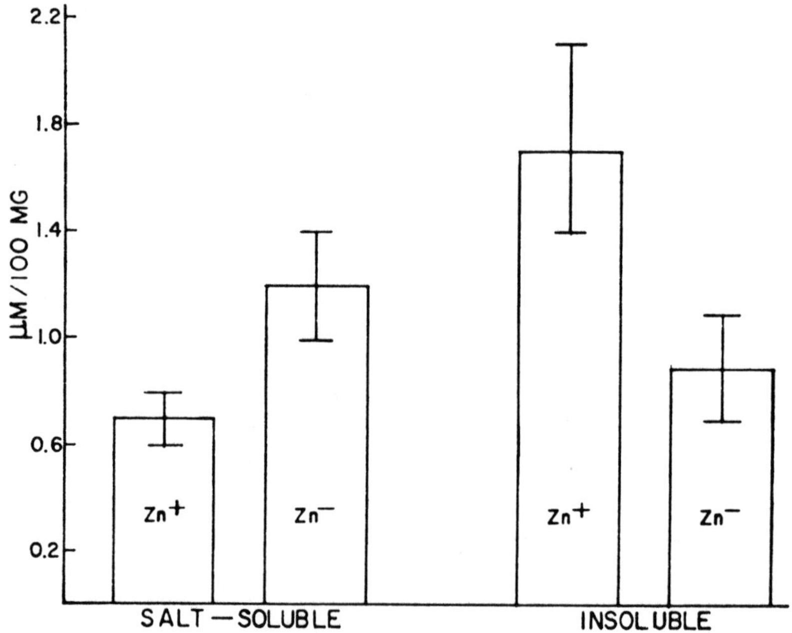

Figure 4. Aldehyde content of salt-soluble and insoluble skin collagen.

TABLE XV

AMINO ACID COMPOSITION OF α-CHAINS FROM SALT-SOLUBLE RAT SKIN COLLAGEN*

Amino acid	Residues per 1000			
	Zinc-supplemented		Zinc-deficient	
	α_1	α_2	α_1	α_2
Aspartic	47	46	47	45
Hydroxyproline	94	92	94	90
Threonine	20	24	20	24
Serine	40	45	40	43
Glutamic	75	75	75	73
Proline	119	105	118	107
Glycine	329	335	329	332
Alanine	110	106	114	108
Valine	20	20	20	20
Methionine	7	4	7	4
Isoleucine	14	10	11	10
Leucine	23	26	20	27
Tyrosine	3	3	3	3
Phenylalanine	12	13	15	13
Hydroxylysine	3.5	9.1	3.3	8.2
Lysine	32.5	25.8	32.6	26.7
Histidine	2	10	2	10
Arginine	49	50	49	53

* Mean of three samples per treatment.

VI. DNA SYNTHESIS

DNA Concentrations in Various Tissues

Zinc deficiency reduces DNA synthesis in rat liver (Fujioka and Lieberman 1964). Our recent studies have demonstrated the need for zinc in the incorporation of cystine-^{35}S into skin protein (Hsu and Anthony 1971) and glycine-2^{14}C into skin collagen (Hsu and Anthony 1973). These findings prompted us to examine the effect of zinc deficiency on DNA synthesis in rat skin. The concentrations of DNA in various tissues are presented in Table XVI. Zinc deficiency had no effect on the total amounts of DNA in the skin, spleen, and thymus. However, DNA contents in the liver and testes of zinc deficient rats were significantly

TABLE XVI

THE CONCENTRATION OF DNA IN VARIOUS TISSUES

Type of Diet	Skin	Liver	Testes	Spleen	Thymus
	mg/g. fresh wt.				
Zn-supplemented (pair-fed)	2.32 ± 0.14[a]	2.30 ± 0.13[b]	3.99 ± 0.12[b]	13.67 ± 0.66	30.66 ± 3.56
Zn-deficient	2.11 ± 0.07	3.10 ± 0.25	4.83 ± 0.32	12.76 ± 0.55	24.62 ± 3.42

a Mean ± SD. Eight or more specimens in each group.

b Difference between Zn-supplemented (pair-fed) and Zn-deficient is statistically significant ($p<0.05$).

higher than that of zinc-supplemented rats.

Thymidine Incorporation into DNA

Measurement of radioactivity in sodium hydroxide-solubilized skin and the specific activity of skin DNA are summarized in Table XVII. There was a significant decrease in the specific activity (CPM/mg DNA) of the DNA in the skin from zinc-deficient rats, 0.5, 1, and 2-hours after thymidine-methyl-^{3}H injection. A decrease in total radioactivity in the skin of zinc-deficient rats was noted at 0.5 hr. post-injection. Only in the 0.5 hr. could the observed decrease in specific activity be a function of isotope available; at 1 and 2 hrs. a difference in the rate of incorporation is indicated.

TABLE XVII

THE SPECIFIC ACTIVITY OF DNA IN THE SKIN AFTER THYMIDINE-METHYL-^{3}H INJECTION

Type of Diet	Hours After Injection	Specific Activity	
		CPM/mg Skin	CPM/mg DNA
Zn-supplemented (pair-fed)	0.5	109 ± 3[a,b]	2184 ± 137[e]
Zn-deficient	0.5	98 ± 3	307 ± 22
Zn-supplemented (pair-fed)	1.0	92 ± 30	6967 ± 935[c]
Zn-deficient	1.0	105 ± 28	2188 ± 394
Zn-supplemented (pair-fed)	2.0	109 ± 3	9023 ± 925[c]
Zn-deficient	2.0	102 ± 7	2180 ± 621
Zn-supplemented (pair-fed)	4.0	394 ± 45	33543 ± 6091
Zn-deficient	4.0	459 ± 39	20064 ± 2902

a Mean ± SD. Five rats in each group.
b Difference between Zn-supplemented (pair-fed) and Zn-deficient rats is statistically significant ($p < 0.05$).
c Difference between Zn-supplemented (pair-fed) and Zn-deficient rats is statistically significant ($p < 0.01$).

TABLE XVIII

THE INCORPORATION OF THYMIDINE-^{3}H INTO TISSUE DNA

Type of Diet	Hours After Injection	Specific Activity CPM/mg DNA			
		Liver	Spleen	Testes	Thymus
Zn-supplemented (pair-fed)	2	1668 ± 81[a]	9904 ± 643[b]	1603 ± 144	683 ± 49
Zn-deficient	2	1459 ± 103	5799 ± 637	1702 ± 174	1283 ± 136
Zn-supplemented (pair-fed)	4	2253 ± 236[b]	15888 ± 2742[b]	4305 ± 766	- - -
Zn-deficient	4	1005 ± 146	2675 ± 392	2919 ± 611	- - -

a Mean ± SD. Five or more rats in each group.

b Difference between Zn-supplemented (pair-fed and Zn-deficient rats is statistically significant ($p<0.01$)

To observe whether the decrease in specific activity of DNA in the skin was a general aberration in the zinc deficient rats, several other tissues were examined. The 2-hr. incorporation of thymidine-^{3}H into DNA decreased in the spleen and increased in the thymus of the zinc-deficient rats significantly (Table XVIII), but no differences were observed in the liver and testes. When these measurements were made 4-hrs following injection, the specific activity of DNA in the liver was significantly decreased in zinc deficiency.

Autoradiographic Analysis

The use of high resolution autoradiography combined with specific labeling of DNA with thymidine-^{3}H permits analysis of certain quantitative histological parameters of cell population kinetics. In this study, pulse labeling was employed to examine the proliferative activity of epidermal cells in zinc-supplemented and zinc-deficient wounded and unwounded rats by determination of the thymidine labeling index, the percentage of cells labeled in the population.

Unwounded epithelium of both zinc-supplemented and zinc-deficient rats showed normal histological appearance. However, there appeared to be fewer cells in the zinc-deficient animals per unit of histological section (Hsu and Hsu 1972). Thymidine-labeling indices for zinc supplemented animals were significantly higher than that for zinc-deficient rats (Table XIX).

Wounding resulted in destruction of a large number of cells at the lesion site. However, two days after wounding, repair was evident; labeled cells, signifying cell proliferation, appeared in the wound site. In histological sections, there were more labeled nuclei in zinc-supplemented specimen (Figure 5A) than in zinc-deficient tissue (Figure 5B). The thymidine labeling index for the zinc-supplemented wounded animals was 29.4 ± 1.2% against 26.1 ± 1.1% for zinc deficient wounded rats.

Four days after wounding, the labeling index for zinc-supplemented animals increased to about 34% whereas it remained at about 25% for zinc-deficient animals. Histological sections revealed that a considerable number of labeling cells were observed in the skin biopsy specimen of zinc-supplemented rats (Figure 5C), but not seen in zinc-deficient animals (Figure 5D). By the sixth day after wounding, the labeling index for zinc-supplemented animals remained about 36%, but the zinc-deficient labeling index also remained around 26%.

The experimental findings conclusively showed that re-epithelialization of wounds is much slower in the zinc-deficient

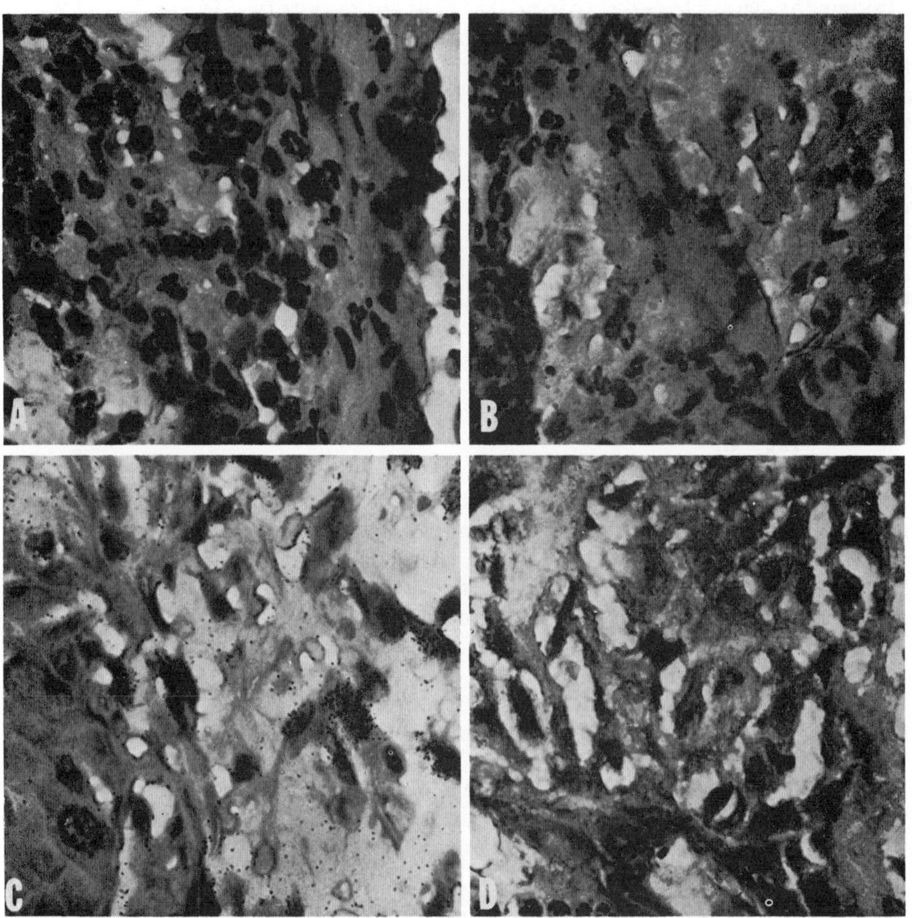

Figures 5A, 5B, 5C and 5D. Autoradiographic analysis of epithelial repair X 1040. A and B are sections 2-days after wounding; C and D, sections 4-days after wounding. Rat A and C are zinc-supplemented; B and D zinc-deficient. Proliferative activity of epithelial cells are much more intense in the zinc-supplemented animals, as shown by the number of labeled nuclei. Such activity leads to a quicker and better organized repair of the tissue.

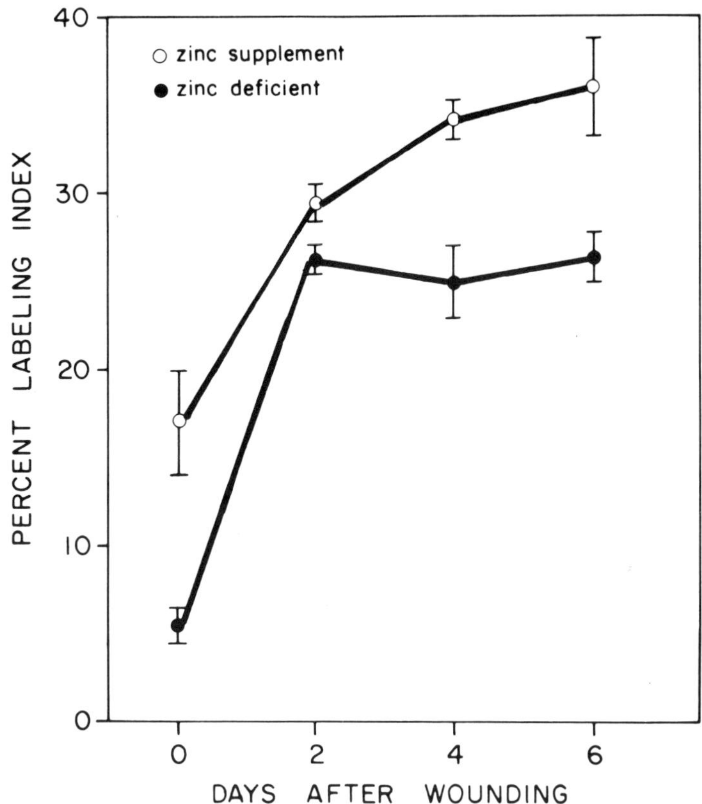

Figure 6. Percent labeling indices for zinc-supplemented and zinc-deficient rats. Labeling index for non-wounded animals are shown at day 0, and represents normal situation. Wounded animals show a much higher labeling index.

animals as demonstrated by isotope counting (Stephan and Hsu 1973), and autoradiography.

TABLE XIX

SKIN UPTAKE OF THYMIDINE C^3H_3

Type of Diet	No. of Rats	Uptake of thymidine C^3H_3	
		dpm/mg Wet weight	Labeling index (%)
Zn-supplemented (pair-fed)	6	607 ± 57[a,b]	17.1 ± 3.4[b]
Zn-deficient	6	537 ± 13	7.4 ± 0.8

a Mean ± SD. Six or more rats in each group.
b Difference between Zn-supplemented (pair-fed) and Zn-deficient rats is statistically significant ($p<0.01$).
c Difference between Zn-supplemented (pair-fed) and Zn deficient rats is statistically significant ($p<0.05$).

The evidence for a zinc requirement for DNA synthesis by rat liver has been shown by several investigators (Fujioka and Lieberman, 1964; Weser, et al., 1970; and Sandstead and Rinaldi, 1969). Recent findings of a zinc-containing DNA polymerase (Slater, et al., 1971) suggest that further study concerning the relationship between zinc and nucleic acid is needed. In zinc-deficient rats the suppression of skin DNA synthesis from ^3H-labeled thymidine has been demonstrated in the present study and also in our autoradiography experiments (Hsu and Hsu, 1972). As observed in our study, the incorporation of radioactive precursor of DNA into skin DNA could result from either direct stimulation of the synthesis of DNA by the zinc or the alteration in permeability transport or pool size of the precursor by zinc. Nevertheless, the decreased incorporation of thymidine into skin DNA in zinc-deficient rats along with the finding that zinc deficiency results in a reduced labeling index in skin cells supports the idea that zinc directly stimulates skin DNA synthesis. This is consistent with the report of Fujioka and Lieberman (1964) indicating a requirement of zinc in DNA synthesis. Although our study does not elucidate the cause of the decreased skin DNA synthesis, one suspects that the association of zinc with specific enzymes involving DNA metabolism may exist. Lieberman et al. (1963) have already shown that zinc-deficient mammalian cells cultured in vitro have decreased activities of DNA polymerase and thymidine kinase compared to control cells. The action of zinc on the synthesis of these enzymes in rat skin in vivo remains to be experimented.

VII. RNA SYNTHESIS

Zinc is a known requirement for RNA synthesis in various microorganism (Wacker, 1962; Winder, 1962; Wegener, 1963) and in plants (Schneider and Price, 1962). Animal studies attempting to show this relationship, however, have not been established. The incorporation of ^{32}P into liver nucleotide was depressed in zinc-deficient rats (Williams et al., 1965). However, there was no consistent effect on nuclear RNA synthesis in rat liver as measured by orotate-^{14}C incorporation (Buchanan and Hsu, 1968; Williams and Chester, 1970). The RNA content per mg DNA of the testes was reduced in zinc-deficient rats but the incorporation of adenine-^{14}C into RNA was unaltered (Macapinlac, et al., 1968). In view of these conflicting reports we felt that a re-study of this area was desirable.

RNA Concentrations in Tissues

Table XX shows that the RNA content of the skin, but not the liver and testes, of zinc deficient rats was significantly reduced compared to zinc-supplemented animals. The normal value of skin RNA presented here is higher in zinc-supplemented rats than the value reported elsewhere (Wannemacher et al., 1965). This difference could be due to the age of the rats used and the analytical procedures employed (Wannemacher et al., 1965).

TABLE XX

THE CONCENTRATIONS OF RNA IN THE SKIN, LIVER AND TESTES

Type of diet	Skin	Liver	Testes
	mg RNA/g fresh wt.		
Zn-supplemented (pair-fed)	3.09 ± 0.67[a,b]	9.53 ± 1.03	4.21 ± 0.21
Zn-deficient	2.13 ± 0.19	9.55 ± 1.08	4.20 ± 0.19

a Mean ± SD. Seven or more rats in each group
b Difference between Zn-supplemented and Zn-deficient rats is statistically significant ($p<0.01$).

Uridine-^{3}H Incorporation into RNA

The Uridine-^{3}H was incorporated to a substantial degree into the skin RNA 1-hour after injection, but the zinc supplemented groups, pair-fed and ad libitum, had higher rates of incorporation than the group that was deficient in zinc (Table XXI). Because of the insignificant difference in the specific activities between the zinc supplemented pair-fed and ad libitum-fed groups. It is concluded that zinc, not food intake, has a stimulating effect on skin RNA synthesis. On the other hand, the radioactivities of liver RNA and testes RNA from zinc-deficient rats were approximately the same as those from zinc-supplemented animals.

TABLE XXI

URIDINE-^{3}H INCORPORATION INTO RNA

Type of Diet	Specific Activity DPM/mg RNA		
	Skin	Liver	Testes
Zn-supplemented (pair-fed)	5523 ± 1406[a,b]	2073 ± 442	2045 ± 335
Zn-deficient	3565 ± 768	2038 ± 268	2314 ± 372

a Mean ± SD. Six rats in each group.
b Difference between Zn-supplemented and Zn-deficient rats is statistically significant ($p<0.05$).

To clarify further on the effect of zinc on liver RNA, additional experiments using orotic acid-^{14}C and liver microsome fractions were conducted and the results are summarized in Table XXII.

The rate of liver RNA synthesis from labeled orotic acid or uridine was again unchanged by zinc depletion. This is in accordance with the observations of other investigators indicating that dietary deprivation of zinc, or zinc deficiency induced by a chelating agent, did not affect liver RNA synthesis (Fujioka and Lieberman, 1964).

Similarly an organ specificity in regard to RNA metabolism may exist during zinc deficiency. Many biochemical defects including impairment of amino acid incorporation and DNA synthesis have been previously described in the skin but not the other tissues of zinc-deficient rats. This organ is apparently a sensitive target in zinc deficiency.

The mechanism by which zinc is involved in RNA synthesis in rat skin is unknown. Studies by Tal et al. (1969) on ethylenediaminetetra-acetic acid-treated E. Coli ribosomes suggest that zinc may function either as a specific stabilizer of the tertiary structure, or that it is involved in the formation of ribosomes as illustrated by the work of Sandstead et al. (1971) in rat liver ribosome and of McClain et al.(1973)

in rat muscle polysome.

TABLE XXII

LIVER RNA SYNTHESIS

Type of diet	18-hrs. incorporation of orotic acid into RNA		24-hrs. incorporation of uridine-^{3}H into RNA	
	CPM/mg fresh wt.	CPM/mg RNA	CPM/mg fresh wt.	CPM/mg RNA
Zn-supplemented (pair-fed)	270 ± 40[a]	22088 ± 3266	58 ± 2	159 ± 8
Zn-deficient	275 ± 44	23339 ± 2598	57 ± 6	166 ± 23

a Mean ± SD. Five rats in each group.

The second possibility is the involvement of zinc in RNA polymerase. Recently RNA polymerase of E. Coli has been demonstrated as a zinc-metalloenzyme (Scrutton et al. 1971). To our knowledge it is not known whether RNA polymerase in mammalian tissues is also a zinc containing enzyme. Nevertheless it is clear that the synthesis of any forms of cellular RNA in the nucleus and the cytoplasm requires RNA polymerase. The findings of a reduced activity of liver nuclear DNA-dependent RNA polymerase by zinc deficiency (Sandstead et al., 1971), strongly suggest that zinc ions participates to a certain degree in RNA metabolism.

The third possibility in regard to the role of zinc in RNA metabolism is its relationship to ribonuclease. Macapinlac et al. (1968) speculated that the reduced content in the testes RNA of zinc-deficient rats are the results of an increase in the catabolism of RNA. This concept is supported by the report of an increased ribonuclease activity in the testes of zinc deficient rats (Somers and Underwood, 1969).

VIII. ELECTRON MICROSCOPIC STUDIES

Fibroblasts

Fibroblasts in zinc-deficient rats were generally smaller and had an increased nucleocytoplasmic ratio. There were fewer organelles, particularly rough endoplasmic reticulum (Figure 7), indicating a derangement in ribosomal formation. Golgi apparatus were also rarely observed. The cell presented an irregular contour with a diminished number of pinocytotic vesicles. The nuclei of zinc-deficient rats also showed irregular contours

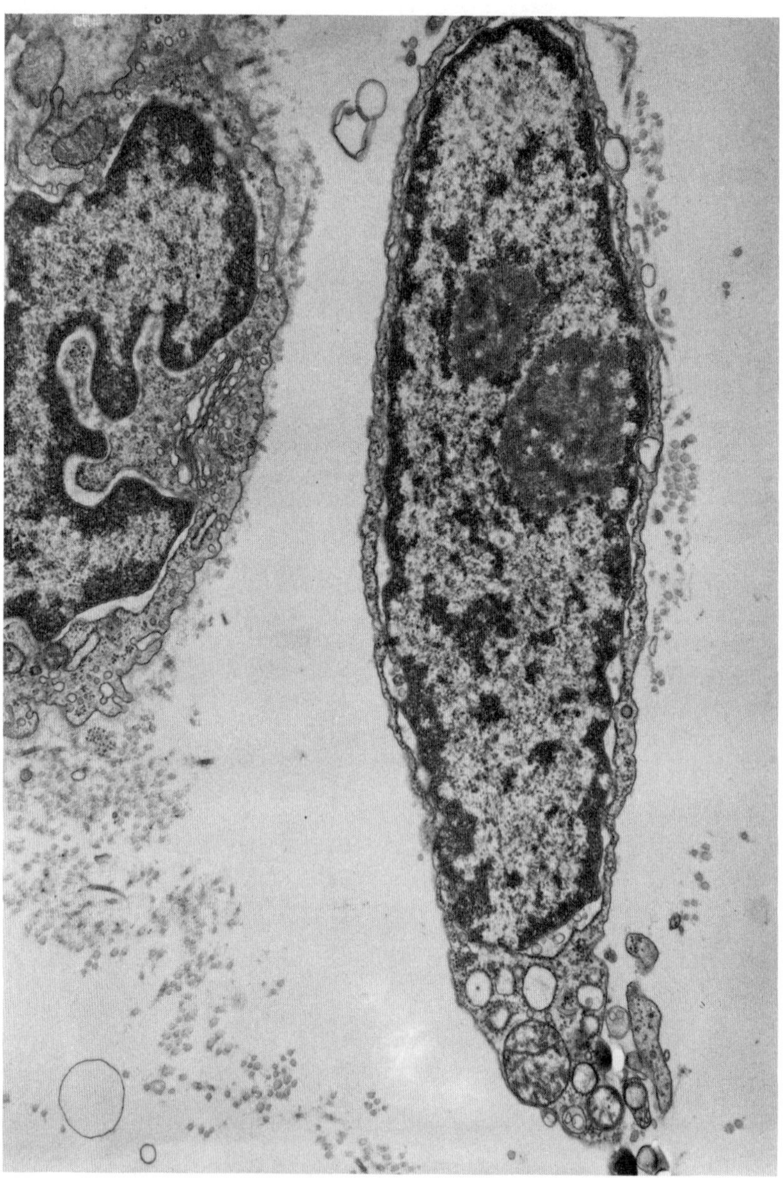

Figure 7. Parts of two fibroblasts from zinc-deficient rats. Nucleo-cytoplasmic ratio is increased. Cytoplasmic organelles particularly the rough endoplasmic reticulum are reduced. The number of pinocytotic vesicles are also reduced. Nucleus shows two prominent nucleoli with well formed nucleolonema (X 17,900).

while chromatin granules had a courser peripheral arrangement. In the nucleoli, nucleolonema were well formed and could be easily identified. Other features of nuclear changes included the numerous nucleolar segregation into fibrillar and granular components (Figure 8). In general, fibroblasts of zinc-deficient rats gave an appearance of cellular immaturity when compared with zinc-supplemented rats (Figure 9).

Collagen Fibers

Although the skin of the zinc-deficient rat appeared grossly thinner, the quantity of connective tissue fibers were difficult to assess morphologically. In acellular and dense collagen bundles, the diameters of collagen fibers were more variable. The interfiber spaces were wide. There were many spur like projections perpendicular to collagen fibers. The cross striations of some fibers partly disappeared and were gradually replaced by amorphous components and spherular materials (Figure 10). In contrast, the collagen fibers from zinc-supplemented rats were smooth, regular in size and uniformly arranged with distinct striations (Figure 11). The interfiber spaces were narrow or almost absent.

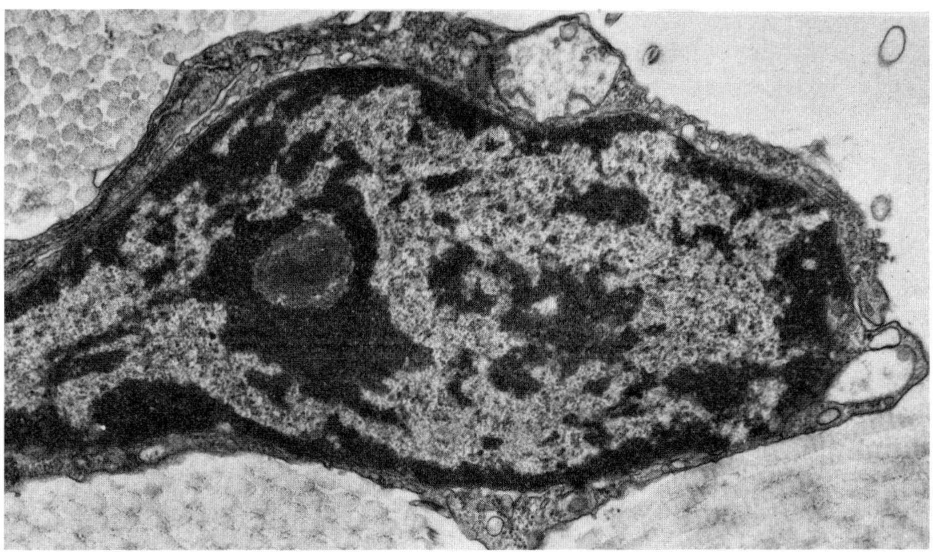

Figure 8. Part of fibroblast from zinc-deficient rat. Cell contour is irregular. Nucleus has course clamping of chromatin along the nuclear membrane. Nucleous is prominent and shows segregation of granular and fibrillar components (X 24,800).

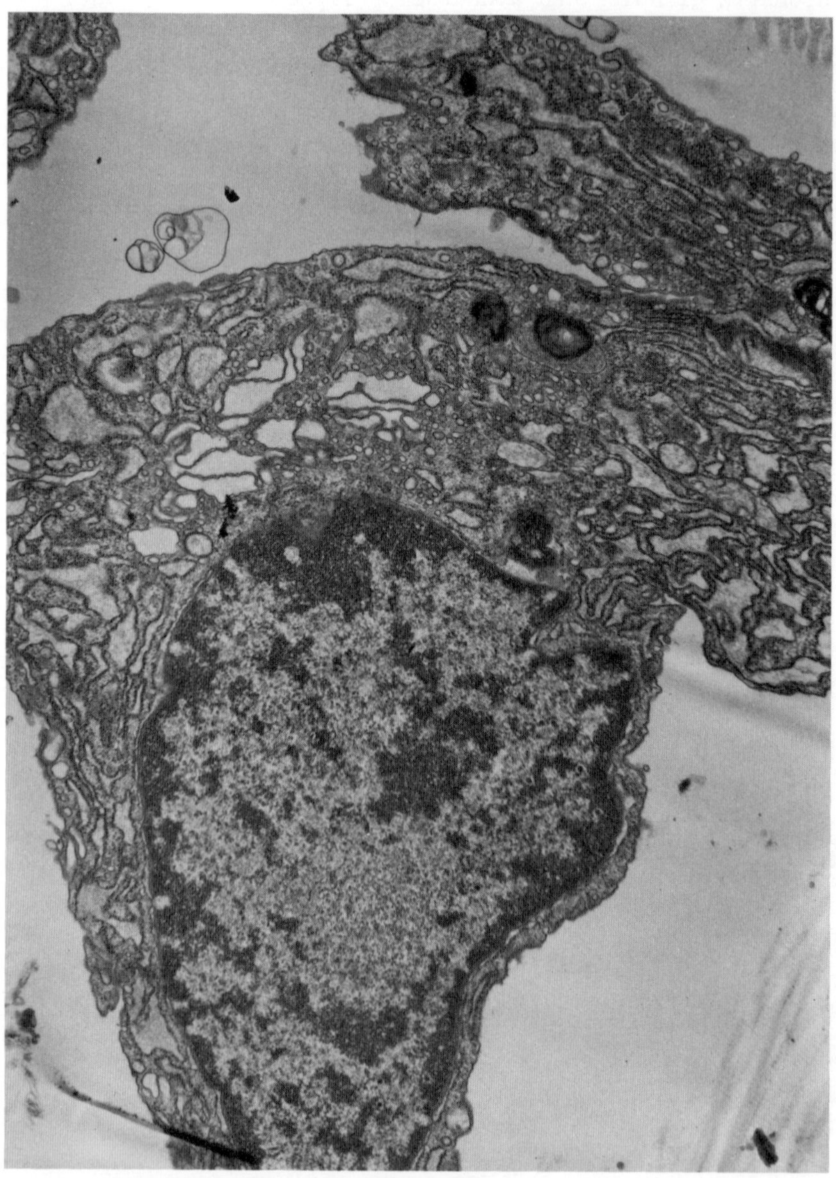

Figure 9. A fibroblast from zinc-supplemented rat. Cytoplasma contains abundant organelles with smaller nucleocytoplasmic ratio. Golgi apparatus are well developed. Numerous pinocytotic vesicles are seen along cell membrane. Two residual bodies containing myelin figures are present. Nuclear chromatin is less coarse without nucleolus (X 17,900).

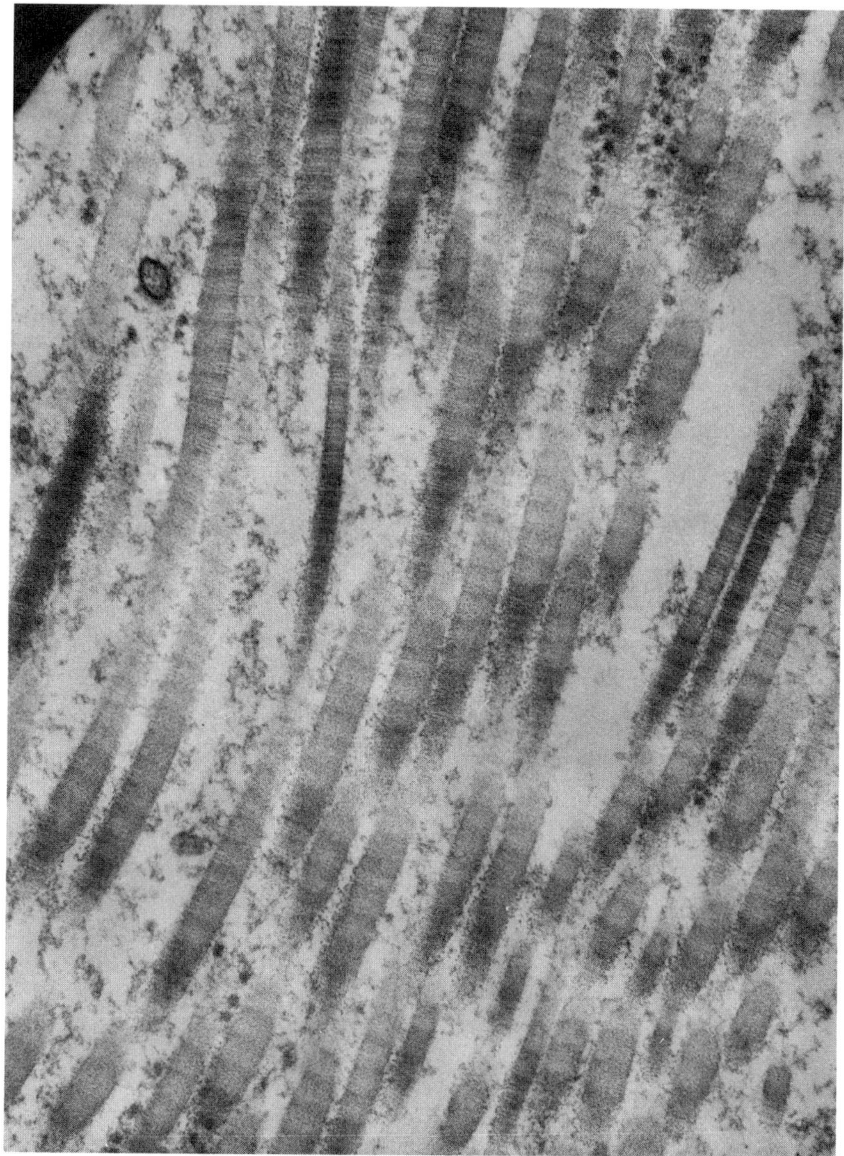

Figure 10. Collagen fibers from zinc-deficient rat show marked variations in diameter of the fibers. Interfiber space is increased. Perpendicular to collagen fibers are many spur-like projections. At the ends of some of the fibers, cross striation disappears and is gradually replaced by amorphous and spherular material (X 71,500).

Figure 11. Collagen fibers from zinc-supplemented rat are regular in size and arrangement. Fibers are tightly packed with small interfiber spaces. There are fine granules along the fibers (X 71,500).

SUMMARY AND CONCLUSIONS

From the preceding discussion, it is evident that zinc is involved in both the transcription and the translation processes which in turn affect protein synthesis. This crude beginning awaits the unveiling of more exciting insights into the alteration in the pathogenesis of skin disease. As more knowledge becomes available, it is hoped that additional theories will be formulated so that a new understanding of how these zinc reactions can be modulated for the benefit of the patient.

ACKNOWLEDGMENTS

This research was supported by General Medical Research Funds from the Veterans Administration. The authors are greatly indebted to Drs. P. E. McClain and G. R. Beecher, Agricultural Research Service, U. S. Department of Agriculture, Beltsville, Md., for their excellent collaboration on this project. The authors also thank Dr. J. K. Stephan and Mr. Thomas H. S. Hsu for their consultation and assistance in completing the DNA study.

REFERENCES

Boquist, L. and Lernmark, A. (1969). Effects on the endocrine pancreas in Chinese hamsters fed zinc deficient diets. Acta Pathol.Microbiol. Scand. 76:215-228.

Buchanan, P.J., and Hsu, J.M. (1968). Zinc deficiency and DNA synthesis in rat liver. Fed. Proc. 27;483.

Carruthers, C. (1962). Biochemistry of skin in health and disease. Charles C. Thomas, Publ., Springfield, Ill. p 27-38.

Day, H.G. and Skidmore, B.E. (1947). Some effects of dietary zinc deficiency in the mouse. J. Nutr. 33;27-38.

Fujioka, M. and Liberman, I. (1964). A Zn requirement for synthesis of deoxyribonucleic acid by rat liver. J. Biol. Chem. 239:1164-1167.

Graham, E.R. and Telle, P. (1967). Zinc retention in rabbits: Effect of previous diet. Science 155:691-692.

Herberg, R.J. (1963). Statistical aspects of liquid-scintillation counting by internal standard technique: A single isotope. Anal. Chem. 35:786-791.

Hsu, J.M., Anthony, W.L. and Buchanan, P.J. (1969). Zinc deficiency and incorporation of ^{14}C-labeled methionine into tissue proteins in rats. J. Nutr. 99:425-432.

Hsu, J.M. and Anthony, W.L. (1970a). Zinc deficiency and urinary excretion of taurine-^{35}S and inorganic sulfate-^{35}S following cystine-^{35}S injection in rats. J. Nutr. 100:1189-1196.

Hsu, J.M., Anthony, W.L., and Buchanan, P.J. (1970b). Zinc deficiency and the metabolism of labeled cystine in rats. In: Mills, C.F. (Ed.): Trace Element Metabolism in Animals. E. & S. Livingston, Edinburgh. p 151-157.

Hsu, J.M. and Anthony, W.L. (1971). Impairment of cystine-^{35}S incorporation into skin protein by zinc-deficient rats. J. Nutr. 101:445-452.

Hsu, J.M. and Anthony, W.L. (1973). Zinc deficiency and collagen synthesis in rat skin. In: Hemphill, D.D. (Ed.): Trace Substances in Environmental Health VI. University of Missouri's Press, Columbia, Missouri, p 137-143.

Hsu, Thomas, H.S. and Hsu, J.M. (1972). Zinc deficiency and epithelian wound repair: An autoradiographic study of ^{3}H-Thymidine incorporation. Proc.Soc.Exp.Biol.Med. 140:157-160.

Husain, S.L. (1969). Oral zinc sulphate in leg ulcers. Lancet 1:1069-1071.

Kernkamp, H.C.C. and Ferrin, E.F. (1953). Parakeratosis in swine. J. Amer. Vet. Med. Ass. 123:217-221.

Larsen, D.L., Maxwell, R., Abston, S., and Dobrkovsky, M. (1970). Zinc deficiency in burned children. Plas.Reconstr.Surg. 46:13-18.

Liberman, I., Abrams, R., Hunt, N. and Ove, P. (1963). Levels of enzyme activity and deoxyribonucleic acid synthesis in mammalian cells cultured from the animal. J. Biol. Chem. 238:3955-3962.

Logothetopoulos, J., Kaneko, M., Wrenshall, G.A. and Best, C.H. (1964). Zinc, granulation and extractable insulin of islet cells following hyperglycemia or prolonged treatment with insulin. In: Brolin, S.E., Hellman, B. and Knutson, H. (Eds): The structure and metabolism of the pancreatic islets. The MacMillan Co. New York, pp 333-344.

Macapinlac, M.P., Barney, G.H., Pearson, W.M., and Darby, W.J. (1967). Production of zinc deficiency in the squirrel monkey (Saimiri sciureus). J. Nutr. 93:499-510.

Macapinlac, M.P., Pearson, W.N., Barney, G.H. and Darby, W.J. (1968). Protein and nucleic acid metabolism in the testes of zinc-deficient rats. J. Nutr. 95:569-577.

Macapinlac, M.P., Pearson, W.N. and Darby, W.J. (1966). Some characteristics of zinc deficiency in the albino rat. In: Prasad, A.S. (Ed.): Zinc Metabolism. Charles C. Thomas, Springfield, Illinois, pp 142-166.

McBean, L.D., Smith, J.C. Jr. and Halstett, J.A. (1972). Zinc deficiency in guinea pigs. Proc. Soc. Exp. Biol. Med. 140:1207-1209.

McClain, P.E., Wiley, E.R., Beecher, G.R., Anthony, W.L. and Hsu, J.M. (1973). Influence of zinc deficiency on synthesis and crosslinking of rat skin collagen. Biochim. Biophys. Acta, 304:457-465.

Millar, M.J., Vincent, N.R. and Mawson, C.A. (1961). Autoradiographic study of the distribution of ^{65}Zn in rat tissue. J. Histochem. & Cytochem. 9:111-116.

Miller, J.K. and Miller, J.W. (1960). Development of zinc deficiency in Holstein calves fed a purified diet. J. Dairy Sci. 43:1854-1856.

Miller, W.J., Pitts, W.J., Clifton, C.M. and Schmittle, S.C. 1964). Experimentally produced zinc deficiency in the goat. J. Dairy Sci. 47:556-558.

Morris, S.C. and McClain, P.E. (1972). Heterogeneity in the cyanogen bromide peptides from striated muscle and heart valve collagen. Biochem. Biophys. Res. Commun. 47:27-34.

O'Neal R.M., Pla, G.W., Fox, M.R.S., Gibson, F.S. and Fry, B.E. Jr. (1970). Effect of zinc deficiency and restricted feeding on protein and ribonucleic acid metabolism of rat brain. J. Nutr. 100:491-497.

Ott, E.A., Smith, W.H., Stob, M. and Beeson, W.M. (1964). Zinc deficiency syndrome in the young lamb. J. Nutr. 82:41-50.

Pories, W.J., Henzel, J.H., Rob, C.G., and Strain, W.H. (1967). Acceleration of healing with zinc sulfate. Ann. Surg. 165:432-436.

Prasad, A.S., Miale, A. Jr., Farid, Z., Sandstead, H., Schulert, A.B., and Darby, W.J. (1963). Biochemical studies on dwarfism, hypogonadism and anemia. Arch. Int. Med. 111: 407-428.

Quarterman, J.C., Mills, C.F., and Humphries, W.R. (1966). The reduced secretion of and sensitivity to insulin in zinc-deficient rats. Biochem. Biophys. Res. Commun. 25:354-358.

Robertson, B.T. and Burns, M.J. (1963). Zinc metabolism and the zinc deficiency syndrome in the dog. Am. J. Vet. Res. 24:997-1002.

Sandstead, H.H., Prasad, A.S., Schulert, A.S., Farid, Z., Miale, A., Bassily, S. and Darby, W.J. (1967). Human zinc deficiency, endocrine manifestations and response to treatment. Am. J. Clin. Nutr. 20:422-442.

Sandstead, H.H. and Rinaldi, R. (1969). Impairment of deoxyribonuecleic acid synthesis by dietary zinc-deficiency in the rat. J. Cell. Physiol. 73:81-83.

Sandstead, H.H., Terhune, M., Brady, R.N., Gillespie, D., and Holloway, W.L. (1971). Zinc deficiency: Brain DNA, protein and lipids and liver ribosome and RNA polymerase. Clin. Res. 19:83.

Schneider, E. and Price, C.A. (1962). Decreased ribonucleic acid levels: a possible cause of growth inhibition in zinc deficiency. Biochim. Biophys. Acta 55:406-408.

Scrutton, M.C., Wu, C.W. and Goldwait, D.A. (1971). The presence and possible role of zinc in RNA polymerase obtained from Escherichia Coli. Proc. Nat.Acad.Sci. U.S.A. 68:2497-2501.

Serjeant, G.R., Galloway, R.E. and Guest, M.C. (1970). Oral zinc sulfate in sickle-cell ulsers. Lancer 2:891-892.

Slater, J.P., Mildvan, A.S. and Loeb, L.A. (1971). Zinc in DNA Polymerase, Biochem. Biophys. Res. Comm. 44:37-43.

Somers, M. and Underwood, E.J. (1969). Ribonuclease activity and nucleic acid and protein metabolism in the testes of zinc-deficient rats. Aust. J. Biol. Sci. 22:1277-1285.

Stephan, J.K. and Hsu, J.M. (1973). Effect of zinc-deficiency and wounding on DNA synthesis in rat skin. J. Nutr. 103: 548-552.

Swick, R.W. and Handa, D.T. (1956). The distribution of fixed carbon in amino acids. J. Biol. Chem. 218:577-585.

Tal, M. (1969). Metal ions and ribosomal conformation. Biochem. Biophys. Acta 195:76-86.

Theuer, R.C. and Hoekstra, W.G. (1966). Oxidation of ^{14}C-labeled carbohydrate, fat and amino acid substrates by zinc deficient rats. J. Nutr. 89:448-454.

Todd, W.R., Elvehjem, C.A., and Hart, E.B. (1934). Zinc in the nutrition of the rat. Am. J. Physiol. 107:146-156.

Traub, W. and Piez, K.A. (1971). The chemistry and the structure of collagen. In: Anfinsen, C.B., Edsall, J.T. and Richards, F.M. (Eds.): Advances in protein chemistry. Academic Press, New York. 25:243-352.

Wacker, W.E.C. (1962). Nucleic acids and metals, three changes in nucleic acid, protein and metal content as a consequence of zinc deficiency in Englena gracilis. Biochemistry 1:859-865.

Wannemacher, R.W. Jr., Bank, W.L. Jr. and Wunner, W.H. (1965). Use of a single tissue extract to determine cellular protein and nucleic acid concentrations and rate of amino acid incorporation. Anal. Biochem. 11:320-326.

Wegner, W.S. and Romano, A.H. (1963). Zinc stimulation of RNA and protein synthesis in rhizopus nigricans. Science 142: 1669-1670.

Weser, D., Seeber, S. and Warnecke, P. (1969). Reactivity of Zn^{2+} on nuclear DNA and RNA biosynthesis of regenerating rat liver. Biochim. Biophys. Acta 179:422-428.

Williams, R.B. and Chester, J.K. (1970). Effect of zinc deficiency on nucleic acid synthesis in the rat. In: Mills, C.F. (Ed.): Trace Element Metabolism in Animals. E. & S. Livingston, Edinburgh, Scotland. p 164-166.

Winder, F. and Dinneny, J.M. (1959). Effects of iron and zinc on nucleic acid and protein synthesis in Mycobacterium smegmatis. Nature 184:742-743.

Wool, I.G., Castles, J.J. and Moyer, A.M. (1965). Regulation of amino acids accumulation in isolated rate diaphragm: Effect of puromycin and insulin. Biochim. Biophys. Acta 107:333-345.

18

NICKEL DEFICIENCY IN CHICKS AND RATS: EFFECTS ON LIVER MORPHOLOGY, FUNCTION AND POLYSOMAL INTEGRITY

F.H. Nielsen, D.A. Ollerich*, G.J. Fosmire and H.H. Sandstead

USDA, ARS, Human Nutrition Laboratory
Grand Forks, North Dakota and
*Department of Anatomy, University of North Dakota
Grand Forks, North Dakota[1]

Nickel is an essential nutrient for animals. It has the following characteristics and biological behaviors which are consistent with essentiality: (1) It is a low molecular weight transition element which forms chelates. Thus it is chemically suitable for biological functions. (2) It is ubiquitous in the earths' crust (0.008%) and in sea water (0.1-0.5 ppb), therefore, it has been generally available to life forms throughout evolution. (3) It is present in relatively constant amounts in plants and animals. (4) It is non-toxic to animals orally except in astringent doses. (5) Homeostatic mechanisms are implied by serum levels, excretion rates, and its tendency not to accumulate. (6) Inadequate amounts of nickel in the diets of experimental animals repeatedly result in an impairment of metabolic processes. These effects correlate with morphologic abnormalities. The first five of these features have been reviewed (Schroeder, et al., 1962; Underwood, 1971; Nielsen, 1971 and 1973; Nielsen and Ollerich, 1973). The last aspect has been the subject of our research for the past several years. This presentation will describe some of these studies.

The first evidence that nickel is essential was obtained from experiments with chicks (Nielsen and Sauberlich, 1970; Nielsen and Higgs, 1971; Nielsen, 1971). Feeding a diet containing less than 40 ppb nickel resulted in an apparently characteristic deficiency syndrome. When compared with controls given a supplement of 3-5 ppm nickel, the deficient chicks showed: (1) pigmentation changes in the

[1]Supported in part by ARS-USDA Cooperative Agreement 12-14-100-11, 178 (61).

shank skin; (2) thicker legs with slightly swollen hocks; (3) dermatitis of the shank skin; (4) a less friable liver which appears to have been related to the fat content; and (5) an enhanced accumulation of a tracer dose of ^{63}Ni in liver, bone and aorta. These findings were observed under conditions which produced suboptimal growth. The abnormalities in leg structure and the dermatitis of the shank skin were inconsistent.

Sunderman et al. (1972) attempted to confirm these findings by feeding a diet containing 44 ppb to chicks raised in a slightly different environment. While they found no gross effects, they did observe ultrastructural changes in the liver. These included dilation of the perimitochondrial rough endoplasmic reticulum in 15 to 20% of the hepatocytes.

In a single experiment with chicks, Leach (1973) observed a growth response with nickel. Later attempts to obtain the growth response were unsuccessful.

Wellenreiter et al. (1970) fed a diet containing 80 ppb nickel to reproducing quail and saw no gross symptoms except an inconsistent positive effect of nickel on breast feathering in birds which were fed a diet relatively deficient in arginine but otherwise adequate in protein.

To clarify and extend the above observations, improvements have been made in the experimental environment and a diet has been formulated with a nickel content of 3-4 ppb. With this diet and environment, we have consistently been able to produce a nickel deficiency in chicks and rats (Nielsen and Ollerich, 1973; Nielsen, 1973).

It has been found that after 3½ weeks, the gross appearances of the deficient and control chicks are similar except for the difference in pigmentation of shank skin which had been observed previously (Nielsen and Sauberlich, 1970; Nielsen and Higgs, 1971; Nielsen, 1971). All chicks at 3½ weeks weigh 350 to 400 g. Leg structure and skin abnormalities which occur are inconsistent. The other gross sign, decreased friability of the liver, which had been observed in our earlier studies is consistently found in the deficient chicks which are raised under the improved experimental conditions.

In contrast to the gross signs, a number of abnormalities in biochemical indices of metabolism are consistently found in the nickel deficient chicks. They include a decreased oxygen uptake by liver homogenates in the presence of α-glycerophosphate, an increase in liver total lipids, and a decrease in liver lipid phosphorus. In addition, a preliminary study has shown an increase in total lipids and lipid phosphorus in the heart.

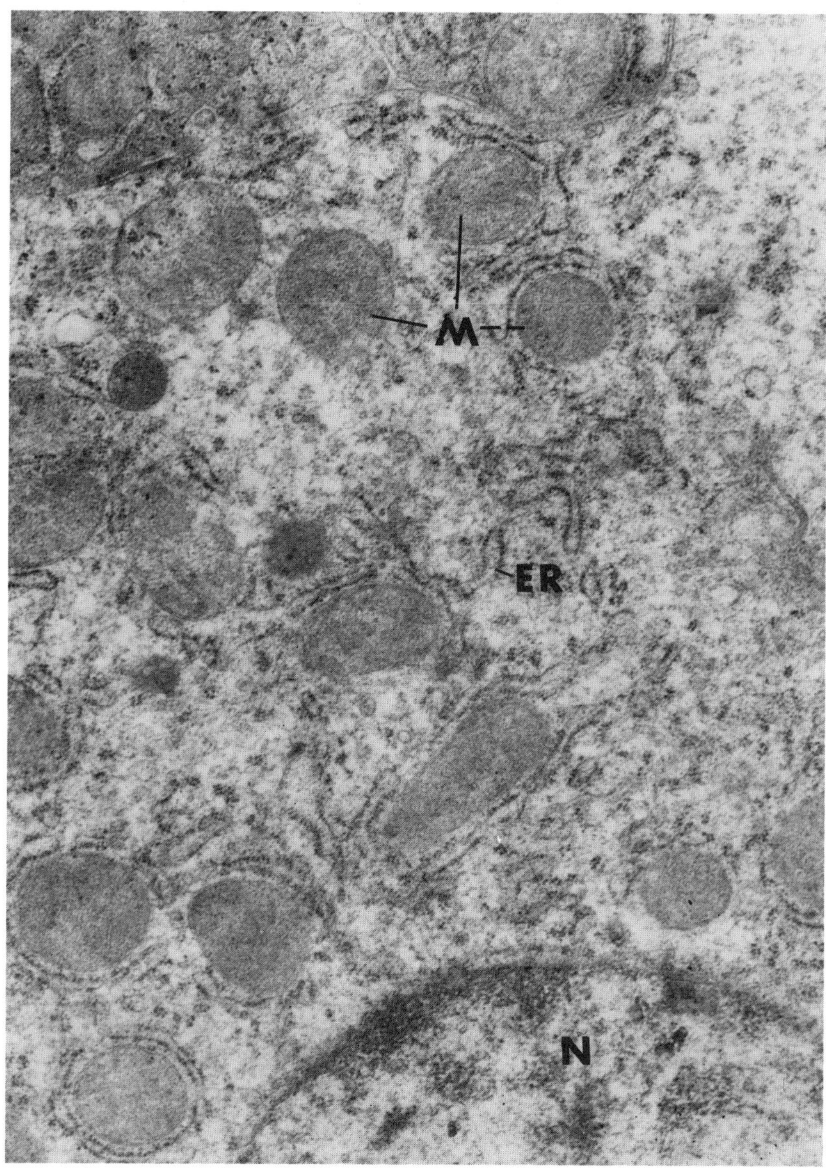

Figure 1. Hepatic cell from a nickel supplemented chick (3 ppm nickel). Compare mitochondria (M) with those in figure 2. Endoplasmic reticulum (ER). Nucleus (N). Uranyl acetate and lead citrate. X 21,600.

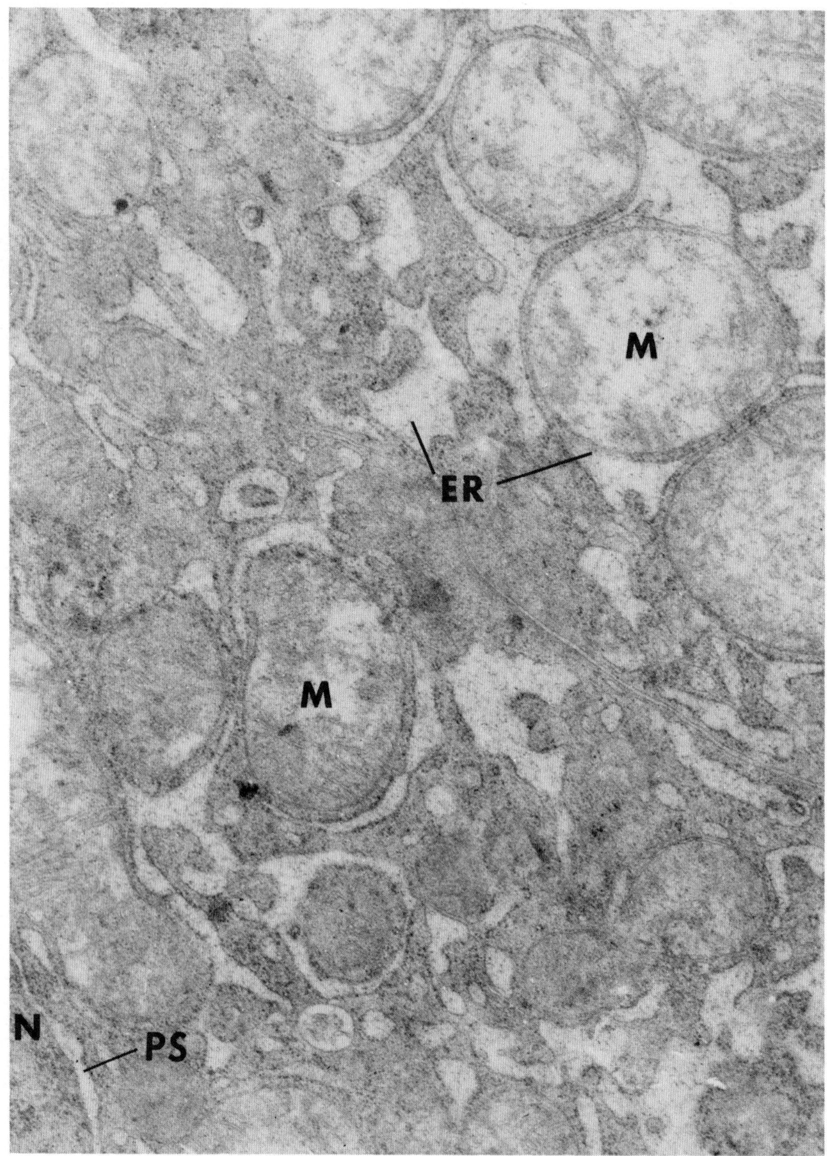

Figure 2. Hepatic cell from a nickel deficient chick (4 ppb nickel). Swelling of mitochondria (M) was evident in numerous hepatic cells. The swelling was in the compartment of the matrix and appeared to cause fragmentation of cristae. Note the dilated cisterns of the rough endoplasmic reticulum (ER) and dilated perinuclear space (PS). Nucleus (N). Uranyl acetate and lead citrate. X 21,600.

Ultrastructural abnormalities in the hepatocytes are also consistently found. Some of these abnormalities are illustrated in figures 1 and 2. These abnormalities are similar to, though more extensive than, those described by Sunderman et al. (1972). They include dilation of the cisterns of the rough endoplasmic reticulum and swelling of the mitochondria. The swelling is in the compartment of the matrix and is associated with fragmentation of the cristae. Other changes include a dilation of the perinuclear space and pyknotic nuclei.

In rats, nickel apparently affects reproduction as seven first generation nickel deficient females which were mated had a significant number of dead pups (15%), while no mortality occurred in the pups of six nickel adequate controls. Nine second generation nickel deficient female rats which were mated had a 19% loss of pups. This finding was somewhat confounded by the fact that the eight controls had a 10% loss. This was, however, roughly only one-half of the loss in the nickel deficient group. During the suckling stage, the nickel deficient pups have a less thrifty appearance and are less active than control pups. To assess this

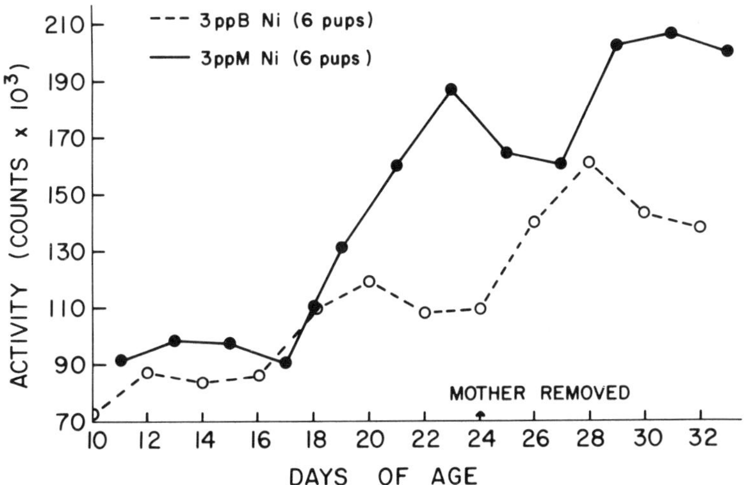

Figure 3. The activity of six each of nickel deficient and supplemented rats from matched litters of the third generation.

last observation, a Stoelting activity monitor[2,3] has been used to measure the activity of matched litters in the second and third generation. Figure 3 indicates that the nickel deficient pups are indeed less active.

Liver oxidative ability has been measured in 12 deficient and 12 control second generation male rats. Grossly, the livers of the deficient rats were a muddy brown color compared with a red brown color of controls. They also had a less distinct substructure and were less friable. As was the case with chicks, the nickel deficient rat liver homogenates oxidized α-glycerophosphate less well than did liver homogenates from control rats.

It has recently been reported that nickel deficiency can be produced in swine (Anke et al., 1973), as well as in chicks and rats. Some of the signs include impaired reproduction and offspring with a less thrifty appearance characterized by a sparse rough hair coat and parakeratosis-like changes in skin. First generation nickel-deficient piglets grow poorly.

Thus, our studies and those of others have established that nickel deficiency can be experimentally produced in three species of animals. The findings do not, however, provide more than a meager insight as to the metabolic function of nickel.

It is known that nickel may complex with macromolecules. For example, it is present in the metalloprotein nickeloplasmin which has been isolated from the serum of rabbits and man (Nomoto et al. 1971 and Sunderman et al. 1971).

Nickeloplasmin is a macroglobulin which has an estimated molecular weight of 7.0×10^5 and contains approximately 0.8 g atoms of Ni/mole. Disc gel and immunoelectrophoresis have shown that purified nickeloplasmin is an α-1 macroglobulin in rabbit serum and an α-2 macroglobulin in man. The functions of this protein are unknown. It is known that nickel can activate numerous enzymes in vitro. Examples include arginase (Hellerman and Perkins, 1935), tyrosinase (Lerner et al. 1950), desoxyribonuclease (Miyaji and Greenstein, 1951), acetyl coenzyme A synthetase (Webster, 1965), and phosphoglucomutase (Ray, 1969). These studies, however, do not necessarily indicate that nickel is a specific activator of a specific enzyme in vivo.

[2]Mention of a trademark or proprietary product does not constitute a guarantee or warranty of the product by the U.S. Department of Agriculture, and does not imply its approval to the exclusion of other products that may also be suitable.

[3]Stoelting Company, Chicago, Illinois.

Nickel also complexes with DNA and RNA. Significant concentrations of nickel have been found in DNA (Wacker and Vallee, 1959; Eichhorn, 1962) and RNA (Wacker and Vallee, 1959; Wacker et al. 1963; Sunderman, 1965) from phylogenetically diverse sources. It has been suggested that nickel and other metals which are present may contribute to the stabilization of the macro-molecular structure of the nucleic acids. Nickel has been shown to stabilize RNA (Fuwa et al. 1960) and DNA (Eichhorn, 1962) against thermal denaturation. In addition, it is extraordinarily effective in the preservation of tobacco mosaic virus RNA infectivity (Wacker, et al. 1963; Cheo et al., 1959). Other studies have suggested that nickel may have a role in the preservation of the compact structure of ribosomes (Tal, 1968, 1969a, 1969b). Nickel will protect ribosomes from in vitro thermal denaturation at low temperatures and will restore the sedimentation characteristics of *E. coli* ribosomes which have been denatured with EDTA.

Because these in vitro studies are consistent with a role for nickel in the metabolism of RNA, we conducted a study of the effects of nickel deficiency on rat liver polyribosomes and the activities of liver RNA polymerase and RNase.

MATERIALS AND METHODS

To bring out the effects of nickel deprivation, successive generations of Sprague-Dawley rats were raised. Thus, the animals were exposed to deficiency throughout fetal, neonatal and adult life. The rats were housed in plastic cages placed inside a controlled environment such as an all plastic rigid isolator[4], or a laminar flow animal rack[5]. The feed and water cups were also plastic. Thus, the only materials the animals came into contact with were plastic, and the air entering the system was filtered to remove contaminating dust.

All equipment used in the experiments was cleaned as follows: (1) washed with detergent and tap water, (2) soaked in a 1:10 solution of radiacwash[6] for 1 hour or more, (3) rinsed with distilled water and (4) rinsed with high purity water[7] (18 megohms-cm resistance).

[4] Germfree Laboratories, Inc., Miami, Florida.

[5] Carworth Division of Becton, Dickinson, and Co., New City, New York.

[6] Atomic Products Corp., Center Moriches, L.I., New York, New York.

[7] Produced by a "Super Q High Purity Water System", Millipore Corp., Bedford, Massachusetts.

The diet used was the same as described previously (Nielsen, 1973). It was based on dried skim milk, ground corn and corn oil. The diet contained approximately 4 ppb nickel on an air dried basis. Control rats were fed the basal diet supplemented with 3 ppm nickel as $NiCl_2 \cdot 6H_2O$. The diets and high purity water were given ad libitum.

The age, sex and generation of rats used in the experimental procedures are given in the results and discussion.

The sucrose density gradients of the liver post-mitochondrial supernatants were obtained by homogenizing the liver after diluting 1:3 (w/v) with a medium containing 0.25 M sucrose, 5 mM Tris · HCl, 25 mM KCl and 5 mM $MgCl_2$, at pH 7.4. The supernatants obtained

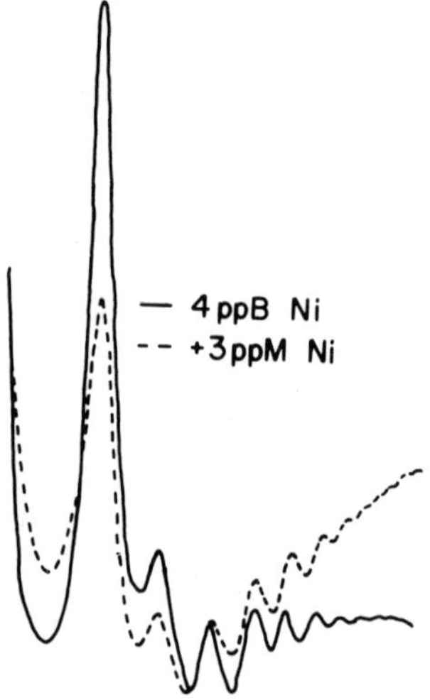

Figure 4. Representative sucrose density gradients of liver post-mitochondrial supernatants obtained from six nickel deficient (4 ppb nickel) and six supplemented (+3 ppm nickel) rats.

after centrifugation at 4400 x g, at 4°, were made 0.5% with respect to deoxycholate and centrifuged at 4687 x g, at 4°. The supernatants were then layered over a linear 15.5-35.5% (w/v) sucrose density gradient and centrifuged for 160 minutes at 284,000 x g, at 4°. The gradients were monitored at 260 nm.

The RNA polymerase was analyzed in intact nuclei by using the high Mg, low salt technique as described by Fosmire et al. (1973).

RNase was determined by the method of Gagnon and de Lamirande (1973) as modified by Fosmire (1973). Protein was determined by the biuret method after samples (0.1 ml) were dispersed with 0.4 ml of 1% NaCl and 0.5 ml of 10% sodium deoxycholate. Statistical analyses was performed by the "t" test (Snedecor and Cochran, 1967).

RESULTS AND DISCUSSION

Sucrose density gradients of liver post-mitochondrial supernatants were prepared from six each of control and deficient rats. Three second generation males, approximately 6 weeks of age from different litters and three third generation females, approximately 6-10 weeks of age from different litters were used. The findings were all similar to the profiles shown in figure 4. It is evident that there is a decrease in polyribosomes with a relative increase in the monosomes in the nickel deficient rat liver.

For the RNA polymerase analyses, two first generation male rats approximately 5 weeks of age, and three third generation male rats approximately 6-10 weeks of age, from each of the nickel deficient and supplemented groups were used. For the RNase analyses, five first generation male rats approximately 4 weeks of age from each the nickel deficient and supplemented groups were used.

The results of the RNA polymerase and RNase analyses are shown in Table I.

These preliminary results indicate that nickel deficiency apparently causes a significant increase in liver nuclear RNA polymerase activity and in the active, and possibly the total, alkaline RNase/protein. These findings substantiate the hypothesis that nickel has a role that is interrelated with certain aspects of nucleic acid, especially RNA metabolism.

TABLE I

Effect of Dietary Nickel on Liver Nuclear RNA Polymerase and RNase Activities

Group	No. of Rats	RNA Polymerase	RNase, pH 7.8 Total	RNase, pH 7.8 Active
		μμM ^{3}H-UTP Incorp./mg DNA	mg/g protein	mg/g protein
Ni Def (4 ppb)	5	7.7[1] ± 1.2[2]	0.99[3] ± 0.04	0.44[4] ± 0.001
Ni Supl (+3 ppm)	5	4.2 ± 1.5	0.82 ± 0.06	0.035 ± 0.002

[1]Significantly different ($P < .05$) from +3 ppm Ni group.

[2]± SEM

[3]Significantly different ($P < 0.10$) from +3 ppm Ni group.

[4]Significantly different ($P < 0.01$) from +3 ppm Ni group.

An increase in these 2 enzymes, and a decrease in polysomes, in nickel deficiency suggests the following hypothesis. The structural stability of the polysome is abnormal, or there is an impairment in the formation of polyribosomes through the combination of ribosomes with messenger RNA. The resultant subnormal quantity of polyribosomes may induce the production of additional messenger RNA via increase RNA polymerase activity. This additional RNA which is not readily incorporated into the polyribosomes stimulates the activity of RNase by its presence in the cytoplasm.

Since a major portion of the polysomes is attached to membranes, it is possible the defect is in this relationship. This is suggested by the ultrastructural abnormalities which occur in the livers of the nickel deficient chicks. The swollen mitochondria may indicate that there is an abnormality in the mitochondrial membrane. The dilation of the perinuclear space and the presence of pyknotic nuclei may be indicative of abnormalities in the nuclear membrane. Finally, the cisterns of the endoplasmic reticulum - a network of membrane bound cavities - to which the polysomes attach, are dilated. This finding is consistent with an abnormality in these membranes. Other supportive evidence includes the finding that the level of phospholipids in liver is decreased by nickel deficiency (Nielsen, 1973; Nielsen and Ollerich, 1973). Phospholipids are an integral part of the various membranes of cells.

Other support for the hypothesis that nickel has a role in the metabolism or function of certain cell membranes is provided by in vitro studies with isolated tissues. Nickel can substitute for calcium in certain steps of the excitation-contraction coupling of isolated skeletal muscle (Frank, 1962; Fischman and Swan, 1967). Other investigators have found that nickel can also substitute for calcium in the excitation process of the isolated nerve cell (Blaustein and Goldman, 1968; Hafeman, 1969). It has been suggested that nickel can substitute for calcium in the complex with a membrane ligand, such as the phosphate of a phospholipid and thus may influence the processes of nerve transmission and muscle excitation and contraction. Though these speculations are attractive, our current knowledge allows us to say little more with confidence than that nickel complexes with macromolecules, and that it thus appears to influence their structure and function.

SUMMARY

Substantial evidence has accumulated which shows that nickel is an essential nutrient for animals. Nickel deficiency repeatedly results in an impairment of a function, or functions, as evidenced by ultrastructural degeneration, by reduced oxidative ability, and by the increased total lipids and decreased lipid phosphorus in the liver of the chicks. Swine deprived of nickel show impaired reproductive performance and the young produced are less thrifty and do not grow as well as nickel supplemented controls. Rats deprived of nickel show a reduced oxidative ability in the liver and abnormalities in the liver polysomal profile. Preliminary results also indicate increased liver nuclear RNA polymerase and increased total and active alkaline RNase/protein.

It is suggested on the basis of available data that nickel, because it can complex with macromolecules, has a role in the metabolism of membranes, and in the metabolism of RNA.

ACKNOWLEDGEMENTS

We wish to thank Dr. Y.Y. Al-Ubaidi and M.A. Fosmire for performing the polysomal profiles; S.H. Givand for performing the RNase analysis; Dr. E. Halas for conducting the activity studies; and G.M. Sutcliffe for his expert technical assistance.

REFERENCES

Anke, M., Grun, M., Dittrich, G., Groppel, B. and Hennig, A. (1973). Low nickel rations for growth and reproduction in pigs. In: Trace Element Metabolism in Animals - 2, edited by W.G. Hoekstra, J.W. Suttie, H.E. Ganther, and W. Mertz. University Park Press, Baltimore, Maryland: In Press.

Blaustein, M.P. and Goldman, D.E. (1968). The action of certain polyvalent cations on the voltage-clamped lobster axon. J. Gen. Physiol. 51: 279.

Cheo, P.C., Friesen, B.S. and Sinsheimer, R.L. (1959). Biophysical studies of infectious ribonucleic acid from tobacco mosaic virus. Proc. Nat. Acad. Sci. 45: 305.

Eichhorn, G.L. (1962). Metal ions as stabilizers or destabilizers of the deoxyribonucleic acid structure. Nature 194: 474.

Fischman, D.A. and Swan, R.C. (1967). Nickel substitution for calcium in excitation-contraction coupling of skeletal muscle. J. Gen. Physiol. 50: 1709.

Fosmire, G.J., Al-Ubaidi, Y.Y., Halas, E.S. and Sandstead, H.H. (1973). The effect of zinc deprivation on the brain. In: Advances in Experimental Medicine and Biology, edited by M. Friedman. Plenum Publishing Corp., New York, New York - This volume.

Frank, G.B. (1962). Utilization of bound calcium in the action of caffeine and certain multivalent cations on skeletal muscle. J. Physiol. 163: 254.

Fuwa, R., Wacker, W.E.C., Druyan, R., Bartholomay, A.F. and Vallee, B.L. (1960). Nucleic acids and metals, II: Transition metals as determinants of the conformation of ribonucleic acids. Proc. Nat. Acad. Sci. 46: 1298.

Gagnon, C. and de Lamirande, G. (1973). A rapid and simple method for the purification of rat liver RNase inhibitor. Biochem. Biophys. Res. Comm. 51: 580.

Hafeman, D.R. (1969). Effects of metal ions on action potentials of lobster giant axons. Comp. Biochem. Physiol. 29: 1149.

Hellerman, L. and Perkins, M.E. (1935). Activation of enzymes. III. The role of metal ions in the activation of arginase. The hydrolysis of arginine induced by certain metal ions with urease. J. Biol. Chem. 112: 175.

Leach, R.M., Personal Communication.

Lerner, A.B., Fitzpatrick, T.B., Calkins, E. and Summerson, W.H. (1950). Mammalian tyrosinase: The relationship of copper to enzymatic activity. J. Biol. Chem. 187: 793.

Miyaji, T. and Greenstein, J.P. (1951). Cation activation of desoxyribonuclease. Arch. Biochem. Biophys. 32: 414.

Nielsen, F.H. (1971). Studies on the essentiality of nickel. In: Newer Trace Elements in Nutrition, edited by W. Mertz and W.E. Cornatzer. Marcel Dekker, Inc., New York, N.Y., p. 215.

Nielsen, F.H. (1973). Essentiality and function of nickel. In: Trace Element Metabolism in Animals - 2, edited by W.G. Hoekstra, J.W. Suttie, H.E. Ganther, and W. Mertz. University Park Press, Baltimore, Maryland: In Press.

Nielsen, F.H. and Higgs, D.J. (1971). Further studies involving a nickel deficiency in chicks. In: Trace Substances in Environmental Health-IV, edited by D.D. Hemphill. University of Missouri, Columbia, Missouri, p. 241.

Nielsen, F.H. and Sauberlich, H.E. (1970). Evidence of a possible requirement for nickel by the chick. Proc. Soc. Exp. Biol. Med. 134: 845.

Nielsen, F.H. and Ollerich, D.A. (1973). Nickel: A new essential trace element. Fed. Proc.: In Press.

Nomoto, S., McNeely, M.D. and Sunderman, F.W., Jr. (1971). Isolation of a nickel α-2 macroglobulin from rabbit serum. Biochemistry 10: 1647.

Ray, W.J., Jr. (1969). Role of bivalent cations in the phosphoglucomutase system. I. Characterization of enzyme-metal complexes. J. Biol. Chem. 244: 3740.

Schroeder, H.A., Balassa, J.J. and Tipton, I.H. (1962). Abnormal trace metals in man - nickel. J. Chron. Dis. 15: 51.

Snedecor, G.W. and Cochran, W.G. (1967). Statistical methods. Iowa State University Press, Ames, Iowa, pp. 91-94.

Sunderman, F.W., Jr. (1965). Measurements of nickel in biological materials by atomic absorption spectrometry. Am. J. Clin. Path. 44: 182.

Sunderman, F.W., Jr., Decsy, M.I., Nomoto, S. and Nechay, M.W. (1971). Isolation of a nickel-α_2-macroglobulin from human and rabbit serum. Fed. Proc. 30: 1274 (abstract).

Sunderman, F.W., Jr., Nomoto, S., Morang, R., Nechay, M.W., Burke, C.N. and Nielsen, S.W. (1972). Nickel deprivation in chicks. J. Nutr. 102: 259.

Tal, M. (1968). On the role of Zn^{2+} and Ni^{2+} in ribosome structure. Biochim. Biophys. Acta 169: 564.

Tal, M. (1969a). Thermal denaturation of ribosomes. Biochemistry 8: 424.

Tal, M. (1969b). Metal ions and ribosomal conformation. Biochim. Biophys. Acta 195: 76.

Underwood, E.J. (1971). Nickel. In: Trace Elements in Human and Animal Nutrition. Academic Press, New York, p. 170.

Wacker, W.E.C., Gordon, M.P. and Huff, J.W. (1963). Metal content of tobacco mosaic virus and tobacco mosaic virus RNA. Biochemistry 2: 716.

Wacker, W.E.C. and Vallee, B.L. (1959). Nucleic acids and metals. I. Chromium, manganese, nickel, iron, and other metals in ribonucleic acid from diverse biological sources. J. Biol. Chem. 234: 3257.

Webster, L.T., Jr. (1965). Studies of the acetyl coenzyme A synthetase reaction. III. Evidence of a double requirement for divalent cations. J. Biol. Chem. 240: 4164.

Wellenreiter, R.H., Ullrey, D.E. and Miller, E.R. (1970). Nutritional studies with nickel. In: Trace Element Metabolism in Animals, edited by C.F. Mills. E & S Livingstone, Edinburgh and London, p. 52.

19

SELENIUM CATALYSIS OF SWELLING OF RAT LIVER MITOCHONDRIA AND REDUCTION OF CYTOCHROME *c* BY SULFUR COMPOUNDS

O. A. Levander, V. C. Morris, and D. J. Higgs

Agricultural Research Service
U.S. Department of Agriculture
Beltsville, Maryland 20705

Schwarz and Foltz (1957) discovered that trace amounts of selenium in the diet could protect vitamin E-deficient rats against liver necrosis. This fundamental discovery led to the recognition of "selenium-responsive" deficiency diseases in several different animal species (Nesheim and Scott, 1961; Schubert *et al.*, 1961). The status of selenium as an essential nutrient in its own right was assured when beneficial effects due to the element were observed even in animals fed sufficient vitamin E (McCoy and Weswig, 1969; Thompson and Scott, 1969).

Since selenium appeared to spare the requirement for vitamin E, much early work concerning the biological role of this element focused on its possible antioxidant properties (Tappel and Caldwell, 1967). More recently, selenium has been shown to be a constituent of the enzyme glutathione peroxidase (Rotruck *et al.*, 1973; Flohe *et al.*, 1973) a function which could explain many of the "antioxidant" effects of the element.

On the other hand, the concept that vitamin E acts solely as an antioxidant is not accepted by all workers in the field (Green, 1970). Also, reports have appeared which suggest that selenium may have a role other than that of an antioxidant (Sprinker *et al.*, 1971; Hurt *et al.*, 1971).

The purpose of the studies reported here was to compare the effects of dietary vitamin E and selenium, alone and in combination, on the *in vitro* swelling of rat liver mitochondria caused by various chemical agents, some of which initiate lipoperoxidation and others of which do not. It was found that dietary selenium, unlike vitamin E, had no protective effect against the mitochondrial swelling induced by compounds which promote lipid peroxidation. Rather, selenium accelerated the swelling caused by certain thiols. Experiments with respiratory inhibitors indicated that the enhanced swelling induced by selenite plus glutathione might be mediated at the cytochrome _c_ level. Indeed, selenium was shown to be an excellent catalyst for the reduction of cytochrome _c_ by sulfhydryl compounds or inorganic sulfide. It is suggested that selenium may act _in vivo_ as selenopersulfide to facilitate the transfer of electrons from sulfur to cytochrome _c_.

MATERIALS AND METHODS

Materials. The animals, diets, and reagents used were described previously (Levander _et al_., 1973a; Levander _et al_. 1973b).

Methods. Techniques used for determination of mitochondrial swelling, estimation of lipid peroxides, and measurement of cytochrome _c_ reduction rates were previously described (Levander _et al_., 1973a; Levander _et al_., 1973b).

RESULTS

Figure 1 (upper curves) shows that there was essentially no effect of dietary vitamin E or selenium on the spontaneous swelling rate of rat liver mitochondria incubated in isotonic KCl-tris buffer at pH 7.4. Addition of 10^{-5} M L-thyroxine to the incubation medium caused a similar increase in swelling rate no matter which diet was fed (Figure 1, lower curves). The extent of lipid peroxidation after swelling was very low in all groups and no differences were seen as a result of diet in either spontaneous or thyroxine-induced swelling. Dietary selenium was not able to protect against either the lipid peroxidation or the mitochondrial swelling caused by 10 μM Fe^{++}, whereas vitamin E was quite effective in this regard (Tables 1 and 2). On the other hand, there was no effect of either dietary vitamin E or selenium against the swelling induced by 15 μM oleate, a reagent which does not promote formation of lipid peroxides. In an analogous experiment, dietary selenium, unlike dietary vitamin E, was shown to have no protective effect against either the mitochondrial swelling or the lipid peroxidation due to ascorbate or a combination of reduced and oxidized glutathiones (Figures 2 and 3, Table 3). In fact, in the latter case selenium actually appeared to stimulate mitochondrial swelling even in the presence of vitamin E. There was no appreciable formation of lipid peroxides during swelling under such conditions. This observation led us to test the effect of selenium on the swelling of rat liver mitochondria induced by a variety of thiols and it was shown that either dietary selenium or selenium added _in vitro_ to the incubation medium could accelerate such

TABLE 1

Effect of Dietary Vitamin E or Se on Fe- or Oleate-induced Swelling of Rat Liver Mitochondria

Supplement to Diet	$-\Delta A_{520} \times 10^3$ (60 min.) Fe^{++} 10 μM	Oleate 15 μM
Vitamin E	53 ± 32	262 ± 56
Vitamin E + Se	54 ± 42	238 ± 71
Se	366 ± 11	218 ± 64

Each value represents swelling of mitochondria after 60 min. at room temperature in 0.125 M KCl-0.02 M tris·HCl, pH 7.4, corrected for spontaneous swelling. Values are means of 3 observations ± standard error.

TABLE 2

Effect of Dietary Vitamin E or Se on Mitochondrial Lipoperoxidation in Fe- or Oleate-induced Mitochondrial Swelling

Supplement to Diet	$\Delta A_{532} \times 10^3$ (60 min.) Fe^{++} 10 μM	Oleate 15 μM
Vitamin E	19 ± 15	11 ± 3
Vitamin E + Se	16 ± 14	1 ± 6
Se	88 ± 8	0 ± 2

Each value represents the level of mitochondrial lipoperoxides after incubation as described in Table 1. Values are corrected for zero time lipoperoxide levels.

TABLE 3

Effect of Dietary Vitamin E or Se on Mitochondrial Lipoperoxidation in Ascorbate- or Glutathione-induced Mitochondrial Swelling

Supplement to Diet	$\Delta A_{532} \times 10^3$ Ascorbate 0.3 mM	$\Delta A_{532} \times 10^3$ GSH + GSSG 10 mM
Vitamin E	16 ± 0	9 ± 0
Vitamin E + Se	9 ± 0	4 ± 4
Se	96 ± 12	59 ± 3

Each value represents the level of mitochondrial lipoperoxides after a 60 min. (ascorbate) or 90 min. (GSH + GSSG) incubation as described in Table 1.

TABLE 4

Effect of Dietary Se on Sulfide-induced Swelling of Rat Liver Mitochondria

Supplement to Diet	$-\Delta A_{520} \times 10^3$ (60 min.) Na_2S 10 mM
Vitamin E	164 ± 37
Vitamin E + Se	323 ± 16

Conditions as in Table 1.

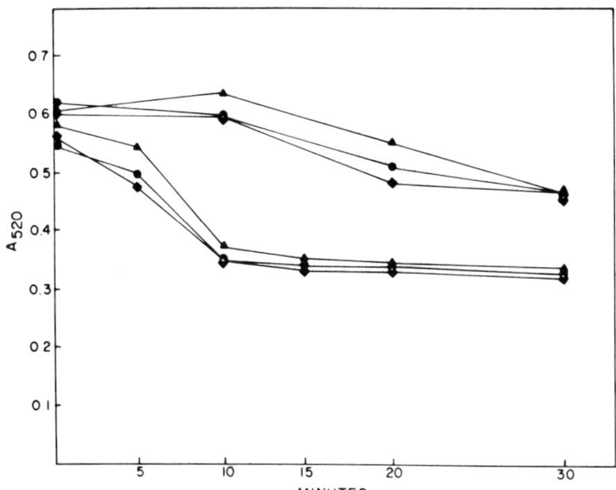

Figure 1. Effect of dietary vitamin E (▲), Se (◆) or both (●) on the spontaneous (upper curves) or thyroxine-induced (lower curves) swelling of rat liver mitochondria. Swelling was carried out in 0.125 M KCl-0.02 M tris·HCl, pH 7.4 at room temperature. Where indicated, vitamin E was added to the diet at 100 ppm as dl-α-tocopheryl acetate and selenium was added at 0.5 ppm as sodium selenite.

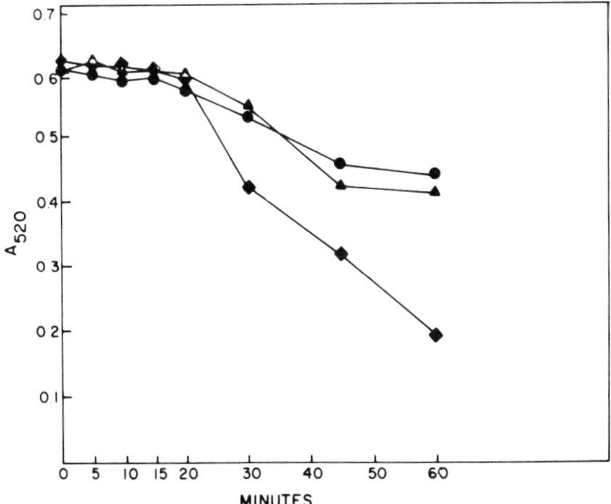

Figure 2. Effect of dietary vitamin E (▲), Se (◆), or both (●) on the swelling of rat liver mitochondria induced by 0.3 mM ascorbate. Conditions as in Figure 1.

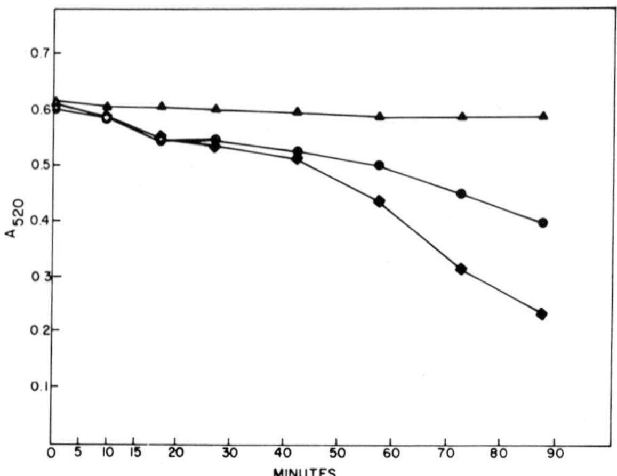

Figure 3. Effect of dietary vitamin E (▲), Se (◆), or both (●) on the swelling of rat liver mitochondria induced by 10 mM GSH plus GSSG. Conditions as in Figure 1.

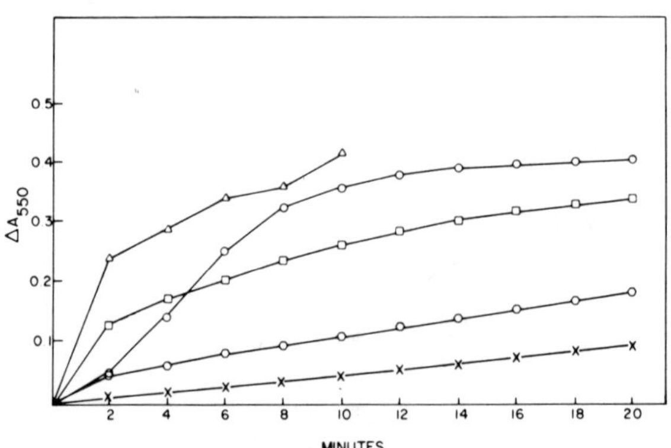

Figure 4. Selenium catalysis of reduction of cytochrome c by GSH. The reaction cuvettes contained 3 x 10^{-5} M cytochrome c and 3.3 x 10^{-4} M GSH in 0.175 M KCl-0.025 M tris·HCl buffer pH 7.45 (basal medium) plus the following additions: none (X); selenocystine, 10^{-6} M (⬡); selenocystine, 5 x 10^{-6} M (□); Na_2SeO_3, 10^{-5} M (○); selenocystine, 10^{-5} M (△). The reaction was initiated by the addition of Se catalyst and was carried out at room temperature. Final reaction volume was 3 ml.

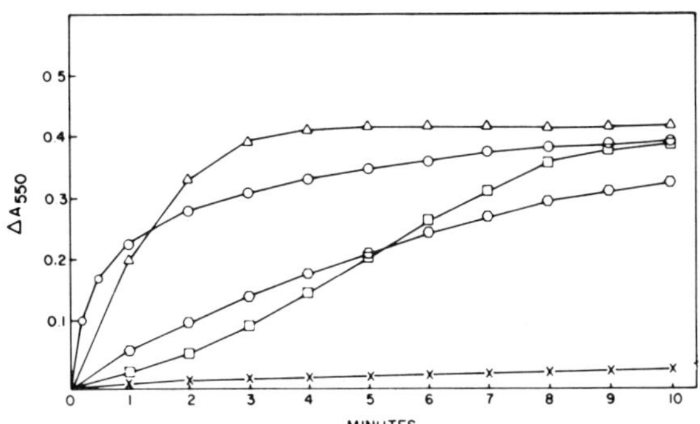

Figure 5. Selenium GSSG (Sigma) or GSSG (Boehringer) catalysis of reduction of cytochrome c by GSH. The cuvettes contained the basal medium (Figure 4) plus the following additions: none (X); Na_2SeO_3, 10^{-5} M (□); GSSG (Boehringer), 3.3 x 10^{-3} M (⬡); GSSG (Sigma), 3.3 x 10^{-3} M (○); or selenocystine, 10^{-5} M (△). Reaction conditions as in Figure 4.

swelling (Levander *et al*., 1973a). Selenium now has also been found to enhance the mitochondrial swelling induced by inorganic sulfide as well (Table 4). The swelling of liver mitochondria prepared from rats fed diets adequate in vitamin E but lacking Se ("Se-deficient mitochondria") caused by glutathione (GSH) plus *in vitro* selenite was found to be totally blocked by cyanide but only partially blocked by amytal or antimycin A, a fact which suggested that this type of swelling might be at least partly mediated at the cytochrome *c* level (Levander *et al*., 1973a). Additional data which are consistent with this concept are derived from experiments that show an inhibitory effect of azide on GSH plus selenite-induced swelling of Se-deficient mitochondria but show little or no inhibitory effect of high levels of dinitrophenol or dicoumarol (Tables 5, 6, and 7).

Further work demonstrated that selenite and especially selenocystine were excellent catalysts for the reduction of cytochrome *c* by GSH in chemically-defined systems (Figure 4 and Levander *et al*., 1973b). In agreement with others (Massey *et al*., 1971) we were able to show that oxidized glutathione (GSSG) was also a good catalyst for the reduction of cytochrome *c* by GSH and that different commercial samples of GSSG exhibited different potencies as catalysts for this reaction (Figure 5). Also of interest was the observation that those samples of GSSG which were the best catalysts for the reduction of cytochrome *c* by GSH were generally the most active in promoting mitochondrial swelling (Table 8).

TABLE 5

Effect of Azide on GSH Plus Selenite-induced Swelling of Se-deficient Rat Liver Mitochondria

GSH 10^{-2} M	$SeO_3^{=}$ 10^{-5} M	NaN_3 M	$-\Delta A_{520} \times 10^3$ (60 min.)
+	-	-	-115 ± 7
-	+	-	60 ± 14
+	+	-	376 ± 29
-	-	10^{-5}	-27 ± 18
-	-	10^{-4}	-45 ± 29
-	-	10^{-3}	-57 ± 19
+	+	10^{-5}	295 ± 53
+	+	10^{-4}	321 ± 18
+	+	10^{-3}	171 ± 32

Conditions as in Table 1.

TABLE 6

Effect of Dinitrophenol on GSH Plus Selenite-induced Swelling of Se-deficient Rat Liver Mitochondria

GSH 10^{-2} M	$SeO_3^{=}$ 10^{-5} M	Dinitrophenol µM	$-\Delta A_{520} \times 10^3$ (60 min.)
+	-	-	53 ± 2
-	+	-	102 ±25
+	+	-	456 ± 4
-	-	250	81 ± 3
-	-	500	143 ± 3
-	-	1000	360 ± 5
+	+	250	486 ± 6
+	+	500	487 ± 4
+	+	1000	454 ±26

Conditions as in Table 1.

TABLE 7

Effect of Dicoumarol on GSH Plus Selenite-induced Swelling of Se-deficient Rat Liver Mitochondria

GSH 10^{-2} M	$SeO_3^{=}$ 10^{-5} M	Dicoumarol µM	$-\Delta A_{520} \times 10^3$ (60 min.)
+	-	-	0 ± 20
-	+	-	117 ± 11
+	+	-	489 ± 4
-	-	10	103 ± 15
-	-	20	194 ± 30
-	-	40	329 ± 15
+	+	10	516 ± 4
+	+	20	538 ± 6
+	+	40	534 ± 5

Conditions as in Table 1.

TABLE 8

Effect of Various Commercial Samples of Glutathione on Swelling of Rat Liver Mitochondria

Commercial Source	$-\Delta A_{520} \times 10^3$ (100 min.) GSH 10 mM	GSSG 10 mM
Brand "B"	31 ± 21	118 ± 18
Brand "S"	171 ± 30	257 ± 22

Conditions as in Table 1 except that animals fed a diet of Purina Laboratory Chow, Ralston Purina Co., St. Louis, Mo. Similar trends were observed with mitochondria taken from rats fed purified diets. Brand "B" is Boehringer-Mannheim Corp., New York, N.Y., and Brand "S" is Sigma Chemical Co., St. Louis, Mo.

TABLE 9

Effect of EDTA on the Se-catalyzed Reduction of Cytochrome c by GSH

EDTA mM	$\Delta A_{550} \times 10^3$ (5 min.) Selenite	Selenocystine
0	223	377
0.5	185	261
1.0	127	245
2.0	54	204

The cuvettes contained the basal medium (Figure 4) and 10^{-5} M Se catalyst plus EDTA as indicated. Reaction conditions as in Figure 4.

TABLE 10

Effect of Ionic Strength on the Se-catalyzed Reduction of Cytochrome c by GSH

KCl	$\Delta A_{550} \times 10^3$			
	Selenite		Selenocystine	
mM	2 min.	10 min.	2 min.	10 min.
0	132	421	341	417
100	132	378	268	365
200	176	371	235	338
300	228	417	279	383

The cuvettes contained 3×10^{-5} M cytochrome c and 3.3×10^{-4} M GSH in 0.025 M tris·HCl buffer, pH 7.45 plus 10^{-5} M Se catalyst and KCl as indicated. Reaction conditions as in Figure 4.

TABLE 11

Effect of Preincubating Se Catalyst with GSH on Reduction of Cytochrome *c*

Preincubation Time (min.)	$\Delta A_{550} \times 10^3$ (5 min.) Selenite	Selenocystine
0	170	342
10	193	271
30	165	279
60	159	215
240	60	101

The reaction cuvettes initially contained 3 x 10^{-5} M cytochrome *c* in 0.175 M KCl-0.025 M tris·HCl buffer pH 7.45. Se catalyst, 10^{-5} M, and 3.3 x 10^{-4} M GSH were mixed and added to fresh cuvettes at the times indicated.

TABLE 12

Effect of Basic pH on Uncatalyzed and Se-catalyzed Reduction of Cytochrome *c* by GSH

pH	$\Delta A_{550} \times 10^3$ (1 min.) -Selenite	+Selenite, 10^{-5} M
7.0	10	12
7.5	7	38
8.0	17	93
8.5	45	216
9.0	355	400

The cuvettes contained 3 x 10^{-5} M cytochrome *c* and 3.3 x 10^{-4} M GSH in 0.175 M KCl-0.05 M tris·HCl buffer at the pH indicated with or without Se. Reaction conditions as in Figure 4.

TABLE 13

Effect of Acidic pH on Se-catalyzed Reduction of Cytochrome c by GSH

pH	$\Delta A_{550} \times 10^3$ (10 min.)
7.0	432
6.5	423
6.0	205
5.5	100
5.0	50

The cuvettes contained 3 x 10^{-5} M cytochrome c and 3.3 x 10^{-4} M GSH in 0.175 M KCl-0.05 M PO_4 buffer at the pH indicated plus 10^{-5} M selenite. Reaction conditions as in Figure 4.

TABLE 14

Se Catalysis of Cytochrome c Reduction by GSH

GSH µM	$SeO_3^=$ 10 µM	$\Delta A_{550} \times 10^3$ (30 min.)
25	-	84
25	+	94
50	-	80
50	+	145
100	-	76
100	+	302

The cuvettes contained 3 x 10^{-5} M cytochrome c in 0.175 M KCl-0.025 M tris·HCl buffer, pH 7.45 plus GSH with or without Se as indicated. Reaction conditions as in Figure 4.

TABLE 15

Non-stoichiometric Relationship Between Se Concentration and Cytochrome c Reduction

GSH μM	$SeO_3^{=}$ μM	Cytochrome c Reduced, μM
25	0	8
25	1	8
25	5	7
25	10	7
50	0	14
50	1	12
50	5	13
50	10	13
100	0	19
100	1	19
100	5	16
100	10	17

The reaction mixtures consisted of 30 μM oxidized cytochrome c in 0.175 M KCl-0.025 M tris·HCl buffer, pH 7.45 plus GSH and selenite as indicated. The reactions were carried out in Thunberg tubes which had been repeatedly evacuated and flushed with nitrogen and then left in the dark at room temperature overnight. The reaction was initiated by adding the selenite from the side arm. Reduction of cytochrome was determined by measuring A_{550} against blanks of 30 μM oxidized cytochrome c directly in the Thunberg tubes and using the extinction coefficient (reduced - oxidized) of 2.12×10^4 cm^2/μmole.

TABLE 16

Se Catalysis of Cytochrome c Reduction by Sulfide

Additions	$\Delta A_{550} \times 10^3$ 2 min.	10 min.
Na_2S, 50 μM	53	290
Na_2S, 50 μM + $SeO_3^{=}$, 10^{-5} M	237	394
Na_2S, 50 μM + Se-cystine, 10^{-5} M	337	364

The cuvettes contained 3×10^{-5} M cytochrome c in 0.175 M KCl-0.025 M tris·HCl buffer, pH 7.45 plus sulfide with or without Se as indicated. Reaction conditions as in Figure 4.

The selenium-catalyzed reduction of cytochrome *c* by GSH could be partially inhibited by 1 mM EDTA (Table 9), but increasing the ionic strength of the medium had no inhibitory effect on this reaction (Table 10). Preincubation of the selenium catalyst with GSH for 10 minutes resulted in a slight enhancement in the cytochrome *c* reduction rate with selenite but with selenocystine there was a slight decrease in the reduction rate (Table 11). Preincubation of either selenium compound with GSH for 4 hours resulted in a significant loss of catalytic activity. Raising the pH of the incubation medium increased the rate of both the uncatalyzed and the Se-catalyzed reactions, although the rate of the latter increased more rapidly (Table 12). Lowering the pH of the medium below 6 markedly decreased the rate of the Se-catalyzed reduction of cytochrome *c* (Table 13 - note differences in incubation periods in Tables 12 and 13).

That selenium acted as a catalyst in stimulating the reduction of cytochrome *c* by GSH is suggested by the data in Tables 14 and 15. The experiment in Table 14 shows that selenite promoted the reduction of cytochrome *c* when GSH was added at concentrations of 50 μM or greater. Although selenite *accelerated* the reduction of cytochrome *c* by GSH under these conditions, the presence or absence of selenite had no effect on the *extent* of the reduction of cytochrome *c* when the reaction was allowed to proceed to completion (Table 15). Finally, selenium has been shown to act as a good catalyst for the reduction of cytochrome *c* by inorganic sulfide as well as by thiol compounds (Table 16).

DISCUSSION

The observations presented here confirm and extend our previous findings that selenium can stimulate the swelling of mitochondria caused by certain sulfur compounds and also that selenium is an efficient catalyst for the reduction of cytochrome *c* by similar sulfur compounds. The fact that dietary selenium, unlike dietary vitamin E, had no protective effect against the mitochondrial swelling associated with lipid peroxidation (i.e., swelling induced by agents such as ascorbate, ferrous ion, or a combination of GSH plus GSSG) would argue against an antioxidant role for selenium under these conditions. The relationship of these results to the newly discovered role of selenium in the enzyme glutathione peroxidase (Rotruck *et al.*, 1973; Flohe *et al.*, 1973) is not clear at present, but earlier work from our laboratory had failed to demonstrate any influence of dietary selenium on the Mg/ATP/serum albumin-driven contraction of mitochondria swollen in 10 mM GSH (Levander *et al.*, 1972). This mitochondrial contraction process had been shown by others to be dependent upon the activity of mitochondrial glutathione peroxidase and catalase (Neubert *et al.*, 1962). The lack of effect of either dietary selenium or vitamin E on swelling induced by thyroxine or oleate is not surprising since the former is in itself a good antioxidant (Bunyan *et al.*, 1961) and neither compound causes swelling that is accompanied by an increased production of lipid peroxides.

The inhibitory effect of azide on GSH plus selenite-induced swelling is consistent with the idea that this type of swelling could be mediated through cytochrome *c*. The interpretation of earlier experiments which showed a similar inhibitory effect of cyanide on GSH plus selenite-induced swelling (Levander *et al.*, 1973a) was confounded by the finding that cyanide was also an effective inhibitor of the selenium-catalyzed reduction of cytochrome *c* by GSH, apparently as a result of destruction of the selenium catalyst via selenocyanate formation (Levander *et al.*, 1973b). The interpretation of the experiment with azide, however, is not compromised in this way since azide at concentrations as high as 0.01 M had no influence on the Se-catalyzed reduction of cytochrome *c* by GSH (Levander, Morris, and Higgs, unpublished observations). In this context, it should perhaps be mentioned that we have not been able to demonstrate any effect of dietary selenium on the activity of rat liver mitochondrial NADH-cytochrome *c* reductase (Unpublished observations). Considerable controversy existed in the early literature as to whether or not vitamin E had any role in cytochrome *c* reductase (Nason and Lehman, 1955; Corwin and Schwarz, 1959; Pollard and Bieri, 1959). It occurred to us that most assays for this enzyme are carried out in the presence of cyanide and therefore if selenium were involved in this enzyme its effect might be masked by the cyanide. Even when azide replaced the cyanide in the assay system, however, no effect of dietary Se was apparent.

The experiments with dinitrophenol and dicoumarol were suggested by the work of Wilson and Merz (1967) who used high levels of these uncoupling agents as respiratory inhibitors. Since these workers concluded that these uncouplers inhibited respiration at a locus before cytochrome *b*, the lack of effect of these compounds against GSH plus selenite-induced swelling is additional indirect evidence that this process may occur through cytochrome *c*.

The fact that certain commercial samples of GSSG as well as certain compounds of selenium are both able to stimulate mitochondrial swelling and are both good catalysts for the reduction of cytochrome *c* by GSH suggests that the activity of GSSG in these systems may be the result of selenium contaminants. Analysis of various lots of commercial GSSG has shown that the trace metal concentration in such preparations varies considerably (Cash and Gardy, 1965). It seems reasonable to consider that the different activities of various GSSG samples might be the result of different levels of selenium contamination. Analyses carried out in our own laboratory have shown different amounts of selenium in different GSSG samples and those samples with the highest catalytic activity also had the highest selenium content (Ferretti and Levander, unpublished observations). However, the absolute levels of selenium present were felt to be too low to account for the catalytic activity of the GSSG. Because of the well-known volatility of many selenium compounds, some of the element may have been lost during the acid digestion used in the analysis.

The GSSG- and selenium-catalyzed reduction of cytochrome c by GSH share some properties in common which is additional circumstantial evidence that the activity of GSSG may be due to selenium contamination. For example, Froede and Hunter (1970) found that the GSSG-catalyzed reduction was only partially inhibited by 1 mM EDTA and we found similar results with the Se-catalyzed reaction. Also, increasing the ionic strength of the medium had no effect on the GSSG-catalyzed reaction (Froede and Hunter, 1970) or on the Se-catalyzed reaction reported here. The lack of effect of EDTA or ionic strength on the GSSG-catalyzed reaction was taken as evidence for the lack of involvement of trace metals in the catalysis by Froede and Hunter, but as seen here the Se-catalyzed reaction is also relatively insensitive to these treatments.

The loss of catalytic activity upon prolonged incubation of the selenium compounds with GSH could be accounted for either by the destruction of a transient intermediate necessary for the catalysis or by the gradual oxidation of GSH to GSSG catalyzed by the selenium itself (Tsen and Tappel, 1958). The mechanism postulated to account for the catalytic activity of selenium in promoting the reduction of cytochrome c by GSH (Levander et al., 1973b) invoked the stepwise reaction of selenite with excess GSH to form the selenopersulfide derivative which was then the species involved in the reduction of the cytochrome:

$$GSSe^- + Cytc^{+++} \longrightarrow GSSe\cdot + Cytc^{++}$$

The fact that basic pH favored the Se-catalyzed reduction of cytochrome c by GSH whereas acidic pH inhibited it is consistent with the concept that the dissociated selenopersulfide is the species involved in the reduction. Of course, the pH effects could also perhaps be explained in terms of efficiency of generating the proper selenium-GSH derivative. The experiment in which it was shown that selenite influenced the rate of reduction of cytochrome c by GSH without influencing the extent of reduction of cytochrome c by GSH (even though the latter was directly related to the concentration of GSH present) was interpreted to mean that selenium was acting as a catalyst in this system.

The experiments which showed that selenium both accelerated the mitochondrial swelling due to inorganic sulfide and catalyzed the reduction of cytochrome c by inorganic sulfide are in agreement with the earlier hypothesis of Diplock *et al*. (1971) that selenium may exert one of its biological effects as part of the active center of an uncharacterized class of catalytically active non-heme-iron proteins. In this manner, selenium could act to facilitate the transfer of electrons from the iron-sulfur center of non-heme-iron proteins to cytochrome. Biochemical precedent for the concept that selenium may play a part in electron transfer now is available from the work of Turner and Stadtman (1973) who have shown that a selenoprotein is involved in the clostridial glycine reductase system. Moreover, Whanger and colleagues (1973) have recently found that their selenium-binding protein in lamb muscle has the spectroscopic properties and amino acid composition of a cytochrome. A role for selenium in electron transport could provide a rationale for many of the physiological consequences seen in vitamin E/selenium deficiency such as the decline in respiration suffered by liver slices prepared from rats fed diets lacking vitamin E and selenium (Chernick *et al*., 1955).

Mention of a trademark or proprietary product does not constitute a guarantee or warranty of the product by the U.S. Department of Agriculture, and does not imply its approval to the exclusion of other products that may also be suitable.

REFERENCES

Bunyan, J., Green, J., Edwin, E.E. and Diplock, A.T.(1961). Some antioxidant properties of L-thyroxine in biological systems. Biochim. Biophys. Acta 47:401-403.

Cash, W.D. and Gardy, M.(1965). Role of metal contaminants in the mitochondrial swelling activities of reduced and oxidized glutathione preparations. J. Biol. Chem. 240:PC3450-PC3452.

Chernick, S.S., Moe, J.G., Rodnan, G.P. and Schwarz, K.(1955). A metabolic lesion in dietary necrotic liver degeneration. J. Biol. Chem. 217:829-843.

Corwin, L.M. and Schwarz, K.(1959). An effect of vitamin E on the regulation of succinate oxidation in rat liver mitochondria. J. Biol. Chem. 234:191-197.

Diplock. A.T., Baum, H. and Lucy, J.A.(1971). The effect of vitamin E on the oxidation state of selenium in rat liver. Biochem. J. 123:721-729.

Flohe, L., Gunzler, W.A. and Schock, H.H.(1973). Glutathione peroxidase: a selenoenzyme. FEBS Lett. 32:132-134.

Froede, H.C. and Hunter, F.E., Jr.(1970). Catalytic effect of GSSG on reduction of cytochrome *c* by GSH--Possible model for facilitation of electron transfer and energy conservation by sulfonium ion formation. Biochem. Biophys. Res. Commun. 38: 954-961.

Green, J.(1970). Vitamin E and the biological antioxidant theory. In: The fat-soluble vitamins, eds. H.F. DeLuca and J.W. Suttie, University of Wisc. Press, Madison, Wisc., pp. 293-305.

Hurt, H.D., Cary, E.E., Allaway, W.H. and Visek, W.J.(1971). Effect of dietary selenium on the survival of rats exposed to chronic whole body irradiation. J. Nutr. 101:363-366.

Levander, O.A., Morris, V.C. and Higgs, D.J.(1973a). Acceleration of thiol-induced swelling of rat liver mitochondria by selenium. Biochemistry, in press.

Levander, O.A., Morris, V.C. and Higgs, D.J.(1973b). Selenium as a catalyst for the reduction of cytochrome c by glutathione. Biochemistry, in press.

Levander, O.A., Morris, V.C. and Varma, R.N.(1972). Effects of dietary selenium or vitamin E on the swelling of rat liver mitochondria induced by glutathione. Fed. Proc. 31:691(Abstract).

Massey, V., Williams, C.H., Jr. and Palmer, G.(1971). The presence of S°-containing impurities in commercial samples of oxidized glutathione and their catalytic effect on the reduction of cytochrome c. Biochem. Biophys. Res. Commun. 42:730-738.

McCoy, K.E.M. and Weswig, P.H.(1969). Some selenium responses in the rat not related to vitamin E. J. Nutr. 98:383-389.

Nason, A. and Lehman, I.R.(1955). Tocopherol as an activator of cytochrome c reductase. Science 122:19-22.

Nesheim, M.C. and Scott, M.L.(1961). Nutritional effects of selenium compounds in chicks and turkeys. Fed. Proc. 20:674-678.

Neubert, D., Wojtczak, A.B. and Lehninger, A.L.(1962). Purification and enzymatic identity of mitochondrial contraction factors I and II. Proc. Nat. Acad. Sci. 48:1651-1658.

Pollard, C.J. and Bieri, J.G.(1959). Studies on the biological function of vitamin E. I. Tocopherol and reduced diphosphopyridine nucleotide-cytochrome c reductase. Biochim. Biophys. Acta 34:420-430.

Rotruck, J.T., Pope, A.L., Ganther, H.E., Swanson, A.B., Hafeman, D.G. and Hoekstra, W.G.(1973). Selenium: biochemical role as a component of glutathione peroxidase. Science 179:588-590.

Schubert, J.R., Muth, O.H., Oldfield, J.E. and Remmert, L.F. (1961). Experimental results with selenium in white muscle disease of lambs and calves. Fed. Proc. 20:689-694.

Schwarz, K. and Foltz, C.M.(1957). Selenium as an integral part of Factor 3 against dietary necrotic liver degeneration. J. Am. Chem. Soc. 79:3292-3293.

Sprinker, L.H., Harr, J.R., Newberne, P.M., Whanger, P.D. and Weswig, P.H.(1971). Selenium deficiency lesions in rats fed vitamin E supplemented rations. Nutr. Repts. Int. 4:335-340.

Tappel, A.L. and Caldwell, K.A.(1967). Redox properties of selenium compounds related to biochemical function. In: Selenium in Biomedicine, ed. O.H. Muth. AVI Publ. Co., Westport, Conn., pp. 345-361.

Thompson, J.N. and Scott, M.L.(1969). Role of selenium in the nutrition of the chick. J. Nutr. 97:335-342.

Tsen, C.C. and Tappel, A.L.(1958). Catalytic oxidation of glutathione and other sulfhydryl compounds by selenite. J. Biol. Chem. 233:1230-1232.

Turner, D.C. and Stadtman, T.C.(1973). Purification of protein components of the clostridial glycine reductase system and characterization of protein A as a selenoprotein. Arch. Biochem. Biophys. 154:366-381.

Whanger, P.D., Pedersen, N.D. and Weswig, P.H.(1973). Selenium proteins in ovine tissues. II. Spectral properties of a 10,000 molecular weight selenium protein. Biochem. Biophys. Res. Commun. 53:1031-1035.

Wilson, D.F. and Merz, R.D.(1967). Inhibition of mitochondrial respiration by uncouplers of oxidative phosphorylation. Arch. Biochem. Biophys. 119:470-476.

20

CHILDHOOD EXPOSURE TO ENVIRONMENTAL LEAD

D.K.Darrow and H.A.Schroeder

Trace Element Laboratory, Department of Physiology

Dartmouth Medical School, Hanover, New Hampshire

1 to 5% of urban adults have blood lead values $\geq$ 40 μgPb/100 ml whole blood whereas approximately 25% of urban children have blood leads at this level. The metabolic exposure (μgPb/kg body weight) of children is over three times adult exposure for assumed normal absorption and ingestion of lead and twice the exposure of adults for inhaled lead. The metabolic exposure of children is greater by 13-24% at any given blood lead value in terms of circulating lead per kilogram of body weight. Blood lead values of children corrected for lower hematocrit values increase blood leads by 3 μg to 22 μgPb/100 ml of whole blood. Lead in the air on the Main Street of Brattleboro, Vermont exceeded average yearly values reported for our major cities by 3.6 to 5.6 times. The amount of lead in the air below four feet was tiwce as high as the values above four feet (8.01 μgPb/m^3 and 3.96 μgPb/m^3 respectively). "Exposed" housing adjacent to heavily traveled roads had significantly more lead inside than houses on side streets (956 ppm vs 566 ppm lead in the dirt and dust). Children between the ages of one and three years mouth and ingest foreign matter as part of their natural oral exploratory behavior and are exposed to lead in dirt and dust. Experiments with mice showed both 30 day old and neonate mice had significantly increased DNA and protein in whole brain or cerebrum respectively when exposed to small amounts of lead by feeding the mothers lead at 25 ppm. There is growing evidence that children suffer subclinical neurological impairment as a result of increased blood lead values. Lead fallout from automobile exhaust is partially responsible for the elevated blood lead values in children.

Over 85% of reported cases of overt lead poisoning occur in children between 1 and 3 years of age (1) and two year old children account for 50% of the deaths from lead poisoning (1). Approximately 25% of urban children have blood lead values over 40 μgPb/ 100 ml WB whereas only 1 to 2% of urban adult females and 3 to 5% of urban adult males have blood leads in this range (2). The Surgeon General has designated 40 μg/Pb/100 ml WB as "evidence suggestive of undue absorption of lead either past or present" (3).

From these figures, it is obvious that children are at greater risk from the toxic effects of environmental lead than adults. This could be due to any one, or a combination of three factors. First, as a developing organism children may be more sensitive to the toxic effects of lead. Second, the lead available to children, either inhaled or ingested, could be more readily absorbed than lead ingested or inhaled by adults. Third, children may be exposed to more lead than adults. We will look at each of these in turn to estimate the contribution of each to the toxic effects of lead in children.

Sensitivity to Toxic Effects of Lead

It is puzzling so little experimental work has been done in this area as it is generally recognized that the developmental process of organisms is particularly sensitive to toxins. Yet, in 1973, the International Lead Zinc Research Organization, Inc. in a reply to an EPA position paper on the health effects of lead stated, "There is no evidence that children are more susceptible to lead than adults" (4). They quote Kehoe who stated in 1969 "--it is fair to say that, although our lack of positive evidence of harm to individuals within these special groups (extreme youth or sex) cannot be said to be documented, apprehension concerning the danger of such harm is not well founded, since no such case has been seen and described." And in practice there is no distinction made between acceptable blood lead values for adults, children or pregnant women (2).

Children have a greater metabolic rate than adults and one can calculate the increased exposure to soft tissue at any given blood lead value. Table 1 shows the relative exposure of adults and children in micrograms of lead per kilogram of body weight at any given blood lead value and indicates that a six month old child with a blood lead of 40 μg/100 ml has a soft tissue exposure equivalent to 50 μg/100 ml for an adult male. A newborn child with 30 μgPb/100 ml has an exposure comparable to 36 μgPb/100 ml in an adult male.

A second measure of the greater metabolic exposure of children may be calculated from the amount of lead ingested or inhaled in relation to body weight. Table II compares lead absorption per

TABLE 1

Soft Tissue Exposure from Blood Lead: Child vs Adult

	New* Born	6** Months	12** Months	Adult** Male	Adult** Female
Blood volume, ml	268.5	650	800	5000	4200
Body weight, kg	3.17	7.3	9.8	69.9	59.8
Ml blood/kg body wt.	84.7	89.0	81.6	71.5	70.2
Blood Lead Values		μgPb/kg body weight			
20 μg Pb/100 ml WB	16.9	17.8	16.3	14.3	14.0
30	25.4	26.7	25.0	21.5	21.1
40		35.6	33.3	28.6	28.1
50		44.5	41.7	35.8	35.1
60		53.4	50.0	42.9	42.1

* Blood volume and body weight from: "Lead Concentrations in the Newborn Infant" B.K. Rajegowda et al, The J.of Pediatrics, Jan.1972, Vol.80, No.1,p.116-118.

** Blood volume and body weight from: Biological Handbook 1962, F.A.S.E.B., Hematopoictic Tissues from Infancy to Maturity: Man,p.348.

TABLE II

Comparative Lead Exposure: Child vs Adult

	70 kg Adult	12.6 kg Child
Ingested Lead		
Ingested μg/day	300	130
*Absorbed μg/day	30	13
μg Pb/kg/day	0.43	1.03
Inhaled Lead		
Air inhaled (m^3/day)	15	6
Air Lead μg Pb/m^3	2.0	2.0
**Lead absorbed μg/day	15.0	6.0
μg Pb/kg/day	0.21	0.48
Comparable Adult Exposure	Absorbed	Exposure
Ingested	72.1 μg Pb/day	721.0 μg Pb/day
Inhaled	33.5 μg Pb/day	4.5 μg Pb/m^3

*Assuming 10% absorption

** Assuming 50% absorption

kilogram of body weight for children and adults (5). The soft tissue exposure for a two year old child is approximately 2.5 times the exposure for a 70 kg adult ingesting 300 μg of lead per day.

Dr. Kehoe reported in 1960 (6) on a series of experiments involving ingestion of lead salts by four human subjects. The dosages added to the lead naturally occurring in the diet were 0.3, 1.0, 2.0, and 3.0 mg per day. The 0.3 mg dosage was discontinued after 420 days as there was no rise in blood lead values and a barely significant rise in urine lead levels after this length of time. The 1.0 mg dosage continued for four years resulting in a blood lead of 46 μgPb/100 ml. After two years, the 2.0 mg dosage resulted in a blood lead of 65 μgPb/100 ml. Four months after commencing the 3.0 mg dosage the blood lead value of the fourth subject had risen to 50 μgPb/100 ml.

These results were extrapolated by Dr. Kehoe to the point in time when the blood lead of the subjects would reach 80 μgPb/100 ml and be potentially dangerous. The addition of 1 mg of lead to the diet would raise the blood lead values to a dangerous level in about 8 years. Ingestion of 2.35 mg of lead (2 mg administered, 0.35 mg in food and beverages) would result in 80 μgPb/100 ml blood in about four years. The dosage of 3 mg of administered lead would reach the point of danger in approximately 8 months.

This data can be applied to children if we take into account their greater metabolic rate. Table III presents the dosages administered by Kehoe in terms of μg of lead per kilogram of body weight. I am assuming a body weight of 70 kg for his subjects as it is not given in his papers and a weight of 12.6 kg for a two year old child.

TABLE III

Comparative Lead Exposure of Administered Lead

Adult vs Child

		Micrograms of Lead		
*Ingested Lead	70 kg Adult	1,270	2,350	3,270
**Absorbed Lead	70 kg Adult	127	235	327
Exposed Lead μg Pb/kg body weight		1.81	3.36	4.67
Absorbed Lead	12.6 kg Child	22.8	42.3	58.8
**Ingested Lead	12.6 kg Child	228	423	588

* Administered plus dietary Lead
** Assuming 10% absorption

The normal values for total lead intake based on fecal lead in children range between 106 μg/day and 206 μg/day (5). Thus a child need only ingest 3 to 6 times the amount of lead normally ingested to reach a level expected to raise his or her blood leads to 50 μgPb/100 ml blood in four months. In contrast, an adult would have to ingest 8 to 10 times his normal lead intake to arrive at a similar level. The margin for children appears to be considerably narrower.

The consequences of the lower hematocrit in children has never been properly evaluated. Since 90% of the lead is attached to the red blood cells, any reduction in their amount will lower the observed blood lead values. Dr. King has stated "If a child has anemia the health consequences at any given level of blood-lead may be greater than those affecting a child with a normal hemoglobin value. For example, a lead level of 40 μg/100 ml in the blood of a child whose red blood cell volume is reduced to half of its normal level may be equivalent in clinical effect to a higher lead level in the blood of a child with a normal red blood cell volume" (5). The Surgeon General in his Policy Statement on Medical Aspects of Childhood Lead Poisoning also states "--seemingly low blood lead levels in children with anemia may be misleading" (3).

It is possible to detail the difference between adult blood lead values and blood leads of children correcting for the lower hematocrit values. Table IV indicates the corrected blood lead values ranging from 6% to 28% higher in children. The blood lead values of 40 μg/100 ml adapted by the Surgeon General (3)has indicative of excessive lead absorption" is in reality closer to 50 μgPb/100 ml and the value of 80 μg/100 ml which would require hospitalization and treatment is closer to 100 μg/100 ml.

TABLE IV

Hematocritic Corrections for Blood Lead

Age	Normal Hematocrit ml RBC/100 ml WB*		Blood Lead Values						
6 mos-6 yrs	36	observed	20	30	40	50	60	70	80
Adult,male	47	corrected**	26	38	51	69	77	89	102
Adult,female	42	corrected**	23	35	46	57	69	81	92

*Normal hematocrits from Clinical Hematology, Wintrobe.

**Corrected values assume 90% of blood lead is associated with the red blood cell. Corrected value =[(Observed Blood Lead X0.9 x $\frac{47}{36}$)0.1 Observed Blood Lead] when comparing with Adult Male Hematocrit Value, $\frac{42}{36}$ used to compare with Adult Female Values.

The hematocrit values for children used in the calculations for Table IV are based on normal values reported by Wintrobe in Clinical Hematology. However, the Ten States Nutritional Survey 1968-1970 (8) indicates that over 45% of the children two years old and younger have values less than the 36 ml RBC per 100 ml whole blood used in my calculations. Consequently the corrected blood lead values in Table IV may be even higher.

Nutritional factors may also play a role in lead toxicity. Experiments with rats show that deficiencies in either calcium or iron enhance the toxic effect of lead. Low calcium diets increased lead absorption and altered the distribution of lead between the hard (bone) and soft tissue (7), and a deficiency of iron increased the lead retained in liver, kidney and bone and decreased liver iron and liver protein (9).

Although there may be isolated instances of calcium deficiency in children, the Ten States Nutritional Survey indicates sufficient calcium intakes for all ages. So calcium would not appear to be a factor increasing the toxicity of lead in children.

Iron deficiency, however, is prevalent among children between 1 year and 3 years of age and as a group they receive less than 50% of adequate iron in their diets (8). Thirty percent of the children between the ages of 24 months and 36 months were ingesting less than 1/3 of the recommended daily allowance for iron (8).

Absorption of Lead in Children

The absorption of ingested lead from the gastrointestinal tract in children is assumed to be 10% and is based on observations in adults (2,6,10,11). The assumed 10% absorption has not to my knowledge been tested, though an EPA report (2) suggests "that healthy children may absorb as much as 50% of their oral lead intake." This figure (50%) would be more in line with the results obtained by Forbes and Reina (12) showing lead absorption of 75-80% in rats up to weaning, after which the adult pattern of 10% absorption is manifested.

A value of 50% absorption would be difficult to uphold on the basis of known fecal lead excretion (5,11) and assumed dietary intake of lead (5,13). Mean fecal lead values in normal children are approximately 130 μgPb/day and 50% absorption would require the total ingestion of 260 μgPb/day. Since dietary lead for adults is estimated to be between 100-500 μg/day (6,13,14) with a mean of 300 and the metabolic requirements for a two year old child are about 50% of adult values, lead ingestion from dietary source, for a child should range between 50-250 μg/day with a mean of 150 μgPb/day. The higher absorption rate can be supported if we assume that "normal" children have an additional source of exposure and this excess lead shows up in the feces.

Absorption of inhaled lead by adults is between 25% and 50% (5,6,13,14,16) depending on particle size. Kehoe (6) found that 64% of inhaled lead with a particle size of 0.05 μ was exhaled. The 36% retained was absorbed and subsequently excreted in the urine. He assumes "virtually complete absorption of particles of lead from lung membrane." When the lead particle size was increased to 2 μ, only 54% of the inhaled lead was exhaled. Of the 46% retained, part was absorbed in the lungs and part was swept into the alimentary tract and subjected to gastrointestinal absorption. Since only about 10% of this lead was absorbed, the amount of lead excreted in the feces was elevated and amounted to more than the total ingested dietary lead.

A study by DuPont (17) indicates about 51% of the particulate emissions from automobiles as measured in a proportional sampling system coupled to an automobile exhaust are greater than 1.0 μ in size and only about 20% are less than 0.3 μ. This size distribution is confirmed indirectly by many studies indicating reduced lead concentrations as the distance increases from highways (18,19,20,21) and by the "unnaturally high" lead content of dirt and dust adjacent to roads (2,13).

The exhaust gas from an automobile burning leaded gasoline contains 1000 to 8000 $\mu gPb/m^3$ while four feet behind the car the lead content drops to 12 to 113 $\mu gPb/m^3$ (24). Part of this decrease is dilution and part can be attributed to larger particles settling out. These larger particles, if inhaled, almost certainly would not reach the lungs but be deposited before reaching the alveolar tissue and swept out to be expectorated or swallowed. If swallowed, they would be subjected to gastrointestinal absorption. A two year old child is approximately 35 inches tall, and an adult 70 inches. The child would be more likely to inhale the larger particles coming directly from an automobile exhaust or which result from dust mobilization by moving automobiles.

It is possible therefore that the fecal lead determinations which correlate so nicely with assumed lead ingestion from assumed dietary intakes are in fact the summation of lead ingested from dietary sources and large inhaled particles. We, in fact, do not know how much lead is absorbed by a two year old child either from ingested lead or inhaled lead.

Lead Exposure of Children

Children exhibit a natural oral exploratory behavior which from birth to one year of age is mostly "mouthing--but not ingesting--almost anything the infant can place in his mouth" (25). Between 12 months and 18 months, the child will not only "mouth" but ingest many foreign materials. "As many as 50 percent of children carefully studied in both middle class and poverty groups habitually and selectively ingest objects other than food" (25).

The ingestion of chips of paint which have a high lead content has and will continue to be a major source of exposure for children in this age group. Dr. Guinee of the Bureau of Lead Poisoning Control in New York City reported in 1971 that of 2,344 apartments of children with lead poisoning inspected "two-thirds of them had at least one location where lead paint was found in excess of the legal limit for lead content" (26). The legal limit for lead content is 1% lead. One third of the apartments tested, however, had no available source of lead paint even at the 1% level. A study in Connecticut of 230 rural children found no statistically significant association between blood lead levels and history of pica, history of pica for paint or plaster, presence of flaking paint in the home, or age of dwelling (15). Though lead paint is implicated in excessive exposure of children to lead and subsequent lead poisoning, it does not explain entirely the epidemic proportions of this disease. This same oral behavior by small children which leads to ingestion of lead paint chips will result in the ingestion of much dirt and dust.

Dirt collected from the sidewalks of the Bedford-Stuyvesant area of New York City averaged 1717 μgPb/gram of dirt (27). Adjacent to heavily traveled roads in Washington, D.C. lead in dirt and dust ranged between 4,000-12,000 μgPb/gram of dirt. Dirt taken near an incinerator in Washington had 32,890 μgPb/gm of dirt; adjacent to the White House the level was 4,340 ppm and near the EPA 5,120 ppm (28). A report from England found 13 of 35 <u>indoor</u> samples of dust and greater than 5,000 ppm lead, three had greater than 10,000 ppm and one was as high as 30,700 ppm (27). Dust <u>inside</u> homes in the Boston area were mostly in the range between 1,000-2,000 ppm (2). The lead found in dirt and dust near roadways and in urban situations is predominantly from automobile exhaust (2,13,21). There are approximately 89 million cars emitting an estimated 4 lbs of lead yearly in the U.S.

The maximum daily permissible intake (M.D.P.I.)level for children suggested by an AdHoc committee to estimate the M.D.P.I. for children and composed of King, Challop, Chisolm, Lin-Fu, Stokinger, Stokinger and Tepper (5) is 300 μg of lead per day from all sources. They assume a daily dietary intake of 130 μg and 4 to 8 μg from inhaled lead which leaves a margin of 160 to 170 μg from extraneous sources. A child need only ingest 160 to 170 milligrams of dirt containing 1,000 ppm lead to exceed the M.D.P.I. The ingestion of a half gram of dirt containing 500 ppm would result in lead exposure in excess of the daily permissible intakes. Considering the oral habits of children between the ages of 6 months and 3 years, and the amount of lead found in dust and dirt both inside and outside, it is difficult to imagine how any urban child could avoid exceeding the M.D.P.I. The only way to avoid exceeding this limit is to raise the limit.

Lead exposure to children from inhaled lead may also be in excess of adult values. A 2 or 3 year old child playing on the sidewalks or street is closer to the ground and thus more exposed to the larger particles of exhaust fallout. The values given for air lead after many years of study in our major cities (2,13,23) are all generated from air sampling stations located considerably above street level;none of the stations was at the one or two foot level. One can only assume that children playing at street level have a greater exposure to air borne lead than adults.

There is then, evidence to suggest that children are more sensitive to the effects of lead; they may have a higher absorption rate and also, a higher exposure to lead than adults. All of these factors may be contributing to the large numbers of children with elevated blood lead values.

Sub-Clinical Effects

It has been well documented that large numbers of urban children have elevated levels of lead in their blood. These children are asymptomatic of the clasical signs of lead intoxication. The question of the sub-clinical effect of long term exposure to elevated blood leads remains unanswered. Either there is no risk from elevated blood lead in children or we have not been asking the right questions. It is difficult to ask the "right questions" of a two or three year old child, who is at best only marginally communicative, has very little basis for comparing "feeling sick from feeling well," and is an organism developing from an unknown genetic makeup. There is evidence accumulating, however, which suggests there are subclinical effects.

Since acute lead poisoning in children affects the brain often causing permanent damage (1,2,13) and lead is a known neuro-toxin (25 31), it would seem reasonable to look for subclinical effects in the brain. Two reports have recently come out which are highly suggestive and also demonstrate how difficult it is to detail "subclinical" effects in the brain.

In Virginia, seventy children (29), with elevated blood lead but no obvious signs of lead intoxication were tested at four years of age using a series of psychological tests. The test results were compared with the results obtained from testing a second group of 72 four year olds with no known lead exposure. The lead exposed group showed significant abnormalities in fine-motor control and behavior. The most frequent combination of abnormal behavior characteristics was negativism, distractibility and constant need of attention.

A second study (30) compared blood and urine lead levels of hyperactive and non-hyperactive children. It was found that hyperactive children with no known cause for the condition had significantly raised blood leads and excreted significantly more lead in their urine when challenged with the chelating agent penicillamine. The blood lead values were a measure of circulating lead and the urine lead after challenge with penicillamine was considered an indication of previous lead exposure. It is estimated there are between 5-6 million hyperactive children in the U.S. today.

Delta aminolevulinic acid dehydratase (ALAD) activity is known to be inhibited by lead (33) and this inhibition can be detected at blood lead values as low as 20 μgPb/100 ml (2). This enzyme is concerned with heme production and apparently there is a considerable reserve capacity, at least in adults. However, a study by Millar et al (32) has shown that lead intoxicated suckling rats have not only reduced ALAD activity in the circulating blood but also in the brain. Therefore, this biochemical abnormality induced by lead and which can be demonstrated at low blood lead values, can be seen in the brain of a developing organism.

If we now look at Kehoe's statement mentioned earlier, we see there is, meager as it is, documentation "of positive evidence of harm" and there are cases which are being "seen and described." The answer to the question "Is there evidence of subclinical toxicity in developing organisms (i.e. children) with marginally elevated blood leads?" can only be answered by experimentation with developing organisms. In the past, we have tried to answer the above question by answers provided through experimentation with mature organisms; if you ask the wrong question you get the wrong answer.

Dr. Schroeder has completed many investigations dealing with the recondite toxicity of lead and other metals in rats and mice. Lead added to the drinking water at 25 ppm resulted in increased mortality, decreased longevity, and fewer tumors in mice (34); a similar effect was observed in rats (35). However, a later study found lead was not toxic to male rats in terms of growth and survival when chromium (1 ppm) was added to the drinking water (36). The animals in these investigations were started on lead after weaning and thus represent recondite toxicity for adult and not developing organisms.

A study on the toxic effects of lead on reproduction of mice and rats shows a much more dramatic effect on developing organisms. Lead fed to mice at 25 ppm resulted in the loss of the strain after two generations. In the F_1 generation, there were 9 deaths of young mice and 69 runts out of 72 offspring while there were no young deaths and no runts in controls. The rats were able to breed through three generations but at the termination of the experiment there were 70 runts and 49 young deaths out of 506 offspring while controls produced 348 offsprings with only one runt and no young deaths (37).

Dr. Schroeder estimates the amount ingested was approximately 100 times the normal human exposure in terms of μgPb/100 g body weight. Retention in soft tissues was about 1/10 the human retention. Concentrations of lead in the soft tissues were comparable to concentrations in human tissue after 50 years. Therefore, lead fed to mice and rats at 10 to 100 times normal human ingestion rates (in the absence of chromium) resulted in a pronounced effect on developing animals and a minimal but significant effect on adult animals. We are presently evaluating a second reproductive study with added chromium (5 ppm) and lead (25 ppm).

Experimental Results

Studies by Winick have shown that cell proliferation in the rat brain continues until about day twenty with the various regions of the brain exhibiting different rates of hyperplasia or cell division (40). The rat brain continues to increase in weight after day 20 by hypertrophy or increase in cell size. Malnutrition during hyperplasia can result in permanent reduction in cell number whereas malnutrition during hypertrophy appears to be reversible (40). Results from similar investigations on human brains indicate cell division has been completed by the 6th month of postnatal growth (38) and that malnourished children have reduced brain weights and a reduction in the number of cells in the brain (39).

To investigate the possibility that lead may have an observable effect on mice, we began looking at the DNA, RNA and protein in the brains of mice. The mice were offspring of female Charles River CD mice maintained on an adequate diet developed and used in this laboratory for many years (36). Lead was added to the drinking water as lead nitrate at 25 ppm with 5 ppm chromium as chromium acetate. At 30 days of age, the mice were sacrificed, brains removed and frozen until analysis could be performed. Protein, RNA, and DNA were extracted and assayed essentially as in Winick and Nobel (41) using the Schmidt and Thanhauser (42) method for extraction, the Lowry method for proteins (43), RNA by the Dische orcinol reaction and DNA by the diphenylamine reaction (44).

The data are summarized in Table V and given in terms of wet weight. There was a slight but not significant reduction in brain weight of the males but not of the females. The amount of DNA in the males increased significantly whereas the female offspring of the lead fed mice showed only a slight increase which was not significant. RNA was increased slightly for both males and females in the experimental group but not significantly. Protein for both male and females was increased significantly over controls. Variability in all parameters measured was increased for the experimental animals.

TABLE V

Brain weight, DNA, RNA and Protein of 30 day old Mice fed Lead

Males *		Controls	P ***	Lead Fed
Brain wt	(gm)	0.5278±0.0166	NS	0.4831±0.0264
DNA/Gram	(mg)	1.251±0.038	< .05	1.512±0.101
RNA/Gram	(mg)	2.677±0.153	NS	2.964±0.231
Prot/Gram	(mg)	33.77±1.5	< .005	48.47±3.41

Females**		Controls	P ***	Lead Fed
Brain wt.	(gm)	0.5271±0.0121	NS	0.5199±0.0175
DNA/Gram	(mg)	1.386±0.131	NS	1.485±0.135
RNA/Gram	(mg)	2.636±0.224	NS	3.004±0.333
Prot/Gram	(mg)	36.38±2.54	< .025	53.23±4.82

* 6 determinations of pooled brains from 5,6,6,5,6,4 animals were used from the control group and 6 determinations from 7,4,4,1,1,1 were used in the lead mice.

** 5 determinations from pooled brains of 4,4,6,3,6 control animals and 5 determinations from 5,4,1,1,1 lead animals.

*** *P* is significance of the difference in the means of contiguous groups by Student's "*t*".

All values are in terms of wet weight.

A second experiment (45) using just the cerebrum of neonate littermate mice shows similar increases in DNA and protein. These mice were raised in a different laboratory, were not sexed, did not have chromium added to the diet, and were fed a different diet (which probably has sufficient chromium) so are not strictly comparable. However, they were mice of the same strain (Charles River CD), fed lead at the same level (25 ppm PB), and analyses were performed in our laboratory.

Table VI summarizes the results and compares lead exposed animals with controls and methyl mercury exposed animals (1 ppm Hg as methyl mercury acetate). Methyl mercury is a known neurctoxin and produces observable neurological sequelae in children (46,47) and animals (48). It seemed useful to compare a known neurotoxin with a suspected neurotoxin.

TABLE VI*

Neonate Cerebrum: Weight, DNA, RNA, Protein

	Mercury Fed**	P***	Control**	P***	Lead Fed**
Cerebrum wt(g)	0.0515±0.0097	< .01	0.0642±0.0097	NS	0.0618±0.0081
DNA/Gram(mg)	0.854±0.418	NS	0.808±0.197	< .01	1.063±0.224
RNA/Gram(mg)	16.3±3.9	NS	13.96±3.7	NS	14.05±2.98
Protein/gm(mg)	16.8±3.7	< .005	12.51±2.30	< .05	15.68±4.34

* From: Darrow and Clark "Variations in DNA, RNA, and Protein in Neonatal Mouse cerebrum born of lead and mercury fed mothers".

** 10 animals used in each group.

*** P is significance of the difference in the means of contiguous groups by Student's "t".

All values are in terms of wet weight.

The cerebrum of these neonate lead exposed mice exhibited a significantly increased DNA and protein content with no increase in RNA or brain weight, while the mercury exposed animals showed reduced brain weight and increased protein. An increase in DNA and protein with no increase in brain weight would be consistent with a proliferation of smaller brain cells. Whereas the reduced brain weight with no reduction in DNA per gram would suggest a fewer total number of brain cells (this is indicated also by considering DNA per brain, 0.041±0.016 mg DNA per brain for mercury animals, 0.051$\pm$0.012 mg DNA per brain for controls, $P < .10$).

The increased numbers of cells suggested by increased DNA in both the 30 day old male mice and neonate mice is consistent with a proliferation of astrocytes and microglial cells which has been reported in both human lead encephalopathy and experimental lead encephalopathy in rats (49). To induce a lead toxicity in suckling rats resembling the acute human condition and including proliferation of astrocytes and microglia, requires feeding 4% lead carbonate (31,000 ppm Pb) to their mothers. Our data suggests that though there are no obvious external signs of lead intoxication in mice fed 25 ppm lead there may be subtle neurological effects. We are continuing our investigations to determine if this condition will be manifest in animals exposed from birth to 30 days.

To evaluate the possibility that children have a greater exposure to lead than adults, we undertook two investigations: First, an estimate of the vertical distribution of air lead on Main Street, Brattleboro, Vermont and second, an estimate of the exposure a child would have in terms of a natural oral exploratory behavior in their own homes.

Three Low Volume Air Samplers (50) were used fitted with plastic tubing, which allowed the sampling heads to be placed at various heights, and glass fiber filter paper (0.3 μ pore size)(51). The sampling heads were approximately 2 meters from the curb. Air flow was 16.5 l/min. and maintained for 60 min., 100 min., or 1108 min. Traffic during the day, 8:30 a.m. to 4:30 p.m., averaged 600 cars per hour. The total number of cars for the longer time, 4:35 p.m. 9/12/72 to 1:02 p.m. 9/13/72, is not known. Weather conditions were clear and sunny with little wind until the second day when it became overcast and rained about 12 noon. Air filters were washed with two washes of nitric acid, filtered, then washed with two washes of distilled water and brought up to a constant volume. Lead determinations were made using the dithizone method as previously reported (53).

Table VII shows the lead distribution at various times of the day and at different heights. Though there is some direct evidence for a stratified air-lead content, there is so much variation it does not lend itself to statistical analysis. However, there are two things that are apparent: first, on September 12 thru September 13, 1972 on the Main Street of Brattleboro air lead values exceeded the daily averages for Cincinnati, Los Angeles, Philadelphia, New York and Chicago (2,22) by 363% to 556%, and second, the combined air lead values below the four foot level averaged 8.01 μgPb/m^3 whereas combined values above four feet averaged 3.96 μgPb/m^3.

Either we picked an unusual sampling site or there are other variables which should be taken into consideration. The lower height certainly is one factor. However, the "Three Cities Survey" (22) was conducted using either 3.0 μ or 2.0 μ pore size filter and we were

TABLE VII

Lead in Air on Main Street, Brattleboro, $\mu g/m^3$

Height Above Street \ Time	8·36 10:15 AM	10:20 AM 12:00 Noon	1:20 3:00	3:25 4:25	4:35 PM 9/12 1:02 PM 9/13
1 ft.	8.94	7.85	3.85		
2 ft.			5.31	17.38	
3 ft.	8.22	4.40			
4 ft.			7.56	21.11	
5 ft.	7.13	8.15			
7 ft.				2.40	
8 ft.					3.01
30 ft.					4.25

Average Air Lead Value for 54.38 m^3 8.34 μg Pb/m^3

1', 2', 3', 4', Average Air Lead Values for 13.53 m^3 air 8.01 μg Pb/m^3

5', 7', 8', 30', Average Air Lead Values for 40.85 m^3 air 3.96 μg Pb/m^3

using a filter with 0.3μ pore size. The study by DuPont cited earlier (17) indicates that about 49% of the lead emissions as measured in their proportional sampling system are less than 1.0 μ in size and Daines (18) has shown that "65% of the lead in the air between 30 and 1750 feet from a highway consists of particles under 2 μ." Lee reported in 1968, that "about 75% of the lead in Cincinnati air was associated with particles having equivalent diameters below 1 μ" even at the four foot level (20). The higher values reported for Brattleboro are probably a combination of greater collecting efficiency and the lower sampling locations. The contribution of each of these factors needs to be evaluated.

Indirect confirmation of the values reported here is presented in a study by Konopinski in 1967 (52) of commuter exposure to air borne lead using a car mounted filtering system and a 3.0 μ pore size filter. Mean concentrations of lead were found ranging from

9 μg/m^3 to 38 μg/m^3 in weekday samples from downtown and freeway routes and on rural roads were from 2.8 μg/m^3 to 5.8 μg/m^3. These air lead concentrations, which are greatly in excess of values reported in either the "7 cities" or "3 cities" survey even using the 3.0 μ pore size filter, could be the result of the mobilization of the heavier lead particles by the moving automobiles. Any location adjacent to moving traffic would probably have higher air-lead values than those reported for our major cities monitored at higher elevations removed from heavy traffic.

To estimate childhood exposure in their own homes, we used pre-weighed self-adhesive labels (Avery #5356, 17.5 cm^2) to pick up dirt and dust. Six locations were selected, 2 near major roads and 4 on side streets. All housing was in roughly similar condition, and of similar age (older than 25 years). All had children in the home ranging in age from 1-1/2 to 20 years old. Two had 6 children; two had four children, and 2 had three children. All samples were taken from floors or low window sills and roughly 1/3 of the samples were from outside locations, porch, sidewalks or driveways, where a child might play. After exposure, the labels were reweighed, subjected to "bomb" ashing (54) and assayed for lead using the dithizone method.

Table VIII compares lead values found in and around "exposed" houses within 5 meters of roads averaging 7500 cars/day and "unexposed" houses located on side streets with only local traffic.

TABLE VIII

Lead in Dirt and Dust--Brattleboro, Vt.

2 exposed houses*	All Samples	Inside	Outside
No. of Samples	15	11	4
Average dirt picked up(g)	0.0074±0.0058	0.0062±0.0052	0.0107±0.0058
Average lead picked up(μg)	8.75±9.56	5.93±3.12***	16.51±15.29
μg Pb/gram dirt (ppm)	1182	956	1543
4 unexposed houses**			
No. of Samples	37	25	12
Average dirt picked up(g)	0.0286±0.0769	0.0070±0.0061	0.0734±0.1231
Average lead picked up(μg)	6.08±7.18	3.96±2.88	10.51±10.63
μg Pb/gram dirt (ppm)	213	566	143

*Within 5 m of road with yearly average of 7500 cars per day.
**Side street with only local traffic.
***Significance of the difference in the mean of lead picked up inside "exposed" houses and "unexposed" houses by Student's t is $P < .05$.

Only the inside dirt had significantly higher lead values in the exposed locations. The labels used outside picked up sand and small rocks in addition to contaminated dirt. This resulted in great variability in weight and lead content. While this is far from a definitive study, it does indicate a correlation between lead inside houses and heavily travelled roads.

The question of the effect of airborne lead on the segment of the population at greatest risk has not been answered. There is reason to believe that children are more sensitive to the toxic effects of lead; they may absorb more lead and be exposed to more lead than adults. It is possible there are subclinical neurological effects in children with increased body burdens of lead. There is need for more information related to childhood exposure to lead, not adult exposure.

REFERENCES

1. Lin-Fu, Jane S. "Lead Poisoning in Children" U.S.D.H.E.W., P.H.S. No. 2108, 1970.

2. "EPA's Position on the Health Effects of Airborne Lead" Publications Section, E.P.A., 401 M St. S.W. Room W 238, Washington, DC 20460.

3. The Surgeon General's Policy Statement on Medical Aspects of Childhood Lead Poisoning, August, 1971.

4. Cole, J.F., Lynam, D.R. "Response to Request for Comment on Notice of Proposed Rule Making on Regulation of Fuels and Fuel Additives." F.R. Vol. 38, No.6 pp 1249-1261, Jan. 10, 1973. International Lead Zinc Research Organization, Inc. Lead Industries Association, Inc. March 9, 1973.

5. King, B.G., "Maximum Daily Intake of Lead without Excessive Body Lead-Burden in Children" Amer. J. Dis. Child, Vol. 122 Oct. 1971, pp. 337-340.

6. Kehoe, Robert A. "The Metabolism of Lead in Man, in Health and Disease" The Harben Lectures, 1960, reprinted from the J. of the Royal Institute of Public Health and Hygiene, 1961.

7. Six, K. and Goyer, R. "Experimental Enhancement of Lead Toxicity by Low Dietary Calcium," J. Lab. and Clin. Med. 76, 933-942, 1970.

8. Ten State Nutrition Survey 1968-1970, U.S.D.H.E.W. Health Service and Mental Health Administration, Center for Disease Control, Atlanta, Ga. 30333, DHEW Pub. No. (HSM) 72-8131.

9. Six, K.M., Goyer, R.A. "The Influence of Iron Deficiency on Tissue Control and Toxicity of Ingested Lead in the Rat" J. Lab and Clin. Med. Vol 79, No.1, Jan. 1972.

10. Hursh, J.B., Suomela, J. "Absorption of ^{212}Pb from the Gastro-intestinal Tract of Man" Acta Radiologica, Vol. 7 Fasc. 2, p. 108-120 April, 1968.

11. Barltrop, D., Killala, N.J.P. "Faecal Excretion of Lead by Children" The Lancet, pp. 1017-1019, Nov. 11, 1967.

12. Forbes, G.B., Reina, J.C. "Effect of Age on Gastrointestinal Absorption (Fe,Sn,Pb) in the Rat" The J. of Nutr. Vol. 102,#5 p. 647, May 1972.

13. "Airborne Lead in Perspective" committee on Biologic Effects of Atmospheric Pollutants of the Division of Medical Sciences Nat. Res. Council, N.A.S.

14. Schroeder, H.A., Tipton, I.H. "The Human Body Burden of Lead" Arch. Environ. Health, Vol. 17, p. 965-978, Dec. 1968.

15. Cohen, A.B., Lipow, M.L. "Blood Lead Levels of 230 Rural Children" Graduation requirement thesis, University of Connecticut School of Medicine, Dept. of Pediatrics, U. Conn. Health Ctr, McCook Hosp. 2 Holcomb St.,Hartford CT(accepted for publication in J.A.M.A.)

16. Goldsmith, J.R., Hexter, A.C. "Respiratory Exposure to Lead: Epidemiological and Experimental Dose-Response Relationships" Science, Vol. 158, No. 3797, pp. 132-134, Oct. 6, 1967.

17. Habibi, K., Jacobs, E.S., Kung, W.G. Jr., Pastell, D.L. "Characterization and Control of Gaseous and Particulate Exhaust Emissions from Vehicles." Presented at the Air Pollution Association, West Coast Section, Fifth Technical Meeting, Oct. 8-9, 1970, San Francisco, California, E.I. duPont de Nemours and Company, Inc. Petroleum Laboratory, Wilmington, Delaware 19898.

18. Daines, R.H., Motto, H., Chilko, D.M. "Atmospheric Lead: Its Relationship to Traffic Column and Proximity to Highways" Environ. Science and Technology Vol.4, pp. 318-322, April 1970.

19. Schuck,E.A., Locke, J.K. "Relationship of Automotive Lead Particulates to Certain Consumer Crops" Environmental Science and Technology, Vol.4, pp. 324-330, April 1970.

20. Lee, R.E., Jr., Patterson, R.K., Wagman, J. "Particle-Size Distribution of Metal Components in Urban Air" Environmental Science and Technology, Vol. 2 pp. 288-290, April 1968.

21. Ault, W.U., Senechal, R.G., Erlebach, W.E., "Isotopic Composition as a Natural Tracer of Lead in the Environment" Environmental Science and Technology, Vol 4, pp 305-313, April, 1970.

22. The Working Group on Lead Contamination, "Survey of Lead in the Atmosphere of Three Urban Communities" U.S.D.H.E.W., Public Health Service, Division of Air Pollution, Cincinnati, Ohio 45226 Jan 1965.

23. Tepper, L., Levin, L. "A Survey of Air and Population Lead Levels in Selected American Communities" Report submitted to EPA, June 1972.

24. Brief, R.S. "Air Lead Concentrations from Automotive Engines" Arch Environ. Health, Vol. 5 pp 527-531, Dec. 1962.

25. Chisolm, Jr., J.J., Kaplan, E., "Lead Poisoning in Childhood--Comprehensive Management and Prevention" J. Pediatrics, Vol 73, No.6, pp. 942-950, Dec 1968.

26. Guinee, V.F. "Lead Poisoning in New York City" Transactions of the New York Academy of Sciences, Series II, Vol.33, No.5 pp. 539-545, May 1971.

27. Letter to William D. Ruckelshaus, Administrator, U.S. E.P.A. dated March 9, 1973 from Louis V. Lombardo, President, Public Interest Campaign commenting on E.P.A.'s gasoline-lead removal program.

28. Center for Science in the Public Interest Newsletter, Vol. 2 #3.

29. de la Burde, B., Choate, Jr. M.S. "Does Asymtomatic Lead Exposure in Children have Latent Sequelae?" The J of Pediatrics, Vol. 81, No.6 pp. 1088-1091, Dec. 1972.

30. David, O., Clark, J., Voeller, K., "Lead and Hyperactivity" The Lancet, pp. 900-903, Oct. 28, 1972.

31. Bryce-Smith, D. "Behavioral Effects of Lead and Other Heavy Metal Pollutants." Chemistry in Britain, Vol.8, No. 6, June 1972.

32. Millar, J.A., Battistimi, V., Cumming, R.L. et al: "Lead and Delta Aminolevalinic Acid Dehydratase Levels in Mentally Retarded Children and in Lead-poisoned Suckling Rats" The Lancet 2, pp. 695-698, Oct. 3, 1970.

33. Weissberg, J.B., Lipschutz, F., Oski, F.A. "Delta-Amino-levulinic Acid Dehydratase Activity in Circulating Blood Cells" New Eng. J. of Med. 284: 565-569, March 18, 1971.

34. Schroeder, H.A., Balassa, J.J. and Vinton, Jr., W.H. "Chromium, Lead, Cadmium, Nickel and Titanium in Mice: Effect on Mortality, Tumors and Tissue Levels." J. Nutr. 83: 239-250, July 1964.

35. Schroeder, H.A., Balassa, J.J. and Vinton, Jr., W.H. "Chromium, Cadmium and Lead in Rats: Effects on Life Span, Tumors and Tissue Levels." J. Nutr. 86: 51-66, May 1965.

36. Schroeder, H.A., Mitchener, M. and Nason, A.P. "Zirconium, Niobium, Antimony, Vanadium and Lead in Rats: Life Term Studies." J. Nutr. 100: 59-68, Jan 1970.

37. Schroeder, H.A. and Mitchener, M. "Toxic Effects of Trace Elements on the Reproduction of Mice and Rats" Arch Environ Health 23: 102-106, Aug 1971.

38. Winick, M. "Nucleic Acid and Protein Content during Growth of Human Brain" Pediatrics Res. 2: 352, 1968.

39. Rosso, P., Hormazabal, J. and Winick, M. "Changes in Brain Weight, Cholesterol, Phospholipids, and DNA content in Marasmic Children. Amer J Clin Nutr Vol. 23, pp 1275-1279, Oct. 1970.

40. Fish, I. and Winick, M. "Cellular Growth in Various Regions of the Developing Rat Brain". Pediatrics Res. 3: 407-412, 1969.

41. Winick, M. and Noble, A. "Quantitative Changes in DNA, RNA and Protein During Prenatal and Post natal Growth in the Rat." Developmental Biology, Vol. 12, No. 3 Dec. 1965.

42. Schmidt, G. and Thanhauser, S.I. "A Method for the Determination of Desoxyribonucleic Acid, Ribonucleic Acid, and Phosphoproteins in Animal Tissue." J. Biol Chem 161: 83-89, 1945.

43. Lowry, D.H., Rosebrough, N.J., Farr, H.L., and Randall, R.J. "Protein Measurement with the Folin Phenol Reagent." J Biol Chem 193: 265-275, 1951.

44. Dische, Z. In: "The Nucleic Acids" (E. Chargaff and J.N.Davidson, eds) Vol.1, Academic Press, New York.

45. Darrow, D.K. and Clark, G. "Variations in DNA, RNA and Protein in Neonatal Mouse Cerebrum born of Lead and Mercury fed Mothers (submitted for publication).

46. Takeuchi, T. "Biological Reactions and Pathological Changes of Human Beings and Animals under the Conditions of Organic Mercury Contamination" International Conference on Environmental Mercury Contamination in Ann Arbor, 1970.

47. Snyder, R.D., "Congenital Mercury Poisoning" New Eng J of Med 284: 1014-1016, May 6, 1971.

48. Spyker, J.M., Sparker, S.B., Goldberk, A.M. "Subtile Consequences of Methyl Mercury Exposure: Behavioral Deviations in Offspring of Treated Mothers" Science 177: 621-622, Aug. 18 1972.

49. Goyer, R.A., Rhyne, B.C. "Pathological Effects of Lead" International Rev Exp Path Vol 12, pp. 1-77, 1973.

50. Gelman Bantam Model #19100, Gelman Instrument Co., Ann Arbor, Michigan.

51. Gelman Instrument Co., Ann Arbor, Michigan, Type E. DOP Test (0.3 μ) 99.7% retention efficiency, no detectable lead in an 8" x 10" sheet.

52. Konopinski, V.J., Upham, J.B. "Commuter Exposure to Atmospheric Lead" Arch Environ Health Vol. 14, pp. 589-593, April 1967.

53. Schroeder, H.A., Balassa, J. J. "Abnormal Trace Metals in Man: Lead" J Chron Dis 14: 408, 1961.

54. Sample placed in polyethylene 2 oz sample bottle, 3 ml sulfuric and 2 ml nitric added and allowed to sit at room temperature overnight. Caps were screwed tight and bottle inserted in tightly fitting metal containers made from standard 1-1/2 I.D. galvanized pipe nipple and 2 galvanized caps. The container with plastic sample bottle inside was brought up to 70° C in a drying oven for two hours, allowed to cool, and caps of sample bottles removed. Samples were left in oven overnight at 70° C and then brought up to 10 ml with H_2O.

21

CELLULAR EFFECTS OF LEAD

R. A. Goyer and J. F. Moore

Department of Pathology

University of North Carolina at Chapel Hill

The particular level of exposure to lead which results in an adverse health effect is related to many factors including "critical organ content," route of administration and a number of metabolic and nutritional factors (Goyer and Mahaffey, 1972). Important considerations related to toxic manifestations of a heavy metal are its distribution among subcellular components and interaction with cellular function and whether it is complexed to a diffusible ligand or is bound to a non-diffusible complex. This paper is a review of available information regarding the subcellular distribution of lead, effects on cellular organelles and the relationship of these effects to symptoms of lead toxicity, and is modified from an earlier working paper submitted to a task group on metal accumulations (Goyer, 1972).

SUBCELLULAR LOCALIZATION OF LEAD

A number of techniques have been employed for measuring the subcellular distribution of lead. Uptake of radioactive lead isotopes at different time intervals following intravenous administration provides a dynamic model whereby animals can be sacrified, and the subcellular distribution of labelled lead measured in cellular compartments following separation by differential centrifugation. Castellino and Aloj (1969) determined the distribution of lead in nuclei, mitochondria, microsomes and cell sap of liver and kidney of rats for 9 days after the intravenous injection of ^{210}Pb. Lead entered the cells within one hour after injection. In the liver within the first 24 to 72 hours there was a decrease in radioactivity in the nuclear fraction and an increase in mitochondria; microsomes and soluble fractions remained stable. These relative distributions remained constant for nine days. On the other hand, in the kidney relative increases of lead in both nuclear and mitochondrial fractions occurred during the first 24 hours with a decrease in microsomes and little change in the soluble fraction. After 24 hours the distribution reached a steady state. Mitochondrial binding of lead was remarkably strong and resisted many washings. None of the rats had been pre-exposed to excessive amounts of lead.

A similar experiment was reported by Barltrop and coworkers (1971). They measured the subcellular distribution of lead in heart, liver, kidney and spleen following intraperitoneal administration of ^{203}Pb. Cells were fractionated into nuclear, mitochondrial-lysosomal, microsomal and soluble fractions between 2 hours and 14 days after administration of the isotope. Renal tissue was further separated into lysosomal and mitochondrial fractions; most of the lead was in the mitochondrial fraction. However, pretreatment of the rats with lead diminished the in-vivo uptake of radioactive lead by the mitochondria suggesting that lead binding sites on the mitochondrion may be saturated. However, isolated rat liver mitochondria incubated with lead form lead-containing granules in the inner compartment (Walton, 1973). Formation of these granules requires the presence of substrate and ATP and is reduced in the presence of metabolic inhibitors such as actinomycin A or dinitrophenol suggesting that lead accumulation in mitochondria is an active, energy-dependent process.

Direct measurement of lead content of subcellular components of the rat kidney has shown that the major fraction of the lead is present in the nucleus of both control and lead toxic rats but during excessive exposure to lead by far the greatest increment is in the nucleus (Goyer et al., 1970b).

NUCLEAR INCLUSION BODIES

A characteristic cellular reaction in lead intoxication is the formation of discrete, dense staining, intranuclear inclusion bodies. These were initially observed by Blackman (1936) some 37 years ago in hepatic parenchymal cells and renal tubular lining cells of children dying of acute lead encephalopathy. More recently they have been reported in livers and renal tubular lining cells of swine (Watrach and Vatter, 1962), dogs (Stowe et al, 1973), rabbits (Hass et al., 1964), fowls (Simpson et al., 1970) and in kidneys only of rats (Goyer, 1968).

The ultrastructure of a typical inclusion body in the nucleus of a renal tubular lining cell of a lead-intoxicated rat consists of a dense central core and an outer fibrillary zone. They are always independent of nucleoli. Inclusion bodies vary considerably in size and in ratio of core to outer fibrils (Bracken et al., 1958; Beaver, 1961; Richter et al., 1968). Richter et al. (1968) estimate the thickness of the fibrils to range from 100 Å to 130 Å, although the size of fibrils varies with fixation and staining procedures. A fine structure to the fibrils (periodicity) has not been observed. The composition of lead-induced intranuclear inclusion bodies has been further studied by histochemistry, autoradiography and by direct analysis after isolation. Histochemical studies have been summarized by Richter et al. (1968). Inclusion bodies do not stain with the Feulgen reaction, although they are sometimes surrounded by Feulgen-positive material. They do not stain with fast green after treatment with trichloroacetic acid, but do stain strongly with mercurial bromophenol blue and by basic fuchsin suggesting they contain protein, but probably not histones. The basis of the acid-fast reaction is uncertain; it may simply be a physical phenomenon. Landing and Nakai (1959) showed that the inclusions contain at most small amounts of lipid and suggested the acid-fast properties of these bodies were related to the presence of sulfhydryl groups in the protein.

The presence of lead in the inclusion bodies was suspected but unproved for many years. Finner and Calvery (1939) stained sections of kidneys from lead poisoned rats with hydrogen sulfide and found particles of lead sulfide occurring in nuclei of tubular epithelia presumably corresponding to intranuclear inclusion bodies. Other attempts to identify the metal either histochemically or by x-ray diffraction have been unsuccessful (Watrach and Vatter, 1962). More recently, lead within the intranuclear inclusion body has been positively demonstrated by autoradiography (Dallenbach, 1965) and electron-probe x-ray microanalysis (Carroll et al., 1970).

The structure and composition of the inclusion bodies have been further studied after isolation by differential centrifugation and sonication (Goyer et al., 1970b). Treatment of inclusions in tissue sections (Richter et al., 1968) or isolated inclusion bodies (Goyer et al., 1970b) with proteolytic enzymes results in partial digestion, but they are not altered by incubation with DNAase or RNAase. Direct chemical analysis of isolated inclusion bodies confirms that they are composed of a lead-protein complex containing approximately 50 ug of lead per mg protein. Lead within the inclusion bodies is 60 to 100 times more concentrated than whole kidney lead (Goyer et al., 1970b; Horn, 1970). Amino acid composition and solubility characteristics of the inclusion body protein resemble those of the residual acidic fraction of proteins in normal nuclei (Goyer et al., 1971).

The protein has high content of aspartic and glutamic acids, glycine, cystine, and tryptophan. The inclusion bodies are insoluble in physiological media but may be dissolved in denaturants like 6 M urea and sodium deoxycholate. The lead is removable by incubation in-vitro with chelators like EDTA. In this circumstance the inclusion body loses its typical morphological structure and becomes an amorphous protein residue which continues to be insoluble (Moore et al., 1973). The inclusions also contain lesser amounts of other metals including calcium, iron, zinc, copper and cadmium in order of decreasing concentrations.

A scheme has been proposed for the possible role of the intranuclear inclusion body in lead poisoning (Goyer, 1971). It is suggested that the inclusion body serves as an adaptive or protective mechanism during transcellular transport of lead. In the course of excretion of lead from capillary to bile in hepatic cells, or by transtubular flow in renal tubular lining cells, a portion of the lead enters the nucleus where it becomes bound into a lead-protein complex and is no longer diffusible. This mechanism has the effect of maintaining a relatively low cytoplasmic concentration of lead and, therefore, reducing the toxic effects of lead on sensitive cellular functions, such as mitochondrial respiration and protein synthesis.

Indirect support for this hypothesis is found in experimental studies of the relationship of the formation of intranuclear inclusion bodies and other renal effects of lead. When groups of rats are given different concentrations of lead in their drinking water, intranuclear inclusion bodies are observed at a lower dose than is any other renal effect of lead; in fact, at a lower dose of lead than that which produced signs and symptoms of lead toxicity. Inclusion bodies appear to be related to renal lead content and discernible in the rat models when kidney lead concentration is between 10 and 20 ug/g wet weight.

The lead-induced nuclear inclusion body may also facilitate urinary excretion of lead. They are found in the urine of children with acute lead poisoning (Landing and Nakai, 1959). In a recent study of renal biopsy material from five lead industry workers with varying times of exposure to lead from 16 weeks to more than 20 years, typical lead-induced inclusions were only found in nuclei of workers employed for less than one year (Cramer et al., in press). Urinary excretion of lead was higher in those workers with inclusion bodies than in older workers without such complexes although other symptoms of lead toxicity were more marked in the latter group. Whether the absence of inclusion bodies in these workmen reflects repeated chelation therapy or implies an alternative cellular metabolism of lead is not known.

A similar study of renal biopsies from 23 lead industry workers with varying periods of exposure to lead has been reported by Richet and coworkers (1964, 1966). Although some form of a nuclear body was found in 17 of the 23 renal biopsies, only one showed the typical morphology of the characteristic lead-induced inclusion body and this was the only individual in the group with a history of short (3 months) exposure to lead. Other nuclear bodies occurring in lead toxic workers resembled granular chromatin masses which occur in a number of pathological conditions characterized by cellular hyperactivity (Bouteille et al., 1967).

Inclusion bodies occur in nuclei of osteoclasts of lead poisoned pigs (Hsu et al., 1973). Their formation is augmented by lowering dietary calcium, suggesting that lead is reabsorbed during osteolysis similarly to calcium. Lead becomes bound into the non-diffusible inclusion body.

The lead-protein inclusion body phenomenon is not restricted to animal cell nuclei but also occurs in plant cells. Bryophytes (moss) are known to accumulate heavy metals in concentrations which are toxic for other groups of plants. It has recently been shown that moss cultivated in a greenhouse and watered daily with lead acetate solutions forms lead containing inclusion bodies in nuclei of leaf cells (Skaar et al., 1973). Similar inclusion bodies may occur in root cells of corn grown in soil with added lead (personal communication, Dr. D. E. Koeppe, University of Illinois, Urbana, Illinois).

EFFECTS OF LEAD ON CELL DIVISION AND PROTEIN SYNTHESIS

Although lead may be concentrated in nuclei in the form of a protein complex, the metabolic effect of lead localization on nuclear function, particularly cellular division, is uncertain. The finding of granular chromatin bodies in biopsies of lead workers suggests increased cell turnover time. Recent studies (Choie and Richter, 1972a) have shown that a single intraperitoneal dose of lead will greatly enhance DNA synthesis and proliferation of renal tubular cells within 2 days and repeated injections given in 2-day intervals will stimulate successive waves of nuclear DNA synthesis in tubular epithelial cells (Choie and Richter, 1972b). This finding is of particular interest since lead does have a carcinogenic effect in rats and mice after long-term intoxication. On the other hand, a carcinogenic effect for lead has not been observed in man or other species of animals.

Excessive exposure to lead does produce structural alterations in the chromosomes of man and experimental animals. Nuclear polyploidy and abnormalities in mitosis have been noted in bone marrow cells by a number of authors (Waldron, 1966) but chromosomal aberrations in lead intoxication have only recently been recognized. Chromosomes from leukocyte cultures from mice fed a diet containing 1% lead acetate show an increased number of gap-break type aberrations usually involving only single chromatids (Muro and Goyer,

1969). Similar gap-break types were also found in lead workers with blood lead levels ranging from 62 to 88 ug/100 g (Schwantitz et al., 1970). In addition, a variety of non-specific changes in chromosome morphology was also present such as adhesions and spiralizing defects. Tetraploid mitosis and the mitotic index were also increased in the workers. Moreover, the percentage of abnormal mitoses correlated very well with urinary d-ALA excretion. Similar studies have not been performed on acutely intoxicated children nor is the mechanism for this lead effect known. Lead, like other heavy metals, does have an affinity for nucleic acid binding but whether lead becomes biochemically bound to chromosomes *in vivo* is not known.

The effect of lead binding to nuclear proteins on cell division is also incompletely understood.

Lead is bound to insoluble nuclear proteins as discussed in the formation of nuclear inclusion bodies. Lead is also associated with NaOH soluble acidic proteins. Acidic nuclear proteins are implicated in the regulation of DNA replication and influence DNA-dependent RNA synthesis (Stein and Baserga, 1972).

There is some information to suggest that lead may affect protein synthesis at the levelof the ribosome. Waxman and Rabinovitz (1966) and Ulmer and Vallee (1969) observed that lead causes disaggregation of polyribosomes. The latter workers also reported inhibition by lead of leucine incorporation into t-RNA in *E. coli*. Purified RNA is depolymerized by lead (Farkas, 1968).

EFFECTS OF LEAD ON MITOCHONDRIA

The affinity and relatively strong binding of lead to mitochondrial membranes suggests that this organelle is a particularly susceptible target of the subcellular effects of lead.

Lead toxicity in humans and experimental animals is associated with alterations in ultrastructure and function of mitochondria in hepatic parenchymal cells, renal proximal tubular lining cells and reticulocytes in the bone marrow. The principle clinical manifestations of lead toxicity involve these organs as well as the central nervous system. Although disturbances in mitochondrial structure and function are not the only subcellular effects of lead, it is suggested that the *in vivo* effect of lead on these organelles contributes to the anemia of lead poisoning and impaired renal tubular function. The relationship of the mitochondrial effects of lead to other symptoms of lead toxicity are less well understood.

Mitochondria in proximal renal tubular lining cells of children, adults and experimental animals with lead poisoning show varying degrees of swelling with margination of cristae (Angevine et al., 1962; Cohen et al., 1960; and Goyer, 1968) probably secondary to increased membrane permeability. Such changes are similar to the non-specific swelling that occurs in the early stages of other forms of cellular injury and is probably reversible. The fate of such mitochondria is uncertain but increased numbers of myelin figures and autophagocytosis suggest an increased rate of turnover. Mitochondria in cells of renal biopsy specimens from lead industry workers show evidence of budding and division. Lead induced granules or inclusion bodies similar to those formed in nuclei may be produced in vitro in isolated liver mitochondria (Walton, 1973) but these have not been noted in vivo.

Other types of intramitochondrial inclusions have been observed but are probably not specific for lead poisoning. Watrach (1964) has described an intramitochondrial lamellar type crystalloid in liver parenchymal cells of lead poisoned swine. These are composed of an intersecting band of fine, loosely packed arrays that run parallel and perpendicular to form a lattice pattern. Such lamellar formations are characteristic of chronic cellular injury accompanying a number of metabolic abnormalities such as diabetes mellitus, alcohol intoxication and biliary obstruction as well as experimentally induced hypoxia (Shiraki and Neustein, 1971).

From the observed alterations in mitochondrial ultrastructure some impairment of functional ability is suggested. Nevertheless, mitochondria from cells of subjects with chronic intoxication are expected to have some degree of respiratory ability since cellular necrosis or irreversible cellular injury is not a prominent morphologic feature. A decreased level of mitochondrial function is compatible with the concept of an altered functional steady state or lowered level of cellular activity.

Measurement of respiratory activity of mitochondria from liver or kidney of lead-treated experimental animals confirms this concept. Teras and Kakhn (1967) found decreased respiratory rates in mitochondria of rabbits treated with daily intraperitoneal injections of lead acetate (10 to 20 mg/kg body weight) for seven months. ADP stimulated respiration was lower for mitochondria from lead-treated rabbits in the presence of three different substrates: α-ketoglutarate, pyruvate and succinate. Phosphorylation was also decreased. The authors suggest respiratory impairment is most pronounced using pyruvate as substrate.

Studies with mitochondria isolated from kidneys of lead poisoned rats show a specific impairment in respiratory and phosphorylative abilities in pyruvate-dependent respiration as evidenced by altered respiratory control (RCR) and ADP:O ratios. Slightly higher State IV and State III rates of toxic mitochondria in the presence of pyruvate suggest partial uncoupling of phosphorylation. These functional parameters which are reflected by RCR and ADP:O are not different from the control mitochondria in succinate-dependent respiration (Goyer and Krall, 1969a). Since succinate oxidation and related electron transport by-pass the pyruvate-NAD reductase enzyme complex, it is suggested that the NAD reductase complex is more sensitive to lead. A more recent report indicates that magnesium restores oxidative ability and dinitrophenol sensitivity in lead toxic mitochondria for NAD coupled respiration, but it has no effect on succinate-supported respiration (Krall and Dougherty, 1971). Decreased oxygen uptake rate for both State III and IV in succinate-supported respiration in lead toxic mitochondria may be evidence of decreased succinoxidase enzyme which correlates with decreased mitochondrial protein (Rhyne and Goyer, 1971).

Ulmer and Vallee (1969) conclude that lead acts specifically with dithiol bonds in the lipoamide dehydrogenase in the pyruvate dehydrogenase system, but this was validated by *in vitro* work only. This site is sensitive to other metals such as mercury and cadmium but it is uncertain whether this site is accessible to lead in the *in vivo* system.

The effect of lead on mitochondria of bone marrow reticulocytes is of particular interest because of the resultant impairment in heme synthesis and subsequent clinical anemia. It also serves as a fairly well defined example of a toxic change in an intramitochondrial enzyme system.

At least three steps in heme synthesis may be affected by lead. Inhibition of d-aminolevulinic acid dehydratase (ALA-D) is probably the enzyme in the heme pathway that is most sensitive to lead. Inhibition of this enzyme results in a block in utilization of d-aminolevulinic acid (d-ALA) and subsequent decline in heme synthesis. Under the proposed scheme of negative feed-back control of heme synthesis as proposed by Granick and Levere (1964), d-ALA synthetase activity is derepressed resulting in increased activity and synthesis of d-ALA. This may be regarded as a second effect of lead on heme synthesis. Since utilization of d-ALA is blocked, urinary excretion is increased. Assay of urinary d-ALA serves as a sensitive index of a biochemical lead effect and is widely used as an indication of lead toxicity (Haeger-Aronsen, 1960).

A third abnormality in heme synthesis described in lead intoxication is inhibition of the enzyme, ferrochelatase (heme synthetase) (Otrzonsek, 1967) which is located on the inner membrane of mitochondria (Jones and Jones, 1968). This effect is particularly interesting because of associated ultrastructural changes in the mitochondrion. Ferrochelatase catalyzes the incorporation of the ferrous ion into the porphyrin ring structure. Bessis and Jensen (1965) have shown that iron in the form of apoferritin and ferruginous micelles accumulates in mitochondria of bone marrow reticulocytes from lead poisoned rats.

Gibson and Goldberg (1970) recently reported no inhibition of this enzyme (ferrochelatase) in homogenates prepared from liver of lead toxic rabbits, but when they added lead acetate to the homogenate in concentrations similar to those present in the original unhomogenized tissues, a marked inhibition occurred. Lead associated with the enzyme _in vivo_ may be so loosely bound as to be lost in the isolation procedure. Nevertheless, the experiment does show that ferrochelatase is readily inhibited by lead.

Other steps in the biosynthetic pathway of heme may also be abnormal in lead toxicity but the evidence for these is incomplete. Increase in urinary excretion of coproporphyrin, the degradative product of coproporphyrinogen III, is a sensitive reflection of lead toxicity. Metabolism of porphobilinogen to coproporphyrinogen proceeds unimpaired. However, since ALA-D is an extra-mitochondrial enzyme, and later steps in heme synthesis are intramitochondrial, coproporphyrinogen must re-enter the mitochondrion to be metabolized further. It has been suggested that transport of metabolites like coproporphyrinogen into the mitochondrial matrix might be impaired by altered inner membrane permeability, or decreased efficiency in oxidation and phosphorylation (Haeger-Aronsen et al., 1968). Whether the increased urinary coproporphyrinogen occurring in lead poisoning is a reflection of a non-specific alteration in mitochondrial membrane or reflects a more specific effect of lead on the intramitochondrial enzyme coproporphyrinogenase, is not known.

The principal clinical manifestations of lead toxicity involve three organ systems: the central nervous system, hematopoietic system and the kidney. A relationship between the mitochondrial effect of lead and the central nervous system pathology has not been demonstrated. However, functional impairment of mitochondria of reticulocytes is responsible, at least in part, for the anemia of lead poisoning, and lead toxic mitochondria of reticulocytes are responsible, at least in part, for the anemia of lead poisoning, and lead toxic mitochondria in proximal renal tubular lining cells must decrease the functional ability of these cells. Transport of a number of substances including sodium, potassium, phosphate ions and small organic molecules such as amino acids and glucose occurs in this portion of the renal tubule and is an active process dependent on ATP largely derived from mitochondrial oxidative phosphorylation.

In acute lead poisoning, particularly in children, excessive excretion of amino acids, glucose and phosphate occurs (Chisolm, 1962). Whether reduced reabsorption of these substances in lead poisoning is entirely related to reduced energy production by lead toxic mitochondria or is also a consequence of a lead effect on other cellular membranes is uncertain. Most of the energy synthesized in proximal tubular lining cells is required for sodium transport. It is interesting that Sandstead et al. (1970) have recently shown that men with chronic lead poisoning, following excessive industrial exposure to lead or ingestion of lead-contaminated illicit moonshine, do have reduced sodium reabsorption. This defect, like the aminoaciduria, is reversible by treatment of the lead poisoning with chelating agents (EDTA).

Respiratory rates of bone marrow reticulocytes from lead intoxicated rats are less than normal (Lessler et al., 1968), probably reflecting impaired mitochondrial respiration. Also some of the enzymes associated with heme synthesis are located either within the inner mitochondrial membrane of matrix or are in close association with this organelle and are dependent on intact mitochondrial respiration. Some impairment of heme enzyme function may follow the lead effect on respiration.

The effect of lead on cellular content of other heme proteins such as cytochromes, catalase and peroxidases has received limited study. Prader and Vannotti in an attempt to compare effects of lead poisoning on hemoglobin and other cellular porphyrin moieties measured levels of hemoglobin and cytochrome c in rabbit liver. In contrast to decrease in the hemoglobin values cytochrome c level remained constant (Vannotti, 1955). However, measurement of the total cytochrome content of rat kidney mitochondria by difference spectra showed a selective decrease in cytochrome aa_3 in lead poisoning (Rhyne and Goyer, 1971). Whether this decrease resulted from inhibition of mitochondrial protein synthesis which is reflected by decrease in cytochrome aa_3 or specific inhibition of the cytochrome aa_3 heme moiety has not been resolved. The protein moiety as well as the heme moiety differs for each class of cytochromes so inhibition of either fraction would result in a decrease of the cytochrome.

LEAD AND LYSOSOMES

The radioisotope studies of Barltrop and coworkers (1971) suggest that lead has less affinity for lysosomes than other subcellular organelles, particularly mitochondria and nuclei; nor does lead intoxication in the rat provoke any remarkable increase in lysosomal enzymes (Krigman et al., 1968). The origin of the lead-induced increase in urinary β-glucuronidase noted by these workers is uncertain but may be derived from a lead effect on endoplasmic reticulum (Krigman et al., 1970). Autophagocytosis of lead containing mitochondria may result in lead containing secondary lysosomes as suggested from observation of biopsy specimens from lead workers (Cramer et al., in press). Also, lead has been identified by x-ray microanalysis in dense cytoplasmic concretions in kidney cells of rats administered large doses of lead intraperitoneally (Murakami, 1971). Such concretions have not been found following oral administration of lead in rats and in human biopsy material.

SUMMARY AND CONCLUSIONS

The intracellular localization of lead has been best studied in liver and kidney, the two organs most concerned with transcellular transport and excretion of lead. Within the cell lead has the greatest affinity for mitochondrial membranes and nuclei. Nuclear lead binds with an acidic protein fraction and accumulates to form a discrete inclusion body which has a characteristic ultrastructure. When the renal tubular cell is presented with even larger concentrations of lead intracytoplasmic concretions are formed. The relationship of these concretions to the lysosomal system has not been demonstrated but dense staining bodies do form in secondary lysosomes derived from degenerating lead toxic mitochondria.

The effect of lead on membrane enzyme systems and biochemical activity of subcellular organelles is related to the pattern of intracellular partitioning. The lead effect on mitochondria decreases the ability of cells to form specific functions, e.g. transport functions in proximal renal tubular lining cells. Sensitivity of enzymes concerned with heme synthesis results in anemia as well as increased urinary excretion of intermediate products of the heme synthesizing pathway.

Measurable amounts of lead are present in most cells of persons in the general population without apparent adverse effect. With an increase in lead burden the transition from no effect to toxicity is probably not sharply definable but reflects a spectrum of cellular events, some compensatory, some pathological. Knowledge of the intracellular distribution of lead and its effect on subcellular components should improve the ability to discern between a tolerable and toxic lead burden.

REFERENCES

Angevine, J.M., Kappas, A., Degowin, R.L., Spargo, B.H.: Renal tubular nuclear inclusions in lead poisoning. A clinical and experimental study. Arch. Pathol. 73:486-494, 1962.

Barltrop, D., Barrett, A.J., Dingle, J.T.: Subcellular distribution of lead in the rat. J. Lab. Clin. Med. 77:705-712, 1971.

Beaver, D.L.: The ultrastructure of the kidney in lead intoxication with particular reference to intranuclear inclusions. Am. J. Pathol. 39:195-208, 1961.

Bessis, M.C., Jensen, W.N.: Sideroblastic anaemia, mitochondria and erythroblastic iron. Br. J. Haematol. 11:49-51, 1965.

Blackman, S.S., Jr.: Intranuclear inclusion bodies in kidney and liver caused by lead poisoning. Bull. Johns Hopkins Hospital 58:384-403, 1936.

Bouteille, M., Kalifat, S.F., Delareue, J.: Ultrastructural variations of nuclear bodies in human diseases. J. Ultrastructural Res. 19:474-486, 1967.

Bracken, E.C., Beaver, D.L., Randall, C.C.: Histochemical studies of viral and lead-induced intranuclear bodies. J. Path. Bact. (Lond) 75:253-256, 1958.

Carroll, K.G., Spinelli, F.R., Goyer, R.A.: Electron probe microanalyzer localization of lead in kidney tissue of poisoned rats. Nature (Lond) 227:1056, 1970.

Castellino, N., Aloj, S.: Intracellular distribution of lead in the liver and kidney of the rat. Br. J. Ind. Med. 26:139-143, 1969.

Chisolm, J.J., Jr.: Aminoaciduria as a manifestation of renal tubular injury in lead intoxication and a comparison with patterns of aminoaciduria seen in other diseases. J. Pediatr. 60:1-17, 1962.

Choie, D.D., Richter, G.W.: Cell proliferation in rat kidney induced by lead acetate and effects of uninephrectomy on the proliferation. Am. J. Path. 66:265-275, 1972a.

Choie, D.D., Richter, G.W.: Stimulation of DNA synthesis in rat kidney by repeated administration of lead. Proc. Soc. Exp. Biol. Med. 142:446-449, 1972b.

Cohen, S., Sweet, A.Y., Mautner, W., Churg, J., Grisham, E.: Light and electron microscopy of lead nephropathy. Am. J. Dis. Child. 100:559-560 (abs), 1960.

Cramer, K., Goyer, R.A., Jagenburg, O.R., Wilson, M.H.: Renal ultrastructure, renal function and parameters of lead toxicity in workers with different lengths of lead exposure. Brit. J. Ind. Med., in press.

Dallenbach, F.: Die aufnahme von radioactivem blei 210 durch die tubulusepithelien der niere. Verh. Dtsch. Ges. Pathaol. 49: 179-185, 1965.

Dingwall-Fordyce, I., Lane, R.E.: A follow-up study of lead workers. Br. J. Ind. Med. 20:313-315, 1963.

Farkas, W.R.: Depolymerization of ribonucleic acid by plumbous ion. Biochim. Biophys. Acta 155:401-409, 1968.

Finner, L.L., Calvery, H.O.: Pathologic changes in rats and in dogs fed diets containing lead and arsenic compounds. Arch. Pathol. 27:433-446, 1939.

Gibson, S.L.M., Goldberg, A.: Defects in haem synthesis in mammalian tissues in experimental lead poisoning and experimental porphyria. Clin. Sci. 38:63-72, 1970.

Goyer, R.A.: The renal tubule in lead poisoning. I. Mitochondrial swelling and aminoaciduria. Lab. Invest. 19:71-77, 1968.

Goyer, R.A.: Lead toxicity: a problem in environmental pathology. Am. J. Pathol. 64:167-182, 1971.

Goyer, R.A.: Cellular and subcellular effects of lead. Working paper for task group on metal accumulation. Proceedings of XVIIth International Congress of Occupational Health, Buenos Aires, Argentina, September, 1972.

Goyer, R.A., Krall, A.R.: Further observations on the morphology and biochemistry of mitochondria from kidneys of normal and lead-intoxicated rats. Fed. Proc. 28:619A, 1969a.

Goyer, R.A., Krall, R.: Ultrastructural transformation in mitochondria isolated from kidneys of normal and lead intoxicated rats. J. Cell Biol. 41:393-400, 1969b.

Goyer, R.A., Leonard, D.L., Moore, J.F., Rhyne, B., Krigman, M.R.: Lead dosage and the role of the intranuclear inclusion body. Arch. Environ. Health 20:705-711, 1970a.

Goyer, R.A., Mahaffey, K.R.: Susceptibility to lead toxicity. Environ. Health Perspect. Exptl. Issue No. 2:73-80, 1972.

Goyer, R.A., May, P., Cates, M., Krigman, M.R.: Lead and protein content of isolated intranuclear inclusion bodies from kidneys of lead-poisoned rats. Lab. Invest. 22:245-251, 1970b.

Goyer, R.A., Moore, J.F., Barrow, E.M.: Lead binding protein in the lead-induced intranuclear inclusion body. Am. J. Pathol. 62:96-97 (abs), 1971.

Granick, S., Levere, R.D.: Heme synthesis in erythroid cells. Progr. Hematol. 4:1-47, 1964.

Haeger-Aronsen, B.: Studies on urinary excretion of 5-aminolaevulinic acid and other haem precursors in lead workers and lead-intoxicated rabbits. Scand. J. Clin. Lab. Invest. Suppl. 47:1-128, 1960.

Haeger-Aronsen, B., Stathers, G., Swahn, G.: Hereditary coproporphyria. Study of a Swedish family. Ann. Intern. Med. 69: 221-227, 1968.

Hass, G.M., Brown, D.V.L., Eisenstein, R., Hemmens, A.: Relations between lead poisoning in rabbit and man. Am. J. Pathol. 45:691-728, 1964.

Horn, J.: Isolation and examination of inclusion bodies of the rat kidney after chronic lead poisoning. Virchows Arch. (A) B:Z 6:313-317, 1970.

Hsu, F.S., Krook, L., Shively, J.N., Duncan, J.R., Pond, W.G.: Lead inclusion bodies in osteoclasts. Science 181:447-448, 1973.

Jones, M.S., Jones, O.T.G.: Evidence for the location of ferrochelatase on the inner membrane of rat liver mitochondria. Biochem. Biophys. Res. Commun. 31:977-982, 1968,

Krall, A.R., Dougherty, W.J.: Effects of lead and magnesium on function and on structural changes of rat kidney mitochondria. Fed. Proc. 30:637 (abs), 1971.

Krigman, M.R., Crane, D., Goyer, R.A.: Lysosomal alterations in lead nephropathy. Fed. Proc. 27:410 (abs), 1968.

Krigman, M.R., Shearin, H., Goyer, R.A.: Distribution of selective enzymes in lead nephropathy. Am. J. Pathol. 59:56 (abs), 1970.

Landing, B.H., Nakai, M.D.: Histochemical properties of renal lead inclusions and their demonstration in urinary sediment. Am. J. Clin. Path. 31:499-506, 1959.

Lessler, M.A., Cardona, E., Padilla, F., Jensen, W.N.: Effect of lead on reticulocyte respiratory activity. J. Cell Biol. 39: 171, 1968.

Moore, J.F., Goyer, R.A., Wilson, M.H.: Lead-induced inclusion bodies: solubility, amino acid content, and relationship to residual acidic nuclear proteins. Lab. Invest., in press.

Murakami, M.: Identification of elements in concretions in the kidney of rats with experimental lead poisoning. Jap. J. Ind. Health 9:194-197, 1971.

Muro, L.A., Goyer, R.A.: Chromosome damage in experimental lead poisoning. Arch. Pathol. 87:660-663, 1969.

Otrzonsek, N.: Die aktivitat der protoham ferro-lyase in Rattenleber und Knochenmark bei experimenteller Bleivergiftung. Int. Arch. Gewerbepathol. Gewerbehyg. 24:66-73, 1967.

Rhyne, B.C., Goyer, R.A.: Cytochrome content of kidney mitochondria in experimental lead poisoning. Exp. Mol. Pathol. 14: 386-391, 1971.

Richet, G.C., Albahary, R., Ardaillou, C., Sultan, C., Morel-Maroger, A.: Le rein du saturnisme chronique. Rev. Fr. Etud. Clin. Biol. 9:188-196, 1964.

Richet, G.C., Albahary, L., Morel-Maroger, L., Guillaume, P., Galle, P.: Les alterations renales dans 23 cas de saturnisme professionnel. Bull. Soc. Med. Hop. Paris 117:441-466, 1966.

Sandstead, H.H., Michelakis, A.M., Temple, T.E.: Lead intoxication. Its effect on the renin-aldosterone response to sodium deprivation. Arch. Environ. Health 20:356-363, 1970.

Schwantitz, G., Lehnert, G., Gebhart, E.: Chromosome damage after occupational exposure to lead. Dtsch. Med. Wochenschr. 95: 1636-1641, 1970.

Shiraki, K., Neustein, H.B.: Intramitochondrial crystalloids and amorphous granules. Arch. Pathol. 91:32-40, 1971.

Simpson, C.F., Damron, B.L., Harms, R.H.: Abnormalities of erythrocytes and renal tubules of chicks poisoned with lead. Am. J. Vet. Res. 31:515-523, 1970.

Skaar, H., Ophus, E., Gullvåg, B.M.: Lead accumulation within nuclei of moss leaf cells. Nature 241:215-216, 1973.

Stein, G., Baserga, R.: Nuclear proteins and the cell cycle. Adv. Cancer Res. 15:287-318, 1972.

Stowe, H.D., Goyer, R.A., Krigman, M.R., Wilson, M., Cates, M.:

Clinical and morphological effects of experimental oral lead toxicity in young dogs. Arch. Path. 95:106-116, 1973.

Teras, L.E., Kakhn, K.H.A.: Oxidative metabolism and phosphorylation of the liver in lead poisoning. Vopr. Med. Khim. 12: 41-44, 1966.

Ulmer, D.D., Vallee, D.L.: Effects of lead on biochemical systems, in Hemphill DD (ed): Trace Substances in Environmental Health, II. Columbia, Mo., University of Missouri Press, 1969, pp. 7-27.

Vannoti, A.: Experimental studies of porphyrin metabolism in cytochrome C synthesis, in Wolstenholme GEW and Millar ECP (eds): Porphyrin Biosynthesis and Metabolism. Boston, Little, Brown & Co., 1955, pp. 128-140.

Wachstein, M.: Studies on inclusion bodies. I. Acid-fastness of nuclear inclusion bodies that are induced by ingestion of lead and bismuth. Am. J. Clin. Pathol. 19:608-614, 1949.

Waldron, H.A.: The anaemia of lead poisoning: A review. Br. J. Ind. Med. 23:83-100, 1966.

Walton, J.R.: Granules containing lead in isolated mitochondria. Nature 243:100-101, 1973.

Watrach, H.M.: Degeneration of mitochondria in lead poisoning. J. Ultrastruct. Res. 10:177-181, 1964.

Watrach, H.M., Vatter, A.E.: The nature of inclusion bodies of lead poisoning, in: Electron Microscopy Proceedings International Congress—5th. Philadelphia, 1962, pp. VV-11.

Waxman, H.S., Rabinovitz, M.: Control of reticulocyte polyribosome content and hemoglobin synthesis by heme. Biochim. Biophys. Acta 129:369-379, 1966.

22

DISTRIBUTION, TISSUE BINDING AND TOXICITY OF MERCURIALS

J. T. MacGregor[1] and T. W. Clarkson[2]

U.S. Department of Agriculture, Western Regional Research Center, Albany, Ca. 94710[1] and the University of Rochester, School of Medicine and Dentistry, Rochester, New York 14620[2]

The development of the environmental mercury problem and the toxicity of mercury and its compounds have received attention in numerous recent reviews (27, 28, 29, 39, 42, 43, 48, 76, 81, 98, 103, 125). Mercury and cinnabar have been known since at least 1500 B.C. and mercury was a familiar substance by the first century B.C. (35, 75). The toxicity of mercury was mentioned by a number of ancient writers and the occupational hazards of mercury were described in detail by Paracelsus in 1533 (41, 53). Ramazzini (97), who in 1700 published De Morbis Artificium, the first comprehensive treatise on occupational diseases, vividly describes the maladies afflicting miners, gilders, mirror-makers and physicians who were exposed to mercury in the course of their work. Although there is a continuing problem of occupational exposure to inorganic mercury* (33, 39), possible effects on broader segments of the population due to the widespread dispersal of mercury in the environment has recently become the focus of concern. The Minamata disaster in the 1950's, in which forty-six deaths occurred among the 121 inhabitants poisoned by methylmercury which had accumulated in the fish of Minamata Bay as the result of the mercury discharged from a chemical plant employing a mercury catalyst (59, 60, 77), dramatically exposed the potential hazards to the general population. Several

*In accord with the definitions of the 1969 International MAC Committee (99), "inorganic mercury" refers to mercury in the form of elemental vapor, mercurous and mercuric salts, and those complexes in which mercuric ions can form reversible bonds to such tissue ligands as thiol groups on proteins; "organic mercury" refers to mercury in those compounds in which mercury is directly linked to a carbon atom by a covalent bond.

other outbreaks of organic mercury poisoning among human populations due to industrial discharge or to poisoning resulting from the use of alkylmercury compounds as seed dressing have occurred since the Minamata episode. Forty-seven cases of methylmercury poisoning and six deaths occurred along the Agano River in Niigata, Japan from 1965 to 1970 as the result of methylmercury discharges from a chemical plant; in 1969 seven persons in New Mexico were poisoned by the ingestion of pork from animals which had eaten seed grain treated with methylmercury dicyandiamide (60); in the 1950's and 1960's between 1200 and 1300 cases of alkylmercury poisoning had been reported in Iraq and other parts of the World; and a recent methylmercury epidemic in Iraq, the most catastrophic epidemic yet recorded, involved 6,530 cases admitted to hospitals and 459 hospitals deaths to date (5). During the 1960's accumulation of mercury was noted in wild life and fish in Sweden (35, 81, 117) and it was becoming apparent that the mercury problem was far more extensive than an isolated acute local situation such as Minamata or Niigata. It was subsequently learned that virtually all the mercury accumulated in the fish was the extremely hazardous methylmercury (see 117, 121) and that inorganic mercury was methylated by anaerobic bacteria in the sediment of waterways (58, 125). The finding that fresh water fish in the Great Lakes of the United States were similarly contaminated rocketed the mercury problem to the forefront of national concern and the governments of both the United States and Canada in 1970 banned the sale of fish containing levels of mercury exceeding 0.5 ppm (103). The contamination of fish with methylmercury is a continuing cause of serious concern.

In spite of our long history of concern about mercury toxicity and the recent flood of experimental evidence spurred by the developments discussed above, we are still largely ignorant of the basic biochemical and physiological mechanisms by which mercury compounds interact with the organism and produce a toxic effect. Integrally related to the effects of mercury in the organism are the factors governing its absorption and distribution to the organs and tissues and finally to the specific subcellular sites where the compound chemically interacts with its ultimate target and exerts deleterious effects. The present discussion will be concerned with the extent of our knowledge concerning the interaction of mercury with tissue constituents, the relevance of these interactions to the distribution within the animal and within the target cells, and the relation of these factors to the toxic manifestations of mercury. There are, of course, numerous forms of mercury and although some generalizations may be drawn, each is really a specific case in itself. This discussion will primarily concern itself with three forms of mercury of current toxicological importance: Methylmercury because of its obvious significance as a widespread environmental hazard; mercury vapor because of its continuing importance as an occupational hazard; and Hg^{++} because this form is a potential metabolite in vivo of all the major classes of organic and inorganic mercurials.

I. FACTORS INFLUENCING THE INTERACTION OF MERCURIALS WITH BIOLOGICAL MACROMOLECULES

A. Chemistry

The chemistry of mercury and its compounds has been outlined in standard chemistry texts (32, 44, 100). A few properties of relevance to the present discussion are briefly summarized below. Mercury, along with cadmium and zinc, falls into Group IIb of the periodic table. These elements have two electrons beyond a completed d subshell, and the removal of more than these two electrons almost never occurs because of the stability of the "pseudo-rare-gas structure" which remains after the loss of the outer two s electrons. Although the plus two oxidation state is predominant in Group IIb, mercury also exists in the plus one oxidation state. Hg_2^{++} is readily obtained by reduction of mercuric salts and is easily oxidized to the +2 state. Compared with mercuric chloride ($HgCl_2$), mercurous chloride (Hg_2Cl_2) is relatively insoluble. Hg_2^{++} forms few complexes, partly because the greater ability of Hg^{++} than of Hg_2^{++} to form stable complexes with most ligands favors the disproportionation of Hg_2^{++}. Hg° is unusually volatile for a heavy metal and is the only one which is a liquid at room temperature. The vapor is monatomic and the vapor pressure of 1.7×10^{-3} mm at 24° presents a hazard, for elemental mercury in equilibrium with its vapor at this temperature results in a concentration of 18.3 mg Hg/m^3, 366 times the average permissible concentration of 0.05 mg/m^3 recommended for occupational exposure by the National Institute for Occupational Safety and Health (33). The metal is surprisingly soluble in both polar and nonpolar liquids. The saturated solution in air free water at 25° is approximately 60 micrograms per liter (32) and we shall see that this has important implications for the transport of mercury into brain during vapor exposure. Mercury is not a transition element because its compounds exhibit no valence in which the d shell is not full, but it resembles the transition elements in its ability to form complexes with various ligands. The mercuric ion complexes are usually orders of magnitude more stable than the corresponding Zn^{++} or Cd^{++} complexes. The characteristic coordination numbers are 2-coordinate linear and 4-coordinate tetrahedral. Octahedral coordination is less common and five coordinate complexes have been described. $HgCl^+$, $HgCl_2$, $HgCl_3^-$ and $HgCl_4^=$ are the most important chloride complexes and at the chloride concentration of serum there are almost equal concentrations of $HgCl_2$, $HgCl_3^-$ and $HgCl_4^=$ (118). The first two ligands are bound much more tightly than the next two, but the ability to exhibit higher coordination numbers may well play a role in the stability of chelates formed in biological systems. The Hg-Cl bond has extreme covalent character and $HgCl_2$ has appreciable solubility in organic solvents (4.55 grams dissolve in 100 ml of ether (118)). $Hg(OH)_2$ is an extremely weak base and aqueous solutions of $SO_4^=$ and NO_3^- salts hydrolyze extensively in water. Mercuric ion adds to olefins readily* but complexing agents reverse this reaction and because of the abundance of biological ligands this type of reaction could not play an important role in the biological effects of mercuric ions.

In contrast with zinc and cadmium, mercury forms a large number of organometallic compounds of the type R-HgX which are stable to air and water. This unusual stability is not a function of higher bond strength, since the mercury carbon bond strengths are in the range of 15-20 kcal/mole and are actually weaker than the zinc-carbon and cadmium-carbon bonds, but is attributed to the very low affinity of mercury for oxygen. If X is NO_3^- or $SO_4^=$ the substance tends to be salt-like, but the chlorides are covalent, non-polar compounds more soluble in organic compounds than water. The organomercurial ions tend to form complexes with various ligands in the same manner as mercuric ion but the stability constants tend to be less than those for the mercuric ion with the same ligands. Of course the R group in the organomercurial occupies one of the two most strongly binding coordination sites, so the organomercurials show much less tendency to cross link macromolecular ligand groups.

B. Affinity for Biological Ligands

Because of the strong tendency of both mercuric ion and $R\text{-}Hg^+$ compounds to form complexes, a very important consideration in the distribution and tissue binding is the competition of the numerous ligand groups in the body for the relatively small quantities of mercurial present. Because of the extremely high affinity of mercurials for sulfhydryl groups (101, 118) it is generally assumed that the physiological effects result from alterations in SH containing macromolecules within or on the surface of the target cell (54, 94, 101). Although it is indeed difficult to imagine ligand groups which could successfully compete with SH, one should keep in mind that there are numerous biological ligands capable of complexing mercurials and both Webb (118) and Rothstein (101) have previously stressed that in a biological system numerous ligands will compete for the mercury present, each one influencing to some extent the reaction of mercury with the target site. It should also be noted that there are many small molecular weight compounds containing amine and carboxyl ligand groups that have stability constants greater than 10^{20} (118) and the possibility that the stabilization of a macromolecule by chelation with ligands other than SH, coupled with other stabilizing factors such as local charge interaction, can result in very stable complexes should be at least borne in mind. The present evidence, of course, suggests that mercury is bound predominantly to SH groups, although there is evidence which has

*A similar addition to acetylene is the basis of the catalytic conversion of acetylene to acetaldehyde and is one reason mercury solutions were used in the Minamata chemical plant.

led to the idea that in the kidney the bivalent mercuric cation may react with one SH group and one non-SH group (105, 120). Although one would predict that mercurials should be quite specific for SH relative to other potential ligands, and indeed mercurials are widely used as specific SH reagents, one must realize that practically all proteins contain SH groups and only a minute quantity of mercury is required to produce its physiological effects. As an example, in the case of methylmercury toxicity, a blood level of roughly 2.5 μmoles per liter is the threshold for the onset of symptoms of toxicity (5). Since there are 12 to 20 mmoles of SH groups per liter of blood, it is evident that at the toxic blood level there is a 4800 to 8000 fold excess of SH groups over the available mercury in blood. A similar excess of SH groups over available methylmercury exists in tissues other than blood. Even if the mercury were completely specific for SH groups, the whole body situation is one of a considerable excess of SH ligands competing for the available mercury. Which of the many available SH groups will in fact combine with the mercury is dependent on the reactivity of the SH present in biological molecules, the stability of these complexes, the accessibility of the SH ligands within molecules, and the ability of the mercury to reach intracellular sites.

C. "Geographical Specificity"

Practically all enzymes are inhibited by mercurials in vitro (94, 118), but this is of little consequence in the whole organism, because the mercury will clearly not be equally available to all enzymes. Rothstein (101) has coined the phrase "geographical specificity" to describe specificity dictated by the structural location and accessibility of SH groups at the cellular level. One example of this type of specificity is the action of Hg^{++} on glycerol permeability of the red blood cell (94). On exposure of the red blood cell to Hg^{++}, mercury first binds with ligands in the membranes that are associated with glycerol permeability. As the mercury passes through the membrane and is redistributed to the hemoglobin and other ligands in the interior of the cell, the cell's original permeability to glycerol is regained. This example simply illustrates that one must consider not only the ability of an agent to react with a particular site and to produce a biological effect, and not only the concentration of the agent in a particular organ or even the cellular concentration, but one must know the actual amount bound to the specific site within the cell responsible for the biological effect. Even a knowledge of subcellular distribution is not sufficient, as is illustrated by the fact that p-chloromercuribenzenesulfonate (PCMBS) bound to external sites of the red blood cell membrane has no effect on potassium efflux, while the rate of potassium leakage is roughly proportional to the number of SH groups in an "internal" compartment of the cell membrane combined with PCMBS (94). Conversely, sugar transport is inhibited only by the combination of the organomercurial chlormerodrin with SH sites on the surface of the membrane and is unaffected by chlormerodrin bound to SH groups within the membrane (101). It is evident that a determination of membrane bound mercurial would give no indication of the distribution with respect to the sensitive sites within the membrane.

D. "Reactivity" of Ligand Groups

It should be emphasized that the SH groups even within a single protein are rarely homogeneous. Thus of the eight SH groups of hemoglobin only two react with p-chloromercuribenzoate (PCMB) or PCMBS, six react with chlormerodrin, and all eight react with Hg^{++} (101). The different reactivities of the SH groups in phosphoglucomutase is illustrated by the fact that sequential titration of four of the six SH groups with p-hydroxymercuribenzoate yields successive second order rate constants of 171, 47, 13, and 8 $L.mole^{-1} \cdot sec^{-1}$ (40). That many SH groups are masked by burial within the three-dimensional structure of the protein is evident from the fact that denaturating agents which cause an unfolding of the tertiary structure increase the reactivity of unreactive SH groups in most proteins (40). These "buried" SH groups may not be equally accessible to all mercurials. For example, it has been pointed out by Webb (118) that p-mercuribenzoate reacts with the SH groups of denatured tobacco mosaic virus but not with the native virus, while methylmercury is able to penetrate and react stoichiometrically with the native virus. The binding of the mercurial may itself induce an unfolding of the protein, and this has been proposed as a mechanism of so-called "all-or-none inhibition", in which it is found that after one SH group reacts, all the other groups in the same molecule react rapidly (40). The complexity of the reaction of Hg^{++} with certain enzymes is illustrated by the reaction of Hg^{++} with plasma acetylcholinesterase (118). The kinetics of the slowly developing inhibition deviates so much from classical theory that Webb declared the results uninterpretable. Since methylmercury and p-mercuribenzoate up to 1 mmolar do not inhibit this enzyme, it has been suggested that the interaction is a dimerization caused by the bifunctional Hg^{++}. The bifunctionality of Hg^{++} is well worth keeping in mind because it also provides a possible explanation for the difference in the plasma to red blood cell ratio between Hg^{++} and the alkylmercury compounds.

A progressive irreversibility of mercury binding is noted in many enzyme systems (118), such that less and less mercury can be removed by BAL or glutathione at progressively longer times after mercury treatment. Webb has noted some possible mechanisms by which irreversible damage to an enzyme may develop with time (118). An initial reversible binding followed by a progressive disintegration of the secondary structure, with unfolding of the helical structure of the polypeptide chain which could free groups capable of forming intermolecular bridges, with subsequent polymerization and precipitation, is one such proposal. It seems to be generally accepted that some progressive structural alteration in the protein is responsible for this progressive change and in many cases there is physical evidence of such structural changes (118), but a slow transfer of the mercury to a more stable, less sterically accessible site is also a possibility. Of several hundred studies on the reversal of enzyme inhibition reviewed by Webb (118), 59% were completely reversible by thiols, 30% were partially reversed and 11% were not reversed at all. Webb's discussion includes the example of yeast alcohol dehydrogenase, which is completely blocked by 2.5 x 10^{-7}M Hg^{++} at an enzyme concentration of 4.5 x 10^{-9}M. Glutathione restored 54% of the activity after 30 seconds but only 17% after 10 minutes. In the above case the reactivation by glutathione is quite rapid, but in the case of erythrocyte pyruvate kinase the reactivation is quite slow, requiring 25 minutes to achieve 59% reversal. The reasons for the differing rates of reversal of inhibition by mercury compounds are not known, but the rate of reversal is not correlated with the affinities of the enzyme for the mercurial. In the case of β-fructofuranosidase it is observed that glutathione and cysteine actually increase the inhibition by Hg^{++}, while BAL effectively reverses the inhibition (118). A similar occurrence is known to operate in the kidney, where BAL is able to reverse mercurial diuresis but the monothiols cysteine and glutathione do not retard (14) and in fact may even enhance the diuresis (80). Although an alternative hypothesis based on the release of mercury from the monothiol complex in the renal tubule has been postulated to account for this phenomenon in kidney (31), elucidation of the mechanism by which cysteine and glutathione increase the inhibition of β-fructofuranosidase in solution may well lead to a better understanding of the response of the kidney to Hg^{++} and to the effects of complexing agents.

One can choose many examples to illustrate that one cannot very confidently generalize about the reactivity of the various mercurials with biological macromolecules. The factors determining the reactivity of mercurials with biological ligands are discussed at great length by Webb (118). The reader is referred especially to Webb (118), but also to the articles by Friedman (40), Gurd and Wilcox (47), and Passow et al. (94) for further information on factors influencing the binding of mercurials to SH and other ligand groups in macromolecules.

II. RELATION OF TISSUE DISTRIBUTION TO TOXICITY

A. Organ Distribution

In spite of the complexities described above, certain generalities may be drawn concerning the relation of the organ distribution of mercurials to toxic symptoms they produce. Detailed reviews of mercury distribution have recently been published (29, 88). A number of the important features of the distribution of Hg°, Hg^{++} and methylmercury are illustrated by the data in Table I. After exposure to Hg^{++} or Hg°, the kidneys accumulate much higher mercury concentrations than any other organ. The distribution of mercury after exposure to Hg° is very similar to the distribution after exposure to Hg^{++}, in line with the fact that Hg° is rapidly oxidized to Hg^{++} in blood (24). A most important difference in the distribution of the two forms is the much greater quantity of mercury that penetrates the blood-brain barrier after exposure to Hg°. This was first demonstrated by Berlin et al. (11) in the mouse and has been found to be true also in rats, rabbits and monkeys (12). Because of the greater penetration of Hg° into the brain, exposure to the vapor results in a greater risk of nervous system involvement, while Hg^{++} is generally considered to exert its toxic effect primarily on the kidney.

Table I. Distribution of Different Forms of Mercury in the Rat

A. Distribution 24 Hr After Administration of 0.5 mg Hg/kg.#

Form of Mercury	Reference	Kidney	Liver	Blood	Brain
Hg°	12	25	1.0	0.30	0.80
Hg^{++}	12	25	0.98	0.23	0.05
Hg^{++}	109	14	1.7	0.15(0.03)*	0.04(0.02)*
Methylmercury	109	2.9	1.1	4.1(2.1)*	0.17(0.21)*

Organ values are given as percent of the administered dose per gram of tissue.

* Value 9 days after mercury administration

B. Red Blood Cell to Plasma Mercury Ratio 24 Hr After Mercury Administration.

Form of Mercury	Reference	Dose (mg Hg/kg)	RBC/Plasma Ratio
Hg^{++}	18	1.2	0.44
Methylmercury	89	1.0	271

Methylmercury is distributed much more evenly than Hg^{++} or Hg°, and like Hg°, crosses the blood-brain barrier more readily than Hg^{++}. In the rat, the methylmercury concentration in brain increases for about three days after methylmercury administration and reaches a maximum value between 4 and 10 days after exposure (89, 109). Nine days after i.v. administration of 0.5 mg Hg/kg as methylmercury, the brain mercury concentration is ten times higher than the brain mercury concentration after an equivalent dose of mercuric nitrate (Table I).

In the rat, practically all the methylmercury in blood is contained in the red blood cells (RBC) as is illustrated in Table IB. There are marked species differences in the RBC/plasma ratio of methylmercury (88), and this ratio is higher in the rat than in most other species. Plasma to brain ratios do not show large species differences however, and the rat has one of the lowest brain/blood methylmercury concentrations of any species, while primates have the highest (29). In man 90 to 95% of the methylmercury in whole blood is found inside the red blood cell (2, 63). Hg^{++}, however, is roughly equally distributed between plasma and red blood cells in man (63, 108). Since virtually all the mercury in plasma is protein bound (18, 24), the RBC/plasma ratio is a reflection of the relative binding to ligands in the plasma and red blood cells. The striking difference in the RBC/plasma ratios for Hg^{++} and methylmercury is most interesting because it implies very marked differences in sulfhydryl specificity between these two forms of mercury. The difference could be related to an ability of the bivalent Hg^{++} to form more stable complexes in plasma than methylmercury.

In the light of evidence in humans suggesting that exposure to levels of methylmercury which do not affect the mother may result in damage to the fetus (60, 98), it is surprising that there is so little data on the passage of different forms of mercury across the placental barrier. Berlin and Ullberg (7, 8) used autoradiography to show that methylmercury given to pregnant mice was evenly distributed in the fetus at concentrations comparable to those found in maternal tissues. After the same dose of $HgCl_2$, mercury was present at quite high concentrations in the placenta, but only traces were present in the fetus. Suzuki et al. (106) compared the fetal content of mercury after mercuric chloride administration to pregnant mice on the 14th day of pregnancy to that after administration of methylmercuric acetate. Unfortunately the doses of the two forms of mercury were not the same, but a larger fraction of the administered dose was found in the fetus after methylmercury than after Hg^{++} administration. Further, quite high concentrations of mercury were present in placenta after Hg^{++} administration, but only small quantities had penetrated to the fetus, while after methylmercury administration the mercury concentrations in placenta and fetus were almost equal. Greenwood et al. (46) found that after 2-1/2 minutes of exposure to mercury vapor the fetal mercury content was 40 times higher than that found 2-1/2 minutes after the i.v. administration of $HgCl_2$ at a dose equivalent to the quantity of absorbed mercury vapor, even though the blood concentration of mercury was 26 times higher after i.v. administration of $HgCl_2$ than after exposure to the vapor.

Consistent with the above data on the brain and fetus, both methylmercury and mercury vapor pass through other lipid membranes much more readily than mercuric mercury. Methylmercury has been shown to produce signs of toxicity after absorption through skin (6) and is much more efficiently absorbed from the gastrointestinal tract than Hg^{++}. Oral doses of methylmercury are almost completely absorbed vs. about 15% or less for Hg^{++} (28). Hughes (54) has postulated that mercury vapor exists transiently in solution in blood lipids and is transferred in this form to the brain and other tissues where it is oxidized and fixed within the cells. Although the oxidation of Hg° to Hg^{++} in blood is quite rapid, Magos (65) has demonstrated that 0.5 to 1.3% of mercury vapor absorbed by blood in vitro is still in the elemental form 15 minutes after the end of exposure. Further, Magos (64), in an ingenious experiment, has shown that mercury vapor generated in the blood stream is rapidly exhaled, while practically no mercury is exhaled after the injection of Hg^{++}. It is thus clear that mercury is indeed transported in the body in the elemental state, in which form it readily crosses the pulmonary membrane. Mercury vapor is essentially completely absorbed from the lung (49, 82) and Hughes (54) has pointed out that the solubility characteristics of mercury vapor favor its uptake from air and transfer into the lipid material of the body.

While Hg° and methylmercury chloride both have much greater lipid solubility than $HgCl_2$ and thus would be expected to cross lipid membranes more rapidly, it has already been noted that mercury chloride has appreciable lipid solubility. However, even though the chloride concentration in plasma is about 0.1 M the affinity of the plasma SH groups for Hg^{++} and methylmercury is so much greater than that of chloride that only minute quantities of the chloride complexes can be available for diffusion. That the chloride complex of Hg^{++} can penetrate membranes rapidly is illustrated by the rapid penetration of Hg^{++} into the erythrocyte when the cells are suspended in NaCl solution (119). Although it is possible for the minute quantities of chloride complex which are theoretically present to be the species which penetrates lipid membranes, it seems more likely that small molecular weight sulfhydryl complexes would be the penetrating species, especially since 5-10% of the total thiol content of human plasma is in the form of small molecules (54). In line with this thinking is the fact that BAL, which forms a small molecular weight complex with mercury, accelerates the penetration of methylmercury and Hg^{++} into the brain (9, 10). BAL, of course, will further decrease the concentration of any free chloride complex in plasma. In a later section we will further examine the effects of endogenously formed small molecular weight sulhydryl complexes of mercury on the tissue distribution of mercury.

B. Biotransformation

Published data indicates that the methylmercury radical is quite stable in animal tissues, although a limited degree of cleavage of the carbon-mercury bond does occur (29, 88). Norseth and Clarkson (89) showed that in the rat practically all the mercury in brain was in the form of methylmercury up to 28 days after injection of a single dose of methylmercury, and pointed out that the delayed appearance of central nervous system toxicity following exposure to methylmercury is not due to an accumulation of inorganic mercury in brain. Cleavage of the carbon-mercury bond of methylmercury occurs in the liver and possibly in other organs, but unchanged methylmercury is the major form present in all organs studied, except large intestine, after single or multiple doses of methylmercury (88, 89). Of the mercury excreted in the feces, which is the major route of mercury excretion after methylmercury exposure, about 50% was in the form of inorganic mercury.

It has been known for some time that Hg° is rapidly oxidized to Hg^{++} in blood (24), but the mechanism by which this oxidation takes place and its importance regarding mercury vapor uptake is only now coming to light. It was originally suggested that the oxidation was simply a chemical phenomenon (24), possibly catalyzed by thiol compounds (86). The observation of Nielsen Kudsk that mercury vapor uptake by blood was inhibited by ethyl and methyl alcohols but not by propyl and butyl alcohols (84), and that ethanol decreased the pulmonary absorption of mercury vapor in man (83), suggested that an enzymatic mechanism, possibly involving a hydrogen peroxide-catalase complex and/or glutathione peroxidase, was involved in the oxidation of mercury vapor. These findings have been summarized by Nielsen Kudsk (86) and the main points are given below.
(1) Ethanol inhibits the uptake of mercury vapor by blood in vitro.
(2) Methylene blue and menadione, both of which cause hydrogen peroxide production, stimulate mercury vapor uptake by blood in vitro.
(3) Ethanol and methanol have a high affinity for the hydrogen peroxide-catalase complex.
(4) Liver catalase in the presence of hydrogen peroxide causes a marked increase in the uptake of mercury vapor by glutathione solutions and this uptake is inhibited by ethanol and aminotriazole.
(5) Aminotriazole, a strong inhibitor of catalase, did not influence the in vitro uptake of mercury vapor by human erythrocytes. It has recently been shown, however, that after a 3-hr pre-incubation with

aminotriazole plus methylene blue, 60% of erythrocyte catalase was inactivated and mercury vapor uptake was decreased by 50% (71).

It was suggested by Nielsen Kudsk that the deposition of mercury in the body after exposure to mercury vapor is related to the activity of catalase in the tissues (86). In the case of blood, glutathione peroxidase may play the dominant role in the oxidation, hence uptake, of mercury vapor (86). Nielsen Kudsk has pointed out that the normal content of oxidized glutathione (formed from H_2O_2 and glutathione peroxidase) in red blood cells is sufficient to account for the in vitro uptake of mercury vapor (85) and has postulated how the effects of F^-, iodoacetate, hydroxylamine, ascorbate, and varying O_2 tension might relate to the levels of oxidized glutathione. The actual mechanisms operating in vivo are still not clear.

Magos et al (69) administered mercury vapor intravenously to rats and measured the tissue distribution and pulmonary elimination of mercury with and without ethanol pretreatment. Mercury exhalation continued for over a minute from ethanol treated rats but for only 10-15 seconds from controls. Much more mercury vapor was exhaled from the ethanol treated animals in spite of the fact that the blood levels were lower in this group, consistent with the idea that impaired oxidation and not a decreased ability of mercury vapor to cross the pulmonary membranes is responsible for the decreased pulmonary uptake after ethanol treatment. The difference in mercury distribution between control and ethanol treated animals was most interesting and is presumably a result of the relative ability of ethanol to alter the capacity of the various tissues to oxidize mercury vapor to Hg^{++}. The decreased tissue concentration was most striking in lung and heart. Lung normally accumulates much higher concentrations of mercury after exposure to the vapor (either by inhalation (49) or i.v. injection (64)) than after injection of mercuric chloride. The brain, significantly, accumulated only about half as much mercury after ethanol as the control group. Ethanol must therefore have decreased the oxidation of elemental mercury in brain, because the fact that free vapor was available in blood for a longer time after ethanol should otherwise have resulted in an increased mercury uptake in brain. While most other tissues accumulated less mercury after ethanol treatment, accumulation in liver increased 7 fold and accumulation in the gastrointestinal tract increased 2 fold. It is very difficult to explain these results unless one assumes that dissolved vapor is rather freely accessible to all body compartments, that mercury retention is dependent primarily on the rate of oxidation in the tissues, and that ethanol inhibits the oxidation of mercury vapor in tissues.

Experiments using aminotriazole to inhibit catalase gave very similar results (71), except in this case the blood level of total mercury was higher in the aminotriazole treated animals than in the controls 3 minutes after the mercury vapor injection, the opposite effect of that observed after ethanol treatment. This would be consistent with a more pronounced effect of aminotriazole on tissue than on blood catalase. Thus, because the tissues do not oxidize the Hg^o as rapidly after aminotriazole, there is a higher concentration of Hg^o in plasma, resulting in a higher concentration in red blood cells and a greater exhalation of mercury vapor. Likewise, the increase in liver mercury implies that the liver oxidase system is less inhibited than other tissue oxidases by aminotriazole. It may be that liver oxidases other than catalase are involved in the oxidation of mercury vapor in this organ, since liver catalase would be expected to be sensitive to aminotriazole.

Future work should serve to clarify the pathways responsible for the accumulation of mercury vapor in the animal. The further elucidation of the mechanisms involved in elemental mercury absorption, oxidation, and accumulation in tissues should provide an excellent example of how the interplay of physiochemical properties, metabolism, and organ accumulation relate to the toxic effects in the whole animal.

C. Distribution Within the Organ

The organ distribution of the mercurials is certainly important and some generalizations have been made above concerning the relation between organ concentration and toxicity. The whole organ concentration of mercury is of course a very crude measure of the mercury concentration at critical sites and indeed the same specificities of accumulation that are observed between different organs are seen between different cells within the same organ, between different compartments within the cell, and finally between different macromolecules within a cell compartment. With regard to the toxic effects discussed above, the organs of primary importance are brain, kidney, blood and liver. The fetus is also important to consider, but very little information seems to be available on this subject. It appears methylmercury concentrations in cord blood are about 20% higher than maternal blood (112) and this may be related to the known difference of fetal from adult hemoglobin (122). Tejning (113) has indicated that methylmercury binds with fetal hemoglobin more readily than with adult hemoglobin. Yang et al. (127) administered approximately 0.5 mg Hg/kg as methylmercury chloride to pregnant rats on day 16 of gestation and determined the mercury distribution in fetal and maternal brain 1,2,4 and 5 days after dosing. The fetal brains contained higher concentrations of mercury than the maternal brains and displayed a different pattern of mercury distribution. Mercury concentrations in fetal cerebellum and cerebrum were 2-4 and 1.5-2.5 times, respectively, higher than in the corresponding parts of the maternal brains. Matsumoto et al (73) administered methylmercury chloride, at a does of 2.0 or 20 mg/kg, to pregnant rats between the 9th and 11th day of pregnancy. Half the animals were given 1 gm/kg penicillamine hydrochloride 4 to 5 hours later. Nine days later the animals were sacrificed. Methylmercury concentrations were higher in fetal than in maternal brain. In the fetus, degeneration of nerve cells in the cerebellum, but more often gross malformations of the cerebellum, were observed in animals not treated with penicillamine, while after penicillamine only a few cases of nerve degeneration and no gross malformations were observed in the cerebellum. Their results also indicated a marked reduction in the methylmercury concentrations in both fetal and maternal brain due to penicillamine administration.

The distribution of mercurials within liver, kidney and blood has been recently reviewed by one of the present authors (29) and will not be considered further at this time.

Of primary interest from the point of view of toxic effects is the distribution and state of binding of mercury in the nervous system. The important symptoms of toxicity arising from chronic occupational exposure to inorganic mercury and exposure to alkyl mercury compounds result from damage to the nervous system. A knowledge of the role mercury distribution plays in the production of the observed symptoms requires a careful correlation of data on the distribution of mercury, distribution of morphologic lesions, and functional disturbances observed in the nervous system after exposure to different doses and time schedules of the form of mercury in question. Because of the complex interrelations between the many parameters relating distribution to toxicity and because of differences in both toxicity and distribution arising from factors such as species variations and differing dosage schedules used by various investigators, it would be beyond the scope of the present discussion to attempt a comprehensive review of the relationship of the distribution of mercury in brain to the neurotoxicity of Hg°, Hg^{++}, and methylmercury. Because of the recent advances in our knowledge of the distribution and toxicity of methylmercury, and because of the toxicological importance of this compound, only a brief discussion of the distribution of inorganic mercury will be presented. The relation of methylmercury distribution within brain to its toxic effects will be made the primary focus of the present discussion.

The distribution of mercury in brain has been investigated in mice and rats using autoradiography by Cassano et al (15) after exposure to radioactive mercury vapor. The pattern of mercury distribution did not vary markedly as a function of time after exposure. The pattern of mercury distribution was similar from 1 to 50 days following ten daily exposures of 6 hr/day to 8 mg/m^3 Hg°. The brain stem nuclei, spinal cord, cerebellar nuclei and cerebellar cortex contained more mercury than cerebral cortex, basal ganglia and the septum. High concentrations of mercury were found in the epithelium of the choroid plexi and ependyma and in nerve and glial cells near the ependyma. The most radioactivity was present in certain nuclei of the midbrain, pons, medulla and most of all in cerebellum. In the cerebellar cortex, mercury was localized chiefly in Purkinje cells with much less in the granular layer. The nucleus dentatus was heavily labeled. In the spinal cord relatively high concentrations were observed in the neurons of the anterior and posterior horns and the postero-medial nuclei.

Berlin et al (7, 11, 12) have investigated the mercury distribution in brain following exposure of mice, rats, rabbits and monkeys to radioactive mercuric salts and mercury vapor. Rats, rabbits and monkeys exposed to mercury vapor for 4 hours at a concentration of 1 mg Hg/m^3 accumulated brain levels of mercury about ten times higher than animals injected with the same dose of $HgCl_2$ (12). After the exposure of each animal to mercury vapor, the amount of mercury absorbed by the animal was determined and a second animal was given a dose of $HgCl_2$ corresponding to the amount of mercury absorbed by the animal exposed to vapor. Except for brain, there were no marked differences in the organ distribution of mercury between animals exposed to mercury vapor and those given the same dose of $HgCl_2$. Shortly after exposure, animals exposed to vapor had a much higher mercury content in red blood cells than the animals injected with $HgCl_2$. Gray matter contained more mercury than white after both forms of mercury. After mercury vapor exposure the nucleus dentatus in the cerebellum, nucleus olivarius inferior in the brain stem and nucleus subthalamicus showed a marked accumulation of mercury. The choroid plexi and calliculus superior also appeared dense in the autoradiograms. In the cerebellum, the granular and Purkinje cell layers initially had more mercury than the molecular layer and with time the concentration in the Purkinje layer was greater than that in the granular layer. In general, mercury distribution in brain after i.v. injection of $HgCl_2$ was similar to that after vapor exposure, except the overall concentration of mercury in brain was much higher after exposure to mercury vapor. Relative to the concentrations in the other brain regions, the concentration in the choroid plexi and pia mater were higher after $HgCl_2$ injection. The uptake of mercury in the area postrema was of the same order of magnitude regardless of the form of mercury administered. The high uptake of mercury in the area postrema after $HgCl_2$ administration, relative to other areas of the brain, is probably due to the high vascular permeability of the area postrema (21).

It has been observed by Nordberg and Serenius (87) that certain cells in the brain stem, which they designated mercuriphilic cells, not only contain relatively higher quantities of mercury after inorganic mercury exposure, but also eliminate the accumulated mercury much more slowly than other neighboring cells. Further, the rate of elimination of mercury from different regions of the brain varies from region to region (87).

With regard to methylmercury, a general picture of the distribution of methylmercury within the nervous system and its relationship to the toxic effects of this compound has begun to emerge during the past few years, although the data is still surprisingly limited. Grant (45), who has carried out extensive correlative studies on the relationship between mercury distribution and the pathological changes observed in monkeys, rats and cats after methylmercury exposure in collaboration with Berlin et al (13), and Kurland (60) have summarized the general pathological picture observed in both man and animals after methylmercury exposure. The important lesions seem to involve the calcarine (visual) cortex, granular layer of the cerebellum and the peripheral sensory nerves and ganglia, together or in various combinations.

In man, methylmercury intoxication results in selective damage to the visual cortex and cerebellum, with disintegration of cells in the calcarine cortex and severe degeneration of cells in the granule cell layer of the cerebellum (56, 59, 111). Matsumoto et al (74) used histochemical techniques to determine the mercury distribution in the brains of autopsied Minamata disease victims and found that the regions of most severe cellular damage exhibited the strongest histochemical reaction for mercury. The distribution of mercury in two cases exhibiting the strongest histochemical reactions is illustrated in figure 1. It can be seen that the highest density of granules was found in the cerebellum and the calcarine fissure. The calcarine fissure is, of course, the cortical visual area, upon which the retina is projected (102). Constriction of the visual field, which is one of the prominent features of methylmercury intoxication in man (55), presumably results from selective damage to the visual cortex.

Grant (45) observed in monkeys and Yoshino et al (128) observed in dogs that the most obvious histological changes after methylmercury exposure occurred in the calcarine cortex and that these changes were associated with a selective accumulation of mercury in this area. Yoshino et al (128) found neither a high mercury content in cerebellum, nor selective cerebellar damage in his dogs, but Hunter et al (55), in their classic description of methylmercury intoxication in humans and experimental animals, noted that the degree of cerebellar damage is related to the dosage schedule. Rats receiving smaller doses and surviving for longer periods exhibit degeneration of the granular layer of the cerebellum, while rats exposed to higher doses (Hunter et al gave roughly 35 mg/rat over a period of 29 days) develop severe peripheral neuropathy and become moribund with no observable changes in the cerebellum.

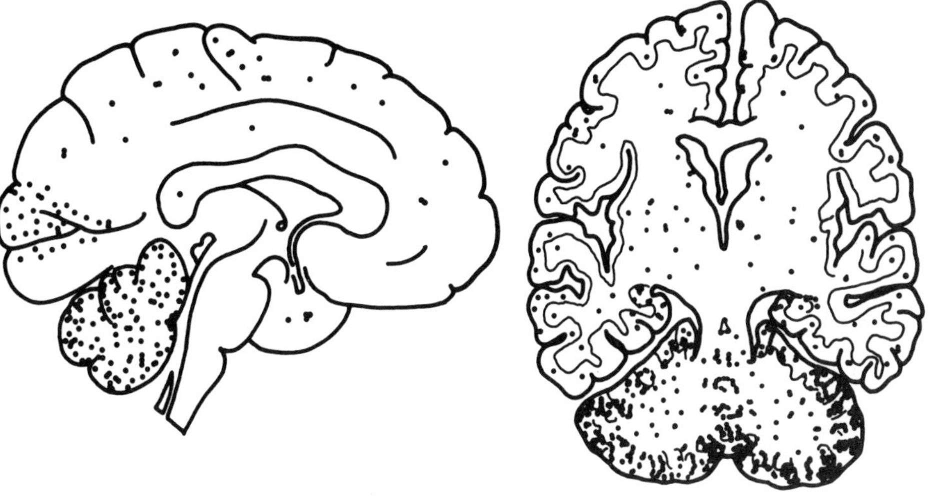

Figure 1. Pattern of Brain Distribution of Mercury Derived From Histochemical Studies of Two Minamata Disease Victims. (Redrawn from Reference 74, with permission.)

Miyakawa and Deshimaru (78) fed rats 10mg/kg/day methyl methylmercuric sulfide for 20 days and observed the changes in brain from 12 to 150 days after initiating treatment, using light and electron microscopy. At 12-13 days no changes were visible with the light microscope but the granule cells bordering the deep sulcus of the cerebellar vermis contained electron dense lysosomes visible with the electron microscope. By 19-20 days there were marked changes in the granule cell layer bordering the deep sulcus of the cerebellar vermis, while the telencephalon and brain stem appeared normal. There There was a conspicuous disappearance of ribosomes from the granule cells and two types of cellular change. Type 1 changes were most numerous and were characterized by the disappearance of ribosomes and Golgi from the cytoplasm, an increased nuclear membrane density and a coagulated appearance of the intranuclear substance. Type 2 changes were characterized by degeneration of the nuclear membrane with streaming of the nuclear substance into the cytoplasm. Chang and Hartmann (22) have recently studied the subcellular distribution of methylmercury using the electron microscope and a histochemical technique for visualizing the mercury. It is interesting that after methylmercury exposure the mercury in the granule cells of the cerebellum is associated with the membranous structures; endoplasmic reticulum, Golgi apparatus and nuclear membrane. Practically no mercury was found inside the nucleus, but large amounts were bound to the nuclear membrane.

Chang and Hartmann (22) have related their findings, using the above histochemical techniques, to the ultrastructural changes observed after methylmercury and $HgCl_2$ exposure (19, 20). Among the cell populations examined, the distribution of mercury was (22):

After methylmercury: dorsal root ganglion neurons > neurons of calcarine cortex > cerebellar Purkinje cells > ventral horn motoneurons > cerebellar granule cells.

After $HgCl_2$: dorsal root ganglion neurons > cerebellar Purkinje cells > ventral horn motoneurons > neurons of calcarine cortex > cerebellar granule cells.

With regard to ultrastructural changes, the cerebellar changes observed were very similar to those observed by Miyakawa and Deshimaru (78) except some changes were also observed in the Purkinje layer. In spite of the far greater degree of degeneration observed in the granular layer, more mercury was present in the Purkinje layer after both methylmercury and $HgCl_2$ exposure. This observation was attributed to the fact that the Purkinje cells are very rich in SH groups, which may offer a protective effect. The most characteristic lesion of methylmercury was a focal cytoplasmic degeneration of the dorsal root ganglia. This characteristic effect of methylmercury in rats was first reported by Hunter et al (55) and has been confirmed by others (16, 79). At a daily dose of 1 mg/kg/day of methylmercury, degeneration of the sensory fibers of the dorsal root was observed beginning after the second week of treatment, while only moderate changes were ever observed in the motor neurons of the ventral root. In a study of the topographical differences in the vascular permeability of the peripheral nervous system, Olsson (93) demonstrated that dorsal root ganglia accumulate large concentrations of fluoresceinisothiocyanate labeled albumin outside the blood vessel walls of the capsule, in the endoneurium and in the spaces between adjacent neurons in the cortex. Chang and Hartmann (22) found the greatest mercury accumulation of any of the neurons they examined to be in the neurons of the dorsal root ganglia after both methylmercury and $HgCl_2$ administration. The high accumulation of mercury in these ganglia and the associated lesions are probably a result of the unusually high vascular permeability in the dorsal root ganglia. Although degeneration of the granule cells of the cerebellum is a characteristic finding after methylmercury exposure, after both methylmercury and $HgCl_2$ these cells contained less mercury than cerebellar Purkinje cells and ventral horn motoneurons, both of which are relatively unaffected during methylmercury intoxication. In sciatic nerve, it was found that only certain nerve fibers contained electron dense granules after methylmercury exposure and degenerative changes seemed to be associated with these mercury containing neurons. This observation fits well with the idea that only certain neurons in the sciatic nerve are affected and that these are the sensory fibers of the dorsal root.

The chronology of the peripheral electrophysiological and morphological changes due to methylmercury intoxication reported by Herman et al (51) correlates well with the general picture of methylmercury associated pathology drawn above. Coincident with the first structural changes some sensory fibers ceased to conduct. By the time animals became ataxic, the morphological changes in the dorsal roots were marked and the compound action potential in spinal sensory (dorsal) roots was depressed by at least 75% in amplitude and area. Ventral (motor) roots were almost normal. The sciatic nerve showed the same changes as dorsal roots, but the changes were "diluted" by healthy efferent axons. It is worthy of note that many sensory fibers failed to conduct before animals showed any obvious neurological disturbances. In this connection Grant (45) observed clinically inapparent morphological damage in monkeys and rats exposed to methylmercury. Such clinically "silent" damage should certainly be taken into account when considering "safe" levels of exposure for human populations.

Somjen et al (104) administered 2 mg Hg/kg as methylmercury labeled with 203Hg to rats 5 days per week and determined the mercury distribution from 8-29 days after initiating treatment using radioassay and autoradiography. In the nervous system, the spinal dorsal root ganglia contained the highest concentration of mercury followed closely by the cerebral cortex and cerebellum, then the subcortical part of the forebrain. On autoradiograms, more grains were observed over gray matter than white and more in neurons than glia, satellite or Schwann cells. There was no difference in the mercury content of the dorsal and ventral root fibers and it was suggested that the data was consistent with the dorsal ganglia being the primary targets, with subsequent degeneration of the sensory axons due to damage to the parent cells in the ganglion. Chang and Hartmann (20), however, report changes in the axons to precede alterations in the cell bodies, but they also reported dorsal root fibers to contain more mercury than ventral root fibers, in contrast to Somjen et al (104). In general, the patterns of distribution and damage observed by Somjen et al (104) are in agreement with the results discussed above.

Some of the important studies relating methylmercury distribution and nervous system toxicity are summarized in Table II.

Although some features of methylmercury intoxication appear to correlate well with mercury distribution, it should be clear that distribution is only one of several factors which must be understood to fully explain the development of physiological lesions after exposure to mercury compounds. In fact, the differences in mercury concentration between different areas of the nervous system are in general not quantitatively very large, and the regions of highest mercury concentration are not in every case those in which lesions occur. In addition to the mercury concentration, the ability to repair structural damage, the content of protective metal binding substances, and other functional parameters of a particular class of cells play an equally important role in determining whether damage occurs in that particular cell population.

III. MOLECULAR BINDING SITES

A. Kidney and Liver

It has been emphasized that a knowledge of mercury concentrations in a particular organ, cell, or subcellular compartment may have little meaning unless one has some knowledge of molecular binding in the compartment of interest. Since a knowledge of mercury biocomplexes present in the target cells, where the biological effects of these compounds are exerted, is essential if we are to develop an understanding of the mechanisms of mercury toxicity, it is somewhat disappointing that our information in this important area is so scanty. It is of course very difficult to define the nature of mercury biocomplexes in vivo. The multiplicity of biological molecules which contain SH and other ligand groups with a high affinity for mercurials results in a staggering number of potential targets. Further, the biocomplexes of mercury are generally reversible, so there is always the possibility of redistribution of the mercury when the normal compartmentation of the cell is disrupted. The general lack of knowledge concerning the chemical basis of most physiological disturbances makes it difficult to relate information determined on isolated biochemical systems to the toxicological manifestations in the whole animal. Some of the problems of relating the interaction of mercurials with particular ligand groups to the subsequent functional disturbances have been discussed by Rothstein (101).

From the above discussion (i.e. affinity for biological ligands) mercurials would be expected to bind to thiol ligands in proteins and smaller molecular weight compounds in vivo if purely physicochemical forces at the molecular binding sites were involved. The weight of evidence indicates that mercury in plasma (18, 24, 29), red blood cells (101), kidney (57, 126), liver (57), and brain (129) is protein bound. Beyond this crude distinction between diffusible and non-diffusible forms of mercury, further definitive information on molecular binding sites is difficult to obtain. Studies based on in vitro addition of mercury compounds to biological preparations frequently give results at variance with those seen after in vivo administration. For example, Piotrowski and associates (95) found that the bulk of mercury (Hg^{++}) when added in vitro to kidney homogenates was bound to high molecular weight proteins, while Hg^{++} administered in vivo was associated with mainly the smaller molecular weight fractions (probably metallothionein, see below). Massey and Fang (72), in studies of liver slices, reported that the subcellular distribution of mercury differed after in vitro versus in vivo uptake. These results are not unexpected. Mercury, added in vitro as a pure compound (usually the chloride or acetate salt) is presented to the tissue in a different chemical form than in vivo. Equilibrium distribution may not be established in vitro, and mercury may bind to a variety of ligands depending upon the experimental conditions. In vivo, true equilibrium distribution is unlikely. Magos and Clarkson (for review, see 29) found that 2,4-dinitrophenol, a classic uncoupler of oxidative phosphorylation, reduced the selective renal accumulation of mercury without changing renal blood flow or glomerular filtration. They concluded that an input of metabolically derived energy was needed for accumulation of mercury by this organ.

Nevertheless, evidence is now beginning to point to the importance of metallothionein, a protein of molecular weight about 10,000 and containing 25 to 35% cysteine, as an important site of binding of inorganic mercury in kidney tissue. Clarkson and Magos (25) used equilibrium dialysis methods to study binding of Hg^{++} in kidney and liver homogenates in the presence of various selected concentrations of penicillamine. Their results were the same whether Hg^{++} was added in vitro or given in vivo to the rats, as would be expected under true equilibrium conditions. Three classes of binding sites were identified, distinguished according to their affinity for mercury. The class of binding sites having the highest affinity for mercury accounted for only a small fraction of the total protein-bound sulfhydryl in the homogenate. It was noted that the amount of Hg^{++} needed to saturate this class of sites corresponded to kidney levels of mercury that just elicited toxic effects in vivo. These authors, however, did not present supporting evidence that the classes of binding sites identified in the in vitro studies were important in the in vivo disposition of Hg^{++} in renal tissue. Recent evidence from Piotrowski's laboratory indicates that these sites probably correspond to metallothionein (57, 95, 123, 124). Metallothionein derived from human kidneys has been known for a long time to contain mercury (96). Piotrowski et al (95), using gel filtration separation of homogenates

Table II. Some Important Findings Relating Methylmercury Distribution and Methylmercury Neurotoxicity.

Reference	Signs, Symptoms, Pathological Changes	Species	Mercury Distribution
Hunter et al. (55)	Ataxia, dysarthria, constriction of visual field in man. Degeneration of peripheral nerves, trigeminal nerves, dorsal (sensory) spinal roots, dorsal columns and granular layer of cerebellum in animals. No change in ventral (motor) spinal roots.	Man, rat, monkey.	
Hunter and Russell (56), Takeuchi (111) and Kurland et al. (59)	Paresthesia, constriction of visual field, hearing loss, dysarthria, ataxia, other symptoms. Degeneration of granule cells of cerebellum, disintegration of cells in calcarine and precentral cortex, foci of cortical atropy in cerebrum.	Man	
Matsumoto et al. (74)		Man	Highest mercury content in cerebellar cortex and calcarine cortex of cerebrum corresponding with regions of severe pathologic changes.
Yoshino et al. (128)	Prominent changes in area of calcarine fissure.	Dog	Highest mercury content in area of calcarine fissure.
Grant et al. (45) and Berlin et al. (13)	Constriction of visual field, impaired motor coordination, possible sensory disturbances. Visual impairment dominated. Damage limited to cerebral cortex. Demonstrated dependence of damage on dosage schedule. Demonstrated clinically silent damage.	Monkey	Corresponded with sites of cerebral cortical damage.

Table II.(Continued)

Reference	Signs, Symptoms, Pathological Changes	Species	Mercury Distribution
Chang and Hartmann (19,20,22)	Morphological changes in dorsal root ganglia, dorsal root fibers, certain fibers in sciatic nerve, granule cells in cerebellum.	Rat	Dorsal root ganglion neurons> neurons of calcarine cortex> Purkinje cells of cerebellum> ventral horn motoneurons> granule cells of cerebellum. In sciatic nerve, fibers with high mercury content were those exhibiting degeneration.
Somjen et al (104) and Herman et al (50), 51)	Morphological changes in dorsal roots coincided with depression of the compound action potential. Ventral roots were almost normal. In sciatic nerve certain fibers showed same morphological changes as those in dorsal root. In more advanced stages cerebellar changes were observed.	Rat	Mercury content quantitatively determined. Highest content in dorsal root ganglia. Spinal cord, spinal roots, perpipheral nerve contained significantly less. No difference in content of dorsal and ventral roots. Cerebral cortex, cerebellum contained only slightly less mercury than dorsal root ganglia. More mercury in neurons than glia, satellite or Schwann cells; more in cell bodies than axons.

demonstrated that, after in vivo dosing of rats, the bulk of the mercury (Hg^{++}) in kidney tissue was associated with the metallothionein fraction. The amount of Hg^{++} needed to saturate the metallothionein sites was the same as that required to saturate the high affinity sites described by Clarkson and Magos, and corresponded to a dose of $HgCl_2$ (0.3 to 0.5 mg Hg/kg body weight) needed to initiate toxic effects. Thus, the idea arose that metallothionein played the role of "shielding factor" against the potential toxicity of inorganic mercury (123). In this context the finding of Piotrowski et al. (95) that pre-treatment with inorganic mercury can increase metallothionein levels in the kidney is most interesting (see Table III) and may account for the development of tolerance to inorganic mercury.

The mechanism by which mercury causes an increase in metallothionein levels is not unequivocally established. Pretreatment with cadmium salts is known to increase metallothionein levels in both liver and kidneys (37). The effect of inorganic mercury is restricted to the kidneys, where it also causes an increased incorporation of ^{14}C from ^{14}C-cysteine given to rats (95). Mercury (Hg^{++}) may induce metallothionein synthesis in kidney tissue but the results published to date do not exclude the possibility that these effects could be due to increased uptake of circulating metallothionein by the kidney. The increased metallothionein levels shown in Table III were produced by repeated dosing at 0.5 mg Hg/kg; doses which begin to produce changes in kidney function.

Fang (36) has demonstrated uptake of mercury into the metallothionein fraction in kidneys of rats after doses of phenyl and ethylmercury salts. However, these compounds are known to dissociate in vivo to inorganic mercury which accumulates in the kidneys. Thus, there is still no definitive evidence for the attachment of organomercurial molecules to metallothionein in vivo.

B. Body Fluids

Cember et al. (18) studied the distribution of mercury among serum proteins following the i.v. administration of $HgCl_2$ to rats at a dose of 0.12 mg/kg or 1.2 mg/kg. Virtually all the mercury in serum was bound to serum proteins. Most was associated with the globulins. Initially, the α_2 globulins contained the most mercury, but at the higher dose of mercury there was a shift of the mercury to the α-globulin and albumin fractions with time. At the lower dose there was less redistribution of the mercury among the serum proteins with time. It is noteworthy that mercury added to whole blood in vitro was bound mainly to albumin, again illustrating the importance of conservatively interpreting data obtained in an in vitro situation. Jakulobski et al (57), using gel filtration,

Table III[a]. The Level of Metallothionein in the Kidney and Liver of Rats, With and Without Exposure to Metals.

Exposure to Metals	Metallothionein, mg/g of Tissue			
	No. of Rats	Kidney	No. of Rats	Liver
None	9	0.39±0.08 (0.26–0.52)	5	0.12±0.05 (0.07–0.20)
Cadmium 7months	3	3.1 (2.6-4.8)	4	4.3 (2.6-9.0)
Mercury 3weeks	6	2.5 (2.1-2.7)	5	0.17 (0.12-0.26)

Mean values and standard deviations; high and low values in parentheses.

---

[a]Table reproduced from reference (95), with permission of the author and publisher.

---

found the highest concentration of serum mercury to be in a fraction containing mainly lipoproteins and benzidine positive globulins one hour after the intravenous administration to rats of 1 mg Hg/kg as $HgCl_2$.

Weiner et al (120) examined the urinary excretory products of a number of different mercurials using polarography and concluded that the main urinary products were cysteine complexes of either the intact organomercurial or of mercuric ion. Jakubowski et al (57) studied the binding of mercury in urine using gel filtration after removal of the urine sediment by centrifugation. It was found that the fraction of dialyzable mercury varied considerably from animal to animal and values from 1% to more than 50% were observed. Gel filtration on Sephadex G-75 revealed that mercury in urine could be resolved into three fractions; fraction one represented relatively high molecular weight compounds, fraction two represented compounds with a molecular weight of approximately 11,000, and fraction three represented low molecular weight compounds. In some cases, all three fractions were present, in other cases fraction two or three, or even both fractions two and three were absent. Fraction three corresponded to dialyzable mercury and the organic components of this fraction appeared to have a molecular weight of 100 to 300. The differences in distribution found in different urine samples did not appear to be related to dose or time after administration of mercury or to the presence of bacteria in the urine samples. In addition to the above complexes of mercury in urine, 20 to 50% of the total mercury in urine was present in the urine sediment (57). Several workers have noted that the desquamation of renal tubular cells after mercury administration may result in a very large excretion of mercury associated with the sloughed cell debris (17, 34, 68, 114). Davies and Kennedy (34) measured the rate of cell excretion in urine after Hg^{++} administration and reported that 0.8 mg/kg increased the cell excretion from a control value of 3,000 to 5,000/hr to roughly 265,000/hr two days after mercury administration. Under some conditions almost all the mercury excreted in urine is associated with the debris removed by centrifugation (17). It is possible that some of the variability observed by Jakubowski et al (57) is related to a variable release of cellular constituents from cell debris in the urine.

Norseth (90,91) has pointed out that the biliary complexes of methylmercury may be important both in the excretion and organ distribution of mercury. Methylmercury is excreted in rat bile in two forms (90); one is a low molecular weight complex originally thought to be a dipeptide (91,23), the other is a protein complex with a molecular weight greater than 50,000. Norseth (92) has reported that the mercury excreted in bile after doses of 1 ug Hg/kg or 1 mg Hg/kg as methylmercury is found predominantly in the low molecular weight complex independently of the dose. The low molecular weight complex is reabsorbed much more rapidly from the intestine than the higher molecular weight complex. When the bile complex is administered to normal rats it is observed that more mercury is accumulated in the kidney than after administration of the same dose of methylmercury chloride and the red blood cell to plasma ratio is only 80 ± 33 compared to 271 ± 22 after methylmercury chloride administration (91). This indicates that the complexed form doesn't enter the red blood cell as readily as the normal plasma form but that it accumulates in the kidney more readily. It is known that Hg^{++} accumulates in the kidney more readily after the administration of the cysteine complex than after mercury chloride administration (68). In bile duct ligated rats given methylmercury, less mercury accumulated in the kidney and more accumulated in the blood, liver and brain than in normal rats (90). The above data strongly suggests that reabsorption of the biliary complex plays a role in determining the distribution of methylmercury within the animal.

Cherian and Vostal (23) have investigated the forms of methylmercury and Hg^{++} in rat bile as a function of dose and time after administration. The biliary excretion of Hg^{++} and methylmercury were similar, although Hg^{++} reached its maximum excretory rate more quickly. The rate of biliary excretion of methylmercury, expressed as a percentage of the injected dose, was independent of dose between 0.075 and 2 mg/kg. At a relatively low dose of 0.1 mg/kg almost all the methylmercury in bile was in the high molecular weight complex. At higher doses of methylmercury the fraction present in the low molecular weight complex increases with increasing dose and increasing time after methylmercury administration. At doses of 1 and 2 mg/kg of methylmercury the majority of mercury in bile was in the low molecular weight complex from one to five hours after administration. These results apparently contradict those of Norseth (92) referred to above, which indicate that there is no dose dependence of the distribution between the high and low molecular weight fractions. The low dose used by Norseth (92) is considerably lower than the lowest dose used by Cherian and Vostal (23), but the trend toward a higher proportion in the high molecular weight complex at lower doses observed by Cherian and Vostal would predict that all the mercury would be in the high molecular weight fraction at this dose. Norseth (92) reported a shift of mercury to the high molecular weight fraction, expecially at very low doses of methylmercury, if the bile was stored frozen before fractionation, but Cherian and Vostal (personal communication) did not find such a shift at the doses they employed. The reason for this apparent discrepency is not clear. After $HgCl_2$ administration, cumulative biliary excretion of mercury is similar to that after similar doses of methylmercury, but the low molecular weight complex does not play such an important role (23). Immediately after i.v. administration of $HgCl_2$, over 50% of the excreted mercury was in the form of a low molecular weight complex, but after about one hour only binding to the higher molecular weight component was important. The distribution of Hg^{++} between the high and low molecular weight complexes is relatively independent of dose between 0.1 mg/kg and 1 mg/kg, and after 2 mg/kg a slightly greater fraction was in a low molecular weight form. The reabsorption of the different methylmercury complexes from the intestine was also investigated. The low molecular weight complex was absorbed most efficiently (there was almost complete absorption from jejenum in 60 minutes), in agreement with Norseth (90). The high molecular weight complex was also absorbed fairly rapidly (46% was absorbed from jejunum in 60 minutes). As pointed out by these authors, it is unlikely that the high molecular weight complex comprises a significant fraction of the mercury excreted in feces because the half time of excretion for methylmercury is independent of dose, while there is a marked dose dependence of the distribution between the biliary complexes.

C. Nervous System

Because the brain is of obvious interest in view of the toxic manifestations of mercury compounds, it is unfortunate that there is so little information on the state of binding of mercury compounds in the brain.

Cassano et al (15) performed successive extractions on homogenates of brain from rats exposed to mercury vapor. No mercury was associated with the lipid, phosphatidopeptide or nucleoprotein fractions. Mercury was associated principally with either the insoluble residue fraction or the water extract. Over half the mercury present in brain 10 days after mercury vapor exposure was extractable with water but the proportion of mercury which could be removed in this manner rapidly diminished with time after exposure. By forty days after the exposure almost all the mercury in brain was associated with the insoluble residue fraction.

Yoshino et al (128) fractionated brain homogenates from rats treated with methylmercury and determined that the mercury was associated exclusively with the proteins. Very little mercury was found in the lipid and nucleic acid fractions. The methods used involved extraction of the lipids with chloroform-methanol, extraction of nucleic acids with hot trichloroacetic acid and precipitation of proteins with trichloroacetic acid. Differential centrifugation showed that most of the mercury was fairly uniformly distributed through the mitochondrial, microsomal and supernatant fractions and that little mercury was found in the nuclear fraction. The purity of the centrifugal fractions was not determined however, and it was difficult to tell how much mercury was actually associated with the red blood cells which were not separated from the nuclei. The work of Chang and Hartmann (22) which demonstrated methylmercury accumulation in the nuclear membrane and other membraneous structures with practically no mercury inside the nucleus has already been noted. It should be obvious, however, that any mercury not firmly bound to a macromolecule which is insoluble throughout the fixation process which is essential for electron microscopic studies would very likely be lost during fixation.

Hirayama and Takahashi (52) separated brain and liver homogenates from methylmercury treated rats into supernatant and granule fractions by centrifugation at 100,000 X G for one hour after preliminary removal of nuclei and cell debris and found roughly 70% of the mercury in brain and 50% of the mercury in liver in the soluble protein fraction. This suggests that mercury binding to the soluble proteins in brain may be at least as important as the mercury binding to the soluble protein fraction of liver and kidney. Unfortunately, no detailed study on the binding of methylmercury or Hg^{++} to the different classes of brain soluble proteins have been reported. It would be most interesting to determine if there are protective, inducable binding proteins in brain which are analogous to the binding protein for Hg^{++} in kidney.

IV. TREATMENT OF MERCURY POISONING

The successful treatment of any type of poisoning is generally directed at one or a combination of three basic goals. These are: 1) a reduction of the concentration of toxin at its molecular target sites; 2) the negation of the deleterious biochemical effect caused by the combination of the toxin with its molecular target(s); 3) support or enhancement of physiological compensation for the biochemical lesion. There is very little knowledge of the biochemical changes which are the primary cause of damage in mercury intoxication. The biochemical changes associated with mercury toxicity are outside the scope of this article, but the limited information available has not yet led to the development of clinically useful antidotes based on a counteraction of the biochemical lesions. Aside from general supportive therapy, efforts to treat mercury intoxication have been mainly based on the use of antidotes which reduce the concentration of mercurial at its target sites in tissues either by forming an inactive complex with the mercurial or by increasing its elimination from the tissue.

The development of 2,3-dimercaptopropanol as an antidote to the toxicity of compounds with a high affinity for SH groups is a classic example of the practical application of a knowledge of the biochemical association between a toxin and a particular type of ligand. Originally developed by the British during World War II as an antidote to lewisite and other arsenical war gases which were known to have a high affinity for SH groups, 2,3-dimercaptopropanol, designated British Anti-Lewisite, or BAL, by American scientists during the war, was also found to be an effective antidote to heavy metal poisoning (61). The ability of BAL, which is a vicinal dithiol compound, to form very stable chelates with mercury permits it to effectively remove mercury from the cellular constituents to which it is normally bound. Although BAL is established as an effective antidote in acute poisoning from soluble inorganic mercury compounds (62) and is still the most commonly used antidote in cases of acute inorganic mercury poisoning, it does not effectively protect from alkylmercury compounds and it does not alleviate neurological disorders caused by mercury vapor exposure (28).

Of the many types of chelating agents which have been tested as antidotes to mercury poisoning since the development of BAL, only the penicillamines appear to offer a substantial advantage over BAL. Aposhian and Aposhian (3) demonstrated that both D-penicillamine and N-acetyl-D,L-penicillamine were orally effective antidotes to the lethal effects of mercuric chloride and were relatively nontoxic. N-acetyl-D,L-penicillamine appears to be effective in increasing the excretion of mercury after mercury vapor exposure and in relieving the symptoms of chronic mercury vapor exposure (28). The prevention of fetal brain damage in the offspring of rats treated with methylmercury by D-penicillamine has been mentioned above. Both D-penicillamine and N-acetyl-D,L-penicillamine mobilize mercury from the tissues and increase the excretion of mercury in cases of human methylmercury poisoning (5,107). Thus it appears penicillamines offer the advantages over BAL that they are orally effective, less toxic, and are effective in treating mercury vapor poisoning and probably alkylmercury poisoning, although it is not yet established if the neurological effects of alkylmercury poisoning, once clinically developed, are improved by penicillamine treatment.

Unithiol (2,3-dimercaptopropanosulphonate), a drug developed in Russia (4) and so far studied only in Eastern Europe, is an analogue of BAL which is apparently even more effective than BAL in mobilizing mercury (115). Unithiol is very effective in removing mercury from the kidneys after exposure to mercury chloride and has been shown to remove mercury primarily from the metallothionein fraction of kidney proteins (115). Unithiol did not cause redistribution of mercury to brain (although there was a shift in organ distribution) when given to animals exposed to mercury chloride (115), as has been observed after BAL treatment (9). It should be noted that this redistribution to brain caused by BAL can be avoided if large enough doses of BAL are given (66). Unithiol is apparently effective in the treatment of occupational mercurialism (38) but we are aware of no reports on its effects in alkylmercury poisoning.

The relatively high efficacy of α-thiola (2-mercaptopropionylglycine) in increasing the excretion of methylmercury and lowering tissue methylmercury levels (including brain) in experimental animals should be noted (92a).

Aaseth (1) has recently demonstrated that mercaptodextran is an effective antidote when administered to mice immediately following exposure to $HgCl_2$. Because mercaptodextran is distributed in the extracellular space and does not penetrate the blood-brain barrier, there was no redistribution of mercury into brain. Unfortunately, when treatment was delayed for 2 hours following mercury administration, it was completely ineffective. Apparently the rate of redistribution of mercury, once it has penetrated to intracellular sites, is so slow that mercaptodextran is no longer effective once the mercury has become intracellular. BAL, on the other hand, still prevented death when given 2 hrs. after the mercury and BAL reduced the kidney level of mercury to about the same extent whether given immediately or 24 hrs. after mercury administration. The diffusibility of the BAL-mercury complex has been mentioned above. It seems that the maximum advantage of both agents might be realized if relatively small doses of BAL (or other diffusible thiols) were given in conjunction with an agent like mercaptodextran, which is distributed extracellularly, has a relatively low toxicity, and favorable mercury-binding properties. The diffusible thiol would serve as a mercury "carrier" to increase the rate of redistribution, while the extracellular thiol served as a "trap" to accumulate and eliminate the mercury.

Although the present approaches to the therapy of mercury poisoning, aside from general supportive therapy, are in the main directed toward preventing mercury absorption or increasing its excretion, one should be aware that an increased excretion of mercury, per se, is not necessarily always beneficial. The redistribution of mercury after BAL administration with increased brain uptake of mercury under certain conditions has already been mentioned. The extent of the hazard associated with redistribution of mercury after BAL treatment is not clear, but it is rather worrisome in light of the rapid metabolism of BAL in vivo (61). As another example, thioacetamide is one of the most effective compounds in increasing mercury release after mercury chloride exposure, but one of the major causes of this release seems to be a synergistic kidney damage caused by the combination of thioacetamide and mercury, with increased sloughing of renal tubular cells (114).

The increased urinary excretion and decreased kidney levels of mercury observed after administration of sodium maleate was originally explained as a cellular disturbance leading to the release of endogenous small molecular weight mercury complexes from the cell (26). Subsequently, it was shown that low doses of maleate, which do not affect mercury excretion or affect kidney function markedly, enhance the effect of D-penicillamine in increasing urinary mercury excretion and decreasing renal mercury after mercury chloride, but not after methylmercury exposure (27, 67, 105). It has been suggested that the combination of maleate and penicillamine might have a therapeutic advantage over BAL in treating Hg^{++} poisoning because this combination produced higher excretion rates of mercury but less redistribution within the animal than treatment with BAL (67). Evidence has been presented that at these low doses maleate was bound to a non-SH mercury binding site adjacent to the high affinity mercury binding site in kidney, reducing the affinity of the renal sites for mercury and creating a favorable situation for the efficient removal of mercury by penicillamine (105). While these observations are interesting and may in the future lead to some improvement in the treatment of mercury poisoning, at present no useful clinical applications of these observations has been demonstrated.

The development of thiol chelating agents to remove mercury from its binding sites in vivo is one example of the development of successful therapeutic agents as a result of some knowledge of the biocomplexes of mercury. Recently, a new approach to the treatment of alkylmercury poisoning based on a knowledge of methylmercury distribution and excretion has been proposed. The rapid biliary excretion of methylmercury complexes which are to a large extent reabsorbed and retained in the body suggested the possibility that this mercury might be trapped in the intestine as a nonabsorbable complex, with a consequent increase the fecal excretion of methylmercury. Takahashi and Hirayama (110) fed a diet containing 3.3% reduced human hair, which contains cysteine SH groups to bind the mercury and is resistant to trypsin digestion, to mice given methylmercury intramuscularly and found the body burden of methylmercury was reduced in the group receiving reduced hair. Clarkson et al (30) tested the effect of feeding a polystyrene resin containing fixed sulfhydryl groups on the rate of excretion of methylmercury and on the gastrointestinal absorption of methylmercury in food. Table III illustrates the effectiveness of a polythiol resin in increasing the elimination of methylmercury administered as a single i.p. dose and in reducing the absorption of dietary methylmercury. Because the resins are not absorbed, there is unlikely to be any systemic toxicity associated with the antidote. Further, there is no possibility of a potentially hazardous redistribution of the mercury into sensitive tissues, as may result after systemic administration of a chelating agent. One of these polythiol resins and two penicillamine derivatives have been used to treat victims of the recent outbreak of methylmercury poisoning in Iraq and it appears that all three agents effectively increased methylmercury excretion in some cases (5), although the data is still too limited to permit a comparison of the relative efficacy of the different treatments. It is possible that the effectiveness of the resin may be improved by combining its use with other therapeutic measures designed to increase the biliary excretion of methylmercury. For example, BAL causes a redistribution of methylmercury from red blood cells to plasma and increases the biliary excretion of methylmercury (116), suggesting that chelating agents might be used to increase the biliary excretion in conjunction with the resin to trap the excreted mercury in the intestine. Phenobarbital, which has recently been shown to increase the biliary excretion of methylmercury, possibly by increasing the formation of methylmercury conjugates suitable for biliary excretion (70), is another agent which might be beneficially used in conjunction with the polythiol resins.

Table III. Excretion and Absorption of Methylmercury in Mice After Polythiol Treatment.[a]

	Biologic Half-Life (Whole Body) (days)	Rate of Absorption From Food (cpm/day)	Predicted Steady State Body Burden (cpm)
Control	7.3	10,450	110,000
Resin Treated	3.6	5,015	26,050

[a]Values were calculated from data in reference (30). Methylmercury was administered as a single i.p. dose of 0.5 mg Hg/kg or was present in food at a concentration of 0.5 mg Hg/gm (sp. act. = 4uCi/mg Hg). The resin was present in the diet at a concentration of 1% by weight.

The above examples were chosen to illustrate some practical applications of our knowledge of the binding, distribution and excretion of mercurials to the treatment of mercury poisoning. As the gaps in our knowledge concerning the particular molecular target sites and the primary biochemical lesions resulting from the interaction of mercurials with these sites are filled in, this knowledge should result in the development of new approaches to the reduction of the hazards of mercury toxicity and to the treatment of mercury poisoning.

Acknowledgement

The authors would like to express their appreciation to Dr. Daniel Gould for helpful discussions regarding the pathology of methylmercury poisoning, to Dr. Mendel Friedman for his invitation to include the present discussion in this symposium, and to Amy Noma for translating reference 74.

References

1. Aaseth, J. (1973). The effect of mercatodextran on distribution and toxicity of mercury in mice. Acta Pharmacol. Toxicol. 32:430-441.
2. Aberg, B., Ekman, L., Falk, R., Grietz, U., Persson, G. and Snihs, J.-O. (1969). Metabolism of methylmercury (^{203}Hg) compounds in man. Arch. Environ. Health 19:478-484.
3. Aposhian, H. V. and Aposhian, M. M. (1959). N-acetyl-d,l-penicillamine, a new oral protective agent against the lethal effects of mercuric chloride. J. Pharmacol. Exp. Ther. 126: 131-135.
4. Ashbel, S. I. (1964). Intoksikatsi Rtut-organitsheskimi jadokhimikatami. Medicina, Moskwa, cited by Piotrowski, J. K. Further investigations on binding and release of mercury in the rat, in Mercury, Mercurials and Mercaptans, ed. Miller, M. W. and Clarkson, T. W., C. C. Thomas, Springfield, pp. 247-260.
5. Bakir, F., Damluji, S. F., Amin-Zaki, L., Murtadha, M., Khalidi, A., Al-Rawi, N. Y., Tikriti, S., Dhahir, H. I. and Clarkson, T. W., Smith, J. C. and Doherty, R. A. (1973). Methylmercury poisoning in Iraq. Science 181:230-241.
6. Berglund, F. and Berlin, M. (1969). Risk of methylmercury cumulation in man and mammals and the relation between body burden of methylmercury and toxic effects, in Chemical Fallout, ed. Miller, M. W. and Berg, G. G., C. C. Thomas, Springfield, pp. 258-269.
7. Berlin, M. and Ullberg, S. (1963). Accumulation and retention of mercury in the mouse I. An autoradiographic study after a single intravenous injection of mercuric chloride. Arch. Environ. Health 6:589-601.

8. Berlin, M. and Ullberg, S. (1963). Accumulation and retention of mercury in the mouse III. An autoradiographic comparison of methylmercuric dicyandiamide with inorganic mercury. Arch. Environ. Health 6:610-616.
9. Berlin, M. and Lewander, T. (1965). Increased brain uptake of mercury caused by 2,3-dimercaptopropanol (BAL) in mice given mercuric chloride. Acta Pharmacol. Toxicol. 22:1-7.
10. Berlin, M., Jerksell, L.-G. and Nordberg, G. (1965). Accelerated uptake of mercury by brain caused by 2,3-dimercaptopropanol (BAL) after injection into the mouse of a methylmercuric compound. Acta Pharmacol. Toxicol. 23:312-320.
11. Berlin, M., Jerksell, L.-G. and von Ubisch, H. (1966). Uptake and retention of mercury in the mouse brain : A comparison of exposure to mercury vapor and intravenous injection of mercuric salt. Arch. Environ. Health 12:33-42.
12. Berlin, M., Fazackerley, J. and Nordberg, G. (1969). The uptake of mercury in the brains of mammals exposed to mercury vapor and to mercuric salts. Arch. Environ. Health 18:719-729.
13. Berlin, M., Nordberg, G. and Hellberg, J. (1973). The uptake and distribution of methylmercury in the brain of Saimiri sciureus in relation to behavioral and morphological changes, in Mercury, Mercurials and Mercaptans, ed. Miller, M. W. and Clarkson, T. W., C. C. Thomas, Springfield, pp. 187-206.
14. Cafruny, E. J. (1968). The site and mechanism of action of mercurial diuretics. Pharmacol. Rev. 20:89-116.
15. Cassano, G. B., Viola, P. L., Ghetti, B. and Amaducci, L. (1969). The distribution of inhaled mercury (Hg^{203}) vapors in the brain of rats and mice. J. Neuropathol. Exp. Neurol. 28:308-320.
16. Cavanaugh, J. B. and Chen, F. C. K. (1971). The effects of methylmercury-dicyandiamide on the peripheral nerves and spinal cord of rats. Acta neuropathol. (Berl.) 19:208-215.
17. Cember, H. (1962). The influence of the size of the dose on the distribution and elimination of inorganic mercury, $Hg(NO_3)_2$, in the rat. Am. Ind. Hyg. Assn. J. 23:304-313.
18. Cember, H., Gallagher, P. and Faulkner, A. (1968). Distribution of mercury among blood functions and serum proteins. Am. Ind. Hyg. Assn. J. 29:233-237.
19. Chang, L. W. and Hartmann, H. A. (1972). Ultrastructural studies of the nervous system after mercury intoxication I. Pathological changes in the nerve cell bodies. Acta Neuropathol. (Berl.) 20:122-138.
20. Chang, L. W. and Hartmann, H. A. (1972). Ultrastructural studies of the nervous system after mercury intoxication II. Pathological changes in the nerve fibers. Acta Neuropathol. (Berl.) 20:316-334.
21. Chang, L. W. and Hartmann, H. A. (1972). Blood-brain barrier dysfunction in experimental mercury intoxication. Acta Neuropathol. (Berl.) 21:179-184.

22. Chang, L. W. and Hartmann, H. A. (1972). Electron microscopic histochemical study on the localization and distribution of mercury in the nervous system after mercury intoxication. Exp. Neurol. 35:122-137.
23. Cherian, M. G. and Vostal, J. J. (1974). Biliary excretion of mercury compounds: II. The effect of dose and time after administration on the forms of mercury in rat bile. Toxicol. Appl. Pharmacol. (in press).
24. Clarkson, T., Gatzy, J. and Dalton, C. (1961). Studies on the equilibration of mercury vapor with blood. Atomic Energy Commission Research and Development Report #UR-582, University of Rochester, New York.
25. Clarkson, T. W. and Magos, L. (1966). Studies on the binding of mercury in tissue homogenates. Biochem. J. 99:62-70.
26. Clarkson, T. W. and Magos, L. (1967). The effect of sodium maleate on the renal deposition and excretion of mercury. Br. J. Pharmacol. Chemother. 31:560-567.
27. Clarkson, T. W. (1971). Epidemiological and experimental aspects of lead and mercury contamination of foods. Fd. Cosmet. Toxicol. 9:229-243.
28. Clarkson, T. W. (1972). Recent advances in the toxicology of mercury with emphasis on the alkylmercurials. CRC Crit. Rev. Toxicol. 1:203-234.
29. Clarkson, T. W. (1972). The pharmacology of mercury compounds. Ann. Rev. Pharmacol. 12:375-406.
30. Clarkson, T. W., Small, H. and Norseth, T. (1973). Excretion and absorption of methylmercury after polythiol resin treatment. Arch. Environ. Health 26:173-176.
31. Clarkson, T. W. and Vostal, J. J. (1973). Mercurials, mercuric ion and sodium transport, in Modern Diuretic Therapy in the Treatment of Cardiovascular and Renal Disease, ed. Wilson, G. and Laut, A. F., Exerpta Medica Foundation, Amsterdam, pp. 221-230.
32. Cotton, F. A. and Wilkinson, G. (1966). Advanced Inorganic Chemistry - A Comprehensive Text. Wiley-Interscience, New York.
33. Criteria Document: Recommendations for an Occupational Exposure Standard for Inorganic Mercury. (1973). U. S. Dept. of Health, Education and Welfare, Public Health Service, National Institute for Occupational Safety and Health.
34. Davies, D. J. and Kennedy, A. (1967). Course of the renal excretion of cells after necrosis of the proximal convoluted tubule by mercuric chloride. Toxicol. Appl. Pharmacol. 10:62-68.
35. D'Itri, F. M. (1962). The Environmental Mercury Problem. CRC Press, Cleveland.
36. Fang, S. C. (1973). The in vivo kinetics of mercury binding of kidney soluble proteins from rats receiving various mercurials, in Mercury, Mercurials and Mercaptans, ed. Miller, M. W. and

Clarkson, T. W., C. C. Thomas, Springfield, pp. 277-291.

37. Fassett, D. W. (1972). Cadmium, in Metallic Contaminants and Human Health, ed. Lee, D. H. K., Academic Press, New York, pp. 97-124.
38. Fesenko (1969). Vrachebnoe Delo 10:85-87.
39. Friberg, L. and Vostal, J. J., eds. (1972). Mercury in the Environment. CRC Press, Cleveland.
40. Friedman, M. (1973). Chemistry and Biochemistry of the Sulfhydryl Group in Amino Acids, Peptides, and Proteins. Pergamon Press, Oxford, pp. 311-363.
41. Goldwater, L. J. (1936). From Hippocrates to Ramazzini: Early history of industrial medicine. Ann. Med. Hist. 8:27, cited by D'Itri, F. M. (1972), in The Environmental Mercury Problem, CRC Press, Cleveland, p. 6.
42. Goldwater, L. J. (1971). Mercury in the environment. Scientific Am. 224:15-21.
43. Goldwater, L. J. and Clarkson, T. W. (1972). Mercury, in Metallic Contaminants and Human Health, ed. Lee, D. H. K., Academic Press, New York, pp. 17-55.
44. Gould, E. S. (1962). Inorganic Reactions and Structure. Holt, Rinehart and Winston, New York.
45. Grant, C. A. (1973). Pathology of methylmercury intoxication: some problems of exposure and response, in Mercury, Mercurials and Mercaptans, ed. Miller, M. W. and Clarkson, T. W., C. C. Thomas, Springfield, pp. 294-310.
46. Greenwood, M. R., Clarkson, T. W. and Magos, L. (1972). Transfer of metallic mercury into the fetus. Experientia 28:1455-1456.
47. Gurd, F. R. N. and Wilcox, P. E. (1956). Complex formation between metallic cations and proteins, peptides, and amino acids. Adv. Protein Chem. 11:311-427.
48. Hartung, R. and Dinman, B. D., eds. (1972). Environmental Mercury Contamination. Ann Arbor Science Publishers, Ann Arbor.
49. Hayes, A. D. and Rothstein, A. (1962). The metabolism of inhaled mercury vapor in the rat studied by isotope techniques. J. Pharmacol. Exp. Ther. 138:1-10.
50. Herman, S. P., Klein, R.,Talley, F. A. and Krigman, M. R. (1973). An ultrastructural study of methylmercury induced primary sensory neuropathy in the rat. Lab. Invest. 28:104-118.
51. Herman, S. P., Krigman, M. R., Klein, R. and Somjen, G. G. (1974). The progression of clinical signs, structural lesions and functional deterioration in sensory neuropathy caused by methylmercury poisoning. J. Pharmacol. Exp. Ther. (in press).
52. Hirayama, K. and Takahashi, H. (1970). Studies on the treatment for methylmercury poisoning. "Lowering of the methylmercury content in the poisoned animal brain." Kumamoto Med. J. 23:56-64.
53. Holmstedt, B. (1967). The toxicology and metabolism of the different mercury compounds, in The mercury problem, Oikos

Suppl. 9:25, cited by D'Itri, F. M. (1972). The Environmental Mercury Problem, CRC Press, Cleveland, p. 6.

54. Hughes, W. L. (1957). A physiochemical rationale for the biological activity of mercury and its compounds. Ann. N.Y. Acad. Sci. 65:454-460.

55. Hunter, D., Bomford, R. R. and Russell, D. S. (1940). Poisoning by methyl mercury compounds. Q.J. Med. 9(New Series): 193-213.

56. Hunter, D. and Russell, D. S. (1954). Focal cerebral and cerebellar atrophy in a human subject due to organic mercury compounds. J. Neurol. Neurosurg. Psychiat. 17:235-241.

57. Jakubowski, M., Piotrowski, J. and Trojanowska, B. (1970). Binding of mercury in the rat: Studies using $^{203}HgCl_2$ and gel filtration. Toxicol. Appl. Pharmacol. 16:743-753.

58. Jernelöv, A. (1973). A new biochemical pathway for the methylation of mercury and some ecological implications, in Mercury, Mercurials and Mercaptans, ed. Miller, M. W. and Clarkson, T. W., C. C. Thomas, Springfield.

59. Kurland, L. T., Faro, S. N. and Seidler, H. (1960). Minamata disease. Wld. Neurol. 1:370-395.

60. Kurland, L. T. (1973). An appraisal of the epidemiology and toxicology of alkylmercury compounds, in Mercury, Mercurials and Mercaptans, ed. Miller, M. W. and Clarkson, T. W., C. C. Thomas, Springfield, pp. 23-47.

61. Levine, W. G. (1970). Heavy-metal antogonists, in The Pharmacological Basis of Therapeutics, Goodman, L. S. and Gilman, A., 4th ed., Mac Millan, London, pp. 944-957.

62. Longcope, W. T. and Luetscher, J. A. (1946). Clinical uses of 2,3-dimercaptopropanol (BAL). XI. The treatment of acute mercury poisoning by BAL. J. Clin. Invest. 25:557-567.

63. Lundgren, K.-D., Swensson, A. and Ulfvarson, U. (1967). Studies in humans on the distribution of mercury in the blood and the excretion in urine after exposure to different mercury compounds. Scand. J. Clin. Lab. Invest. 20:164-166.

64. Magos, L. (1968). Uptake of mercury by the brain. Br. J. Ind. Med. 25:315-318.

65. Magos, L. (1968). Transport of elemental mercury by blood. Naunyn-Schmiedeberg's Arch. Pharmacol. 259:183-184.

66. Magos, L. (1968). Effect of 2,3-dimercaptopropanol (BAL) on urinary excretion and brain content of mercury. Br. J. Ind. Med. 25:152-154.

67. Magos, L. and Stoytchev, T. (1969). Combined effect of sodium maleate and some thiol compounds on mercury excretion and redistribution in rats. Br. J. Pharmacol. 35:121-126.

68. Magos, L. (1973). Factors affecting the uptake and retention of mercury by kidneys in rats, in Mercury, Mercurials and Mercaptans, ed. Miller, M. W. and Clarkson, T. W., C. C. Thomas, Springfield, pp. 167-184.

69. Magos, L., Clarkson, T. W. and Greenwood, M. R. (1973). The

depression of pulmonary retention of mercury vapor by ethanol: Identification of the site of action. Toxicol. Appl. Pharmacol. 26:180-183.

70. Magos, L., MacGregor, J. T. and Clarkson, T. W. (1974). The effect of phenobarbital and sodium dehydrocholate on the biliary excretion of methylmercury in the rat. Toxicol. Appl. Pharmacol. (in press).

71. Magos, L., Sugata, Y. and Clarkson, T. W. (1974). Effects of 3-amino-1,2,4-triazole on mercury uptake by in vitro human blood samples and by whole rats. Toxicol. Appl. Pharmacol. (in press).

72. Massey, T. H. and Fang, S. C. (1968). A comparative study of the subcellular binding of phenylmercuric acetate and mercuric acetate in rat liver and kidney slices. Toxicol. Appl. Pharmacol. 12:7-14.

73. Matsumoto, H., Suzuki, A., Morita, C., Nakamura, K. and Saeki, S. (1967). Preventive effect of penicillamine on the brain defect of fetal rat poisoned transplacentally with methylmercury. Life Sciences 6:2321-2326.

74. Matsumoto, H., Kameda, T. and Takeuchi, T. (1969). Pathological studies on organic mercury poisoning--Supplement to histochemical demonstration of mercury in the brain of Minamata disease. Adv. Neurol. Sci. (Jap.) 13:270-278.

75. Mellor, J. W. (1940). A Comprehensive Treatise on Inorganic and Theoretical Chemistry, Vol. IV. Longmans, Green and Co., London.

76. Miller, M. W. and Clarkson, T. W., eds. (1973). Mercury, Mercurials and Mercaptans. C. C. Thomas, Springfield.

77. Minamata Disease (1968). (Study Group of Minamata Disease), Kumamoto University, Japan,338 pp.

78. Miyakawa, T. and Deshimaru, M. (1969). Electron microscopial study of experimentally induced poisoning due to organic mercury compound. Acta Neuropathol. (Berl.) 14:126-136.

79. Miyakawa, T., Deshimaru, M., Sumiyoshi, S., Teraoka, A., Udo, N., Hattori, E. and Tatetsu, S. (1970). Experimental organic mercury poisoning. Changes in peripheral nerves. Acta Neuropathol. (Berl.) 15:45-55.

80. Mudge, G. H. and Weiner, I. M. (1958). The mechanism of action of mercurial and xanthine diuretics. Ann. N.Y. Acad. Sci. 71:344-354.

81. Nelson, N., Byerly, T. C., Kolbye, A. C., Kurland, L. T., Schapiro, R. E., Shibko, S. I., Stickel, W. H., Thompson, J. E., Van Den Berg, L. A. and Weissler, A. (1971). Hazards of Mercury. Env. Res. 4:1-69.

82. Nielsen Kudsk, F. (1965). Absorption of mercury vapour from the respiratory tract in man. Acta Pharmacol. Toxicol. 23:250-262.

83. Nielson Kudsk, F. (1965). The influence of ethyl alcohol on the adsorption of mercury vapour from the lungs in man. Acta Pharmacol. Toxicol. 23:263-274.

84. Nielsen Kudsk, F. (1969). Uptake of mercury vapour in blood in vivo and in vitro from Hg-containing air. Acta Pharmacol. Toxicol. 27:149-160.
85. Nielsen Kudsk, F. (1969). Factors influencing the in vitro uptake of mercury vapour in blood. Acta Pharmacol. Toxicol. 27:161-172.
86. Nielsen Kudsk, F. (1973). Biological oxidation of elemental mercury, in Mercury, Mercurials and Mercaptans, ed. Miller, M. W. and Clarkson, T. W., C. C. Thomas, Springfield, pp.355-369.
87. Nordberg, G. F. and Serenius, F. (1969). Distribution of inorganic mercury in the guinea pig brain. Acta Pharmacol. Toxicol. 27:269-283.
88. Nordberg, G. F. and Skerfving, S. (1972). Metabolism, in Mercury in the Environment, ed. Friberg, L. and Vostal, J. CRC Press, Cleveland, pp. 29-91.
89. Norseth, T. and Clarkson, T. W. (1970). Studies on the biotransformation of ^{203}Hg-labeled methyl mercury chloride in rats. Arch. Environ. Health 21:717-727.
90. Norseth, T. and Clarkson, T. W. (1971). Intestinal transport of ^{203}Hg-labeled methyl mercury chloride. Arch. Environ. Health 22:568-577.
91. Norseth, T. (1973). Biliary complexes of methylmercury: A possible role in organ distribution, in Mercury, Mercurials and Mercaptans, ed. Miller, M. W. and Clarkson, T. W., C. C. Thomas, Springfield, pp. 264-272.
92. Norseth, T. (1973). Biliary excretion and intestinal reabsorption of mercury in the rat after injection of methyl mercuric chloride. Acta Pharmacol. Toxicol. 33:280-288.
92a. Ogawa, E., Suzuki, S., Tsuzuki, H., Tobe, M., Kobayashi, K. and Hojo, M. (1970). Experimental studies on the distribution and excretion of methylmercury chloride. Proc. of the International Symposium on Thiola, Nov. 25-26, 1970, pp.238-253.
93. Olsson, Y. (1968). Topographical differences in the vascular permeability of the peripheral nervous system. Acta Neuropathol. (Berl.) 10:26-33.
94. Passow, H., Rothstein, A. and Clarkson, T. W. (1961). The general pharmacology of the heavy metals. Pharmacol. Rev. 13:185-224.
95. Piotrowski, J. K., Trojanowska, B., Wisniewska-Knypl, J. M. and Bolanowska, W. (1973). Further investigations on binding and release of mercury in the rat, in Mercury, Mercurials and Mercaptans, ed. Miller, M. W. and Clarkson, T. W., C. C. Thomas, Springfield, pp. 247-260.
96. Pulido, P. Kägi, J. H. R. and Vallee, B. L. (1966). Isolation and some properties of human metallothionein. Biochemistry 5: 1768-1777.
97. Ramazzini, B. (1713). Diseases of Workers. Translated from the Latin text De Morbis Artificium of 1713 by Wright, W. C. Hafner Publishing Co., New York (1964).
98. Report of an Expert Committee. (1971). Methylmercury in fish. Nord. Hyg. Tidskr. Suppl. 4.

99. Report of an International Committee. (1969). Maximum allowable concentrations of mercury compounds. Arch. Environ. Health 19:891-905.
100. Rochow, E. G., Hurd, D. T. and Lewis, R. N. (1957). The Chemistry of Organometallic Compounds. John Wiley and Sons, New York.
101. Rothstein, A. (1973). Mercaptans, the biological targets for mercurials, in Mercury, Mercurials and Mercaptans, ed. Miller, M. W. and Clarkson, T. W., C. C. Thomas, Springfield, pp. 68-92.
102. Ruch, T. C. (1965). Binocular vision and central visual pathways, in Physiology and Biophysics, Ruch, T. C. and Patton, H. D., W. B. Saunders Co., Philadelphia, pp. 441-454.
103. Skerfving, S. (1972). Mercury in fish - Some toxicological considerations. Fd. Cosmet. Toxicol. 10:545-556.
104. Somjen, G. G., Herman, S. P., Klein, R., Brubaker, P. E., Briner, W. H., Goodrich, J. K., Krigman, M. R. and Haseman, J. K. (1973). The uptake of methyl mercury (^{203}Hg) in different tissues related to its neurotoxic effects. J. Pharmacol. Exp. Ther. 187:602-611.
105. Stoytchev, T., Magos, L. and Clarkson, T. W. (1969). Studies on the mechanism of the maleate action on the urinary excretion of mercury. Eur. J. Pharmacol. 8:253-260.
106. Suzuki, T., Matsumoto, N., Miyama, T. and Katsunuma, H. (1967). Placental transfer of mercuric chloride, phenyl mercury acetate and methyl mercury acetate in mice. Ind. Health 5:149-155.
107. Suzuki, T. and Yoshino, Y. (1969). Effects of D-penicillamine on urinary excretion of mercury in two cases of methyl mercury poisoning. Jap. J. Ind. Health 11:487-488.
108. Suzuki, T., Miyama, T. and Katsunuma, H. (1970). Mercury contents in the red cells, plasma, urine and hair from workers exposed to mercury vapour. Ind. Health 8:39-47.
109. Swensson, A. and Ulfvarson, U. (1968). Distribution and excretion of mercury compounds in rats over a long period after a single injection. Acta Pharmacol. Toxicol. 26:273-283.
110. Takahashi, H. and Hirayama, K. (1971). Accelerated elimination of methyl mercury from animals. Nature (Lond.) 232:201-202.
111. Takeuchi, T. (1972). Biological reactions and pathological changes in human beings and animals caused by organic mercury contamination, in Environmental Mercury Contamination, ed. Hartung, R. and Dinman, B. D., Ann Arbor Science Publishers, Ann Arbor, pp. 247-289.
112. Tejning, S. (1970). The mercury contents in blood corpuscles and in blood plasma in mothers and their new-born children. Rep. 70-05-20, Dept. Occup. Med., University Hospital, Lund, Sweden.
113. Tejning, S. Cited by Kurland, L. T. (1973). An appraisal of the epidemiology and toxicology of alkylmercury compounds, in Mercury, Mercurials and Mercaptans, ed. Miller, M. W. and Clarkson, T. W., C. C. Thomas, Springfield, pp. 23-47.
114. Trojanowska, B., Piotrowski, J. K. and Szendzikowski, S. (1971). The influence of thioacetamide on the excretion of mercury in rats. Toxicol. Appl. Pharmacol. 18:374-386.

115. Trojanowska, B. and Szendzikowski, S. Mobilizing effects of Unithiol upon mercury deposits in the rat. (In press).
116. Vostal, J. J. and Clarkson, T. W. (1971). The effect of BAL and related thiols on plasma/red blood cell distribution of methyl mercury. The Pharmacologist 13:289. (Abstract).
117. Vostal, J. (1972). Transport and transformation of mercury in nature and possible routes of exposure, in Mercury in the Environment, ed. Friberg, L. and Vostal, J., CRC Press, Cleveland.
118. Webb, J. L. (1966). Enzyme and Metabolic Inhibitors. Volume II. Academic Press, New York, pp. 729-1070.
119. Weed, R., Eber, J. and Rothstein, A. (1962). Interaction of mercury with human erythrocytes. J. Gen. Physiol. 45:395-410.
120. Weiner, I. M., Levy, R. I. and Mudge, G. H. (1962). Studies on mercurial diuresis: Renal excretion, acid stability and structure-activity relationships of organic mercurials. J. Pharmacol. Exp. Ther. 138:96-112.
121. Westöö, G. (1966). Determination of methylmercury compounds in foodstuffs. I. Methylmercury compounds in fish, identification and determination. Acta Chem. Scand. 20:2131-2137.
122. White, A., Handler, P. and Smith, E. L. (1964). Principles of Biochemistry. McGraw-Hill, New York, pp. 193-199.
123. Wisniewska, J. M., Trojanowska, B., Piotrowski, J. and Jakubowski, M. (1970). Binding of mercury in the rat kidney by metallothionein. Toxicol. Appl. Pharmacol. 16:754-763.
124. Wisniewska-Knypl, J. M., Trojanowska, B. B., Piotrowski, J. K. and Jablonska, J. K. (1972). Binding of mercury in rat liver by metallothionein. Acta Biochim. Pol. 19:11-18.
125. Wood, J. M. (1971). Environmental pollution by mercury, in Advances in Environmental Science and Technology, Vol. 2, ed. Pitts, J. N. and Metcalf, R. L., Wiley-Interscience, New York, pp. 39-56.
126. Yagi, K. and White, H. L. (1958). Comparison of ammonium sulfate fractionation of proteins and of protein-bound mercury in kidney soluble fraction of chow-fed and sucrose-fed rats. Am. J. Physiol. 194:547-552.
127. Yang, M. G., Krawford, K. S., Garcia, J. D., Wang, J. H. C. and Lei, K. Y. (1972). Deposition of mercury in fetal and maternal brain. Proc. Soc. Exp. Biol. Med. 141:1004-1007.
128. Yoshino, Y., Mozai, T. and Nakao, K. (1966). Distribution of mercury in the brain and its subcellular units in experimental organic mercury poisonings. J. Neurochem. 13:397-406.
129. Yoshino, Y., Mozai, T. and Nakao, K. (1966). Biochemical changes in the brain in rats poisoned with an alkylmercury compound, with special reference to the inhibition of protein synthesis in brain cortex slices. J. Neurochem. 13:1223-1230.

23

INTERACTIONS OF MERCURY COMPOUNDS WITH WOOL AND RELATED BIOPOLYMERS

Mendel Friedman and Merle Sid Masri

Western Regional Research Laboratory, Agricultural Research Service, U.S. Department of Agriculture, Berkeley, California 94710

I. ABSTRACT

Mercury compounds, which constitute environmental health hazards, can be taken up by wool, and other keratins and their derivatives, and by other agricultural products and by-products. Wool can bind mercury to about half of its weight from concentrated mercuric chloride solutions and can quickly recover a substantial proportion from very low, but biologically important, concentrations in the parts-per-billion range. It binds both naturally occurring and manufactured inorganic and organic mercury compounds. Binding capacity can be further increased by chemical modification, binding efficiency, by nondestructive recovery. Factors that influence binding of mercury compounds to keratins are discussed.

II. INTRODUCTION

Mercury is a well publicized, a strong, and subtle health hazard. This element has no known desirable effect, even in traces, in normal human nutrition. Widespread and persistent contamination of land and water from manufacturing and agricultural uses of mercury has increased the general level of unintended consumption (Table 1).

General exposure to ingested mercury has increased in the last ten to fifty years. Local episodes of acute poisoning have occurred. Environmental contamination by mercury is persistent and cannot easily be cleaned up by simple, direct methods, because mercury is firmly bound to organic sediments and soils, where it is metabolized by bacteria and persists and is recycled and concentrated in food chains. (Goldwater, 1971; Miller and Clarkson, 1973; McGregor and Clarkson, this volume). However, vegetation typically has a much lower content of mercury than the soil in which it grows (Malyuga, 1964). Evidently soil plants select against mercury by a factor of about ten. The corresponding evidence for water plants is not available, as far as we know. (See also, Anelli, Pelosi, and Galoppini, 1973).

We have proposed to take advantage of the affinity of wool for mercury compounds to recover them from contaminated water. The proposal affords an attractive possible use for waste wool too short for textiles, with the prospects of recovering a valuable product (mercury) and removing it as a health hazard. Our surveys (Friedman and Waiss, 1972; Masri and Friedman, 1972; Friedman, Harrison, Ward, and Lundgren, 1973; Friedman and Masri, 1973; Masri, Reuter, and Friedman, 1974) indicate that, besides wool, other agricultural protein and polyphenolic products such as feathers, blood meal, soy meal, walnut hulls, and peanut skins bind mercuric chloride and methylmercuric chloride effectively. A large fraction of adsorbed mercurials can be recovered from wool with either citrate, ethylenediamine tetracetate, or chloride at pH 6. Part, presumably combined with sulfhydryl groups, remains more firmly bound. These studies are intended to develop improved scavengers for toxic and industrial metals and to utilize the potential value of keratin proteins and related biopolymers as instructive models for mercury binding *in vivo*.

The following key references and the references cited therein form a useful entry to the literature on protein-mercury interactions: Leach, 1960, 1966; Webb, 1966; Friberg, 1972; Tratnyek, 1972; Miller and Clarkson, 1973; Friedman, 1973; Friedman et. al., 1973; and MacGregor and Clarkson, this volume. Related studies of interactions of wool (Brady et. al., 1973) and animal hair (Michelsen, 1973) with various metal ions are currently under way in other laboratories.

III. WOOL

A. Special Characteristics. Some characteristics that may give wool and other hairs special utility are (1) low solubility; (2) accessibility to water and solutes in aqueous media; (3) physical form, as crimped and resilient fibers with diameters of the order of 20 to 100 μm; (4) relatively high content of particular reactive groups (Table 2) that may serve as binding sites for mercu-

Table 1. Significant Concentrations of Mercury

Source	Dissolved Concentration as cited	Dissolved Concentration moles/liter	Reference
Sea water	0.03 ppb	1.5×10^{-10}	Monier-Williams, 1949; Löfroth, 1969; Sillen, 1963
	0.1 ppb	5×10^{-10}	Robertson, 1970
Rain	0.2 ppb	1×10^{-9}	Monier-Williams, 1949; USGS, 1970
Human Blood plasma[a]	2.3 ng/g	1×10^{-8}	Löfroth, 1969
Public water supply, maximum permissible[b]	5 ppb	2.5×10^{-8}	USPHS, 1962
Maumee River, Antwerp, Ohio[c]	6 ppb	3×10^{-8}	USDI, 1970
Merced River, Happy Isles, Yosemite National Park[d]	6 ppb	3×10^{-8}	Durum and others, 1971
Chlorine plant effluent, with efficient sludge removal	0.1 mg/liter 0.3 mg/liter	5×10^{-7} 1.5×10^{-6}	Bouveng and Ullman, 1969
Chlorine plant waste, industrial sample	138 ppm	6.9×10^{-4}	Our research

Source	In solids Concentration as cited	In solids Concentration moles/kilogram	Reference
Food, range commonly present[e]	0.005 ppm to 0.05 ppm	2.5×10^{-8} 2.5×10^{-7}	Monier-Williams, 1949
Food, provisional Maximum allowable in total diet[e]	50 ng/g 0.05 ppm	2.5×10^{-7}	FAO/WHO, 1965, 1970
Ocean Fish, usual upper limit	150 ng/g	7.5×10^{-7}	Löfroth, 1969
Soil, normal	0.1 ppm to 0.3 ppm	5×10^{-7} 1.5×10^{-6}	Monier-Williams, 1949; USGS, 1970
Ocean sediments (n = gram-atoms per 0.6 kg of sediment)	- log n = 5.8	2.6×10^{-6}	Sillen, 1963
U.S. canned tuna[f]	0.37 ppm	1.8×10^{-6}	New York Times, 1970
Fish, provisional maximum allowable			
1. USFDA, 1970	0.5 ppm	2.5×10^{-6}	New York Times, 1970
2. Sweden, 1969	1. mg/kg	5×10^{-6}	Miettinen, 1969
Fish, 'contaminated'			
1. Tuna, maximum	1.12 ppm	5.6×10^{-6}	New York Times, 1970

Table 1. continued

Source	In solids Concentration as cited	moles/kilogram	Reference
Fish 'contaminated'			
2. Swordfish	1.3 ppm	6.5×10^{-6}	S.F. Chronicle, 1970
3. Halibut	6.0 ppm		Spinelli *et al.*, 1973
Human hair, normal	1,350ng/g	6.7×10^{-6}	Löfroth, 1969
	0.2 ppm	1×10^{-6}	Jervis, 1970
	to 15 ppm	7.5×10^{-5}	
	5-100mg/g		Nord *et al.*, 1973
Human hair, (ab)normal[g]	150 ppm	7.5×10^{-4}	Jervis, 1970
Fish, Minimata Bay, toxic[e]	5 mg/kg[h]	2.5×10^{-5}	Miettinen, 1969
	to 20 mg/kg	1×10^{-4}	
	27 mg/kg[i]	1.3×10^{-4}	Löfroth, 1969
	to 102 mg/kg	5×10^{-4}	
Wheat flour from methylmercury-treated grain	3.7 mg Hg/gm	1.85×10^{-2}	Bakir *et al.*, 1973
	to 14.9 mg Hg/gm	7.45×10^{-2}	
Wool, residual binding[j]	4.26 mg/g	2.1×10^{-2}	Our research
Wool, maximum recorded binding at 25°C.[k]	235 mmoles/100g	2.35×10^{0}	Speakman and Coke, 1939

[a]Presumably normal.

[b]"Mandatory maximum;" higher concentrations are grounds for rejecting water supply

[c]Sample take June 1970; this was the only station with more than the "mandatory maximum" among 430 tested.

[d]Sample taken October 1970 including sediment; highest value in publication cited.

[e]Based on fresh weight. Contents in fish are not affected importantly by method of cooking.

[f]Average of 138 samples.

[g]Estimated upper limit for absence of toxic symptoms.

[h]Estimated contents in seafood that poisoned frequent consumers.

[i]Analyses related to 88 cases of poisoning per 10,000 population.

[j]Wool was saturated with mercuric chloride, which was then removed as much as possible by washing several times with 0.01 M ethylenediaminetetraacetic acid adjusted to pH 6.

[k]From 0.2 M mercuric acetate in 0.1 M acetic acid.

References to Table 1

Bakir, F., Damluji, S. F., Amin-Zaki, L., Murtadha, X. M., Khalidi, A., Al-Rawi, N. Y., Tikriti, S., Dhahir, H. I., Clarkson, T. W., Smith, J. C., and Doherty, R. A. (1973). Methylmercury poisoning in Iraq. Science 230, 131.

Bishop, J. E. Wall Street Journal (December 14, 1970) Pacific Coast Edition 83 (117): 1,9.

Bouveng, H. O., and Ullman, P. "Reduction of Mercury in Waste Waters from Chlorine Plants" (1969). Chlorine Institute Pamphlet Number R-10. New York.

Durum, W. H., Hem, J. D., and Heidel, S. G. (1971) "Reconnaissance of Selected Minor Elements in Surface Waters of the U.S. October 1970," U.S.D.I. Geological Survey Circular 643, Washington.

Jervis, R. E. Cited in Chem. Eng. News, October 8, 1970, page 8.

Löfroth, G. (1969) "Methylmercury." Chlorine Institute Pamphlet Number R-101, New York.

Miettinen, J. K. "On the methylmercury problem and the present status of the mercury investigations by the Radiochemistry Department." Presented at the Finnish Chemists Meeting in Helsinki, October 14-16, 1969; transcript from the Chlorine Institute, New York.

Monier-Williams, G. W. "Trace Elements in Foods." Wiley, New York, 1949.

New York Times Service as reported in the San Francisco Chronicle, December 16, 1970, page 1.

Nord, P. J., Kadaba, M. P., Sorenson, J. R. J. (1973) Mercury in human hair. A study of the residents of Los Alamos, N.M., and Pasadena, California by cold vapor atomic absorption spectrophotometry. Arch. Environ. Health 27, 40-41.

D. Robertson as quoted by Bishop in the article cited above.

San Francisco Chronicle, December 18, 1970, page 3.

Sillen, L. G. (1973) How has sea water got its present composition? Svensk Kem. Tid. 75, 161-177.

Speakman, J. B., and Coke, C. E. (1939) The action of mercuric chloride on wool and hair. Trans. Faraday Soc. 35, 246-262.

Spinelli, J., Steinberg, M. A., Miller, R., Hall, A., and Lehman, L. (1973). Reduction of mercury with cysteine in comminuted halibut and hake fish protein concentrate. J. Agr. Food Chem. 21, 264-268.

United States Department of the Interior News Release, October 4, 1970. United States Public Health Service. "Drinking Water Standards." USPHS Publication 956. Washington, 1962.

WHO/FAO as reported by Jervis and Bishop in the articles cited and cited by R. Christoll, L. G. Erwall, K. Ljunggren, B. Sjöstrand, and T. Westermark, "Methods of activation analysis for mercury in the biosphere and in foods." Presented at the 1965 International Conference on Modern Trends in Activation Analysis. College Station, Texas.

rials or that can be chemically modified to provide binding sites or a more favorable binding environment; and (5) variety and juxtaposition of reactive sites that may allow cooperative reaction to bind substances more effectively than by the different kinds of sites acting individually.

B. Effect of pH. In the absence of strongly bound ions other than hydrogen, wool has a region of very small net charge between about pH 5 and 9. Below this range it has an increasing net positive charge as carboxyl groups bind hydrogen. Above this range imidazole, sulfhydryl, phenol, amine, and guanidine groups successively release hydrogen so that wool is increasingly negative. The net charge on wool will be specifically affected by selective uptake of other ions.

The pH's and times used in our first survey were chosen on the basis of preliminary tests with wool. Figure 1 shows the relative binding of mercuric chloride and methylmercuric chloride at various pH's. Table 3 gives also the binding in moles per gram of wool and the residual concentrations. Under the test conditions, mercuric chloride is taken up in substantial amounts at all pH's from 2 to 10, and best near or below 2 and near 9. Methylmercuric chloride is less adsorbed, and its pH dependence differs; its maximum binding is near pH 10. Effects of time and concentration on the extent of binding are illustrated in Tables 4-7. The wool-liquid ratio in these studies was usually 1:100.

C. Salt Effects. Webb (1966) has reviewed factors governing interaction of mercury with proteins because of its wide use in studying enzyme reactivity. Mercuric ion has a strong, well-defined tendency to form complexes with chloride and especially hydroxyl ions. Hydroxyl complexes tend to precipitate well on the acid side of neutrality, especially if the salt is well dissociated. Consequently, most sorption measurements have been made in acid media.

Mercuric nitrate in strong acid and in the absence of complex-forming species gives the divalent cation as the main form of mercury. Mercuric chloride is very little dissociated. In the presence of excess chloride, the anions $HgCl_3^-$ and $HgCl_4^{2-}$ form successively. In 0.1M chloride at low pH, about 40% of the mercury is present as $HgCl_2$ (un-ionized), 25% as $HgCl_3^-$, and 35% as $HgCl_4^{2-}$. Methylmercuric ion forms analogous complexes, but the pK's for dissociation are roughly 1 unit less than for the Hg^{2+} complexes.

Mercuric chloride tends to associate with chloride ions to form negatively charged complexes. For this reason, it was expected that chloride ion (e.g., in lakes, sea water, and industrial effluents) would interfere with mercuric chloride binding to wool (Leach, 1960, 1966). Systematic studies, summarized in Tables 8 and 9 show that this phenomenon operates for both native and various chemically modified wools. Chloride interference due to complexing with $HgCl_2$ should decrease with increasing pH leading to hydrolysis as Cl^- is replaced by OH^-. This is indeed the case.

It is noteworthy (Table 10) that (a) sulfate ion does not inhibit mercury binding since $MgSO_4$, $ZnSO_4$, and Na_2SO_4, did not affect mercury uptake by wool; (b) the chloride ion effect appears to be at a maximum limiting value at about one molar chloride concentration; (c) the chloride effect is independent of the particular cation used since our studies show that NaCl, KCl, $CaCl_2$, $FeCl_2$, or $CuCl_2$ all inhibit mercury binding to wool; and (d) we observe a similar halide ion effect with methyl mercuric chloride and with sodium iodide (Table 11).

D. Effect of Chemical Modification. We wish to explore further how the mercury-binding capacity of wool can be increased by chemical modification, as would be expected if additional binding sites are introduced. We find that it is indeed feasible to increase the binding capacity of wool by such chemical modification and that modified wools appear to be useful for recovering mercurials from contaminated water.

Mercury binding by wool, wool derivatives, and ion exchange resins is compared in Tables 9, 12-17. The results show that (a) chemical modification usually increased mercury binding by wool; (b) both native wool and the chemically modified wools are more effective mercury scavengers than the synthetic resins tested; and (c) the enhanced effectiveness takes place at both low and high initial mercury concentration.

Although most derivatives bound more mercury than native wool, Table 9 shows that binding by S-(p-nitrophenethyl) wool is less than that by native wool. Leach (1960) found that either esterifying wool (removing carboxyl) or reducing and then carboxymethylating it greatly increased binding of $HgCl_2$. These increases may be due in part to making the wool more accessible. However, our results show that reduction of wool followed by alkylation with a reagent that does not introduce potential ligand sites does not necessarily improve its mercury-binding properties. In contrast, reduced wool (W-SH), oxidized wool ($W-SO_3^-$), and thiosulfate wool ($W-S-SO_3^-$) all show greater binding than with a native wool (Table 14). Thus, SH, SO_3^-, and $S-SO_3^-$ appear especially effective for mercury binding. Further improvement in the binding capacity of wool could undoubtedely be achieved by thiolation, that is, by incorporating SH groups in addition to those that can be made by reducing disulfide bonds.

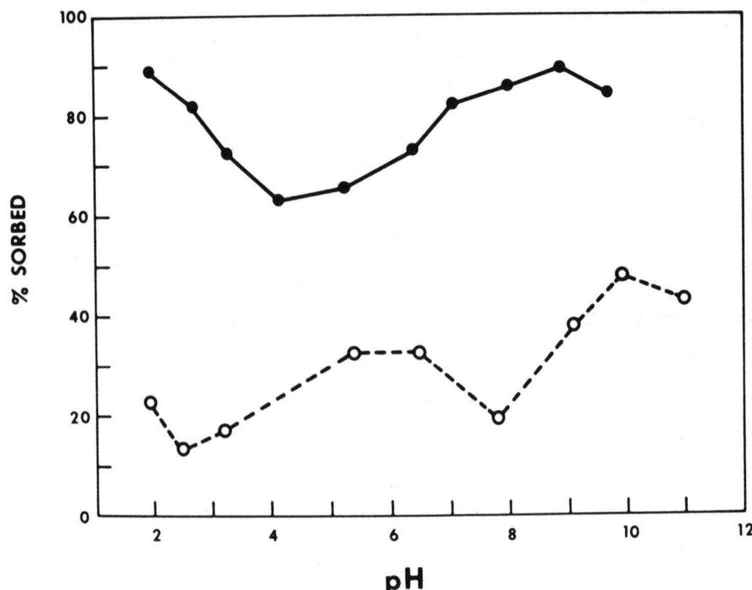

Fig. 1. Effect of pH on sorption of mercuric chloride (upper plot) and methylmercuric chloride by wool (lower plot). (Friedman et al., 1973)

Results of a series of experiments on the binding of mercuric chloride over a wide range of concentrations to native, reduced, and S-(2-pyridylethyl) wools (Tables 15-17) suggest that the modified wools have a greater capacity for $HgCl_2$ at the parts-per-million mercuric ion levels. However, at the parts-per-billion level, the three wool types studied appear equally effective, possibly because covalent binding to residual SH groups takes place.

IV. MECHANISM OF BINDING

A. Rates of Sorption. Our initial observations suggested that 90% or more of sorption from $HgCl_2$ (in HCl at pH 2) or from methylmercuric chloride at pH 6 might be reached in 30 to 60 min, but that sorption from $Hg(NO_3)_2$ (in HNO_3 at pH 2) was much slower (Tables 18, 19). At the higher concentrations, binding from $HgCl_2$ may reach essentially its equilibrium value in 6 hr, from $Hg(NO_3)_2$ not before

Table 2

Potential Reactive Groups in Wool

Kind[a]	Concentration, moles/kg
Peptide (secondary amide)	8.8
Aliphatic hydroxyl	1.47
Half-disulfide	0.86
Total base	0.86
arginine	0.55
lysine	0.22
histidine	0.07
terminal amine	0.02
Free carboxyl	0.84
Primary amide	0.76
Phenolic hydroxyl	0.29
Tryptophan	0.04
Methionine	0.04
Sulfhydryl	0.04

[a]The disulfide content is variable and can be influenced by nutrition and weathering. The carboxyl content can be increased at the expense of primary amide by hydrolysis. The sulfhydryl content can be increased by copper deficiency in the diet and is also affected by weathering. (Friedman et. al., 1973)

one or two weeks. At the low concentration, the rates and amounts of binding from these two salts are much more alike: about 90% of the available mercury was taken up in 1 hr or less. However, these measurements do not define the residual concentration in equilibrium as clearly as desirable for theoretical analysis.

The main point of practical interest is that as the concentration is decreased, sorption of inorganic mercury, either the chloride or the nitrate, from acid solution becomes very efficient.

Table 3

Effect of pH on Sorption Mercuric Chloride and Methylmercuric Chloride by Wool[a]

Mercuric chloride			Methylmercuric chloride		
pH	Sorption, micromoles Hg/g wool	Residual concentration, micromoles/l.	pH	Sorption, micromoles Hg/g wool	Residual concentration, micromoles/l.
1.95	93.6(88.4%)	0.110	1.95	36.0(22.8%)	1.097
2.7	86.4(82.1%)	0.169	2.5	19.4(13.2%)	1.147
3.3	76.4(72.6%)	0.259	3.2	30.5(17.2%)	1.321
4.15	66.5(63.2%)	0.349	5.4	50.9(32.6%)	0.947
5.25	69.2(65.8%)	0.324	6.5	47.1(32.7%)	0.872
6.4	77.0(73.2%)	0.254	7.8	27.7(22.2%)	0.872
7.05	86.4(82.1%)	0.169	9.1	50.9(38.0%)	0.748
8.0	90.3(85.8%)	0.135	9.95	63.7(47.9%)	0.623
8.9	94.2(89.5%)	0.100	11.0	49.9(42.9%)	0.598
9.7	88.6(84.2%)	0.150			

[a]For each measurement, 1.35g(dry weight) of wool was shaken at room temperature (21°C) in 150 ml of buffer for 2hr (mercuric chloride) or for 30min (methylmercuric chloride). The initial concentration of the mercuric chloride was 190 micrograms Hg/ml; initial concentrations of the methylmercuric chloride varied from 210 to 320 micrograms Hg/ml. (Friedman et. al., 1973)

Table 4

Effect of Time on Absorption of $HgCl_2$ (Two Initial Concentrations) by Wool at pH 2.0

Time (minutes)	Concentration (ppm.) as Hg	Time (minutes)	Concentration (ppm.) as Hg
0	1975	0	17.5
5	1525	5	8.5
10	1335	10	7
15	1225	15	5.5
30	1105	30	3.25
61	1105	62	2.5
360	1075	420	.5

Table 5

Effect of Time on Absorption by Wool of CH_3HgCl from an Initial Concentration of 315 ppm at pH 6.0

Time (min.)	ppm as Hg
0	315
3	250
5	235
10	225
15	220
30	215
70	210

Table 6

Effect of Concentration of $HgCl_2$ on Extent of Absorption by Wool at pH 2.0 After Shaking for 48 Hours

Original concentration (mg/ml) as Hg	Final concentration (mg/ml) as Hg	Percent left in solution
1840	1000	54.4
580	120	20.7
200	28	14
58	1	1.7
38	1	2.6
20	1	5

Table 7

Effect of Concentration of CH_3HgCl on the Extent of Absorption by Wool at pH 6.0 After Shaking for 30 Minutes

Original concentration (mg/ml) as Hg	Final concentration (mg/ml) as Hg	Percent left in solution
290	200	69
80	30	37.5
51	10	19.6
26	4	15.4
15.4	2	13

Table 8

Effect of NaCl on Mercury Uptake by Wool

	Solution pH 2		Solution pH 7	
Concentration of added NaCl, M	Residual concn., mg Hg/l. (ppm)	Binding, mg Hg/g	Residual concn., mg Hg/l.	Binding, mg Hg/g
0	170	43	320	28
0.15	270	33	450	15
0.5	380	22	520	8
1	460	14	520	8
2	480	12	540	6
Initial concn.	600		600	

[a]One gram of wool was shaken for 1 hr at room temperature with 100 ml of 0.01N HCl (pH2) or water (pH7) with added sodium chloride as indicated. The initial mercury content was 600 mg/l. Mercury binding was calculated from residual concentration of Hg in solution. (Friedman and Masri, 1973)

Table 9

Effect of Chloride on Hg Uptake by Modified Wools[a]

Wool	NaCl, M	Wt. increase, mg/g original wool
Native	0.0	510
Reduced plus vinylpyridine[b]	0.0	810
Reduced plus vinylpyridine	1.0	213
Reduced plus vinylimidazole[b]	0.0	605
Reduced plus vinylimidazole	1.0	285
Reduced plus vinylpyrrolidone[b]	0.0	680
Reduced plus vinylpyrrolidone	1.0	307
Reduced plus p-NO_2-styrene(HCl)	0.0	202
Reduced plus p-NO_2-styrene(HCl)	1.0	107
Reduced plus p-NO_2-styrene(H_2O)	0.0	430
Reduced plus p-NO_2-styrene(H_2O)	1.0	81

[a]One gram of native or modified wool was shaken for 24 hr at 25°C in 50 ml 0.2M $HgCl_2$ in 0.1N HCl without or with 1M NaCl as shown. The sample was washed in water and the uptake determined as the increased weight after drying.

[b]See Table 10 for structures.

(Friedman and Masri, 1973)

Table 10

Chloride Ion Effect on Mercury Uptake by Wool from Aqueous $HgCl_2$

Salt	Conc. (M)	Hg Uptake mg/g
NaCl	0.0	76
	.01	75
	.5	50
	2.	26
KCl	.01	73
	.5	46
	2.	20
$CaCl_2$	0.1	73
	.5	35
	2.	22
$CuCl_2$	.01	72
	.1	60
	1.	28
$FeCl_2$	.1	59
	1.	30
$ZnSO_4$	1.	67
$MgSO_4$	1.	60

1 gram of wool was shaken in 50 ml of 0.01M $HgCl_2$ in H_2O for 24 hrs at 25°C.

Table 11

Effect of NaCl on CH_3HgCl Binding

Control	pH 7	pH 2
Native wool	600 μg/ml	600 μg/ml
Native wool + 1.0M NaCl	350	500
S-β-(4-Pyridylethyl)wool	500	500
S-β-(4-Pyridylethyl)wool + 1.0M NaCl	275	525
Reduced wool	40	40
Reduced wool + 1.0M NaCl	40	35

Table 12

Mercury Uptake by Wool Fibers and Chemically Modified Wool Fibers[a]

Sample	Modifying agent	Mercury absorbed, mg Hg/g wool
Control	None (untreated wool)	500
1	N-vinylimidazole	650
2	N-vinylpyrrolidone	750
3	2-vinylpyridine	850

[a]One g of material was treated with 100 ml of 0.2M aqueous mercuric chloride solution for 24 hr at room temperature. The keratin material was then removed from the solution, washed thoroughly, and analyzed for mercury by atomic absorption spectroscopy. (Friedman and Waiss, 1972)

TABLE 13

Mercury Uptake by Native and Reduced Wool[a]

Hg Concn., mg/ml		Calcd uptake		Wt. incr. washed wool, mg/g
Initial	Final	mg/g	%	
0.1(100 ppm)	0.002(0.002)	4.9	98(100)	12(19.0)
1.0	0.2(0.0012)	40	80(100)	50(48.5)
2	0.8(0.0014)	60	60(100)	73(139)
4	1.7(0.042)	115	58(100)	116(148)
8	3.4(1.5)	230	57(81)	177(257)
16	8.6	370	46	280
24	13.0	550	46	374
32	16.5	775	49	444
40	21.5(19.5)	925	46(49)	510(625)

[a]Values for reduced wool are in parenthesis; 1 gram of wool was shaken in 50 ml aqueous $HgCl_2$ solutions at 25°C for 24 hr. (Friedman and Masri, 1973).

TABLE 14

Mercury Uptake by Modified Wools and Ion Exchange Resins[a]

Material	Residual concn., mg Hg/1., (ppm)	Mercury adsorbed, %	Structures
None (original mercury concentration	600		
Reduced wool	10	98.3	W—SH
Reduced wool plus N-vinylpyrrolidone	5.5	99.3	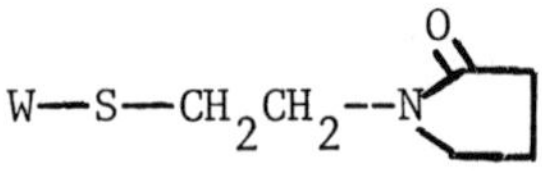 W—S—CH_2CH_2--N(pyrrolidone ring)
Reduced wool plus N-vinylimidazole	18	97	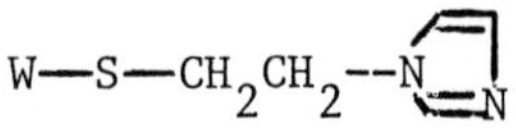 W—S—CH_2CH_2--N(imidazole ring)
Wool plus $NaHSO_3$	50	91.6	W—S—SO_3^-
$KMnO_4$-oxidized wool	180	70	W—SO_3^-
Wool plus N-vinylimidazole	190	68.3	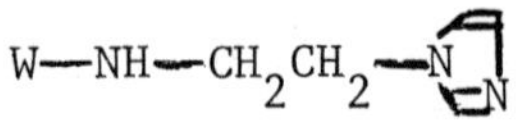 W—NH—CH_2CH_2—N(imidazole ring)
Reduced wool plus 2-vinylpyridine	190	68.3	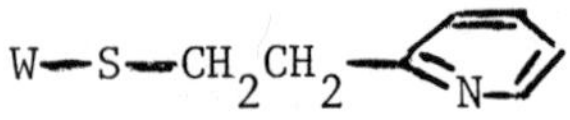 W—S—CH_2CH_2—(pyridine ring)
Waste wool	270	55	W—S—S—W
Dowex 2 X8 resin	320	46.6	R—CH_2—SO_3^-
Dowex 1 X8 resin	350	41.6	R—CH_2—SO_3^-
Dowex 1-A chelating resin	470	38.3	R—CH_2—$N(CH_2COO^-)_2$

[a] Reaction conditions are as follows: 1.5-g samples of adsorbent were shaken at 21°C for 30 min in 150 ml 0.01N HCl containing mercuric chloride to give 600 mg mercury/liter. (Friedman and Masri, 1973).

TABLE 15

Comparison of Hg^{2+} Uptake by Native (N), Reduced (R), and S-β-(2-Pyridylethyl (P)) Wool ($HgCl_2$-H_2O, 50 ml/g, 24 hours, 25°)

	Hg^{2+} conc. (mg/ml)		Hg^{2+} uptake		
Wool	Initial	Final	Mg/g	%	Partition ratio[a]
N	0.1(100 ppm)	0.0017	4.9	98	3,000
R	0.1	0.0002	5	99.8	25,000
P	0.1	0.0003	40	99.7	17,000
N	1	0.2	40	80	200
R	1	0.001	50	99.9	50,000
P	1	0.016	49.2	98.4	3,100
N	2	0.8	60	60	75
R	2	0.0016	99.9	99.9	62,000
P	2	0.218	89.1	89.1	410
N	4	1.7	115	58	67
R	4	0.042	198	99	5,000
P	4	0.79	160.5	80.2	202
N	8	3.4	230	57	67
R	8	1.5	325	81	210
P	8	2.5	276	69	110
N	40	21.5	925	46	43
R	40	19.5	1025	49	53
P	40	25.2	740	37	29

[a](Mg Hg^{2-}/g wool)/[mg Hg^{2+}/ml (final)]. (Masri and Friedman, 1973).

TABLE 16

Uptake by Reduced Alkylated Wool (P) (S-pyridylethyl wool)
($HgCl_2$-H_2O, 20 hours, 25°)

Hg^{2+} conc. initial	Final	Conditions	Partition ratio[a]	
1 ppm	0.616 ppm	141. medium + 200 mg P[b] 20 hr. 25°	44,000	(47,000)[b]
100 ppb	61 ppb	141. medium + 20 mg P[c] 20 hr. 25°	447,000	(360,000)[c]

[a]Partition ratio: (μg Hg^{2+}/g/wool)/[μg Hg^{2+}/ml (final)].

[b]Direct analysis →29μg Hg^{2+}/mg wool (5.8 mg Hg^{2+}/samples vs. 5.4 mg theoretical).

[c]Direct analysis of wool →22 μg Hg^{2+}/mg wool (0.44 mg Hg^{2+}/samples vs. 0.55 mg Hg^{2+} theoretical).

(Masri and Friedman, 1972).

TABLE 17

Hg^{2+} Uptake by Reduced Wool

($HgCl_2$-H_2O, 20 hours, 25°)

Conc. range Hg^{2+} in medium			
Initial	Final	Conditions	Partition ratio[a]
1016 ppb	423 ppb	141. medium + 200 mg wool 20 hr, 25°	98,000
103 ppb	58 ppb	141. medium + 200 mg wool 20 hr, 25°	543,000

[a]Partition ratio: (μg Hg^{2+}/g wool)/[μg Hg^{2+}/ml (final)]. (Masri and Friedman, 1973).

TABLE 18

Extent of Absorption of $Hg(NO_3)_2$ by Wool at pH 2.0 As a Function of Time of Shaking. Studies were Done at Two Initial Concentrations, 2075 ppm. and 20 ppm.[a]

Time	Concentration (ppm.) as Hg	Time (min.)	Concentration (ppm.) as Hg
0	2075	0	20
30 minutes	2006	3	10.75
6 hours	1950	5	7.5
17 hours	1905	15	5
24 hours	1830	30	2
30 hours	1790	60	.5
48 hours	1730		
76 hours	1670		
360 hours	1485		

[a](Friedman *et. al.*, 1973)

In the parts-per-million range, the partition coefficient reaches values of several thousand, and then increases further in the parts-per billion range. It is this property as much as any other that leads us to suggest that wool may be a practical adsorbent for mercury.

Sorption of methylmercuric chloride appears less favorable. At pH 6, its partition coefficient is roughly one tenth of that of the inorganic mercury compounds at low pH, and the amount bound is also less by a factor of about 10. However, useful binding is rapidly achieved, and the amounts that can be bound may be higher by 20% than those observed after 30 min.

If the uptake is graphed against the square root of the time as in Figure 2, the initial rise will be linear as long as the uptake is limited by the rate of diffusion. As saturation is approached, the graph departs from linearity and approaches a limiting value (ideally). This limit is determined in part by the residual concentration of mercury in solution. As the concentration is increased, the amount of mercury bound per gram of wool increases, but the proportion of the total mercury that is bound decreases.

B. Binding Mechanism. Detailed interpretation of the binding processes and stoichiometry presents a substantial challenge. If the mercury-containing units were taken up independently on a single type of binding site in such a way that adsorption of the first unit did not affect adsorption of the next, adsorption would follow a Langmuir isotherm. In this case, as Scatchard (1949) shows, a straight line would result if x/C is plotted against x, where x (for this discussion) is the gram atoms of mercury taken up per kilogram of adsorbent (or millimoles per gram) and C is the residual concentration in moles per liter. The intercept of such a line on the x-axis indicates the concentration of binding sites. The intercept on the x/C axis defines the classical first association constant. This can be regarded as a limiting partition coefficient.

However, when the sorption data for wool are graphed according to Scatchard, the resulting Figure 3 shows curved isotherms without plausible intercepts. This curvature may result from the existence of two or more types of binding sites, which may react at different rates with mercuric chloride.

Scatchard shows that the curvature in the case of proton binding to proteins can be very nearly accounted for by allowing for the changing electrostatic interaction between the protein and the species being bound as successive units are bound. We have tried unsuccessfully to find an empirical exponential correcting factor that will make the graph for $HgCl_2$ linear. The observed x/C shows no indication of approaching zero at high uptake. This circumstance suggests that part of the mercury may perhaps be taken up by preferential solubility without localized binding as described by Raoult's law.

When the sorption results are graphed logarithmically (Figure 4), they are seen to be roughly represented by isotherms according to Freundlich. For the inorganic salts in acid media (pH 2-4),

$$\log_{10}x = 0.33 \log_{10}C + 1.9_4. \quad (1)$$

This gives the mercury sorption x (milligrams of Hg bound per gram of wool) for a given residual concentration C (gram of Hg per liter), usually within a factor of 2 within the range of C from 0.001 to 40. We confirm that wool can bind more than half of its weight of mercury from concentrated mercuric acetate or mercuric chloride.

For methylmercuric chloride at pH 6, the corresponding Freundlich relationship is

$$\log_{10}x = 0.4 \log_{10}C + 1.3_0. \quad (2)$$

This holds with about the same precision as above in the range of C from 0.001 to 0.2 g Hg per liter. In this range, the binding varies from one seventh to one fifth that from $HgCl_2$.

The Freundlich relationship (extrapolated) can be used to make a rough estimate of the amount of wool needed to remove mercury from water under given conditions. For instance, to remove 138 mg Hg from a liter of industrial effluent to bring it to the mandatory maximum level of 5 micrograms per litter, permitted in a public water supply, would require about 0.1 kg wool at pH 2.

The vertical dashed arrow at 0.1 ppb, a representative value for mercury in sea water, indicates the accumulation of mercury in the animal part of the sea food chain; X indicates roughly the concentration in plankton; and the head of the arrow indicates the content expected in large predatory fish. The latter content is a few times higher than the tentative allowable maximum in individual foods. Note that the concentrations in fish are less than would be expected for wool in equilibrium with sea water as indicated by Freundlich lines. However, the difference is roughly consistent with the actual protein content of fish. The idea that the mercury content of fish protein approximates the value expected from sorption equilibrium needs to be evaluated critically in comparison with evidence for food chain accumulation.

The following mathematical analysis is based on adding a small amount of wool δw, leaving it until equilibration, then removing it and replacing it with another δw. The concentration of mercury in a solution into which a weight of wool dw has been placed is given by:

$$C = Co + \left(\frac{\partial C}{\partial w} \right) dw \tag{3}$$

If the equilibrium concentration of Hg in wool is C_w and if it is assumed that changes in solution weight due to sorption of mercury and other causes are negligible, then the weight of Hg removed from the solution is $WCo - C_w dw$, where W is the initial solvent weight and Co the initial Hg concentration in the solution. The final solution Hg concentration is found by dividing by the solution a dilution weight:

$$C = Co - \frac{Cw\ dw}{W} \tag{4}$$

Thus comparing (3) and (4) we have: $$\frac{\partial C}{\partial w} = -\frac{Cw}{W} \tag{5}$$

Integrating from Co to C:

$$\frac{w}{W} = -\int_{Co}^{C} \frac{dC}{Cw} = \int_{C}^{Co} \frac{dC}{Cw} \quad \text{or } w = W\int_{C}^{Co} \frac{dC}{Cw} \tag{6}$$

If $Cw = \alpha C^n$ (assuming the Freundlich isotherm operates); $\log Cw = \log \alpha + n \log C$ (7)

and

$$w = \frac{W}{\alpha\ (1-n)} \left(Co^{(1-n)} - C^{(1-n)}\right) \tag{8}$$

A plot of C_w vs. log C should be linear (Figure 4). Such plots, however, for the three wool types in (Figure 5) deviate from linearity at low concentrations. The deviation may be due to the fact that a different binding mechanism operates at the low levels. At these levels, all the wool types appear equally effective, possibly because covalent binding to residual SH groups takes place.

C. Activation Energies. Hojo (1959) estimated the activation energy of mercury adsorption from rates of uptake from mercuric nitrate at 10, 20, and 30°C. Taking x, the amount of mercury taken up at a given time, t, as a fraction of the amount adsorbed at saturation (70 hours), he assumed:

$$\frac{dx}{dt} = \frac{K}{x^n}$$

and derived

$$\log x = \frac{1}{n+1} \log K + \frac{1}{n+1} \log t + C.$$

In these equations, K is the rate constant, n is a constant, and C is the constant of integration. Plots of log x against log t for the three temperatures were approximately linear and parallel (Figure 6, Table 20). The ratio log (K_2/K_1) is the vertical difference between two of these lines divided by the slope $\frac{1}{n + 1}$ (i. e. $(n + 1)(\log x_2 - \log x_1)$). The activation energy is then:

$$E = \frac{RT_1T_2}{T_2 - T_1} \ln(K_2/K_1).$$

In this way values of 7 to 10 kcal/mole were found.

Hojo's first equation can be challenged because it implies a finite rate of binding (K) when the wool is saturated (x = 1). An alternative expression to describe the rate of uptake may be:

$$\frac{dx}{dt} = K(1 - x)^n, \text{ for } n \geq 0.$$

In this case integration gives (except for n = 1)

$$Kt = \frac{1}{n - 1} \quad (1 - x)^{(1 - n)} -1 \quad \text{and for } n = 1,\ Kt = \ln(1 - x).$$

Designating the right-hand side of these equations as F(x) we may plot the data in Table 20 in the form F(x) versus t for four values of n (Figure 7).
Data from Hojo's graphs replotted according to these equations give the best linear relations when n is assigned the value 3. These results then allow (Figure 7) revised estimates of K and, in turn, E: about 6 kcal/mole. Since the value by either treatment is one-fourth to one-third of the activation energy found for dye binding, Hojo correctly concludes that sorption of Hg(II) under the conditions of his experiments is energetically favored. That is, it occurs more readily than dye binding as reported (Vickerstaff 1954). The rate of uptake of mercury is less affected by change in temperature in this range.

D. Model Studies. Reactions of mercury compounds with amino acids containing sulfhydryl (Arnon and Shapira, 1969), sulfide (Natusch and Porter, 1971), disulfide (Brown and Edwards, 1969), amino and pyridino (Fish and Friedman, 1972), imidazole (Brooks and Davidson, 1966), and indole (Ramachandran and Witkop, 1964) groups lead to the expectation that these in wool may bind mercury under suitable conditions (not necessarily in dilute aqueous media at room temperature or in presence of complexing ions). Much study has been devoted to developing conditions under which

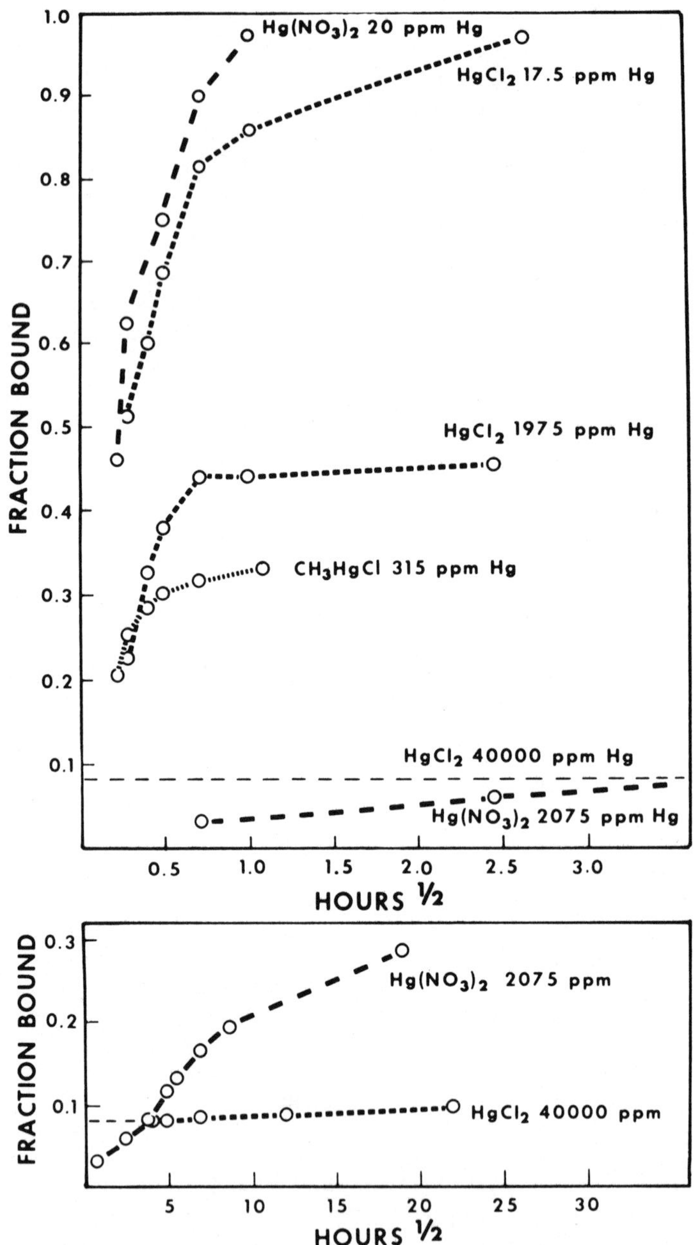

Fig. 2. Rates of uptake of mercury by wool from various salts. (Friedman et. al., 1973)

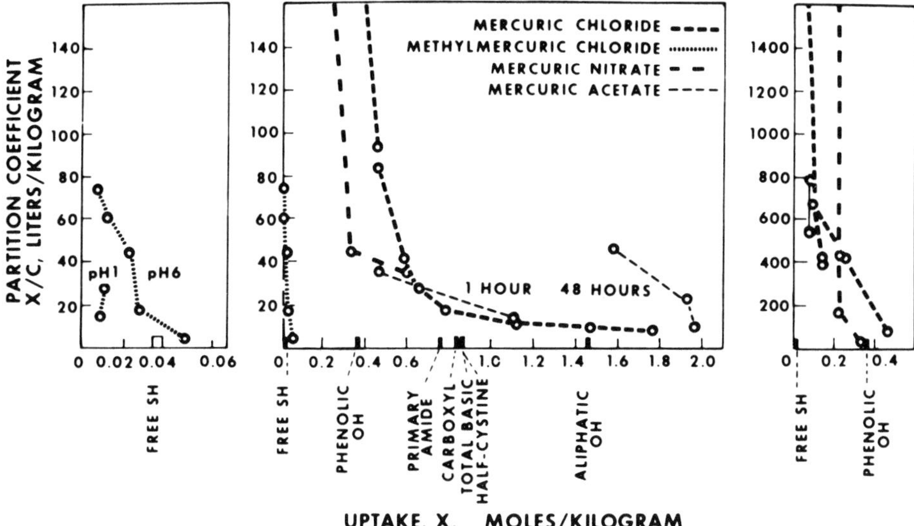

Fig. 3. Sorption of mercury by wool from various salts at various concentrations (Scatchard plot). Contents of the various reactive groups in wool are indicated. (Friedman et. al., 1973)

reactions of mercuric salts or organic mercury derivatives can be used to determine sulfhydryl contents of simple mercaptans and proteins (Leach, 1960; Friedman, 1973). Thus, reaction of various organic mercurials, but especially methylmercuric iodide at pH 7 is judged to measure free sulfhydryl reliably. Stoichiometric binding by SH depends on having complexing ions in solution that compete effectively with all other protein binding sites for the Hg species present. This is accomplished in various cases by chloride, ammonium salts, sulfite, and by adjusting pH to favor reactions of sulfhydryl (as RS^-) over amino (as RNH_3^+) groups.

The binding stoichiometry of inorganic mercuric salts is less certain because of the possibilities of one mercury atom combining with two sulfhydryl groups or only one. When mercury is not in large excess and the substrate is mobile, approximate 1:2 combining ratios may be observed, especially when denaturing agents are present. Under other conditions, binding can greatly exceed the small amount corresponding to the free sulfhydryl content of wool. Leach (1960, 1966) has shown that the binding from mercuric chloride in acid is scarcely affected by previously blocking the free sulfhydryl groups. Speakman and Coke (1939) assumed that (besides sulfhydryl) the basic groups of lysine and arginine residues in wool would bind mercury because of the recognized formation of mercury-amino com-

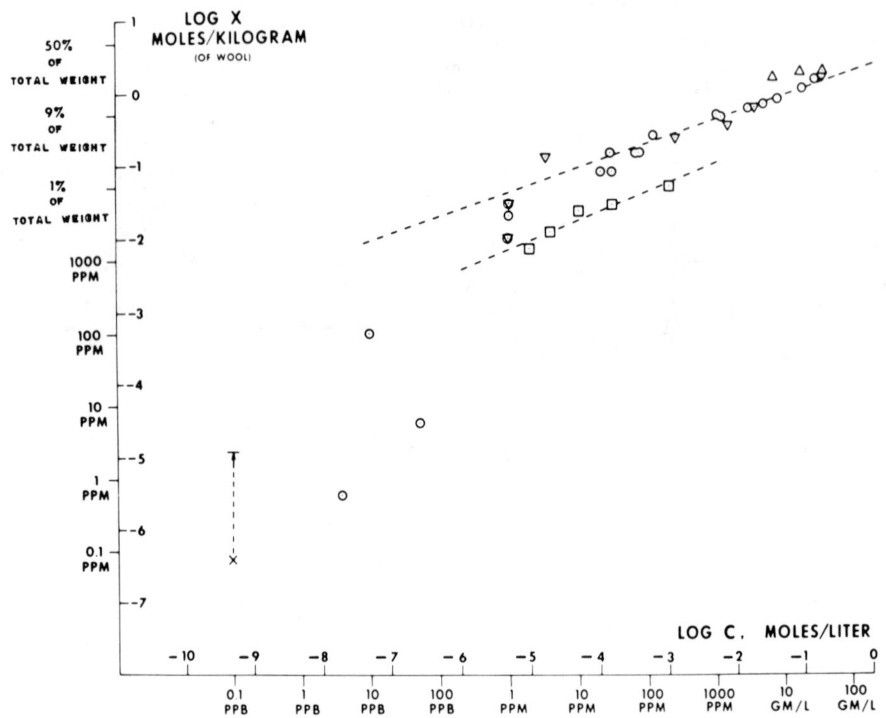

Figure 4. Sorption of inorganic salts and methylmercuric chloride to wool: (◯) mercuric chloride; (▽) mercuric nitrate; (△) mercuric acetate (these in acid media with the corresponding anions); (□) methylmercuric chloride pH 6. The vertical dashed arrow indicates the accumulation of mercury in the animal part of the sea food chain. (Friedman et. al., 1973).

plexes. For supporting evidence, they measured mercury uptake of wool that had been treated with nitrous acid to destroy primary amino groups. Binding was, indeed, decreased by this treatment, but because the drop in mercury binding was less than proportional to the drop in acid binding, existence of additional binding sites was inferred. Combination with peptide bonds was therefore proposed. This suggestion could not be confirmed (Fish et. al., 1973).

These studies do not clearly distinguish effects on rates of adsorption from effects on total uptake at equilibrium. A satisfactory theory should consider electrical charge of the protein physical constraints on uptake by the limited swelling of wool, the ionic states of the mercurial and of the reactive sites in wool, the binding constants of the reactive sites, and the presence of competing complex-forming reagents with higher or lower binding

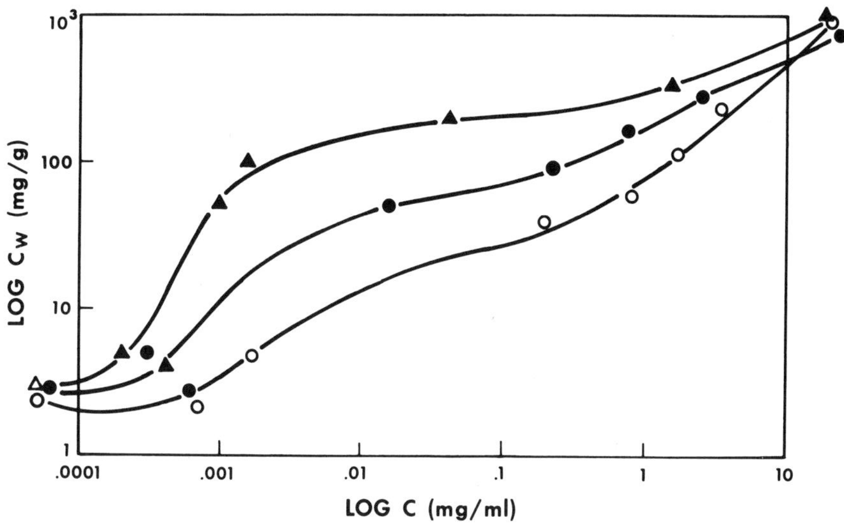

Fig. 5. Plot of log C_w vs. log C for (▲) reduced wool; (●) S-β-(2-pyridylethyl) wool; and (O) native wool.

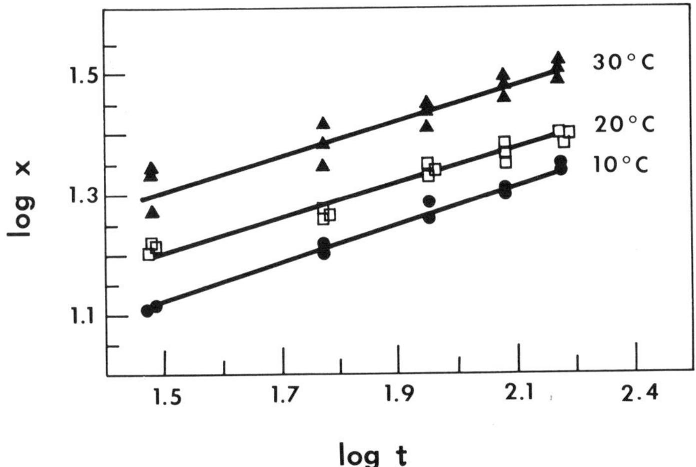

Fig. 6. Plot of log x vs. log t for $Hg(NO_3)_2$ binding to wool at three temperatures. (Redrawn from Hojo, 1959).

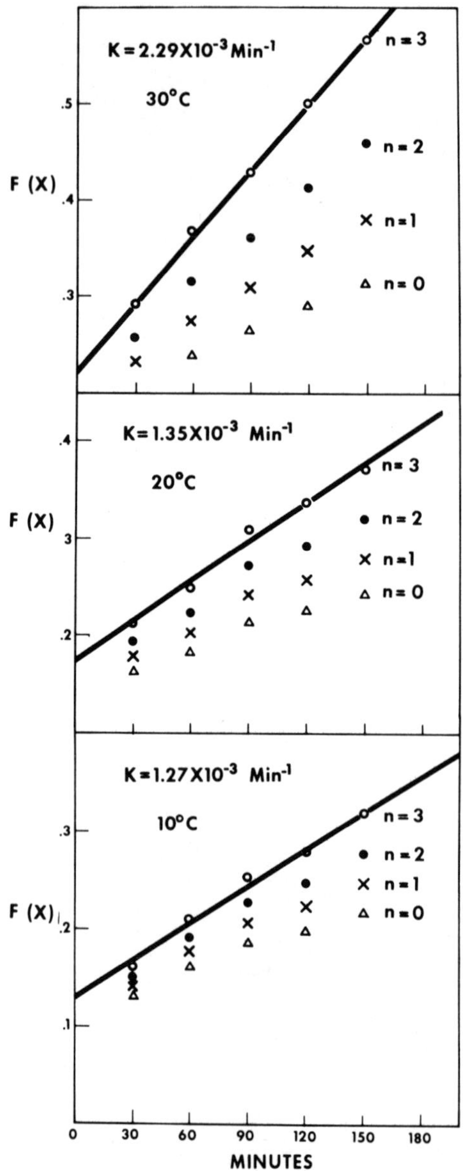

Fig. 7. Plot of F(x) versus time for $Hg(NO_3)_2$ binding to wool at three temperatures. See text.

TABLE 19

Effect of Concentration of $Hg(NO_3)_2$ on Extent of Absorption by Wool at pH 2.0 After Shaking for 48 Hours

Original concentration (μg/ml) as Hg	Final concentration (μg/ml) as Hg	Per cent left in solution
2050	1460	71.3
660	250	37.9
230	3.5	1.5
60	1	1.7
20	1	5

[a](Friedman *et*. *al*., 1973)

TABLE 20

(Data from Figure 6)

t, min	x at 10°C	x at 20°C	x at 30°C
30	0.131	0.163	0.205
60	0.161	0.183	0.240
90	0.186	0.214	0.266
120	0.199	0.227	0.293
150	0.218	0.243	0.315

TABLE 21

Mercury Sorption by Model Substances[a]

Material	Monomer unit weight	Mercury absorbed: Milli-grams/g	Mercury absorbed: Micromoles per monomer unit	Residual concn., milligrams Hg/l.
None				600
Polyamide-6,6	125.2	8	0.24	520
Poly(methyl glutamate)	143.1	6	0.14	540
Poly(glutamic acid)	129.1	7	0.19	530
Polytyrosine	163.2	9	0.21	510
Polyglycine	57.1	15	1.18	445
Poly(arginine hydrobromide)	237.1	22	0.42	380
Poly(lysine hydrochloride)	164.6	56.7	1.66	33
Polyhistidine	137.1	56.2	2.01	28
Egg albumin		38		220
Bovine serum albumin		33		290
Horse γ-globulin		22		380

[a]Fifty-mg samples of polymer in 2 ml 0.01M HCl in dialysis bags were equilibrated for 24 hr against 3 ml 0.01M HCl without the polymer. The initial $HgCl_2$ concentration in both compartments was 600 ppm. Calculations of bound Hg are based on the residual concentrations in the total volume (5 ml) as measured for the solutions outside of the bags. (Friedman and Masri, 1972).

constants at any particular pH. Clear understanding is important in designing chemical treatments for improving the adsorption characteristics of wool and in selecting conditions for adsorption and recovery.

Data on relative affinities of model poly(amino acids) for mercuric chloride (Table 21) show that although all protein functional groups may take part in mercury binding, amino groups of imidazole groups of histidine and lysine side chains residues appear most effective (presumably after SH groups). Arginine residues are moderately effective. Polyglycine appears to bind a

TABLE 22

Desorption of Mercury from Wool by Repeated Rinsing at pH 6

Rinse number	Volume recovered, ml	Mercury concentration micrograms/ml	Mercury recovered, micrograms	Fraction recovered,
A. With 0.01 M Ethylenediaminetetraacetate				
1	22.5	2260	50900	0.576
2	24	800	19200	0.809
3	25	270	6750	0.887
4	24	105	2520	0.916
5	25	60	1500	0.935
Hydrolyzed residue	50	115	5750	1
			86620	
B. With 0.1 M Citrate				
1	22.5	2030	45700	0.533
2	24.5	705	17300	0.733
3	25	310	7780	0.825
4	24.5	130	3180	0.862
5	25	60	1500	0.879
6	25	30	750	0.888
7	24.5	25	610	0.895
Hydrolyzed residue	50	180	9000	1
			85820	

Half-gram samples of wool with adsorbed mercury were shaken 15 minutes at 21°C with 25 ml portions of the indicated solutions.

(Friedman _et_. _al_., 1973)

TABLE 23

Desorption of Sorbed Mercury from Wool

Reagent and Conditions	Desorbed mercury %	Extract con centration moles/liter	Residual sorption moles/kg	Partition coefficien liters/kg
Citrate (Na), 0.1M				
1 g/100 ml	78	0.0038	0.12	31
1 g/50 ml	67	0.0065	0.18	28
	69	0.0108	0.27	25
Diphenylthiocarbazone in $CHCl_3$	0	--	0.49	--
1,2-Dithioglycerol	0	0	0.87	--
1,4-Dithiothreitol	0	0	0.87	--
Ethylenediaminetetraacetate (Na)				
1 g/50 ml	63	0.0034	0.11	32
(duplicates, same result)	70	0.0068	0.16	24
	72	0.0113	0.24	22
(15 min., also 16 hr.)	74	0.0117	0.22	19
1 g/20 ml	43	0.0104	0.31	30
Hydrochloric acid, 0.1 M, not neutralized, 1 g/500 ml, 30 min.	45	0.00027	0.19	700
Mercaptoacetic acid, not neutralized, 1 g/500 ml				
0.03 M	33	0.00020	0.23	1200
0.03 M	35	0.00022	0.22	1000

Table 23 (cont.)

Mercaptoacetate (Na), 0.3 M	72	0.0070	0.15	21
	80	0.0127	0.17	14
	78	0.0076	0.12	16
by direct analysis of the residue in the preceding experiment				
	73	(0.0076)	0.15	20
Mercaptoethanol, 0.3 M	54	0.0027	0.25	93
(Thiola) 2-Mercaptopropionylglycine	69	0.0108	0.27	25
Teorell-Stenhagen buffers, 1 g/500 ml				
pH 2.1	44	0.00020	0.14	700
pH 5.0 (most extraction)	76	0.00035	0.06	170
pH 10.8	56	0.0026	0.11	420

The wool used for desorption studies contained mercury bound from 0.01 M or, usually, 0.1 M $HgCl_2$ at pH 2, 21°C, and air-dried. Single extracts were made, usually by shaking 15 minutes in 50 ml of 0.01 M aqueous solutions of the various reagents adjusted to pH 6, per gram of air-dried wool. Exceptions are noted.

Mercaptoacetate at pH 6, but not mercaptoacetic acid, gave an extract that deposited a green precipitate on standing. A white precipitate formed in the mercaptoethanol solution during and after extraction. These precipitates were removed before the extracts were analyzed.

(Friedman, et. al., 1973).

surprisingly large amount of mercuric chloride. Apparently either peptide bonds or aliphatic (hydrophobic) parts of the polymer may participate in mercury binding. However, the situation is not straightforward since the other poly(amino acids) as well as nylon, a polyamide, exhibit a lesser tendency to bind mercury. Possibly conformational and proximity factors are important here. Furthermore, since poly(glutamic acid), poly(methyl glutamate), and polytyrosine show a low affinity for mercuric chloride, evidently carboxyl, ester, and phenolic groups are less important in mercury binding by proteins at low pH.

Since at high initial mercury concentration more mercury is bound to wool than can be accounted for in terms of free functional groups, additional mechanisms must operate in the binding of mercurials to proteins. One possibility is that the protein could act as a solid solvent for mercurials. This possibility is supported by our observations that the extent of binding can be quantitatively accounted for in terms of a partitioning of the mercurial between the liquid phase and the solid (wool) phase. Another possibility is that mercuric chloride forms polymeric lattices at some places within the wool structure (Sikovski, Simpson and Wood, 1960; Simpson, 1973). The nature of the aggregated deposits from concentrated reagent is presently unknown. In the case of $HgCl_2$, polymer formation could result from interactions between mercury and chloride atoms, as illustrated in Figure 8.

V. DESORPTION OF BOUND MERCURY FROM WOOL

A. Experimental Results. Two separate lots of wool with adsorbed mercury were used. The first (Hg-wool A) was made by shaking 7.5 g of wool in 750 ml of 0.01M $HgCl_2$, pH about 2.5, for 20 hr; it was then washed three times with distilled water and air dried at room temperature. It contained 46 mg of mercury per gram of wool. This lot was used to survey effects of pH.

The second lot (Hg-wool B) was made similarly except that 0.1M $HgCl_2$, pH about 1.5, was used. In this case the wool was cut into short pieces to aid uniformity, shaking was continued for 60 hr, and the wool was not rinsed. Excess liquid was absorbed by pressing between filter papers. This lot contained 160 mg of mercury per gram of wool. It was used to study desorption by solutions of various mercury binding reagents.

To measure desorption, 0.5-g samples of Hg-wool (air-dry weight) were shaken 15 min at 21°C with 25-ml portions of test solution. The mercury content of the liquid was then found as described. Mercury remaining on the wool was found after hydrolyzing a weighed sample in a measured volume of 6N HCl under reflux for 22 hr by analyzing the resulting solution. To determine possible loss of mercury during hydrolysis, 6N HCl containing 108 mg of mercury per ml was boiled under similar condition. Recovery was 98.3%.

The following reagents were tested, as 0.01M aqueous solutions adjusted to pH 6, for their ability to remove mercury from Hg-wool B: mercaptoacetic acid, mercaptoethanol, dithiothreitol, dithioglycerol dimercaprol (BAL), 2-mercaptopropylglycine, ethylenediaminetetraacetic acid (EDTA), and citric acid. Attempts to recover as much mercury as possible from Hg-wool B were made by repeated extractions with 0.01M EDTA and with 0.1M citrate, both at pH 6.

The pH dependence of binding shown in Figure 1 suggests that a slightly acid pH, 4, would be best for removing inorganic mercuric ion from wool, and a lower pH, 2.5, for the methylmercuric ion. Actual trial in 0.01M phosphate suggests that pH 6 is most favorable for desorbing inorganic mercury. We then attempted to increase the amount desorbed at pH 6 by using various complexing agents as 0.01M solutions. The most useful reagents appear to be aqueous ethylenediaminetetraacetate (EDTA) and citrate. More than 90% of the mercury is recovered in concentrated form from the wool (Tables 22,23). The amount of mercury remaining with the wool in each case is roughly in the proportion of one mercury atom to two sulfhydryl groups.

Of the large amount of mercury taken up by reduced wool from $HgCl_2$, only 17% could be extracted by EDTA. The remainder from either $HgCl_2$ or CH_3HgCl indicates a possible content of 0.3 to 0.4 mole/kg of available sulfhydryl sulfur in the reduced wool (as 1 SH:1 Hg). Regardless of the mechanism by which the mercury is originally adsorbed, it seems likely to move to sulfhydryl groups, to the extent that these exist, when circumstances are favorable (given time enough and a high enough pH).

The very small dissociation of mercury mercaptides may make it hard to discover an effective soluble complexing reagent that can successfully compete with sulfhydryl groups in proteins, especially if the mercury is bound in proteins by multiple interaction (See MacGregor and Clarkson, this volume). The very slow excretion of mercury after it has become established in the human body suggests that comparable firm binding has occurred. (In this case, the rate of excretion may possibly be a measure of the catabolism of the proteins to which mercury is bound.) Thus, studies of desorption from wool may be useful in medicine.

B. Competitive Binding by Wool and Polyethylene or Glass Containers. During the course of our studies it became apparent that some mercuric salt was being lost to the containers on prolonged storage of mercury salt solutions in the parts-per-billion range. Since such losses have wide-range practical and theoretical implications, a series of experiments was carried out to delineate this problem. These experiments are summarized in Table 23, and Figure 9.

The mercuric chloride balance during the described events, shown in the right-hand column of Table 24 indicates that a total of 4375 μg of mercuric ion was lost to the container.

Additional studies, summarized in Figure 9, demonstrate that nearly all of the mercuric chloride that has been lost to the container can be recovered with the aid of NaCl and/or HCl. Chloride ions in aqueous and in acid solution appear to prevent binding of Hg^{2+} to the container.

We also examined the sorption of mercuric chloride by glass containers. In typical experiments, the Hg^{2+} concentration of a 12-ppb mercuric chloride solution in a 10-liter glass carboy dropped to 10-ppb after 24 hr. This corresponds to a 17% loss or 0.008 μg Hg^{2+}/cm^2. Similarly, a 30-ppb mercuric chloride solution a a 1-liter volumetric flask lost about 35% of its Hg^{2+} after 9 days, corresponding to 0.02 μg Hg^{2+}/cm^2. Additional experiments have shown that it is relatively safe to store mercuric chloride in the ppm range, since the total loss falls within the experimental error of the analyses.

Precleaning the glassware with 5% to 10% HNO_3, followed by rinsing of the container with a 0.1N HCl solution containing 1N NaCl, effectively removes any adsorbed mercury from the surface. Similarly, addition of NaCl and/or HCl to dilute mercuric ion solutions prevents loss of mercury to container.

Our results show that it is possible to desorb nearly all of the mercuric salt from the container with the aid of chloride ion and hydrochloric acid and that wool effectively competes with the container for mercuric ions. This observation is not only of theoretical interest but of obvious practical importance.

VI. OTHER BIOPOLYMERS

A. Carbohydrate Derivatives. Chitosan (deacetylated chitin), other polyamines derived from cellulose, polyamines derived from diadehyde starch (Figure 10,11), and poly(aminostyrene) bind mercury in large amounts from water solutions of $HgCl_2$. In contrast, unmodified starch and cellulose adsorb very little mercury, while chitin (with acetylamino groups) binds much less than chitosan. In several instances the adsorbents bound more than one atom of mercury per nitrogen and more than their own weight of mercury (Table 25).

These results suggest that polyamino derivatives of carbohydrates may be useful to remove and recover mercury compounds from

TABLE 24

Loss of $HgCl_2$ to Polyethylene Tank Surface

Event	Sampling, time, hr	Hg^{2+}, ppb	Balance Hg^{2+}, μg
Start 501.	t_0	10	+500
	t_{72}	3	
	t_{120}	2	
$+Hg^{2+}$ to 30 ppb	t_0	30	+1400
	t_{43}	10	
$+Hg^{2+}$ to 30 ppb	t_0	90	+4000
	t_{17}	57	
	t_{64}	18	
+ 100 mg P	t_{24}	12	
Filter	Wool has 300 g Hg^{2+}		-300
Empty, rinse			-600
Refill 501. H_2O	t_{18}	1.5	
+ HNO_3(0.05N)	t_2	1.5	
	t_{24}	4	
+ 500 mg P	t_{66}	2.5	
Filter	Wool has 500 g Hg^{2+}		-500
Empty, rinse			-125
			+4375 g
(4375 g Hg^{2+})/(surface 6900 cm^2) = 0.63 g Hg^{2+}/cm^2			

(Masri and Friedman, 1973).

TABLE 25

Mercury Uptake by Carbohydrates and Derivatives[a]

Solid polymer tested	Aqueous medium $HgCl_2$ solution: ml/g polymer	Aqueous medium $HgCl_2$ solution: Initial concn, mg Hg/ml	mg Hg/g polymer	Initial Hg in medium, %	g-atom Hg/ g-atom N
Cellulose, starch, or DAS	25	4	5–12	5–12	0.23
DAS-MDA[b]	17	4	64	94	
DAS-MDA	25	40	280	28	0.23
DAS-MDA (0.1N HCl)	25	40	970	97	0.80
DAS-MDA (0.1N HCl)	50	40	1,050	52	0.86
DAS-MDA, reduced	25	40	617	62	0.55
DAS-MDA, reduced (0.1N HCl)	25	40	830	83	0.69
DAS-PDA[c]	25	4	96	96	
DAS-PDA	30	40	705	59	0.46
DAS-PDA (0.1N HCl)	30	40	999	83	0.65
DAS-PDA, reduced	33	40	1,033	78	0.68
DAS-PDA, reduced (0.1N HCl)	33	40	993	75	0.65
DEAE-cellulose	25	40	290	29	2.26
Aminoethyl cellulose	25	40	55	5	0.98
Aminoethyl cellulose (0.1N HCl)	25	40	128	13	2.25
p-Aminobenzyl cellulose	25	40	90	9	1.58
p-Aminobenzyl cellulose (0.1N HCl)	25	40	120	12	2.11
Chitin	25	4	33	33	
Chitin	25	40	175	17	0.17
Chitosan	25	4	100	100	
Chitosan	25	40	948	95	0.76

TABLE 25 (continued)

Chitosan	50	40	1,425	71	1.15
Poly(p-aminostyrene)	25	4	100	100	
Poly(p-aminostyrene)	25	40	912	91	0.54
Poly(p-aminostyrene)	50	40	1,450	72	0.86
Wool fiber	50	4	116	58	
Wool fiber	50	40	510	25	d

[a]Unmodified celluloses used were absorbent cotton and chromatography grade cellulose powder; starch was purified from potato, DAS was the same used to prepare DAS-MDA and DAS-PDA resins. Mercury concentrations or uptake are on basis of Hg, not $HgCl_2$; thus 4 mg Hg/ml refers to 0.02M $HgCl_2$. The medium was prepared by dissolving $HgCl_2$ in H_2O or, when indicated, in 0.1N HCl (aqueous). Usually tests were performed in both media (H_2O or 0.1N HCl) but results in 0.1N HCl are listed only when they appeared different from those in H_2O.

[b]Dialdehyde Starch-Methylenedianiline (DAS-MDA) Resin.

[c]Dialdehyde Starch-p-phenylenediamine (DAS-PDA) Resin.

[d]$\sim$3 g-atom Hg/g-atom N of basic amino acid residues (arginine, histidine, lysine); 0.75 g-atom Hg/g-atom N of lysine.

(Masri and Friedman, 1972).

TABLE 26

Mercury Sorption by Agricultural By-Products[a]

Material	Sorption from mercuric acetate	
	Mercury adsorbed, mg/g	Residual concn, of Hg/l.
Walnut expeller meal	880	11
Peanut skins	820	12
Wool	580	14
Rice straw	280	17
Plum pit shells	240	18
Peanut hulls	220	18
Rice hulls	180	18
Sugar cane bagasse	180	18

[a]In these tests, half-gram air-dried samples were shaken with 50 ml of 0.1M aqueous mercuric acetate (20 g Hg/l.) at pH 3.4–3.7 for one day at room temperature. The supernatant liquid was filtered through glass wool and analyze. (Friedman and Waiss, 1973)

TABLE 27

Mercuric Ion Uptake by Various Substrates[a]

Substrate	Hg concn, mg Hg/ml Initial	Residual	Calctd Hg Uptake[b] mg Hg/g
Milorganite (H_2O)	40	22	460
Milorganite (0.1NHCl)	40	31	225
Bark, Douglas fir	4	2.9	28
Bark, Douglas fir	40	36	100
Bark, redwood	40	30	250
Bark, black oak	40	24	400
Dry Redwood leaves	20	13	175
Dry pine needles	20	13	175
Senna leaves	40	30	250
Sulfuric acid lignin	40	38.5	150
Peat moss	20	10.4	240
Orange peel (white inner skin)	4	3.1	
Orange peel (white inner skin)	20	15	125
Orange peel (outer skin)	4	1.8	55
Orange peel (outer skin)	20	9	275
Chitin	40	36	100

[a]With all substrates except lignin, 1 g of substrate was equilibrated for 1 day in 25 ml of $HgCl_2$ (in H_2O with one indicated exception; in 0.1N HCl); with lignin a ratio of 1 g per 100 ml solution was used.

[b]Values calculated by difference between initial and residual concentrations determined by atomic absorption. The treated lignin sample was also analysed directly by atomic absorption on a portion digested with sulfuric acid--potassium permanganate. Result of this analysis showed 10.8% Hg. compared to a value of 13.0% calculated by difference between initial and residual Hg concentration.
(Masri, Reuter, and Friedman, 1974).

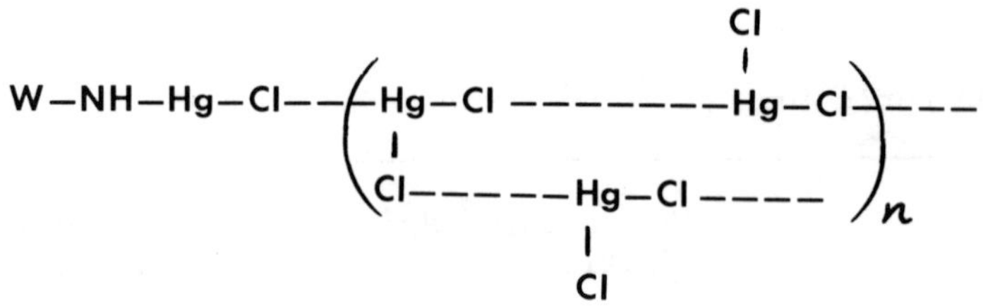

```
                                        Cl
                                        |
W–NH–Hg–Cl––(Hg–Cl ––––––––––––––––––Hg–Cl)–––
             |
            (Cl–––––––Hg–Cl––––– )n
                       |
                       Cl
```

Mercuric chloride "polymer" bound to wool

Fig. 8

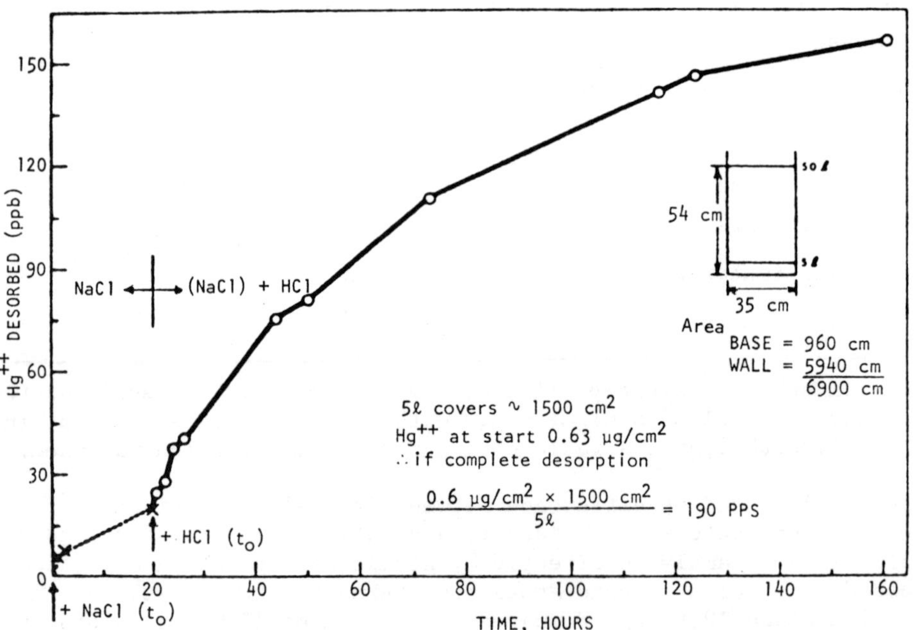

Fig. 9. $HgCl_2$ desorption from polyethylene tank by NaCl-0.05M HCl. (Masri and Friedman, 1973).

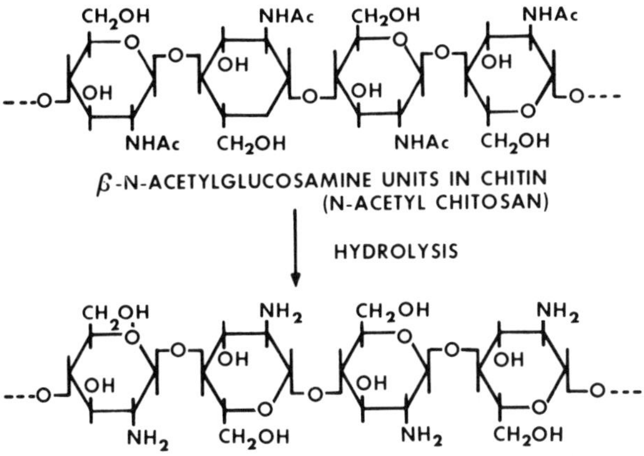

Fig. 10. Structure of Chitosan.

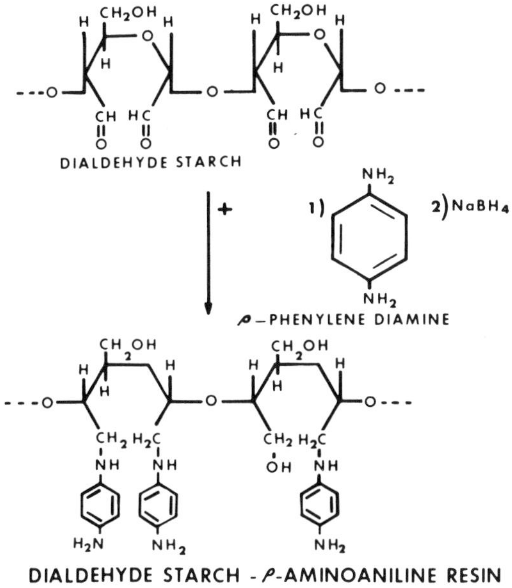

Fig. 11

water (Muzzarelli and Isolafi, 1971; Masri and Friedman, 1973). They suggest also that free amino groups and possibly other basic groups in proteins and other biological materials are important in the natural accumulation and distribution of mercury in the biosphere and that the natural occurrence of mucopolysaccharides in the N-acetylated or N-sulfated forms--e.g., chitin, chondroitin, hyaluronic acid, heparin--may provide a mechanism for the exclusion of nitrogen-metal complex formation in certain biological structures and functions.

A number of other carbohydrate derivatives have been found to bind heavy metal ions. These include thiol-substituted dextrans (Gaber and Fluharty, 1971), cotton fabrics containing polyethylene and polypropylene sulfides (Avny et. al., 1972), aminoethyl cotton (Roberts and Rowland, 1973), thiolated aminoethyl cellulose (Lee and Richardson, 1973), and starch xanthate-cationic polymer complex (Swanson et. al., 1973).

B. Other Agricultural Products and By-Products. Sorption of mercury compounds from water by various agricultural products was surveyed by atomic adsorption and X-ray fluorescence spectroscopy (Friedman and Waiss, 1972; Masri, Reuter, and Friedman, 1974). The results summarized in Tables 26 and 27 show that polyphenolic materials - e.g., tannins as in walnut expeller meal, peanut skins, and wood barks and leaves, as well as milorganite-are highly effective scavengers.

VII. ACKNOWLEDGMENTS

It is a pleasure to thank Dr. W. H. Ward for preparing Table 1 and for constructive suggestions; Dr. E. Menefee for assistance with the interpretation of the activation parameters and for mathematical derivations; and C. S. Harrison and E. Marshall for excellent technical assistance.

REFERENCES

Anelli, G., Pelosi, P. and Galoppini, C. (1973). Influence of mercury on the amino acid composition of tobacco leaves. Agr. Biol. Chem., 37, 1579.

Arnon, R. and Shapira, E. (1969). Crystalline papain derivative containing an intramolecular mercury bridge. J. Biol. Chem., 244, 1033.

Avny, Y., Leonov, D. and Zilkha, A. (1972). Adsorption of heavy metal salts by cotton fabrics containing polyethelene and polypropylene sulfide. Israel J. Chem., 11, No 1, 53-61.

Brady, P. R., Freeland, G. N., Hine, R. J. and Hoskinson, R. M. (1973). CSIRO Division of Textile Industry, Victoria, Australia, private communication.

Brooks, P. and Davidson, N. (1966). Mercury (II) complexes of imidazole and histidine. J. Am. Chem. Soc., 82, 2118.
Brown, P. R. and Edwards, J. O. (1969). Reaction of disulfides with mercuric ions. Biochemistry, 8, 1200.
Fish, R. H., Scherer, J. R., Marshall, E. C. and Kint, S. (1972). A column chromatographic and laser Raman Spectroscopy study of the interaction of mercuric chloride with wool. Chemosphere, No 6, 267-272.
Fish, R. H. and Friedman, M. (1972). A novel mercury (II) chloride complex of S-β-(2-pyridylethyl)-L-cysteine. J. Chem. Soc. Chem. Commun., 812.
Friberg, L. (1972). "Mercury in the Environment." CRC Press, Cleveland, Ohio.
Friedman, M. and Waiss, A. C., Jr. (1972). Mercury uptake by agricultural products and by-products. Environ. Sci. Technol., 6, 457-458.
Friedman, M. (1973). "Chemistry and Biochemistry of the Sulfhydryl Group in Amino Acids, Peptides, and Proteins," Pergamon Press, Oxford, England and Elmsford, New York.
Friedman, M., Harrison, C. S., Ward, W. H. and Lundgren, H. P. (1973). Sorption behaviour of mercuric and methylmercuric salts on wool, (a) J. Applied Polym. Sci., 17; 377-390, (b) presented at the Division of Water, Air, and Waste Chemistry, 161st American Chemical Society Meeting, Los Angeles, California, March 28-April 2, 1971. Preprints, (1971). 11, No 1, 109-14.
Friedman, M. and Masri, M. S. (1973). Sorption behaviour of mercuric salts on chemically modified wools and polyamino acids. J. Applied Polym. Sci., 17, 2183-2190.
Goldwater, L. J. (1971). Mercury in the environment. Scientific American, 224, 15-21.
Leach, S. J. (1960). The reaction of the thiol and disulfide groups with mercuric chloride and methylmercuric iodide. Australian, J. Chem., 13, 520-547.
Leach, S. J. (1966). "A laboratory Manual of Analytical Methods of Protein Chemistry," P. Alexander and H. P. Lundgren, eds., Vol. 4, Chapter 1, Permagon Press, New York, N. Y.
Lee, S. Y. and Richardson, T. (1973). Use of thiolated and amino-ethyl cellulose to remove mercury bound to solubilized fish protein. J. Milk Food Sci., 36, 267-273.
Malyuga, D. P. (1964). "Biochemical Methods of Prospecting." Authorized translation from the Russian edition (1963). Consultants Bureau, New York, p. 69.
Masri, M. S. and Friedman, M. (1972). Mercury uptake by polyamine-carbohydrates. Environ. Sci. Technol., 6(8), 745-746.
Masri, M. S. and Friedman, M. (1973). Competitive binding of mercuric chloride in dilute solutions by wool and polyethylene or glass containers. Environ. Sci. Technol., 7, 951-953.
Masri, M. S., Reuter, F. W. and Friedman, M. (1974). Binding of metal cations by natural substances. J. Applied Polym. Sci.,

18, 675-681. Cf. also, Text. Res. J. (1974), 44, 298-300.
Michelsen, D. L. (1973). Virginia Polytechnic Institute, Blacksburg, Virginia, private communication.
Miller, M. W. and Clarkson, T. W. (1973). "Mercury, Mercurials, and Mercaptans." C. C. Thomas, Springfield, Illinois.
Muzzarelli, R. A. A. and Isolati, A. (1971). Methylmercury acetate removal from waters by chromatography on chelating polymers. Water, Air, and Soil Pollution, 1, 65-71.
Natusch, D. F. S. and Porter, L. J. (1971). Proton magnetic resonance studies of metal-complex formation in some sulphur-containing α-amino acids. J. Chem. Soc., A, 2527.
Ramachandran, L. K. and Witkop, B. (1964). The interaction of mercuric acetate with indoles, tryptophan, and proteins. Biochemistry, 3, 1603-1616.
Roberts, E. J. and Rowland, S. P. (1973). Removal of mercury from aqueous solutions by nitrogen-containing chemically modified cotton. Environ. Sci. Technol., 7, 552-555.
Scatchard, G. (1949). Ann. N. Y. Acad. Sci., 51(4),660.
Speakman, J. B. and Coke, C. E. (1939). Trans. Faraday Soc., 35, 246.
Sikorski, J., Simpson, W. S. and Woods, H. H. (1960). Studies of the reactivity of keratins with heavy metals. Proceedings of the 4th International Conference on Electron Microscopy G, Mollenstadt, ed., Springer Verlag, Berlin, Vol. 1, p. 707.
Simpson, W. S. (1973) Private Communication.
Swanson, C. L., Wing, R. E., Doane, W. M. and Russell, C. R. (1973). Mercury removal from waste water with starch-xanthate-cationic polymer complex. Environ. Sci. Technol., 7, 614-19.
Tratnyek, J. P. (1972). "Waste Wool as a Scavenger for Mercury Pollution in Waters." U.S. Government Printing Office, Washington, D.C. Chem. Abstracts, 78, No 61931.
Vickerstaff, T. (1954). "The Physical Chemistry of Dyeing." Second Edition. Interscience, New York.
Webb, J. L. (1966). "Enzyme and Metabolic Inhibitors," Vol. 2, Academic Press, New York, N. Y.

24

INTERACTIONS OF KERATINS WITH METAL IONS: UPTAKE PROFILES, MODE OF BINDING, AND EFFECTS ON PROPERTIES OF WOOL

Merle Sid Masri and Mendel Friedman

Western Regional Research Laboratory, Agricultural Research Service, U.S. Department of Agriculture, Berkeley, California 94710

I. INTRODUCTION

The binding of a number of metal ions by some natural substances and agricultural by-products representing different classes of biopolymers such as proteins (e.g., wool and keratin derivatives), polyphenolic substances (e.g., lignin-humic acid rich substances, bark, activated sewage residue), and polyamine carbohydrates (e.g., chitosan, diamine-modified dialdehyde starch), was measured to test these materials as potential scavengers for toxic and industrial metal ions that may be present in waste industrial effluents or other water supplies. Our results from these studies have been described elsewhere (Masri and Friedman, 1972, 1973; Masri et. al., 1973, 1974 a, b; Masri and Friedman, 1974; Friedman and Masri, 1973); representative uptake data are shown in Tables (1-5). The results afford comparisons among these different insoluble polymers as regards their capacity and rate of binding of the different ions. In general, many instances of relatively substantial and rapid binding of certain metal ions were observed. The bound cations resist washing with water but stability to other more vigorous treatments were not tested further (except in a few experiments with Hg(II) bound to keratins, Friedman and Masri, this volume). Also, competitive binding form a mixture of cations in solution was not tested (i.e., relative affinity series were not determined). The results of this general survey however yield some information about the role of different functional groups of these environmental earth substances in the binding of metal ions. Also, such informa-

TABLE 1

Mercury Ion Uptake by Carbohydrate Derivatives From Aqueous $HgCl_2$[a]

	Hg removed	
	mg Hg/g polymer	%
Cellulose, starch, or DAS[b]	5-12	5-12
DAS-MDA[c]	280	28
DAS-MDA (0.1N HCl)	970	97
DAS-MDA (0.1N HCl)	1,050	52
DAS-PDA[d]	705	59
DAS-PDA (0.1N HCl)	999	83
DEAE-cellulose	290	29
p-Aminobenzyl cellulose	90	9
p-Aminobenzyl cellulose (0.1N HCl)	120	12
Chitin	175	17
Chitosan	948	95
Chitosan	1,425	71
Poly(p-aminostyrene)	912	91
Poly(p-aminostyrene)	1,450	72
Wool Fiber	510	25

[a] 25-50 ml 0.2M $HgCl_2$/g polymer-24 hr.

[b] 25 ml 0.02 M $HgCl_2$/g., DAS= dialdehyde starch.

[c] DAS-MDA = methylene dianiline modified (crosslinked) DAS.

[d] Phenylenediamine modified DAS.

(Masri and Friedman, 1972)

TABLE 2

Metal Ion Content of Substrates Equilibrated with Aqueous Salt Solutions (24 hr)

	Percent Metal in Equilibrated Substrates			
	Oakbark	Milorganite	Chitosan	Poly(aminostyrene)
$HgCl_2$	11.3	13.0	45.5	46.8
$HgNO_3$			25.4	
$CdCl_2$	2.7	3.2	20.0	1.1
$Pb(NO_3)_2$	13.6	8.9	42.0	3.4
$Pb(C_2H_3O_2)_2$			18.8	8.1
$ZnCl_2$	4.2	2.4	16.5	3.4
$Co(NO_3)_2$	2.3	0.85	10.2	0.45
$NiCl_2$	1.4	0.70	12.8	0.1
$Cr(NO_3)_2$		3.5	2.1	0.13
$CuCl_2$	2.3	1.0	13.5	7.2
$Fe(NH_4)_2(SO_4)_2$			6.5	
$FeCl_3$	2.6	5.7	a	0.28
$Mn(NO_3)_2$	0.6		7.1	
$AgNO_3$	7.5	5.5	23.7	16.3
$AuCl_3.HCl$			48.4	39.6
$PtCl_4.2HCl$	1.4	11.0	41.0	36.5
$PdCl_2$	9.6	11.4	37.8	8.6
$(NH_4)_6Mo_7O_{24}$			27.8	3.0

[a] Liquifies.

(Masri et. al., 1974b)

TABLE 3

Metal Ion Binding to Four Substrates from Aqueous Salt Solutions[a/]

Salt	Uptake: mmole metal/g substrate (in 24 hr)			
	Bark	Milorganite	Chitosan	Poly(p-aminostyrene)
Hg	.62	.72	5.6	5.70
Cd	.23	.29	2.78	.10
Pb	.74	.46	3.97	.17
Zn	.63	.37	3.70	.52
Co	.21	.15	2.47	.07
Ni	.22	.12	3.15	.02
Cr	-	.69	.46	.03
Cu	.35	.15	3.12	1.31
Fe	.43	-	1.18	.05
Mn	.10	-	1.44	-
Ag	.73	.54	3.26	1.98
Au	-	-	5.84	5.75
Pt	.06	.57	4.52	3.82
Pd	.90	1.18	6.28	.87
Mo	-	-	3.90	.31

[a/] Calculated from values in Table 2.

(Masri et. al., 1974b)

TABLE 4

Metal Cation Uptake by Native Wool Contacted with Salt Solutions For One Day

Salt solution[a]	Weight increase in treated wool[b]	Metal in treated wool, %[c]	Calculated mequiv metal/g wool	Color of treated wool
$HgCl_2$	33.9	29.8	2.25	White
$HgNO_3$	29.7	21	1.49	Dark purple
$AgNO_3$	14.6	10.9	1.12	Olive, darkens (air, light)
$A_gC_2H_3O_2$	13.3	9.5	1.02	Orange, brown, darkens
CuCl	9.8	6.5	1.10	Turns green
$CuCl_2$	4.4	2.0	0.33	Light green
$SbCl_3$	9.7	7.1	0.65	White
$UO_2(C_2H_3O_2)_2$	18.7	14.9	0.77	Orange
$UO_2(NO_3)_2$	11.4	8.2	0.39	Orange
$SnCl_2$	6.5	6.9	0.63	White
$Pb(NO_3)_2$	13.8	10.3	0.58	White
$PtCl_4 2HCl$	13.6	9.2	0.55	Light Orange
$ZnCl_2$	3.1	3.3	0.52	White
$BiCl_3$	7.7	6.9	0.36	White
$FeCl_3$	3.8	1.7	0.32	Light tan
$CdCl_2$	4.3	3.2	0.30	White
$PdCl_2$	5.7	3.0	0.30	Tan-grey
$MgSO_4$	2.1	0.3	0.13	White
$Ce(SO_4)_2$	0.9	1.4	0.11	Pink-tan
$Cr(NO_3)_3$	4.3	0.5	0.10	Light violet
$Sr(NO_3)_2$	0.3	0.6	0.07	White
$NiCl_2$	1.0	0.2	0.04	Light green
$Co(NO_3)_2$	0.3	0.2	0.04	Light pink

[a]Salts: 10-20 mmol per 1 g wool in 50-100 ml water or water acidulated with nitric (with mercurous nitrate), acetic (with silver and uranyl acetates), or hydrochloric acids (with antimony, bismuth and tin salts). Solutions were filtered if not clear. With CuCl, a 1% suspension in water was used.

[b]Thus 1 g native wool after treatment with $HgCl_2$ weighed 1.510 g; i.e., the treated wool product shows (1.510-1.0)/1.510 x 100 = 33.9% weight increase. The 1.510 g product had 29.8% Hg by atomic absorption.

[c]Determined by atomic absorption except with Ag, Sb, U, Pt, Bi, Pd, Sr, and Ce salts which were done by X-ray fluorescence analysis. (Masri et. al., 1974a)

TABLE 5

Metal Cation Uptake by Native Wool in 1 hr From Aqueous Salt Solutions[a]

Salt	Weight increase of treated wool, %	Metal in wool, %	Mequiv metal/g wool
$HgCl_2$	25.3	18.7	1.26
$HgNO_3$	6.5	4.2	0.23
$AgNO_3$	13.0	9.5	1.10
$AuCl_3 \cdot HCl \cdot 4H_2O$	14.6	10.5	0.63
CuCl	6.2	3.7	0.62
$SbCl_3$	4.3	4.1	0.35
$UO_2(C_2H_3O_2)_2$	9.3	7.4	0.34
$Pb(NO_3)_2$	7.7	6.2	0.33
$ZnCl_2$	3.2	1.9	0.30
$PtCl_4 \cdot 2HCl$	9.5	5.0	0.29
$BiCl_3$	7.4	5.5	0.29
$CdCl_2$	3.4	1.9	0.17
$PdCl_2$	3.0	1.6	0.16

[a]See footnotes a–c in Table 4. With gold chloride, only 1.1 mmol were used per 1 g wool, and the contact time was 10 min, (see text); treated wool is golden yellow.

(Masri et. al., 1974a)

tion may be relevant to the distribution of metal ions in the environment.

Most of our work, however, was done with wool. The studies were aimed at:
(1) testing the effectiveness of this keratin as an insoluble matrix for the removal and recovery of salts from water solutions.
(2) testing the effects of chemical modification of wool on its metal ions binding profile to gain information on the mode of interaction of metal ions with wool (and proteins in general) by comparing the uptake profile of native unmodified wool with those of different wool derivatives in which specific functional groups were modified.

The use of wool as a model for the study of protein-metal interaction is attractive for several reasons. The insolubility of wool greatly facilitates its handling in measurement of metal salts uptake from aqueous (and non-aqueous) media; also direct determination of the bound cations and anions in the treated wool after removing it from a given salt solution and rinsing with water (or other solvent) is facilitated by its insolubility. The use of wool in this way avoids the need for tedious equilibrium dialysis tests. The various protein functional groups in wool can be readilly modified to yield insoluble keratin derivatives which are also suitable for similar uptakes tests as with native wool.

(3) Finally, another objective of the study was to observe effects of the salt treatments of wool on its physical properties, especially those related to its use as a textile fiber, such as shrink-, moth- and flame-resistance; such desirable effects with certain salts have been recorded. In fact, most of the studies in the literature on the interaction of wool with metal salts were mainly directed to discovering such desirable effects, although these studies usually also resulted in an understanding of the nature of binding. Some of these studies on wool-metal ion interactions are summarized in Table 6.

In a companion paper, the interaction of wool with Hg(II) is discussed in detail (Friedman and Masri, this volume). Some of our results on the interaction of wool with other metal ions have been reported (Masri et. al., 1974a; Masri and Friedman, 1974) and highlights of some of these results will be discussed in the present report.

II. KERATIN DERIVATIVES

<u>1. Native Wool</u>. The native wool (WSSW) used for metal ion uptake tests and to prepare the various wool derivatives was a fine white top from Dubois, Idaho, 1961 clip that had been purified by extraction with benzene and methanol and air-dried.

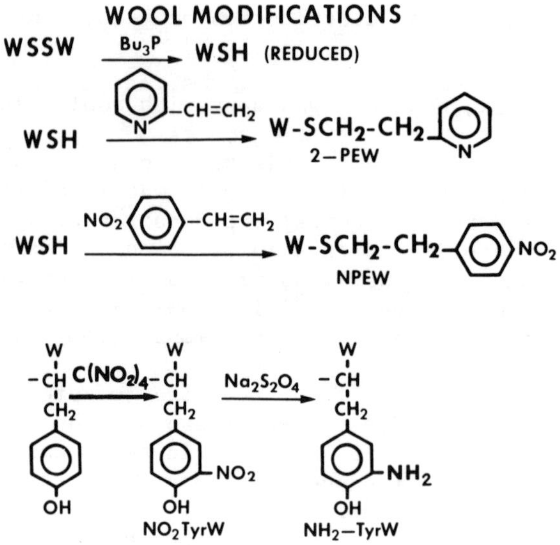

WOOL MODIFICATIONS

Figure 1

2. Wool Derivatives. The wool derivatives (Fig. 1) prepared included: 1) reduced wool (WSH) in which disulfide group of native wool was reduced with n-tributylphosphine (TBP) (Maclaren and Sweetman, 1966); amino acid analysis indicated about 95% reduction of disulfide; 2) S-β-(2-pyridylethyl)-wool (2-PEW), prepared from reduced wool by S-alkylation of WSH with 2-vinylpyridine (2-VP); amino acid analysis showed about 90% conversion of disulfide; 3) S-(p-nito-phenethyl) wool (NPEW), prepared by S-alkylation of WSH with p-nitrostyrene (p-NS); about 90% conversion of disulfide was indicated by amino acid analysis; 4) 3-nitrotyrosyl wool (NTW) prepared by nitration of WSSW with tetranitromethane (TNM); 5) 3-aminotyrosyl wool (ATW), prepared by reducing NTW with sodium dithionite; 6) thiosulfate wool prepared by treating WSSW with aqueous $NaHSO_3$ for 1 hour in air (WSSW $\longrightarrow$ WSH + $WSSO_3$; since no Cu (II) was used to reoxidize the sulfhydryl groups generated, the treated wool presumably had a mixture of sulfhydryl and thiosulfate groups); 7) dialdehyde starch-tanned WSSW, which was prepared as follows: native wool (10g) was treated for two days at room temperature with dialdehyde starch (15g) solubilized in carbonate buffer (400 ml, 0.3M, pH 9.5); isopropanol (50 ml) was also added to wet the wool. The wool (orange-yellow) was then removed and washed with water and methanol and air-dried. Amino acid analysis of a hydrolysed sample showed greately reduced lysine and arginine contents. The methods of preparation and analysis of these wool derivatives have been reported in detail elsewhere (Masri and Friedman, 1974). Figure 1 summarizes the above reactions.

TABLE 6

Binding of Metal Ions to Wool (References)

Metal Ion	Reference	Site of Interaction - Remarks
Hg(II)	Speakman and Coke (1939)	Absorption by side chain amino groups; non-involvement of disulfide at 25° in presence of dilute HCl
	Barr and Speakman (1944)	Production of unshrinkability and increased elastic modulus by cross-linking of peptide chains by mercuric acetate.
	Hojo (1958)	Alkali treatment of wool increased sorption of Hg(II) more than for Ag(I) and Cu(II).
	Leach (1960)	Over pH 1-6, $HgCl_2$ binds (un-ionized) probably to amide and peptide groups. Binding is decreased in presence of complexing anions (e.g. Cl^-, $S_2O_3^=$). $SO_3^=$ produces new sulfhydryl binding sites by reversible splitting of disulfide at pH 7-10 especially with CH_3HgI (basis for analytical method for disulfide). Noninvolvement of free carboxyl and amino groups in binding.
	Hemrajani and Narwani (1963)	Binding of $HgCl_2$ and $Hg(NO_3)_2$ between pH 4-5 by native and modified wool was compared. <u>Binding of $HgCl_2$</u>: a) by simple adsorption to surface (small fraction); b) by coordination with $-CONH_2$; and c) with -CONH- accompanied by H^+ displacement from -NH- of peptide linkage (to give -N(HgCl)-CO- + HCl?) but not of histidine (which does not dissociate below pH 6-8). Noninvolvement of carboxyl or unreduced disulfide in binding of $HgCl_2$. <u>Binding of $Hg(NO_3)_2$</u>: Noninvolvement of -CONH-, $-CONH_2$, or -COOH; binding is by: a) adsorption to surface; b) interaction with disulfide; and c) hydroxyl of tyrosine (cross-linking two residues by replacement of two protons of two hydroxyls).

Table 6. Binding of Metal Ions to Wool (References) - Continued

Metal Ion	Reference	Site of Interaction - Remarks
	Masri and Friedman (1973, 1974b); Friedman and Masri (1973)	Quantitative data on uptake by native and several chemically modified wools and discussion of modes of binding. Significance of natural sulfhydryl content of native wool in its effectiveness of scavenging Hg(II) from dilute sol'ns. Increased capacity and effective binding by reduction of disulfide. Hg(I) and Hg(II) differ in their mode of interaction with wool.
	Fish et. al. (1972)	Study of interaction of wool with $HgCl_2$ using laser Raman spectroscopy.
Ag(I)	Cecil (1950)	Silver ion reacts strongly with thiols to form mercaptides (basis of disulfide estimation after hydrolytic decomposition with sulfite).
	Simpson and Mason (1969)	Binding by: a) major reversible complex formation with carboxyl accompanied by H^+ release; b) minor and slow irreversible reaction with disulfide associated with brown discoloration (~5-8% of total uptake); between pH 3-6.5, maximum uptake is about 1.1 mmol/g wool, almost equivalent to aspartyl and glutamyl free residues.
Cu(II)	Guthrie and Laurie (1968)	Carboxyl is primary site between pH 3-5 (mohair keratin).
	Kokot et. al. (1972, 1973)	In weakly acidic aqueous medium, interaction is: a) mainly by ionic binding to free carboxyl (H^+ competes for carboxyl site), resulting in two esr active paramagnetic species (green color) of wool/copper (II) products; b) minor reaction with disulfide associated with brown discoloration and resulting in a diamagnetic component. This reaction does not occur in propanolic media; i.e., Cu(II) interacts with product of the interaction of water with disulfide of wool.

Table 6. Binding of Metal Ions to Wool (References) - Continued

Metal Ion	Reference	Site of Interaction - Remarks
Cu(II) cont'd	Swan (1961)	Neutral and alkaline cupric sulfite solutions react with sulfhydryl and disulfide of proteins converting them to thiosulfate groups; in reaction Cu(II) is reduced to Cu(I); low Cu(II) concentration is effective if solution has access to O_2 (Cu(I) reoxidized to Cu(II)).
	Hemrajani and Narwani (1967)	Uptake from nitrate and chloride between pH 4-5 is attributed to: a) simple adsorption; b) combination with $-CONH_2$; c) with -CONH-; and d) probably hydroxyl of tyrosine. No reaction with carboxyl between pH 4-5 deduced from results of uptake by esterfied wool. (cf. Guthrie and Laurie, 1968; and Kokot et. al., 1972, 1973)
	Masri et. al. (1974a, 1974b)	Maximum uptake about 1/3 mmol/g wool; practically no increased uptake by reduced wool; minor participation of disulfide. Much higher uptake of Cu(I) (about 1.1 mmol/g native and 2.1 mmol/g reduced wool).
Pb(II)	Hemrajani and Narwani (1967)	Similar comments apply here as for Cu(II) uptake study by same authors.
	Masri et al. (1974); Masri and Friedman (1974b)	Maximum uptake from aqueous $Pb(NO_3)_2$ in 1 day is about 0.6 mmol/g native wool; about the same uptake by reduced wool. Apparently ~2/3 of total uptake by native wool is due to interaction with disulfide.
Zn(II)	McPhee (1965)	Uptake of Zn(II) in presence of sodium sulfite is accompanied by increased resistance to felting shrinkage upon washing if fatty acid soaps are used in wash water. Although uptake of Zn(II) from sulfate or acetate sol'ns is substantial (12-14% weight increase of wool vs. 5-6% weight increase from solutions with sulfite) only minor effects on shrinkage are produced.

Table 6. Binding of Metal Ions to Wool (References) - Continued

Metal Ion	Reference	Site of Interaction - Remarks
Zn(II) cont'd	Hemrajani and Narwani (1967)	Similar comments apply as under Cu(II) study above by same authors.
	Habib et. al. (1967)	About 33% uptake from solution of zinc ammonium hydroxide or zinc ethylenediamine hydroxide. Wool fiber super-contracts (~20%) apparently irreversibly. X-ray shows structure disorganization.
	Masri and Friedman (1974b)	Uptake from water in one day from conc. sol'n is about 0.5 mmol/g native and 1.3 mmol/g reduced wool. Suggesting only minor participation of disulfide in binding to native wool.
Cd(II)	Hemrajani and Narwani (1967)	Similar remarks apply as for Cu(II) study above by same authors.
	Habib et. al. (1971)	Similar uptakes and wool structure disorganization from sol'ns of cadmium ethylenediamine hydroxide as with correspoding zinc salt sol'n. (Habib et. al., 1967)
	Masri and Friedman (1974b)	About 0.3 mmol/g native and 0.8 mmol/g reduced wool. Minor interaction with disulfide of native wool.
Al(III)	Hartley (1968a)	Maximum uptake occurs between pH 3.5-4. Binding is: a) mainly ionic to carboxyl and is accompanied by H^+ displacement. Amino groups oppose uptake due to salt links formation with carboxyl. No evidence of binding to amino or thiol groups; b) by van der Waals forces to non-specific sites (~1/3 total uptake); and c) to sulfonate groups (cysteic acid residues normally present in small conc. due to oxidation of sulfhydryl or disulfide). Bound Al(III) is not washfast (preference by Al(III) of water or hydroxyl ions as ligands to carboxylate)
Ni(II)	Bell and Whewell (1952)	Maximum equilibrium uptake from nickel ammonium hydroxide is extrapolated to ~25% Ni(II). More weight increase is obtained than can be

Table 6. Binding of Metal Ions to Wool (References) - Continued

Metal Ion	Reference	Site of Interaction - Remarks
Ni(II) cont'd	Bell and Whewell (1952) cont'd	accounted for by content of Ni(II) alone; i.e., Ni(II) is incorporated with associated radicals which are probably coordinated NH_3 initially, but later undergo hydrolysis upon washing with water, leaving OH or H_2O as the associated radicals. (No nitrogen content increase in treated and washed wool.) Modification of amino, phenolic, carboxyl or disulfide functions has little effect on equilibrium uptake. It is suggested that the NH_3 coordinated in the nickel ammonium hydroxide salt are exchanged with -NH- groups in main chain of wool. Uptake is accompanied by profound changes in fiber dimension and water-absorbing characteristics.
	Masri et. al. (1974a); Masri and Friedman (1974b)	Uptake from $NiCl_2$ in H_2O is negligible (0.04 mmol/g wool); and is only slightly increased by reduction of disulfide.
Cr(III)	Hartley (1968b, 1970)	Maximum uptake occurs at pH 2.5 and depends on anion, decreasing in this order: fluoride>formate≃sulfate>>chloride. This is due to differences of a) diffusion of Cr(III) species into fiber; and, b) rates of reaction with free carboxyl of wool. Mechanism of reaction with carboxyl involves two steps: 1) <u>rapid</u> pre-equilibrium ionization of carboxylate group; and, 2) <u>slow</u> loss of H_2O ligand from inner coordination sphere of the Cr (III) species (related to b) above). Cr(III) is bound covalently to wool and is washfast (unlike Cr(IV) anions which are bound ionically and can be displaced by excess negative ion). There is no evidence of binding to

Table 6. Binding of Metal Ions to Wool (References) - Continued

Metal Ion	Reference	Site of Interaction - Remarks
Cr(III) cont'd	Hartley (1968b 1970) cont'd	amino groups but binding to sulfonate (cysteic acid) and thiol groups occurs. With Cr(III) chloride (but probably not with sulfate) crosslinks between carboxyl groups of wool are introduced. With Cr(III) sulfate, sulfate-bridged Cr(III) complexes are probably formed in which one chromium atom bound directly to a carboxyl is also bound through a sulfate bridge to another chromium atom. Maximum uptake occurs at pH 2.5, polymeric $Cr(OH)_3$ precipitates above this pH. Below pH 2.5, considerable hydrolysis of the wool at amide groups occurs; also hydrolysis at peptide bonds adjacent to aspartyl residues may occur (Leach, 1955). Hydrolysis of wool may result in new binding sites.
Zr(IV)	Laurie (1966)	Maximum uptake (~12% ZrO_2) was obtained with pre-swollen wool (formic acid) at 70° and pH 3.0, in presence of tartarate as complexing agent to prevent premature precipitation of ZrO_2. Apparently after binding of Zr(IV) to wool at acid pH (Zr(IV) is probably exhausted on amino groups), it becomes fixed within the fiber upon washing with water--the rise in pH causing hydrolysis and precipitation as large aggregates within the fiber. Unlike with Cr(III), apparently there is little crosslinking of wool by Zr(IV). Treatment improves mechanical property of fiber (break at extension).
	Benisek (1971, 1972)	Uptake in the presence of complexing agents (citric, oxalic or tartaric) has fire-retardant effect on wool. Apparently the complexing agent stabilizes the salt in aqueous sol'n (pH~4 or below) and the

Table 6. Binding of Metal Ions to Wool (References) - Continued

Metal Ion	Reference	Site of Interaction - Remarks
Zr(IV) cont'd	Benisek (1971, 1972) cont'd	complex slowly hydrolyses in wool to insoluble basic salts.
Ti(III) and Ti(II)	Laurie (1968b)	Ti(III) (which, unlike Ti(IV) is not polymeric in water even up to pH 3) was used to impregnate wool, then allowed to oxidize to Ti(IV) by exposing the treated wool to oxygen. This results in immediate polymerization within fiber. Treatment (2-3% uptake) had no effect on whiteness of fabric. Fabric had increased affinity to dyes such as Orange II. Some decrease in tear strength and increase in stiffness occurred.
	Benisek (1971, 1972)	Fire-resistance property is imparted by treatment with Ti(IV) from aqueous systems containing complexing agents (e.g., citrate).

3. Silk Derivatives. Some metal ion uptake experiments were done using native silk and two silk derivatives, namely, 3-nitrotyrosyl- and 3-aminotyrosyl-silk. These were prepared in a similar manner as the corresponding wool derivatives. Silk was chosen for its much higher tyrosine content relative to wool. Raw silk spun as a yarn was cut with scissors into short lengths then ground in a Wiley mill (about 80 mesh screen). The coarsely ground (shredded) silk was washed with trichloroethylene, methanol, water then methanol again and air-dried. The silk was then used for uptake tests or derivative preparation; some fine dark brown particulate matter was released from the silk during washing and was discarded.

III. METHODS

A. Uptake from Water Solutions. Uptake tests by native and modified wool top from water solutions have been described (Masri et. al., 1974a; Masri and Friedman, 1974). The wools were soaked at room temperature in salt solution for 1 or 24 hours (using 10-20 mmol salt in 50-100 ml water or acidified water per gram of keratin) then thoroughly washed with water and methanol and air-dried. In these metal ion uptake tests, the amount of bound metal ions was only a small fraction of that available in solution; thus the metal salt concentration was not limiting and we assume the uptake data approximate maximum uptake from infinite concentrations. (See text for tests with gold chloride.)

B. Uptake by Woven Wool Fabric. Some metal salt treatments were performed on woven (native) wool fabric to test effects on physical properties that may enhance its utility as a textile fiber. Specifically, we tested the fabrics for flammability, insect resistance and laundering shrinkage. We did not attempt to optimize uptake conditions in these treatments. Tests of flammability were usually done in duplicate on 10" x 3" cloth swatches suspended vertically according to a modified ASTM procedure D 1230. Fabric circles 3 1/2" diameter cut-out with a die were used in shrinkage tests. Area shrinkage was measured after an Accelerotor wash and compared with shrinkage of untreated control cloth. Moth resistance was determined by measuring mortality and weight of excrement of black beetle larvae exposed to the treated and untreated woolen fabrics in a 14-day CSMA feeding test. We are indebted to Dr. Robert Davis, Director of the U.S. Department of Agriculture, Stored-Products Insects Research and Development Laboratory, Savannah, Georgia, for these tests. Experiments with wool fabric for flame- and insect-resistance were done using either undyed fabric that had not been treated chemically before the salt treatment except for a wash with trichloroethylene and methanol, or a similar fabric that had been rendered shrink-resistant by an interfacial deposition of a polyurea polymer onto the wool (Whitfield, et. al., 1963; Whitfield, 1971a,b). Two lots of the polyurea-fabric were used: one was a commercial sample in which the interfacial polymerization was carried out using a prepolymer with pendent reactive groups (isocyanate). The second lot was treated in our Laboratory by polycondensation of the monomers on the wool fabric. With both lots, the salt treatments were performed _after_ the interfacial polymer deposition step. Tests are in progress to check whether salt treatment during or before the interfacial polymerization step would be efficacious or advantageous.

The salt treatments of fabric were done in either aqueous or aprotic solvents:

1. _Uptake from Aprotic Solvents_. The relatively low uptake by native wool top from water media of some metal ions such as Zr (IV) and Ti (IV) (see below), and the known tendency of these ions to hydrolyze in water and form polynuclear polymeric species even at low pH and concentration (Laurie, 1966; 1968b), which would hinder salt penetration into the fiber, prompted us to carry out some of the salt uptake tests with fabric samples from salt solutions in dimethylformamide (DMF). This solvent also helps penetration into the fiber by causing swelling of the wool. Whether DMF may also form complexes with the salts was not investigated. It was expected that washing the fabric with water after impregnation with the DMF solution of certain salts (e.g., those of Zr (IV) and Ti (IV)), would cause these salts that had penetrated the fiber (regardless of the state of binding) to undergo hydrolysis and form polymeric species that would precipitate and thus be fixed within the fibers. In preliminary experiments, swatches 6" x 11", each weighing 9-10 g were spread flat individually in enamel pans about 2" deep (one swatch to a pan) and were wetted with about 70 ml of DMF in which was dissolved about 4 g of salt. The pans were covered and left in an oven at 70° for 2 hours. The swatches were then squeezed dry, rinsed briefly with DMF then with methanol, water, then methanol again and air-dried. In later experiments, the procedure was modified as follows: The oven temperature was set at 110°, and the samples were left for only 1 hour.

2. _Uptake from Aqueous Media_. In other tests fabrics were treated overnight with salts dissolved in water (with $SbCl_3$) or aqueous 1_N_ HCl (with $ZrCl_4$ and $BiCl_3$) at room temperature. Then they were washed with water then methanol (tests with $SbCl_3$) or with 1_N_ HCl followed by water and methanol (tests with $ZrCl_4$ and $BiCl_3$) and then air-dried. In still other tests in aqueous media, the swatches were treated by boiling for about 45 minutes in solutions of the salts dissolved in water containing citrate as complexing agent (Benisek, 1971).

3. _Other Methods_. Amino acid analyses of the keratin derivatives were performed on hydrolysates (110°, 6N HCl, 20 hrs, in sealed ampules purged with nitrogen gas) by ion-exchange chromatography

on a Phoenix 8000 Amino Acid Analyzer according to Spackman et. al. (1958). Metal ion contents of the treated keratin derivatives were determined by x-ray fluorescence analysis as previously described (Masri et. al., 1974b; also see Reuter and Raynolds, this volume). A weighed aliquot (300 mg) of the ground keratin was pressed in a die into a thin disc (32 mm diameter) and introduced in a plastic holder into a Qanta Metrix energy dispersive x-ray fluorescence spectrometer with exciting energy source from a rhodium x-ray tube. Fluorescent energy from the K- or L- shells of the various metals was detected, analyzed and quantitated. Results from x-ray analysis were spot-checked by atomic absorption spectroscopy applied to nitric acid-potassium premanganate digests of the treated wool. A Perkin Elmer Model 303 spectrometer was used with an acetylene burner and appropriate hollow cathode tubes.

IV. UPTAKE PROFILES AND MODES OF BINDING FROM AQUEOUS SOLUTIONS

1. Native Wool. Native wool top binds substantial amounts of many metal ions from water solutions (Tables 4 and 5). Thus, uptake in 24 hours ranged from more than 2 mmoles to about 2/3 mmole metal /g wool with Hg(II), Hg(I), Ag(I), Au(III) (short contact time), Cu(I), U(VI), Sn(II) and Sb(III); and between 2/3 mmole and 1/3 mmole/g wool with Pb(II), Pt(IV), Zn(II), Bi(III), Cd(II), Pd(II), and Fe(III). Native wool bound (in 24 hrs.) only 0.1 mmole metal ion/g wool with Cr(III) and Ce(IV) and negligible amounts (less than 0.05 mmol/g wool) with Ni(II), Co(II), Th(IV), As(III), Zr(IV), Tl(I), and Ti(IV). Comparison of these 24 hr. uptakes with corresponding 1 hr. uptakes also indicated relatively rapid binding of many of these cations to native wool. Elemental analysis for chlorine and nitrogen performed on some samples also showed some binding of the anion portions of the salts.

This metal binding by native wool was often changed appreciably by specific functional group modification. Differences in uptake profiles of the various keratin derivatives, in terms of capacity and rate of binding, depended on both the particular ion and the particular protein modification. Comparison of these differences yielded information clarifying the role of the different functional groups in binding metal ions and the modes of interaction of wool (and very likely proteins in general) with metal ions. Conclusions from this comparison have been discussed (Masri and Friedman, 1974).

2. Reduced Wool. Comparison of the 24 hour uptakes by WSH with corresponding values for WSSW showed increased uptakes per gram of wool of 0.9, 0.78, 1.06, 0.85, 0.75, and 0.47 mmoles of mercury (II), silver (I), copper (I), antimony (III), zinc (II), and cadmium (II), respectively. These increases indicate that the sulfhydryl polymer has great affinity for these metal cations.

That the observed increases were not due to conformational changes of the protein following reduction of the disulfide crosslinks (which may uncover new sites for interaction) was supported by comparison to the results with NPEW. This conclusion was also supported by the observation that in most instances the increases in metal uptake for reduced wool compared to the native form were roughly equivalent to the sulfhydryl content of WSH (about 0.9 mmol/g). The significance of the natural (though small) sulfhydryl content of WSSW in the highly effective binding by native wool of mercuric ions, at low concentrations, had been pointed out (Masri and Friedman, 1973).

Since uptake increases of these metal ions were roughly equal to the sulfhydryl content of WSH, we suggest that these ions do not interact appreciably with the disulfide function under our test conditions.

Sulfhydryl group content of dry reduced wool, held in air up to 3 weeks at room temperature and humidity did not change significantly. The sulfhydryl content was determined by amino acid analysis after S-alkylation (Masri et. al., 1972). In view of this relative stability to air oxidation, WSH appears to be a practical and effective matrix for removing many metal ions from water systems.

Comparison of metal ion uptakes by WSH and WSSW also shows:
a) While the 24 hr. uptake of mercury (II) by WSH was 0.9 mmol/g more than by WSSW, no change was obtained with Hg(I). This may be interpreted as indicating that uptake of Hg(I) by WSSW (about 1.5 mmol/g) is to a large extent due to interaction with the disulfide bond. Further evidence for this conclusion comes from the observed decreased uptake of Hg(I) by NPEW (0.38 mmol/g), in which the disulfide bond function has been abolished. Also, if it is assumed that reaction of Hg(I) with sulfhydryl groups is rapid while that with disulfide bonds is much slower, then the observed difference between 1 hr. uptakes of 0.44 mmol Hg(I) by WSH and WSSW is readily explicable (Color of WSSW that had been treated with mercurous nitrate was dark purple, while that of NPEW similarly treated was unchanged (white)). Due to the tendency of mercury to undergo disproportionation and the oxidizing tendency of the nitro group, part of the uptake of 1.5 mmol cation/g WSSW from mercurous nitrate solution may represent uptake of the cation as Hg(II) rather than Hg(I).
b) While copper (I) uptake by WSH was increased by 1 mmol/g wool, the uptake of copper (II) was essentially unchanged from that by WSSW. We suggest that this difference may be related to the known ability of Cu(II) to catalyse oxidation of sulfhydryl groups in proteins (Cecil and McPhee, 1959; McPhee, 1963). The uptake by WSH of Fe(III), which similarly catalyses oxidation of sulfhydryl groups (Leach, 1959), also was not increased over the uptake by WSSW. We

suggest that the lack of difference in uptakes by WSH and WSSW of Ce(IV), Cr(III) and perhaps Ni(II), Co(II) and As(III) may reflect a tendency of these cations also to catalyse oxidation of sulfhydryl groups in proteins. Also, Cr(III) sulfide and other sulfides readily hydrolyse in water (Kokot and Feughelman, 1973).

c) The uptake of Pd(II) by WSH was increased over that by WSSW by about 2.2 mmol/g wool (24 hr). This increase is more than twice the stoichiometric amount of sulfhydryl groups generated in WSH. This observation was reproducible. This increase was attributed to increased accessibility of the wool substance due to the absence of disulfide crosslinks, thus making available for interaction sites that were unavailable in the crosslinked WSSW. The result with 2-PEW, in which the disulfide function was also abolished, also showed more than stoichiometric increase in Pd(II) uptake over the 0.9 mmol of pyridylethyl groups/g wool, and can be explained similarly. Results of Pd(II) uptake by NPEW (no disulfide), however, showed no increase from that of WSSW, perhaps due to steric or electrostatic conditions or the potential of the nitro group of NPEW to oxidize Pd(II) (see results of Pd(II) uptake by NTW and ATW).

d) Although the uptakes by WSH compared to WSSW of nickel (II), cobalt (II), thorium (IV), arsenic (III), zirconium (IV), and thallium (I) increased much less than the potential of 0.9 mmol/g wool, these increases were significant in view of the fact that the uptake of these ions by WSSW was essentially nil. The relationship of limited uptake of some of these ions and their probable tendency to reoxidize WSH was mentioned above. The limited uptake of Zr(IV) may be related to the polynuclear (polymeric) nature of Zr(IV), to its tendency to hydrolyze in water even at low pH and low Zr(IV) concentrations (Laurie, 1966), and to salt links between free amino and free carboxyl groups of wool.

e) Uptake of gold (III) by WSH, with a solution of about 1.2 mmoles gold chloride per gram of wool, was almost quantitative within 10-20 minutes of contact. WSH appeared to undergo degradation as with WSSW with excess gold chloride and longer treatment times.

f) The 1 hour uptake of uranium (VI) by WSH was higher than the corresponding 1 hour uptake by WSSW, but the 24 hour uptakes were the same. WSH treated with uranium (VI) for 24 hours lost weight and appeared degraded.

<u>3. S-β-(2-Pyridylethyl)-Wool</u>. In general, the introduced pyridylethyl groups of 2-PEW were not as effective in the modified wools as the sulfhydryl groups for binding the various cations. However, increased uptakes, compared to either WSSW or WSH, were observed with chromium (III), zirconium (IV), titanium (IV), and with gold (III). Also, treating 2-PEW with excess gold chloride for even as long as 24 hours appeared not to degrade the wool (no weight loss). The 1 hour uptake of Sb(III) by 2-PEW was about 0.6 mmoles higher than the 1 hour uptake by WSSW, but the 24 hour

uptake by 2-PEW was nil. This unusual observation was reproducible; it may indicate hydrolysis or degradation of 2-PEW by prolonged treatment with the Sb(III) solution.

4. S-(p-Nitrophenethyl)-Wool. The slightly lower uptakes usually observed with NPEW compared to WSSW may reflect minor participation of the disulfide of WSSW in the binding of the various metal ions. The decreased uptake of Hg(I) has been mentioned earlier. Another outstanding result was the increased uptake of Sn(II) by NPEW compared to WSSW. Similar increases were also seen with other keratins modified to contain nitro groups, for example NTW and 3-nitrotyrosyl-silk. Thus, native silk treated with $SnCl_2$ for 24 hours contained only about 0.5% Sn(II), while 3-nitrotyrosyl silk took up about 10% Sn(II). A possible interpretation is that the nitro groups of the modified keratins oxidize the stannous ion (Sn(II)) to the stannic (Sn(IV)) ion, which in turn may be the species that interacts with potential binding sites in the keratin (in which the nitro groups would become reduced to amino groups). To check this possibility, uptakes of Sn(IV) from aqueous $SnCl_4$ by WSSW, NPEW, NTW, ATW, and by native, 3-nitrotyrosyl- and 3-aminotyrosyl-silk were determined. The uptakes, were not as high as anticipated perhaps due to the high acidity and the greater positive charge of the Sn(IV) ion which may hinder interaction with the positively charged wool. If the Sn(II) ion were oxidized in situ at or near the binding sites, possible hindrance to penetration into the wool structure due to the pH and electronstatic effects would be lessened. Higher uptake of tin ion by the nitro keratins from $SnCl_2$ compared to $SnCl_4$ solutions can be rationalized in this way.

5. 3-Nitrotyrosyl- and 3-Aminotyrosyl Wools. In contrast to the high uptake of Sn(II) by NTW and NPEW, uptake of Sn(II) by 3-aminotyrosyl wool (ATW) was less than by WSSW, again suggesting that oxidation to stannic ion rather than the generation of amino groups in keratin may be the important circumstance responsible for the increased Sn(II) uptake by the nitro derivatives. Similarly, uptake of Sn(II) by 3-aminotyrosyl silk was decreased to about one-eighth the uptake by 3-nitrotyrosyl silk. Uptake of Pd(II) was increased with both NTW and ATW, especially with the latter. In contrast, Pd(II) uptake by NPEW was not increased. The substituted tyrosyl phenolic residue may therefore be an active binding site. Oxidation of Pd(II) by the nitro-keratins may occur.

Results with ATW also showed improved uptakes with Hg(I), and Pt(IV). Therefore, the 3-aminotyrosyl residue appears to be an effective binding site for these cations.

The uptakes by ATW of Sb(III), Pb(II), and U(VI) were less than those by WSSW. The phenolic tyrosyl residues in native wool appear to be binding sites for these cations, so that substitution of these residues may interfere with their reactivity.

6. Dialdehyde Starch-Treated Wool. Dialdehyde starch-treated wool was used in a few tests to measure uptake of Hg(I), Hg(II), and Ag(I) by this modified wool. The dialdehyde starch appears to block the basic amino acid residues of wool. The uptake of Hg(II) in 24 hrs. was greatly decreased to only about 0.25 mmol/g W-DAS in one experiment and 0.3 mmol/g in a second experiment compared to an uptake of about 2.25 mmol/g WSSW. In contrast, the uptake of Hg(I) by W-DAS was unchanged from that by WSSW (1.56 mmol Hg(I)/g W-DAS vs. 1.49 mmol/g WSSW). Also, as with native wool, W-DAS became brown-purple after treatment with mercurous nitrate. These results are consistent with the inferred participation of basic amino acid residues of WSSW in the binding of Hg(II) and our conclusion that Hg(I) binds to WSSW mainly by interaction with disulfide function (which apparently is not affected by the dialdehyde starch), whereas only a slight interaction with disulfide may occur in the binding of Hg(II) to WSSW. Thus, the observed differences between Hg(II) and Hg(I) uptakes with W-DAS further corroborate this conclusion. Another pertinent observation is that formaldehyde solutions desorb appreciable amount of bound mercury ion from native wool that had been treated with $HgCl_2$ but not from native wool that had been treated with $HgNO_3$; presumably formaldehyde competes for the basic amino acid residues of native wool (Fig. 2).

The uptake of Ag(I) by W-DAS was essentially unchanged from that by WSSW; formaldehyde did not desorb Ag(I) from wool. These results are in agreement with the suggested mode of binding of Ag(I) by interaction with free carboxyl groups of wool. (Simpson and Mason, 1969).

The decrease in $HgCl_2$ uptake by W-DAS (compared to WSSW) of about 2 mmoles/g wool, exceeds the basic amino acid residue content of native wool of about 0.85 mmol/g and suggest that inaccessibility of certain binding sites to $HgCl_2$ due to crosslinking of wool by dialdehyde starch may be responsible for some of the decrease. Uptake results with another wool sample that was treated with formaldehyde instead of dialdehyde starch support this view that both crosslinking as well as blocking of basic residues are responsible for the drastic decrease of $HgCl_2$ uptake by W-DAS.

7. Thiosulfate Wool. Metal ion uptakes by the thiosulfate wool preparation used were usually intermediate between those by WSSW and WSH. Higher uptakes than with WSSW were observed especially with Hg(II), Ag(I), Sb(III), Bi(III), Pb(II), Cd(II), Zn(II) and Pd(II). The results, however, were not readily interpretable to be informative on the mode of binding to wool or the role of thiosulfate groups. Perhaps the use of wool in which all disulfide had been converted to thiosulfate (two thiosulfate groups from each disulfide, rather than mixed sulfhydryl and thiosulfate) would have been more instructive. The results therefore will not be discussed

Figure 2. Exchange Experiments

(a) Starting Wools: binding of Ag(I), Hg(I), or Hg(II) to Native Wool (WSSW)

Treatment (1 day)		Wool Product designation	Remarks Color & weight change mg/g starting wool
WSSW + AgOAc	in H_2O	W-AgOAc	Yellow-tan + 111 mg/g
WSSW + $AgNO_3$	in H_2O	W-$AgNO_3$	Yellow-olive + 171 mg/g
WSSW + $HgNO_3$	dil. HNO_3	W-$HgNO_3$	Dark purple + 340 mg/g
WSSW + $HgCl_2$	in H_2O	W-$HgCl_2$	White + 510 mg/g
WSSW + $Hg(OAc)_2$	dil. HOAc	W-$Hg(OAc)_2$	White (yellow cast) + 614 mg/g

(b) Re-equilibration of Wool Products in (a) above

Starting Wool	Re-equilibration medium		Remarks Color & weight change of final wool product (mg/g starting wool)
W-AgOAc	+ $Hg(OAc)_2$	in H_2O	Turns brown in hours + 350 mg/g - 1 day + 560 mg/g - 3 days
W-AgOAc	+ $HgCl_2$	H_2O	White; + 645 mg/g - 1 day
W-AgOAc	+ $HgNO_3$	H_2O	Jet black in minutes! + 130 mg/g - 1 day + 340 mg/g - 3 days
W-AgOAc	+ $HgNO_3$	dil. HNO_3	1 hr.: black 1 day: black with purple streaks 2 days: purple 3 days: purple + 230 mg/g
W-AgOAc	+ $CuCl_2$	H_2O	Green, + 26 mg/g when removed and dried: sensitive to light and/or rapidly turning violet on exposed surface
W-$AgNO_3$	+ $Hg(OAc)_2$	dil. HOAc	Tan-purple - 1 day + 470 mg/g - 1 day
W-$AgNO_3$	+ $HgNO_3$	dil. HNO_3	Minutes: black; 2 hrs.: black with purple streaks

Figure 2. - Continued

Starting Wool	Re-equilibration medium		Remarks Color & weight change of final wool product (mg/g starting wool)
W-$AgNO_3$ cont'd	+ $HgNO_3$ cont'd	dil. HNO_3	1 day: smoky black-gray + 14 mg/g 3 days: dark brown-purple + 290 mg/g
W-$HgNO_3$	+ $HgCl_2$	dil. HCl	Minutes: brown to tan to yellow 66 hrs.: purple-tan - 11 mg/g (loss)
W-$HgNO_3$	+ $AgNO_3$	dil. HNO_3	66 hrs.: lighter purple - 45 mg/g (loss)
W-$HgCl_2$	+ $HgNO_3$	dil. HNO_3	Minutes: chalky white 1 day: lilac 2 days: purple 66 hrs.: purple + 106 mg/g, also white ppt (110 mg) probably calomel
W-$Hg(OAc)_2$	+ AgOAc	dil. HOAc	No color change - 74 mg/g (loss) - 1 day
W-$Hg(OAc)_2$	+ $AgNO_3$	dil. HNO_3	Minutes: black 1 hr.: purple with black streaks and black ppt in medium - 205 mg/g (loss) - 1 day and gray ppt in medium (Ag° or Hg°)
W-$AgNO_3$	+ $Hg(OAc)_2$	H_2O	bright golden yellow in hrs. + 310 mg/g - 1 day + 420 mg/g - 3 days (becomes brittle and crumbly)
W-$Hg(OAc)_2$	+ HCHO[a/]		silvery gray; - 176 mg/g (loss)
W-$HgNO_3$	+ HCHO[a/] (100 ml/1.34 g)		purple-tan (lighter purple than W-$HgNO_3$) - 34 mg/g (loss)
W-$AgNO_3$	+ HCHO[a/]		chestnut brown; + 10 mg/g

[a/] 100 ml of 20 ml isoprop. + 20 ml of a 35% HCHO solution (H_2O) + water balance/ 1.14 g wool.

here further except to indicate that partial sulfitolysis provides an alternate facile method of generating some sulfhydryl content in wool suitable for metal ion binding as with fully reduced wool (WSH). It is possible that this partial reduction of wool may yield a more stable polysulfhydryl polymer than fully reduced wool.

8. Exchange Experiments. Another area of the work that is not clearly understood at this time, will be mentioned briefly. In some experiments, native wool that had been allowed to bind Hg(II), Hg(I), or Ag(I) was re-equilibrated with aqueous solutions of Hg(II), Hg(I), or Ag(I) salts (the salt used for re-equilibration being other than the one originally bound to the wool). In many instances with these re-equilibration-exchange experiments, dramatic and often rapid color changes accompanied by large weight changes of the wool samples took place. Some examples are schematically illustrated in Figure 2. Still more complicated changes occured (not shown) when the re-equilibrated wool products in Figure 2(b) were further subjected to a second re-equilibration in a medium containing a salt different from the one used for the first re-equilibration. General remarks can be drawn from the observations depicted in Figure 2. First, the anions of the salts used for the original uptake or in the re-equilibration affected the observed changes. Second, the observed weight changes of the wool indicated that the main binding sites in native wool for Ag(I), Hg(I), and Hg(II) were not the same. Further interpretation of the data must await more thorough analysis including analysis of the significant metals and anions in the wool and in the medium, changes in pH during treatment and amino acid analysis of the wools at each stage of treatment. (For example, we found that $AgNO_3$ in dilute nitric acid appears to enhance nitration of tyrosyl residues as revealed by presence of 3-nitrotyrosine in hydrolysates of treated wool).

9. Colors of Wools Treated with Salt Solutions. Native and modified wool top acquired characteristic colors after the treatment with the aqueous solutions (Masri et al., 1974a). The colors resist washing with water and methanol. Aside from being esthetically pleasing, the colors sometimes suggested differences of binding mechanisms for the different modified wools. A few examples are pointed out below. Note that the color of the original wools was white except for NTW (orange-yellow), ATW (white to very light tan), and W-DAS (light orange-brown). An orange cast may develop with NPEW on the surface exposed to light and is associated with a free radical signal as reported earlier (Masri et al., 1972), but the colors obtained with this wool derivative due to metal salt treatment were readily distinguishable and easily recognized. Examining the color of the wool samples (a color photograph is available on request) shows the following:

a) Color changes due to treatments of NTW were masked by the original intense color of the wool;

b) Colors resulting from treatment of ATW with most of the salts used were various shades of brown. These colors usually started to develop rapidly as soon as the wool samples were wetted and reached almost full intensity within minutes. These rapid changes may be associated with oxidation of the wool by some of the cations and with crosslinking of the wool mediated through the aromatic amino groups. If such crosslinking occurs, it may limit accessibility of some binding sites; such a process may explain the reduced uptake by ATW compared to WSSW of several metal ions. (Masri and Friedman, 1974).

c) Whereas the color of WSSW treated with $HgCl_2$ was unchanged (white), the color of WSSW treated with $HgNO_3$ was a striking dark purple. The color difference emphasize the contrasting modes of interaction of the two salts with native wool. As mentioned earlier, our uptake data suggest that Hg(I) (as mercurous nitrate), unlike Hg(II) (as mercuric chloride or acetate), binds by interacting mainly with disulfide of wool. Note here that with NPEW (which has no disulfide and shows greatly reduced Hg(I) uptake compared to WSSW), no purple color developed after treatment with $HgNO_3$ solution. While on the other hand W-DAS (intact disulfide) bound appreciable amount of metal ion and turned purple-brown.
Treatment of 2-PEW with mercurous nitrate resulted in only a very light purple color. The slight color may have resulted from interaction with the small residual disulfide content of this preparation (only about 90% of the disulfide was modified to give S-β- (2-pyridylethyl)- groups). The pyridylethyl groups themselves appear to be suitable Hg(I) binding sites in 2-PEW as suggested from the uptake data (Masri and Friedman, 1974). Thus, the absence of purple color with NPEW, the slight color with 2-PEW (which had residual unreacted disulfide), the low uptake of Hg(I) by NPEW but high uptake with 2-PEW, all suggest the importance of intact disulfide rather than merely high Hg(I) uptake for the development of purple color. Another remark on the color change with mercury salts: When mercuric nitrate rather than mercuric chloride or acetate was used with WSSW, a slight purple color did develop slowly (several days at room temperature; faster if uptake medium was heated to 50-60°) but the color intensity was much weaker at comparable metal ion uptake than when binding was from mercurous nitrate.

Perhaps the small tendency of $Hg(NO_3)_2$ to give a purple colored product with WSSW (presumably due to reaction with disulfide in a similar manner as occurs with $HgNO_3$) may be related to the greater dissociation of $Hg(NO_3)_2$ in water compared to $HgCl_2$ and to some conversion of $Hg(NO_3)_2$ to $HgNO_3$.

The purple color observed in tests with $HgNO_3$ might be due to *in situ* dismutation of Hg(I) to Hg(II) and colloidal Hg^o formation and precipitation of red HgS.

d) Although uptake of Hg(I) by WSH is high (comparable to uptake by WSSW), the wool product formed is white (except for a slight color probably related to the small residual content of unreduced disulfide). Apparently, the reaction of Hg(I) with disulfide in WSSW is dissimilar to its reaction with sulfhydryl groups of WSH giving rise to different wool mercury complexes.

e) Colors of WSSW, WSH, 2-PEW, and NPEW wools immediately after removal from medium containing CuCl were light yellow but rapidly started to change to green as soon as the wools were removed and washed with water and methanol. The wools became intense dark green within hours after drying (associated with intensification of paramagnetic Cu(II) esr signal in the wool). Thus, the copper salt is bound originally as the cuprous ion but is rapidly oxidized in air to cupric ion. Whether structural changes of the protein accompany this valence change was not investigated; the cupric ion is easily desorbed from the keratins by dilute acetic acid.

f) Samples of WSH contacted with Pb(II), Bi(III) or Tl(I) salts were distinctly yellow in contrast to WSSW samples that had been treated similarly. These color differences are compatible with conclusions made earlier on the role of disulfide and sulfhydryl groups of native and reduced wool in the binding of these metal ions deduced from comparison of uptake data by native and modified wool.

g) Color of WSSW was light blue-green with Co(II) and white with Ni(II), but the color of WSH treated with either cation was black-brown. The color change with WSH was rapid (within minutes) and may be associated with oxidation or crosslinking of WSH and limited uptake of these cations by the wool. (Also, colors of 2-PEW that had been treated with Co(II) or Ni(II) were light pink or green, respectively.)

h) Treatment with Au(III) resulted in bright golden-yellow to rich yellow-brown colors with the various wool derivatives. The wool products became slightly purple on the surfaces exposed to light and air for several weeks, suggesting reduction to colloidal metallic Au^o (purple), perhaps with change in the wool protein structure. Wool and derivatives that had been treated with Ag(I) also darkened with time on exposed surfaces probably due to photo-reduction to Ag^o.

V. FLAME- AND INSECT RESISTANCE OF TREATED WOOL FABRIC

Many of the studies in the literature on the binding of metal ions to wool, although yielding basic information on the nature of wool- and protein-metal ion interactions, were stimulated by and directed toward a desire to enhance and improve the physical properties of the fiber in its use as textile material. These include such properties as decreased laundering shrinkage, decreased flammability, increased insect resistance, water repellancy, color stability, modified dye uptake and mechanical properties, and desirable hand. We also have tested some practical properties of fabrics treated with salt solution from aqueous or aprotic solvents. The results will be reported in detail in future publication; illustrative examples are discussed below.

1. Flame Retardance. Flammability test on fabric that was treated from solutions in water containing citrate as complexing agent, confirm the reported efficacy of titanium and zirconium salts in decreasing flammability (Benisek, 1971). Our results also showed high uptakes and improved flammability when treatment with $ZrCl_4$ and $TiCl_3$ was from DMF. Treatment with $TiCl_3$ whether from DMF or aqueous system, however, resulted in severe discoloration of the fabric. The higher uptakes from DMF (*vs*. water), as mentioned earlier, may be due to presence of the salts in DMF as monomeric species, a circumstance that would probably favor their penetration into the swollen fiber, where they may become bound to specific sites (amino and peptide functions) or nonspecifically. The subsequent wash with water then can bring about *in situ* hydrolysis, polymerization and precipitation of the metal ion derivatives inside the fiber, where they become fixed (insoluble) regardless of whether the original binding was specific or non-specific. We believe such mechanism may apply to uptake from DMF of other salts that have a tendency to hydrolyse and form polynuclear species in water. Of course, the use of DMF for treating fabrics may not be economically practical unless the solvent is to be reused. Fire retardant effects were also given with other salts (from DMF), especially $BiCl_3$ and (less satisfactorily) $ZnCl_2$ and $SbCl_3$. Fire-resistance to various degrees was also obtained when treatment was from water, with $SbCl_3$, $ZnCl_2$, $ZrOCl_2$, $TiCl_4$, and $BiCl_3$ (Masri et. al., 1974a; Masri and Friedman, 1974). (Illustrative results are shown in Tables 7-9.) The mechanism by which these salts impart fire-resistance to the fabric is not clearly known. Since most of the salts we used were the chlorides, we analysed some of the treated fabrics for chloride content (Tables 7 and 8). The observed chloride content may contribute to the decreased flammability; whether the metal ions also contribute to this effect is not clear.

2. Insect Resistance. Fabrics treated in DMF with $SbCl_3$, $BiCl_3$, $TiCl_4$, $ZrCl_4$, and $ZnCl_2$, all showed a degree of moth-resistance in the CSMA 14-day feeding test. Of these, $SbCl_3$ gave the best results (Table 8) and scored sufficiently effective in the tests. This effect with $SbCl_3$ was obtained although the uptake of this salt was not particularly high as compared with the uptakes of the other

TABLE 7

Flammability of Native Wool Fabric Treated with Salt in DMF at 70° for 2 hrs.[a]

Salt	Weight Add-on % original wool	Flammability[b] After-flame burn time (seconds)	Char length (inches)	% chlorine in treated wool
$ZrCl_4$	3.0	3.8	4.6	2.77
$BiCl_3$	8.6	4.3	5.1	2.62
$ZrCl_2$	1.0	5.9	5.3	1.52
$SbCl_3$	1.1	31.0	Total(10")	[c]
None (solvent only)	-	30.0	Total(10")	

[a] All treated fabrics were washed with DMF, methanol, water then methanol again and air dried.

[b] Modified ASTM procedure D1230; test were run in duplicates; 12 seconds experimental flame; for effective treatments: char length should be less than 6" and after-flame time less than 12 seconds.

TABLE 8

Flame- and Insect-Resistance of Native Wool Fabric Treated with Salts in DMF at 110° for 1 hr.

Treatment[a/] (Salt)	Weight increase (% of original wool)	Flammability[b/] After-flame burn time (seconds)	Char length (inches)	Insect resistance (Excrement per larvae) (mg)	% Cl in treated wool
None (control) (W)		33	total (10")	3.1	0.3
$ZnCl_2$	14	1.4	3.7	- [d/]	3.86
$ZnCl_2$ (W)	6	-	-	1.03	-
$SbCl_3$	9	8.2	5.1	0.39	3.70
$SbCl_3$ (W)	2	14.0	7.0	0.45	-
$ZrCl_4$	13	1.9	3.3	0.86	3.8
$ZrCl_4$ (W)	5.8	2.7	3.5	1.2	2.7
$BiCl_3$	20	1.2[e/]	4.8	0.6	5.4
$BiCl_3$ (W)	17	2.4[e/]	4.7	1.4	-
$TiCl_4$ (W)	0.4	3.2	4.5	0.7	2.4
$ZnCl_2$ + $SbCl_3$ (W)	6.4	3.9	4.9	0.23	-
$ZrCl_4$ + $SbCl_3$ (W)	5.7	1.9	3.8	1.39	-
$BiCl_3$ + $SbCl_3$ (W)	8	5.7	4.9	0.25	-

[a/] All treated fabrics were washed with DMF and methanol; those marked (W) were washed also with water.

[b/] Modified ASTM procedure D1230. (Cloth suspended vertically, 12 seconds experimental flame.

[c/] 14-day CSMA feeding test with black carpet beetle larvae.

[d/] Hyphen indicates not done.

[e/] After-glow was usually observed with bismuth salt treatments.

TABLE 9

Flammability Tests of Wool Fabric Treated with Zirconium and Titanium Salts From Water Containing Citric Acid[a]

Treatment (Salt)	Flammability Tests[b] After-flame burn time (seconds)		Char length (inches)		Remarks
	Mean	SD	Mean	SD	
$ZrOCl_2$	7.8	3.4	5.9	1.5	n = 19; two separate expts.
$TiCl_4$	3.3	1.9	4.5	0.2	n = 18; two separate expts.

[a] According to Benisek (1971). Conditions were: liquor to wool ratio 20:1, 0.01% Igebal D, 45 minutes boiling then 10 minute wash under tap. For zirconium salt: 4.5% $ZrOCl_2 \cdot 4H_2O$ + 5% of conc. HCl (37%) + 6% citric acid all O.W.W. (on weight of wool). For titanium salt: 5% of $TiCl_4$ (50%) + 6.5% citric acid O.W.W.

[b] Modified ASTM procedure D1230 - See Footnote b Tables 7 & 8.

TABLE 10

Shrinkage Tests[a] of Wool Fabric Treated with Salt Solutions in DMF at 110° for 1 Hr.

Treatment (Salt)	Percent Area Shrinkage
None (control) DMF only	47
$ZnCl_2$	41
$SbCl_3$	35
$ZrCl_4$	27
$BiCl_3$	24

[a] A 2 minute Accelerotor wash of treated fabric circles (equivalent to about 20 regular machine washes); 1/2% soap solution, 200 ml wash volume.

salts. The effectiveness of these salts was also tested when the treatments were done from water or water acidified with HCl to minimize hydrolysis or water with citrate as complexing agent. Results were satisfactory with $SbCl_3$. (Masri and Friedman, 1974.) The other salts were only partially effective. The satisfactory insect resistance but inadequate flame-resistance imparted by $SbCl_3$ and the opposite order of effectiveness as regards these two properties in the case of the other salts, prompted us to test the effects on uptake, flammability, and insect resistance for fabrics treated in DMF containing both $SbCl_3$ plus one of the other salts at a time. Results with pairs were satisfactory: both fire- and insect-resistance were imparted to the fabrics.

In general, the finishing treatment for shrinkage control did not interfere with the effectiveness of the salt treatments on flame- and insect-resistance. Also, as was pointed out earlier, the flammability and insect resistance tests were measured after thorough washing under running tap water (except for a few instances indicated in the Tables) in order to remove loosely-bound salt. High uptakes (as weight add-on relative to original weight of wool) were frequently measured (from DMF) when the treated fabrics were washed with DMF and methanol but not water. In future experiments, we intend to further examine wash-fastness of the bound salts (and durability of associated effects on fabric properties) after more thorough machine washing with and without detergents. Skin-sensitization test on the salt-treated fabrics are also planned.

Although some shrink-resistance was conferred on the fabrics due to treatment with some of the salts, the extent of the effect was not outstanding or adequate. The best results which were obtained with $BiCl_3$, $ZrCl_4$, and $SbCl_3$ indicated 25-35% area shrinkage compared to about 45% shrinkage with control untreated fabric in a 2 minute Accelerotor wash (equivalent to about 20 regular machine washes). (Table 10).

VI. CONCLUDING REMARKS

The uptake profiles of a number of metal ions were determined for native wool and a number of wool derivatives in which certain functional groups of the protein were specifically modified. The results show that wool and these derivatives bind substantial amounts of many metal ions efficiently (relatively high capacity and rapid uptake). Comparison of the uptake profiles of native wool and these derivatives yield informative points which suggest modes of interaction of wool (and proteins in general) with metal ions. The comparison also shows that specificity of metal ion binding can be imparted to wool by certain chemical modification and suggest means of imparting further specificity by appropriate choice of protein modifying reactions and reagents. Thus reagent may be chosen to add to the wool new functional groups that are binding sites or block existing binding sites. Also, treatment of wool woven fabric with some salts has been shown to confer upon the wool desirable properties. Specifically, flame- and insect-resistance and some laundering shrinkage control were observed with salts of Bi(III), Zr(IV), Ti(IV), Sb(III), and Zn(II).

The increased binding and modified binding specificity for various metal ions as a result of specific functional group modification of wool (e.g. increased uptake of many ions when disulfide of wool is reduced and increased uptake of Sn(II) by nitro-keratin), suggest several types of potential applications of wool and its derivatives for concentrating, removing, and recovering toxic, precious, radioactive or other metal ions from various liquids. For example: wool and its derivative might be useful in purifying natural water supplies or industrial waste water such as effluents from mining, photographic, electroplating or other manufacturing operations.

Another potential application is for treatment of metal ion poisoning, for example, by Hg(II), Cd(II), Pb(II), As(III), Tl(I), or even Cu(II) and Zn(II). Clarkson and Co-workers (1973) (See also MacGregor and Clarkson, this volume) have shown the value of a sulfhydryl-containing polymer that is not digestible, for enhancing elimination of methylmercury from the body. Takahashi and Hirayama (1971) have already proposed the use of reduced hair in the same way. Given orally, these adsorbents bind mercury in the digestive tract, including mercury excreted in the bile, preventing its absorption or reabsorption from the intestine. Wool derivatives may be used in the same way to bind mercuric and other toxic ions and may have advantages.

Another potential method of using the fibrous sorbents to modify the metal ion content of the body is to pass the blood through a shunt in the circulation over a bed of the sorbent. Administration of metal ion mobilizing drugs such as 2,3-dimercaptopropanol (BAL) or d-penicillamine during the filtration may prove especially useful. The drugs would effect release of the toxic ions across blood-tissue barriers into the circulation where they may become fixed to the insoluble adsorbent in contact with the blood before they are redistributed to critical areas in the body, depending on the effectiveness of the sorbents in competing with the tissues (and the mobilizing ligands themselves perhaps) for binding the metal ions. Rapid and effective binding from the circulation by the sorbent (as may occur with reduced wool for example) may thus prevent build-up of high blood levels of the toxic ions and concomitant redistribution associated with the use of these mobilizing drugs. The insolubility and low allergenicity of wool suggest that immune responses would not be troublesome.

A third kind of potential application of wool or wool derivative as metal ion sorbents is in chemical analysis of dilute solutions (Masri and Friedman, 1973; also see Friedman and Masri, this volume). When the necessary partition coefficients have been established, analysis of wool (or modified wool) that has been equilibrated with a given liquid should establish the metal ion activities in the liquid.

Finally, we envision use of metal ion uptake profiles as probes to obtain information about protein structure. When enough information has accumulated about modes of metal ion interactions with specific binding sites, protein structural details may be deduced from uptakes of various metals.

Acknowledgments

We wish to thank the following from our Laboratory for their help: Eddie C. Marshall, F. William Reuter, Amy T. Noma, Geraldine E. Secor, Buenafe T. Boutz, Iva B. Ferrell, Lona M. Christopher, Linda C. Brown, Robert D. Wong, and Michael J. Beaucage. We are also indebted to Robert Davis and Roy Bry, Stored-Product Insects Research and Development Laboratory, ARS-USDA, Savannah, Georgia, for the insect-resistance tests.

References

Bakir, F., Damluji, S. F., Amin-Zaki, L., Murtadha, M., Khalidi, A., Al-Rawi, N. Y., Tikriti, S., Dhahir, H. I., Clarkson, T. W., Smith, J. C., and Doherty, R. A. (1973). Methylmercury Poisoning in Iraq. _Science_, _181_, 230-241.

Barr, T., and Speakman, J. B. The Production of Unshrinkability by Cross-linkage Formation in Wool, Part I - Woven Fabrics. _J. Soc. Dyers Colour._, _60_, 335-340 (1944).

Bell, J. W., and Whewell, C. S. The Reaction Between Wool and Nickelammonium Hydroxide I. The Absorption of Nickel by Wool From Solutions of Nickelammonium Hydroxide. _J. Soc. Dyers Colour._, _68_, 299-305 (1952).

Bell, J. W., and Whewell, C. S. A Note on the Action of Nickelammonium Hydroxide on Proteins in General. _ibid_, _68_, 305 (1952).

Benisek, L. Use of Titanium Complexes to Improve the Natural Flame Retardancy of Wool. _J. Soc. Dyers & Colour._, _87_, 227-278 (1971).

Cecil, R. Quantitative Reactions of Thiols and Disulfides with Silver Nitrate. _Biochem. J._, _47_, 572-584 (1950).

Cecil, R., and McPhee, J. R. The Sulfur Chemistry of Proteins. _Advan. Protein Chem._, _14_, 255-389 (1959).

Fish, R. H., Scherer, J. R., Marshall, E. C., and Kint, S. (1972). A Column Chromatographic and Laser Raman Spectroscopy study of the Interaction of Mercuric Chloride with Wool. Chemosphere, No 6, 267-272 (1972).

Friedman, M., and Masri, M. S. Sorption Behavior of Mercuric Salts on Chemically Modified Wools and Polyamino Acid. J. Appl. Polymer Sci., 17, (1973).

Friedman, M., Noma, A. T., and Masri, M. S. New Internal Standards for Basic Amino Acid Analyses. Anal. Biochem., 51, 280-287 (1973).

Gurd, F. R. N., and Wilcox, P. E. Complex Formation with Metallic Cations. Advan. Protein Chem., 11, 311- (1956).

Guthrie, R. E., and Laurie, S. H. The Binding of Copper (II) to Mohair Keratin. Aust. J. Chem., 21, 2437-2443 (1968).

Hamrajani, S. N., and Narwani, C. S. Polarographic Study of Metal Ion Complexes with Keratin Fibers (Wool) at pH 4-5 and Temperature 30°. Part I. Chlorides and Nitrates of Hg^{++}, Cu^{++}, Cd^{++} and Pb^{++}. J. Indian Chem. Soc., 44, 704-709 (1967).

Hartley, F. R. Studies in Chrome Mordanting II. The Binding of Chromium (III) Cations to Wool. Aust. J. Chem., 21, 2723-2735 (1968).

Hartley, F. R. Studies in Chrome Mordanting, V. Kinetics and Mechanism of the Interaction of Chromium (III) Salts with Wool. Aust. J. Chem., 23, 275-285 (1970). Also see: Wool Science Review, the Chrome Mordant Process for Dyeing Wool, 37, 54-63 (1969).

Hartley, F. R. The Uptake of Aluminium by Wool. Aust. J. Chem., 21, 1013-1022 (1968).

Hojo, N. The Change of Adsorbability of Hg, Cu and Ag of Wool Through Alkali Treatment. Sen-i Gakkaishi, 14, 953-955 (1958).

Kokot, S., Feughelman, M., and Golding, R. M. An Electron Spin Resonance Study of the Copper (II) Interaction with Wool-keratin, Part I. Textile Res. J., 42, 704-708 (1972); and Part II, ibid, 43, 146-153 (1973).

Laurie, S. H. Uptake of Metal Complexes by Wool and Their Effects, Part I. Uptake of Zirconium from Aqueous Sulfate Solutions. Textile Res. J., 36, 476-480 (1966).

Laurie, S. H. Uptake of Metal Complexes by Wool and Their Effects on its Physical Properties, Part II. A Method of General Application for Impregnating with Titanium Dioxide. Textile Res. J., 38 1140-1141 (1968).

Leach, S. J. The Reaction of Thiol Disulfide Groups with Mercuric Chloride and Methylmercuric Iodide, II. Fibrous Keratins. Aust. J. Chem., 13, 547-566 (1960).

Maclaren, J. A., and Sweetman, B. J. The Preparation of Reduced and S-Alkylated Wool Keratins Using Tri-n-butyl-phosphine. Aust. J. Chem., 19, 2355-60 (1966).

McPhee, J. R. Reaction of Wool with Sodium Sulfite and Zn^{++}. Textile Res. J., 34, 382-384 (1964).

Masri, M. S., and Friedman, M. Mercury uptake by polyamine-carbohydrates. Environ. Sci. Technol., 6(8), 745-746 (1972).

Masri, M. S., Windle, J. J., and Friedman, M. P-Nitrostyrene: New Alkylating Agent for Sulfhydryl Groups in Reduced Soluble Proteins and Keratins. Biochem. Biophys. Res. Comm., 47, 1408-1413 (1972).

Masri, M. S., and Friedman, M. Competitive Binding of Mercuric Chloride in Dilute Solutions by Wool and Polyethylene or Glass Containers. Environ. Sci. Technol., 7, 951-953 (1973).

Masri, M. S., and Friedman, M. Effects of Chemical Modification of Wool on Metal Ion Binding. J. Applied Polym. Sci., 18, in press (1974).

Masri, M. S., Reuter, F. W., and Friedman, M. Binding of Metal Cations by Natural Substances. J. Applied Polym. Sci., 18, 675-681 (1974a).

Masri, M. S., Reuter, F. W., and Friedman, M. (1974). Interaction of Wool with Metal Cations. Textile Res. J., 44, 298-300.

Simpson, W. S., and Mason, P. C. R. Absorption of Silver Ions by Wool. Textile Res. J., 39, 434-441 (1969).

Sokolovsky, M., Riordan, J. F., and Vallee, B. L. Tetranitromethane. A Reagent for the Nitration of Tyrosyl Residues in Proteins. Biochemistry, 5, 3582-89 (1966).

Sokolovsky, M., Riordan, J. F., and Vallee, B. L. Conversion of 3-Nitrotyrosine to 3-Aminotyrosine in Peptides and Proteins. Biochem. Biophys. Res. Comm., 27, 20-25 (1967).

Spackman, D., Stein, D. H., and Moore, W. H. Anal. Chem., 30, 1190-1206 (1958).

Speakman, J. B., and Coke, C. E. Trans. Faraday Soc., 35, 246 (1939).

Swan, J. M. The Reaction of Protein Thiol and Disulphide Groups with Cupric Sulphite Solutions. Aust. J. Chem., 14, 69-83 (1961).

Takahashi, H., and Hirayama, K. Accelerated Elimination of Methyl Mercury from Animals. Nature, 232, 201-202 (1971).

Whitfield, R. E., Miller, L. A., and Wasley, W. L. Wool Fabric Stabilization by Interfacial Polymerization IV. Polyureas, polyesters, polycarbonates. Additional observations on polyamides. Textile Res. J., 33, 440-444 (1963).

Whitfield, R. E. Wool Stabilization with Polyurea Finishes. Applied Polymer Symposia, 18, 559-567 (1971a).

Whitfield, R. E. Some New Polymeric Finishes for Wool Textiles. Textile Chemist and Colorist, 3, 256-259 (1971b).

25

X-RAY PHOTOELECTRON SPECTROSCOPIC STUDIES OF BIOLOGICAL MATERIALS: METAL ION PROTEIN BINDING AND OTHER ANALYTICAL APPLICATIONS

M. M. Millard

USDA, ARS

Berkeley, California

INTRODUCTION

X-Ray Photoelectron Spectroscopy (XPS) is a relatively new type of electron spectroscopy involving ejection of electrons from the core levels of atoms in materials by X-rays and measurement of their energy. Structural as well as analytical information may be obtained using this technique and in a short time XPS has found wide application in chemistry (Siegbahn et al., 1967). The purpose of this article is to review the limited number of applications that exist in the area of biochemistry and hopefully to stimulate further interest in this area. Commercial instruments have become available in the last few years and the capability of instrumentation is rapidly advancing. Initial experiments on instruments of early design indicate that unique structural and analytical information can be obtained and the quality of this information should be greatly improved using newer instruments.

The material covered in this review article will be organized in the following way. The fundamentals of XPS will be presented including examples of structural information obtainable from chemical shifts. The limitations placed on samples of biological interest such as radiation exposure will be considered. Principles of instrumentation will be briefly outlined using the varian IEE spectrometer as an example. Structural applications on the nature of metal ion binding in metallo proteins will be reviewed together with various analytical applications such as the detection of chemical modification on proteins by XPS.

Many excellent review articles on XPS have been published recently emphasing various aspects of the subject. These include Hercules (1970), Jolly (1970), Brundle (1971), Lindberg (1971), Clark (1972), Baker (1972), Hercules (1972), and the Proceedings of the Asilomar Conference on Electron Spectroscopy (Shirley, 1972).

FUNDAMENTALS

In X-ray photoelectron spectroscopy, core electrons are ejected from the surface of materials by x rays. Most work is done using the Mg K_α line at 1253.6 eV or the AlK_α line at 1486.6 eV. The emitted electrons are analyzed in electrostatic or magnetic energy analyzers. A curve is generated consisting of the number of electrons as a function of energy. When a core electron is photo ejected from the ground state and the electron emitted suffers no energy loss processes, a simple relationship exists between the kinetic energy of the emitted electron, the binding energy of the electron, and the energy of the exciting x ray photon. The binding energy of the core electron can be calculated from the expression:

$$E_{BE} = E_{x\text{-ray}} - E_{kinetic} - C$$

where C is a constant correction factor for a given spectrometer and involves quantities such as the work function of the spectrometer material. Binding energies are usually calculated relative to a reference element such as carbon or gold. A schematic drawing of an X-ray photoelectron experiment, the energy conservation equation, and a simple energy level diagram are given in Figure 1. Other processes such as Auger or shake up and shake off can lead to lines in the spectrum of lower intensity and will not be considered further (Shirley, 1972). The X ray Photoelectron Spectrum of vitamin B_{12} (Siegbahn, 1967) is shown in Figure 2 and serves to illustrate a typical spectrum.

Core electron binding energies for the light elements are listed in Figure 3, illustrating the order of magnitude of the electron energies involved. Fortunately for the chemist, the binding energy of core electrons is related to atomic charge or chemical environment and a chemical shift results between atoms with unlike atomic charges. The order of magnitude of the chemical shift for some elements is illustrated in Figure 4 (Jolly, 1970), where the core shifts are given for nitrogen, carbon, and xenon. The 1S binding energy for nitrogen changes by 8 eV going from diethylamine to nitrogen dioxide. The 1S binding energy in carbon varies by 11 eV going from methane to carbon tetrafluoride.

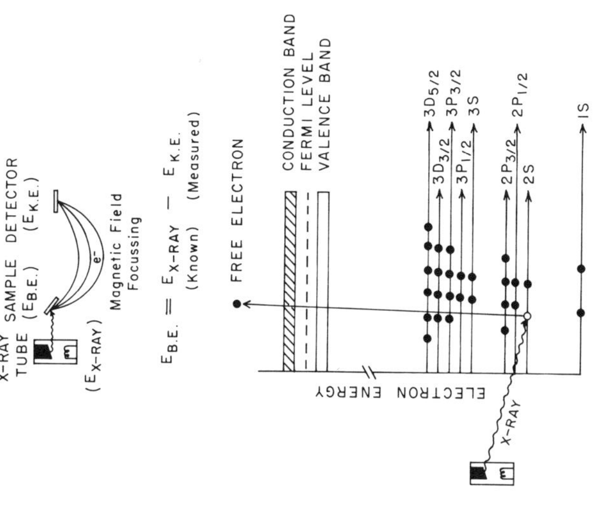

Figure 1. Schematic XPS experiment. (Siegbahn et. al., 1967)

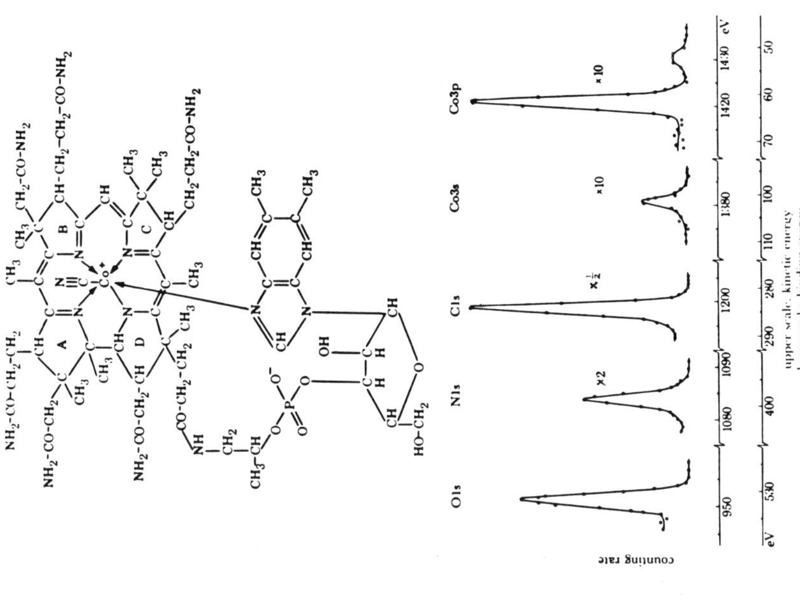

Figure 2. Electron spectrum from Vitamin B_{12}. (Siegbahn et. al., 1967)

	$1s_1/2$	$2s_1/2$	$2p_1/2$	$2p_3/2$	$3s_1/2$	$3p_1/2$	$3p_3/2$	$3d_3/2$	$3d_5/2$
Li	55								
Be	111								
B	188			5					
C	284			7					
N	399			9					
O	532	24		7					
F	686	31		9					
Ne	867	45		18					
Na	1072	63		31	1				
Mg	1305	89		52	2				
Al		118	74	73	1				
Si		149	100	99	8		3		
P		189	136	135	16		10		
S		229	165	164	16		8		
Cl		270	202	200	18		7		
A		320	247	245	25		12		
K		377	297	294	34		18		
Ca		438	350	347	44		26		5
Sc		500	407	402	54		32		7
Ti		564	461	455	59		34		3
V		628	520	513	66		38		2
Cr		695	584	575	74		43		2
Mn		769	652	641	84		49		4

Figure 3. Binding energies of some light elements (B.E. are in eV). (Jolly, 1970).

Atomic Level	Compound	Relative Binding Energy eV
N 1s	$(CH_3)_2NH$	-0.7
N 1s	CH_3NH_2	-0.3
N 1s	NH_3	0
N 1s	HCN	1.2
N 1s	NNO	3.2
N 1s	N_2	4.35
N 1s	NO	5.5
N 1s	N_2F_4	6.8
N 1s	NO_2	7.3
C 1s	CH_4	0
C 1s	CO	5.4
C 1s	CO_2	6.8
C 1s	CF_4	11.0
Xe $3d^5/_2$	Xe	0
Xe $3d^5/_2$	XeF_2	2.95
Xe $3d^5/_2$	XeF_4	5.5
Xe $3d^5/_2$	$XeOF_4$	7.0
Xe $3d^5/_2$	XeF_6	7.9

Figure 4. Relative core electron binding energies. (Jolly, 1970).

Unfortunately the chemical shift seems to be inversly proportional to the atomic radius and therefore is much smaller for the transition elements, is about 1 eV for mercury. Figure 5 gives the 2p binding energy of sulfur in various sulfur compounds. There is a difference of about 8 eV between this quantity in sulfides and sulfonic acids. The line width obtained from solids varies between 2 to 3 electron volts and the resolution of most instruments is between 1 and 1.5 eV. Curve deconvolution techniques can be used to resolve overlapping lines which is a common occurence. An illustrative example of the structural information obtainable from X-ray Photoelectron Spectroscopy is given by the electron spectra of dibenzyldisulfide and its oxidation products shown in Figure 6 (Siegbahn, 1967). When dibenzyldisulfide is oxidized to the thiolsulfinate and thiolsulfonate, a second line appears at higher binding energy and the area under these lines is proportional to the amount of oxidized and reduced product. Siegbahn also studied the oxidation of insulin by this technique. This is probably one of the first examples of the use of XPS to follow chemical changes in proteins.

Another important factor in the study of substances by XPS is the relative sensitivity of detection of the elements. The atomic photoelectric cross section varies with the elements. Wagner (1972) has studied the sensitivity of many of the common elements and the relative atomic sensitivities for the elements obtained by him are given in Figure 7. This can be a very important consideration in dealing with samples containing elements in low concentrations.

SAMPLE CONSIDERATIONS

In the application of XPS to samples of biochemical interest, the following points concerning the nature of the sample and its exposure to radiation should be kept in mind. Samples are usually in the form of powders or solid material although frozen solutions may be accomodated. The sample is maintained under a vacuum of about 10^{-5} or 10^{-6} torr. Attachments on instruments have been devised to allow sample preparation under inert atmospheres or under vacuum followed by direct insertion into the instrument minimizing exposure to the atmosphere. The sample is exposed to fairly intense radiation and the possibility of radiation damage exists. Kramer (1971) estimated the dose rate in the Berkeley spectrometer to be 10^3 rads/hr by placing a film badge in the instrument. Several authors have noted changes in the spectra from samples with time (Siegbahn et al., 1967; McDowell et al., 1972; Carlson, 1973; Ibers et al., 1973; Grunthaner, 1974). Often these changes can be minimized by cooling the sample to

	S2p (eV)	
Na_2S	160.8	
$R-S-H$	162.0	
$R-S-R$	162.3	162.4
$R-S-S-R$	162.7	
S_8	162.2	
$R_2N-S-NR_2$	162.5	
$R-S-OR$	164.2	
$R_2S=O$	164.9	
$R-SO_2^{\ominus}$	165.4	
$SO_3^{2\ominus}$	165.8	
$(RO)_2S=O$	167.0	167.1
$R-SO_2-R$	167.1	
$R-SO_2Cl$	167.3	167.3
$R-SO_2N<$	167.3	
$R-SO_3^{\ominus}$	167.2	
$R-SO_2OR$	167.5	
$SO_4^{2\ominus}$	168.0	
$RO-SO_3^{\ominus}$	168.7	
$(RO)_2SO_2$	168.6	
$[S-SO_3]^{2\ominus}$	160.9	
$R-S-SO-R$	162.3	
$R-S-SO_2-R$	162.8	
$R-S-SO_3^{\ominus}$	162.8	
$R-OS-S-R$	164.8	
${}^{2\ominus}[O_3S-S]$	166.9	
$R-O_2S-S-R$	167.1	
${}^{\ominus}O_3S-S-R$	167.8	

Figure 5. Binding energies for various sulfur compounds (Siegbahn et al., 1967).

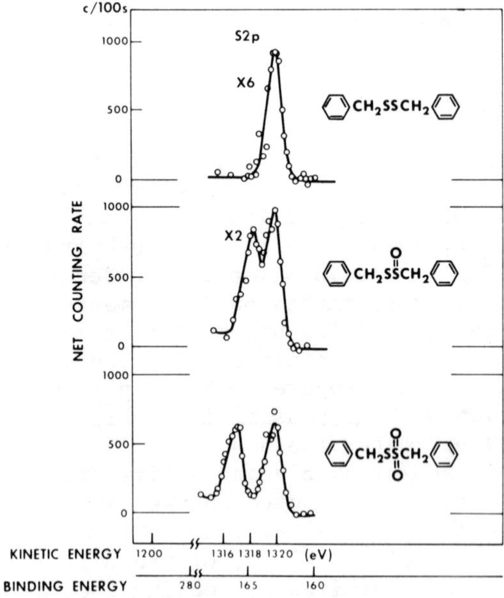

Figure 6. Electron spectrum of dibenzyl disulfide and its oxidation products. (Siegbahn et al., 1967).

F 1s line used as standard

Z	Element	Strong line	Sensitivity by peak height	Sensitivity by peak area
3	Li	1s	0.024	0.022
5	B	1s	0.14	0.14
6	C	1s	0.27	0.24
7	N	1s	0.42	0.41
8	O	1s	0.52	0.61
9	F	1s	1.00	1.00
11	Na	1s	2.14	2.09
12	Mg	1s	2.24	2.24
13	Al	2s	0.28	0.23
14	Si	2p-2p3/2	0.22	0.17
15	P	2p-2p3/2	0.40	0.26
16	S	2p-2p3/2	0.46	0.33
17	Cl	2p-2p3/2	0.55	0.46
19	K	2p3/2	0.94	0.85
20	Ca	2p3/2	1.09	1.01
22	Ti	2p3/2	1.10	1.10
24	Cr	2p3/2	1.69	1.53
25	Mn	2p3/2	1.18	1.55
26	Fe	2p3/2	1.82	1.76
28	Ni	2p3/2	4.10	3.68
30	Zn	2p3/2	3.73	4.24
32	Ge	2p3/2	4.80	5.60
33	As	2p3/2	4.80	5.87
34	Se	3p3/2	0.64	0.86
35	Br	3d-3p3/2	0.54	0.71
37	Rb	3d-3d5/2	1.20	0.95
38	Sr	3d-3d5/2	1.26	1.03
42	Mo	3d5/2	2.37	2.04
45	Rh	3d5/2	2.16	2.00
48	Cd	3d5/2	3.60	3.80
50	Sn	3d5/2	5.75	5.75
51	Sb	3d5/2	6.2	6.7
53	I	3d5/2	5.14	5.03
56	Ba	3d5/2	6.16	6.63
62	Sm	3d5/2	2.80	6.90
72	Hf	4f-4f7/2	1.14	0.85
74	W	4f7/2	1.52	1.37
75	Re	4f7/2	2.11	1.77
77	Ir	4f7/2	1.80	1.72
78	Pt	4f7/2	2.24	1.93
82	Pb	4f7/2	3.92	4.10
83	Bi	4f7/2	3.48	4.16
92	U	4f7/2	5.80	6.45

Figure 7. Relative Atomic Sensitivity of the Elements. (Wagner, 1972).

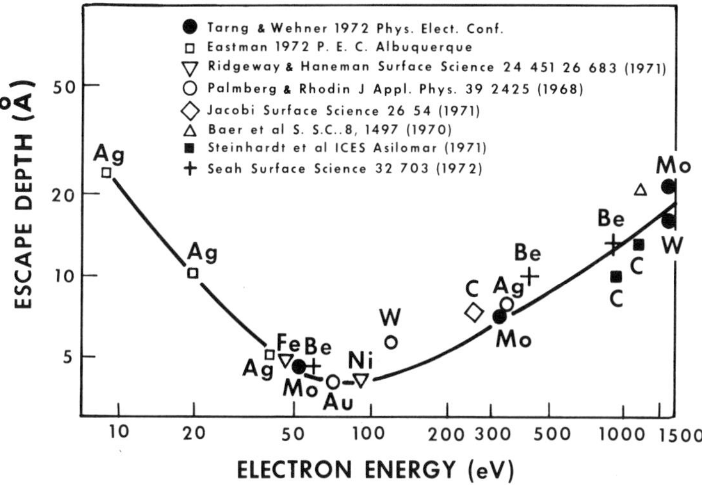

Figure 8. Auger electron escape depths vs energy for various materials. (Palmberg, 1973).

temperatures approaching that of liquid nitrogen while the spectra are being recorded.

Photoejected electrons escape only near the surface without suffering energy loss processes. Consequently, information is obtained from less than 100 Å in depth from the surface. Electron escape depths obtained by Auger spectroscopy for several elements over a range of energies are plotted in Figure 8.

Some precaution should be exercised in extrapolating conclusions concerning data obtained from the surface of materials to bulk properties.

When accurate chemical shift information is required, shifts due to sample charging can be troublesome (Hayes, 1973). Data is usually referenced to a standard line such as gold or graphite. Various methods have been adapted to obtain consistant data such as evaporation of thin gold films on the surface of the sample or intimate mixing of graphite or salts with the sample. In most cases, authors have been satisfied that reproducible data can be obtained by adopting one of these techniques.

INSTRUMENTATION

Since the field of XPS is relatively new, rapid advances in instrumentation are occurring. Several manufacturers are currently marketing instruments and the cost and capability of a given instrument vary considerably. A recent article (Lucchesi and Lesler, 1973) should be consulted for a detailed comparison of the performance of the instruments currently available as well as a coverage of the principles of instrumentation.

The basic components of an x-ray photoelectron spectrometer are the x-ray source, the electron energy analyzer, a detector and recorder, and the associated control devices. Early research instruments employed magnetic deflection systems for the electron energy analyzer. However, commercial instruments available today employ electrostatic energy analyzer and detector systems as well as the sample chamber and associated electronic control equipment.

The author is most familiar with the Varian IEE Spectrometer. This instrument is representative of the instruments currently available and its principles of operation will be briefly reviewed. A schematic drawing of the x-ray source and sample chamber is shown in Figure 9. Electrons are emitted by a heated tungsten filament and bombard the magnesium or aluminum anode which emitts x-rays which then pass through an aluminum foil and strike the sample.

The photoejected electrons exit through a slit and enter a spherical electrostatic energy analyzer shown in Figure 10. The electrons energy is retarded to less than 100 eV, and then analyzed and focused onto an electron multiplier. The retarding voltage is swept in order to obtain a spectrum. The instrument is controlled by a computer whose function is shown by means of a block diagram

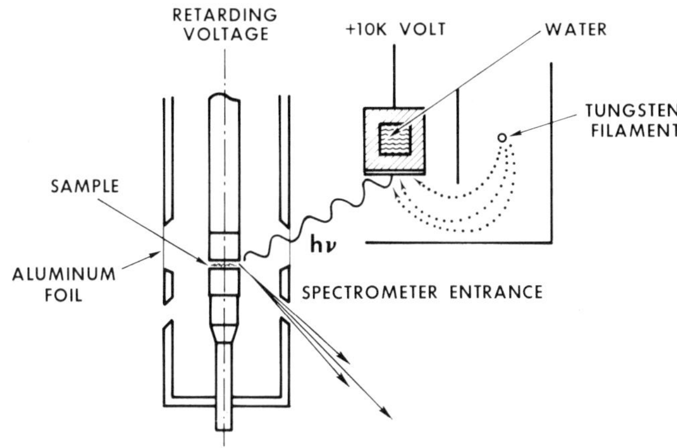

Figure 9. X-ray source and sample (Bodmer, 1971).

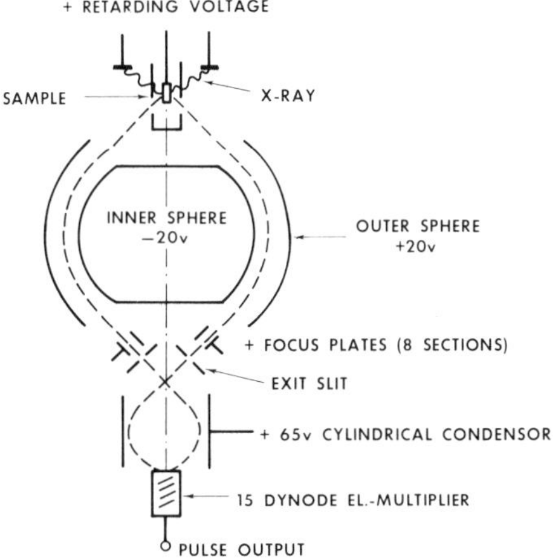

Figure 10. Cross section of the analyzer (Bodmer, 1971).

in Figure 11. The instrument is controlled by a built-in 620 minicomputer. The computer controls the retarding voltage of the spectrometer and gathers the data (electron pulse count) from the multiplier. The commands from the operator are given via teletype to the computer. The accumulated spectra can be analyzed by the operator on the oscilloscope or on the X-Y recorder. The data can also be printed on the teletype if the operator wishes to do so. The operator can enter a program, which allows him to take up to ten different regions of the energy spectrum. Each of these regions can be swept many times in order to get good signal to noise ratio.

APPLICATIONS OF XPS IN BIOCHEMISTRY

The monograph by Siegbahn (1967) contains some early examples of the study of molecules of biological interest by XPS. The electron spectrum from vitamin B_{12} containing lines from oxygen, nitrogen, carbon and cobalt is reproduced in Figure 2. The electron spectra from heparin, insulin, and a number of amino acids is also presented in this monograph. The oxidation of insulin was monitored from the sulfur 2p electron spectrum. A line at higher binding energy appeared when insulin was oxidized and the ratio of the areas under the sulfur 2p line due to insulin and its oxidation product yields the relative amounts of these two substances.

The first serious attempt to relate binding energy studies to structure was a study on nonheme iron proteins reported by Kramer (1971) in a Ph.D thesis. The nonheme iron proteins studied ranged in molecular weight from six to twelve thousand and some contained sulfur in two environments; acid labile and cysteine sulfur, as well as iron. The properties of the substances studied are summarized in Figure 12.

Some of these iron sulfur proteins contain two types of sulfur atoms, acid labile and cysteine type. X-ray crystal structure data is available for some of these compounds. In rubredoxin, isolated from C pasteurianum (Jensen 1972), the sulfur atoms are in a distorted tetrahedral arrangement about the iron atom. The high-potential iron-sulfur protein (HiPIP) of Chromatium Vinosum strain D consists of a polypeptide chain with four cysteines and an inorganic prosthetic group composed of four sulfur and four iron atoms bound to the apoprotein by the cysteine sulfur atoms. The molecule contains two iron sulfur complexes, each containing four iron and four inorganic sulfur atoms with four cysteine sulfur atoms coordinated to the iron atoms. The iron and inorganic sulfur atoms in each complex are arranged in approximate cubes with iron and sulfur atoms at alternate corners of each face (Carter, 1972). A similar configuration

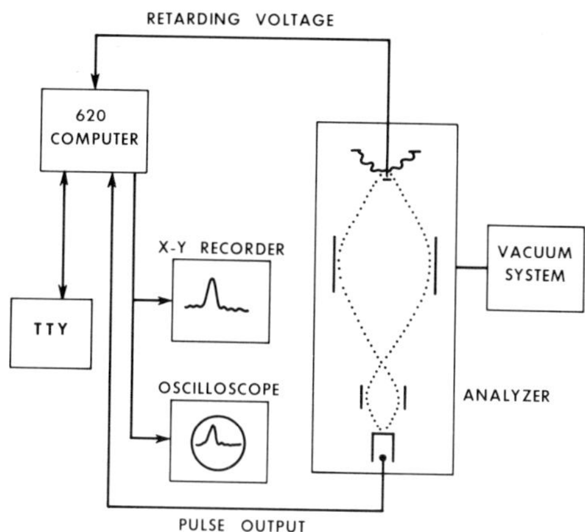

Figure 11. IEE spectrometer block diagram (Bodmer, 1971).

PROTEIN	STOICHIOMETRY OF Fe-S_i MOIETY	Fe3P B.E.	AREA RATIO	S2P B.E.	AREA RATIO
OXIDIZED RUBREDOXIN	1 Fe-0 S_i 4 Cysteines	54.6$\pm$.4eV	1	162.7$\pm$.3eV	1
REDUCED HIPIP	4 Fe-4 S_i 4 Cysteines	53.0$\pm$.4eV	1	162.9$\pm$.3eV 161.5$\pm$.3eV	1 1
OXIDIZED CLOSTRIDIAL FERREDOXIN	X Fe-X S_i 8 Cysteines (x=6 to 8 atoms)	53.4$\pm$.4eV	1	163.2$\pm$.3eV 161.5$\pm$.4eV	(approx) 4 3
OXIDIZED SPINACH FERREDOXIN	2 Fe-2 S_i 5 Cysteines	53.6$\pm$.5eV	1	163.1$\pm$.4eV 161.1$\pm$.4eV	(approx) 5 2
APO CLOSTRIDIAL FERREDOXIN	0 Fe-0 S_i 4 Cystines	— — —	–	163.2$\pm$.2eV	1

Figure 12. Summary of the quantitative results from XPS spectra of nonheme iron proteins. (Kramer, 1971).

has been reported for the iron sulfur complex in bacterial ferredoxin (Adman 1973).

The electron spectra in the sulfur 2p region and the iron 3p region for the proteins studies are shown in Figure 13 and 14, respectively. The binding energies in several model iron sulfur compounds together with the iron sulfur protein data is presented in Figure 15. A binding energy vs. calculated atomic charge correlation for some of these compounds is presented in Figure 16 (Klein, 1971).

The sulfur spectra contain two lines in the lower binding energy region which correlate well with the inorganic and cysteine sulfur, respectively. A line at higher binding energy indicates the presence of oxidized sulfur. The iron data is somewhat poorer and illustrates the sensitivity in low concentration. These data were obtained on a field free magnetic type of instrument of the Siegbahn design. The samples were run as frozen solutions.

Leibfritz (1972) has also reported the results of a study of the electron spectra of iron sulfur proteins. The data was obtained on frozen solutions using a Varian IEE instrument. The iron 2p level was investigated for various oxidized and reduced forms of the protein as well as various model compounds containing iron in oxidation states two and three. The iron 2p data is given in Figure 17.

Data 7a and 7b are from the oxidized and reduced form of the model iron dithiolene dimer whose structure is shown in Figure 18. The structure for compound 8 is also included in this figure. Compound 9 is trimeric iron acetate. Compounds 4 and 9 contain iron surrounded by oxygen atoms. The spectra obtained for compounds 1, 2, and 3 is presented in Figure 19. Although the signal to noise ratio is poor and illustrates the sensitivity difficulty with present instrumentation, a good correlation was obtained between the Fe2p binding energies in the oxidized and reduced form of the model compound and the iron sulfur proteins containing iron in oxidation state two and three. Iron (III) bound to sulfur or nitrogen fell in the range of 710 $\pm$ 0.5 eV. Iron (II) was in the range 708.0 $\pm$ 0.5 eV. Iron III bound to oxygen atoms had a 2p binding energy of about 711.0 $\pm$ 0.5 eV. The sulfur 2p spectra of compounds 1-3 were reported to consist of two lines in the region of 164.0 and 162.8 eV corresponding to cysteine and labile sulfur respectively.

Weser (1972,1973) and co-workers studied the electron spectra of the metallo protein bovine erythrocuprein and several amino

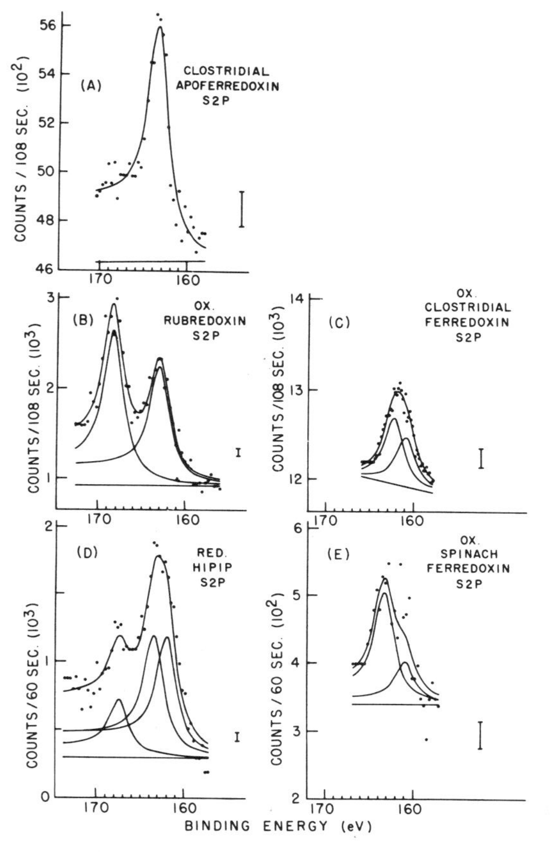

Figure 13. S 2p photoelectron spectra of nonheme iron proteins. (Kramer, 1971).

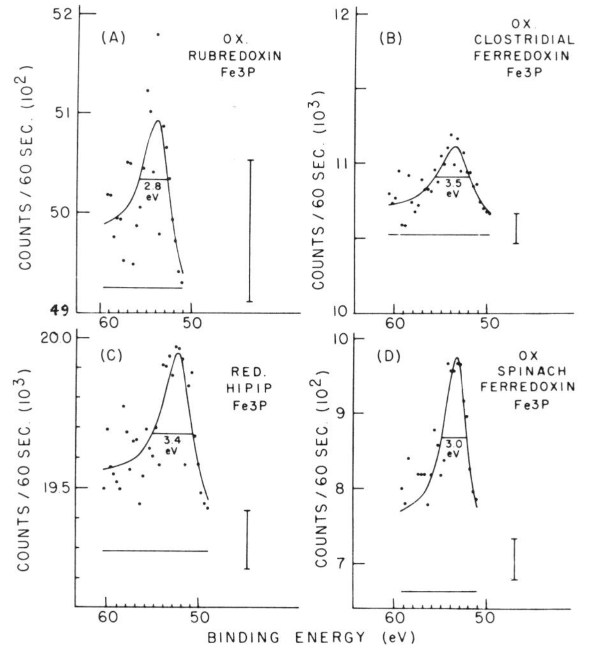

Figure 14. Fe 3p photoelectron spectra of nonheme iron proteins. (Kramer, 1971).

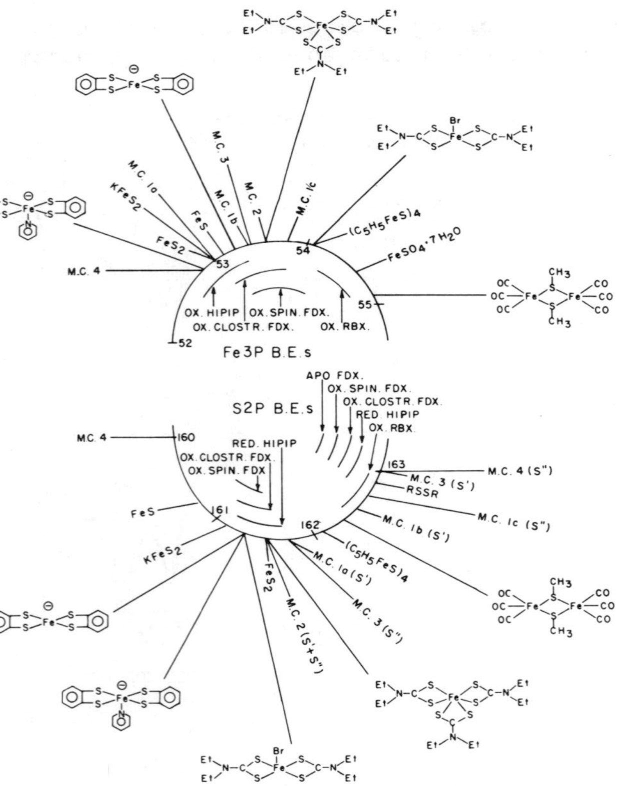

Figure 15. Summary of XPS data for non heme iron proteins and iron compounds. (Kramer, 1971)

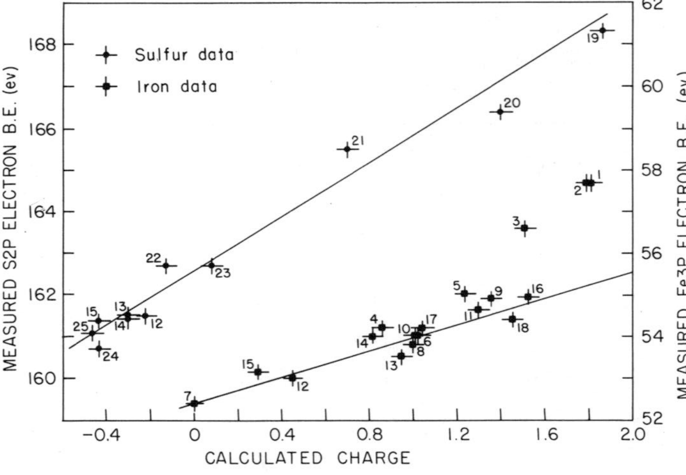

Figure 16. Plot of measured Fe 3p B.E. and S 2p B.E. versus calculated charge on iron and sulfur. (Klein, 1971).

	Compound	IE [eV]	Oxidation State Fe
(1)	*(C. pasteurianum)*	710.0	+3
(2)	*(C. acidi-urici)*	709.9	+3
(3)	(HIPIP)	708.2	+2
(4)	Iron phosvitin	711.0	+3
(5)	Hemin chloride	710.2	+3
(6)	Salen	710.1	+3
(7a)	(oxidized)	710.3	+3
(7b)	(reduced)	707.9	+2
(8)		708.5	+2
(9)		711.8	+3

Figure 17. Iron 2p binding energies. (Leibfritz, 1972).

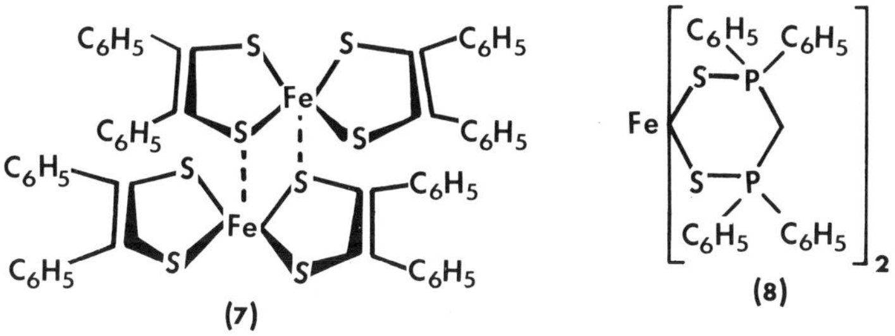

Figure 18. Compound structures from table in Figure 17.

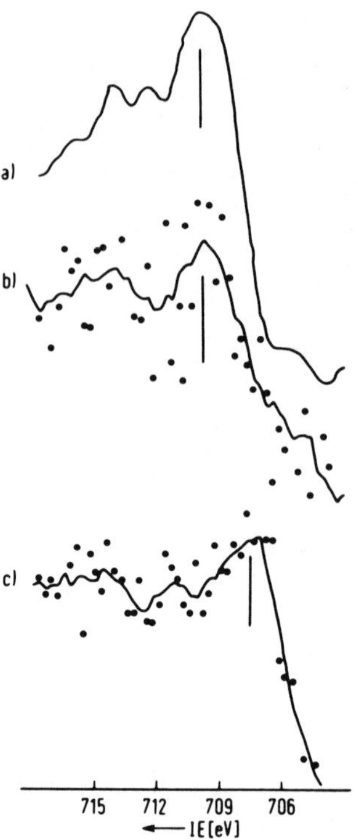

Figure 19. X-ray electron spectroscopy of the iron $2p_{3/2}$ level in the ferredoxins a) (1), b) (2), and c) (3). (Leibfritz, 1972).

acid complexes of copper and zinc. Erythrocuprein contains 2 atoms of copper and two atoms of zinc per 32600 g of the protein. These studies were aimed at gaining information on the nature of the binding site of the metals in this protein. Data was collected using a Varian IEE instrument on samples in the form of powders or frozen solutions. The electron spectra obtained for the protein and some of the amino acid complexes are shown in Figure 20 and the binding energies are tabulated in Figure 21. The signal to noise ratio for the metals in the protein is quite low. The intensity for the line was considered to be anomalously low. On this basis, the authors concluded that this low intensity was due to the fact that the zinc ion was located in the interior of the protein, and that the copper ion was exposed on the surface.

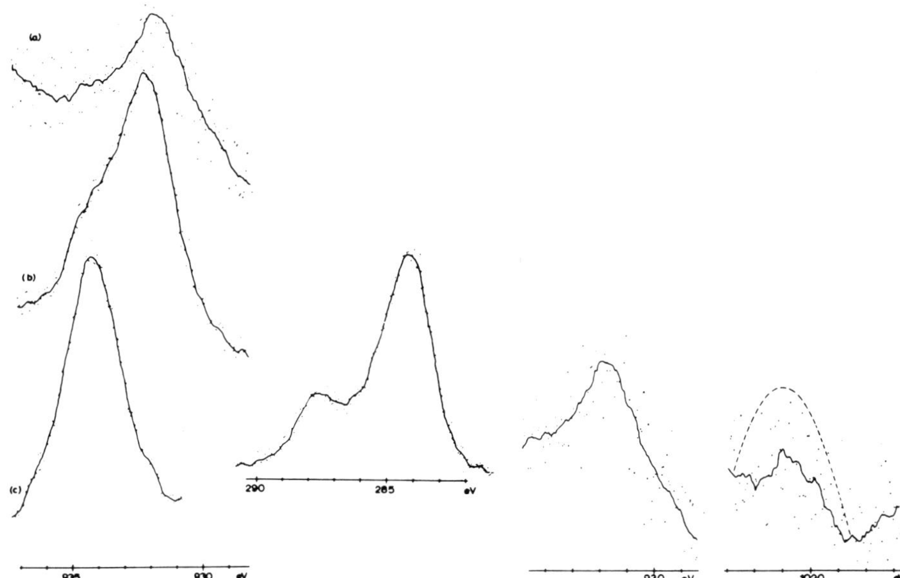

Figure 20. X-ray photoelectron spectra of the Cu $2p_{3/2}$ levels of erythrocuprein (a), $Cu[Cu(Asp)_3]$ (b), and $Cu(Val)_2$ (c). X-ray photoelectron spectra of the $2p_{3/2}$ levels of Cu (d) and Zn (e) ions in erythrocuprein. (---) in (e) indicates the signal expected if Zn is located on the surface of the enzyme. (Weser, 1972, 1973).

BINDING ENERGIES OF N, C AND S ATOMS IN Cu(II) AND Zn(II) COMPOUNDS

	N 1s (eV)	*COO (1s) (eV)*	*CH (1s) (eV)*
$Cu(Lys)_2$	399.4	287.5	284.0
	400.5		
$Cu(Gly)_2 \cdot H_2O$	399.9	288.5	284.0
$Cu(His)_2$	399.5		284.0
$Cu(NH_3)_4SO_4 \cdot H_2O$	399.3		
$Cu[Cu(Asp)_3]$	399.3	287.3	284.0
$Cu(Val)_2$	399.5		284.0
$Cu(Ala)_2$		288.1	284.0
$Zn(Ala)_2$		288.1	284.0
Erythrocuprein	399.1		
	S 2p (eV)	*C 1s (eV)*	
Erythrocuprein	163.2	287.2	
	167.7	284.0	

BINDING ENERGIES OF THE ELECTRONS OF THE $2p_{3/2}$ LEVELS OF SOME Cu(II) COMPOUNDS AND SOME Zn(II) COMPOUNDS

	Binding energy (eV)
$CuSO_4$, water free	933.3
$CuSO_4 \cdot 5H_2O$	934.5
$Cu(NH_3)_4SO_4 \cdot H_2O$	933.9
	932.0
$Cu(Lys)_2$	933.8
$Cu(Ala)_2$	933.9
$Cu(His)_2$	934.1
$Cu(Val)_2$	934.3
$Cu(Gly)_2 \cdot H_2O$	934.6
$Cu[Cu(Asp)_3]$	934.1
	932.0
Cu in erythrocuprein	931.9
$ZnSO_4 \cdot 7H_2O$	1021.5
Zn stearate	1021.0
$Zn(Ala)_2$	1021.3
Zn in erythrocuprein	1020.5

Figure 21. (Weser, 1972, 1973).

A conclusion of this nature should probably be regarded as highly tentative. The copper and zinc 2p binding energies in the metalloprotein were found to be 931.9 and 1020.5 eV, respectively. These values were considerably below those reported for the copper and zinc complexes studied. X-ray photoreduction of Cu(II) to Cu(I) has been observed (McDowell et. al. 1972; Grunthaner, 1974). The anomolous low binding energy for the metals in these proteins suggest that the ions have been reduced.

An extensive study of the application of XPS to metalloproteins containing copper and iron has recently been published in a thesis (Grunthaner, 1974). The problem of x-ray radiation damage to samples and data handling techniques to improve the signal to noise ratio and minimize sample run times was extensively investigated. Data on protein samples were obtained at temperatures selected to minimize radiation damage and data smoothing based on fourier transform techniques was developed and employed. Changes in the spectra of metal proteins as a function of time were studied and spectral difference techniques used to examine these changes. Most of the data was obtained on a Hewlett Packard 5950 A instrument and sample charging was compensated for using a flood gun.

The nature of the copper ion site in stellacyanin, plastocyanin, laccase and hemocyanin was investigated. Copper complexes were studied as model compounds. In agreement with the findings of McDowell et. al., 1972, paramagnetic copper (II) ions exhibit shake up satellites which are absence in the spectra of diagametic copper (I). The satellite structure was found to be sensitive to ligand environment about the copper. Extensive spectral changes during sample irridation were observed and some of these changes were correlated with photo reduction of Cu(II) to Cu(I). Copper 2p lines were assigned to various copper species in these proteins. Sulfur 2p electron spectra were also obtained for the copper model compounds and metalloproteins. Although the sulfur spectra was complex, overlapping lines were deconvoluted and spectral assignments attempted.

The XPS spectra of non-heme iron sulfur proteins and several model iron sulfur complexes was also studied. Sample oxidation and resulting spectral changes seemed to be a severe problem with these compounds. The oxygen 1s and sulfur 2p as well as the iron 2p spectra gave evidence for sample oxidation in the spectrometer. By studying changes in the spectra with time and difference techniques as well as spectral similarities to model iron sulfur compounds, assignments were made for iron and sulfur lines for these compounds. This thesis was obtained too late to allow inclusion of figures. However, this important study should be consulted for pertinent techniques and spectral details.

The binding of magnesium ions to the cell walls of certain bacteria has been investigated by XPS (Baddiley,1973). Cell wall preparations were isolated and exposed to magnesium chloride solution. The cell walls were exposed to Mg^{++} ions in the absence and presence of D alanine esterified to the cell walls. When Mg^{++} ions were bound to the cell walls in the presence of D alanine esterified to the cell walls, a broader Mg 2S line was present. When the D alanine was removed the higher binding energy component of the disappeared. A model was presented to account for the role of the D alanine in modifying the mechanism of Mg^{++} ion binding.

Some porphyrin and phthalocyanine compounds have been studied by XPS (Hayes 1973). Two nitrogen lines in the 1S region were observed for tetraphenylporphine free base at 397.2 and 399.1 eV. Presumably one type of nitrogen contains a proton and the other does not. The N 1S line in most metalloprophyrins appears very close to the low binding energy line above. The binding energy of the iron atom in oxidized and reduced iron prophyrin was considerably different. The two kinds of nitrogen in metal phthalocanines were identical with respect to XPS.

In our laboratory, we have explored several metallo proteins using a Varian IEE instrument with a high intensity x-ray source. The point of interest was to investigate the sensitivity of this instrument with respect to lines from metals in metallo proteins. Cytochrome C and some metal conalbumins were chosen for study. These materials ranged in molecular weight from about twelve to seventy-six thousand.

The iron 2p core electron spectrum is shown in Figure 22 and the iron 3p core electron spectrum in Figure 23 for cytochrome C. As can be seen from the spectrum in Figure 21, the signal to noise ratio is low and the line is broad. The spectrum in the iron 3p region contains a line around 52 eV with a shoulder at higher binding energy. The intesne line at 52 eV is believed to be due to the 2p line from magnesium ion which occurs at about 51 eV. The line at higher binding energy is due to the iron atom in cytochrome C. The intense line at $\sim$ 51 eV was also present in the iron spectrum from iron conalbumin and is probably due to the presence of Mg^{++} ion as an impurity in these samples. This overlap of lines for two elements points out this possibility and one should always consider the origin of all lines in the spectral region under study. The Cu 2p core electron spectrum from copper conalbumin is shown in Figure 24. This spectrum also points out the weak intensity of lines from the transition metals when present in low concentrations in metallo proteins. One must take extreme precautions to exclude the possibility of impurities giving rise to these weak signals.

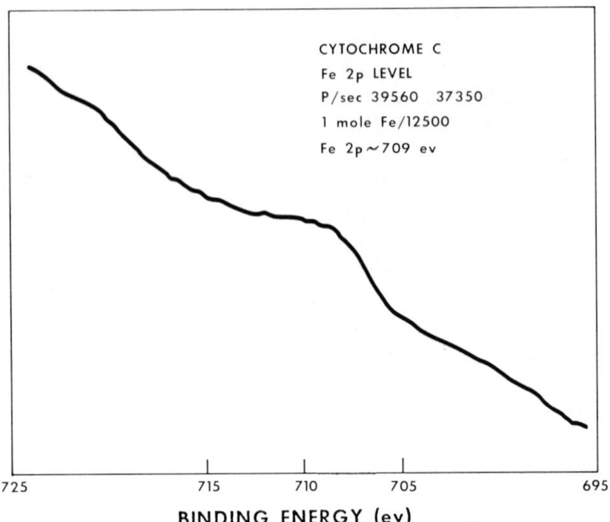

Figure 22. Iron 2p electron spectra from cytochrom C.

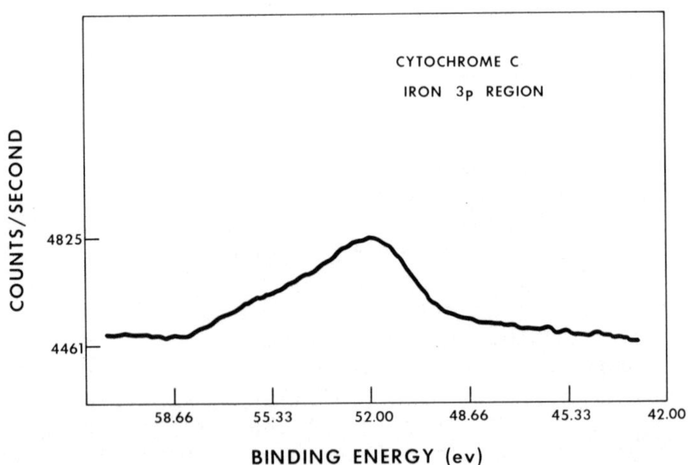

Figure 23. Iron 3p electron spectra from cytochrom C.

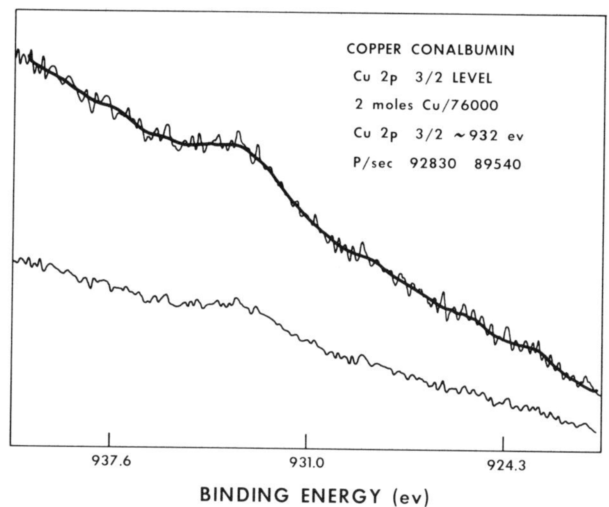

Figure 24. Copper 2p electron spectra from copper conalbumin.

During the course of these studies some mercury compounds were investigated. The mercury line was found to be considerably more intense in comparable concentration than the 3d transition metals studied. Mercury compounds are known to react with the sulfhydryl group of proteins and enzymes and some model mercury sulfur compounds were studied together with some fairly well characterized mercury protein derivatives. The mercury 4f line spectrum obtained from mercury papain, PCMB ovalbumin, and mercuric chloride is shown in Figure 24, 25, and 26, respectively. The intensity of the line in these mercury proteins is quite reasonable. Mercury papain and PCMB ovalbumin are believed to contain mercury sulfur bonds and the mercury line in these compounds agrees quite well with the shift observed for mercury dibenzylmercaptide. The sulfur 2p core line does not seem to shift appreciably. This may be due to the fact that a small percentage of the sulfur present in these molecules

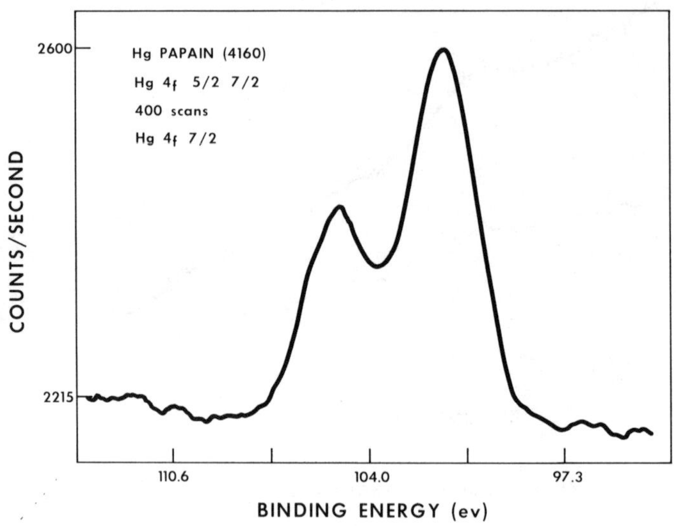

Figure 25. Mercury 4f electron spectra from mercury papain. (Millard and Friedman, 1974).

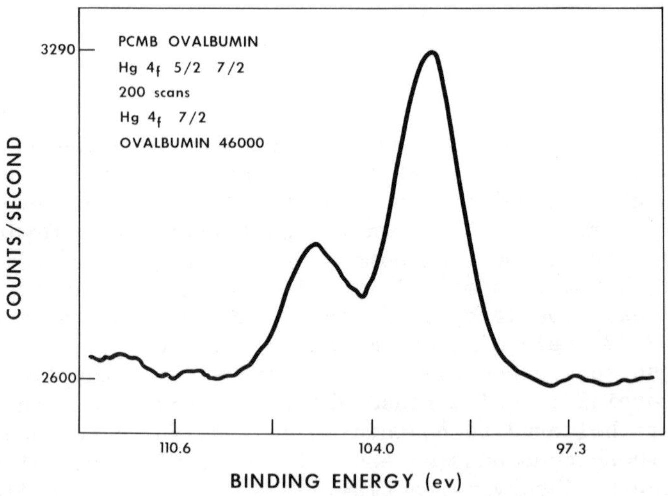

Figure 26. Mercury 4f electron spectra from PCMB ovalbumin. (Millard and Friedman, 1974).

is involved in bonding to mercury as well as the fact that the shift is probably small. This data is summarized in Figure 28. Although mercury has a strong photoelectric cross section, the shift scale for mercury is only about one electron volt. While investigating the binding of mercury ion to some proteins, the binding energy shift and line intensity of the halide ion associated with the mercury ion was studied. It soon became apparent that useful structural information could be gained from the nature of the anion associated with the metal ion. The chlorine 2p line was not resolved into its two components although a shoulder sometimes appeared. Chlorine data from several proteins treated with mercuric chloride as well as chloride ion and mercuric chloride is presented in Figure 29. The deconvalented chlorine line spectra from hemoglobin and bovine serum albumin treated with mercuric chloride are presented in Figures 30 and 31, respectively. The electron spectra in the chlorine region for the two proteins were essentially the same before and after exposure of these proteins to mercuric chloride. Two lines were present and these were apparently present in the original protein sample. The line at lower binding energy, due to chloride ion at 196.7 eV, is in the region of chlorine covalently bound to carbon. Apparently those protein samples contained some impurity containing carbon chlorine bonds. The presence of mercuric chloride can also be ruled out by the absence of a chlorine line around 199 eV. Since the chlorine shift is relatively large between chloride ion and compounds with covalent bonds, it is easy to differentiate between ionic chloride, carbon chlorine, and chlorine associated with metals. This can often be of much help in ellucidating structures in biologically important compounds containing metal ions and halogen in ionic or covalent forms.

Klein and Kramer (1971) have investigated XPS for the estimation of protein quantity and quality in grain protein. Their data is summarized in Figure 32. The nitrogen and sulfur content could be accurately determined in these materials from the area under the electron lines from nitrogen and sulfur. The resolution of the magnetic spectrometer used was not sufficient to accurately resolve the side chain and polypeptide nitrogen lines. However, in principle this should be possible with instruments capable of higher resolution. XPS has great promise as an analytical method for the determination of nitrogen and sulfur in organic materials. The method would be rapid, require small amounts of sample, and be nondestrutive.

XPS has been used to detect the presence of chemical modifications in proteins and estimate the extent of the modification of the functional group. Some feeling for the analytical accuracy obtainable from area ratio's from electron spectra can be

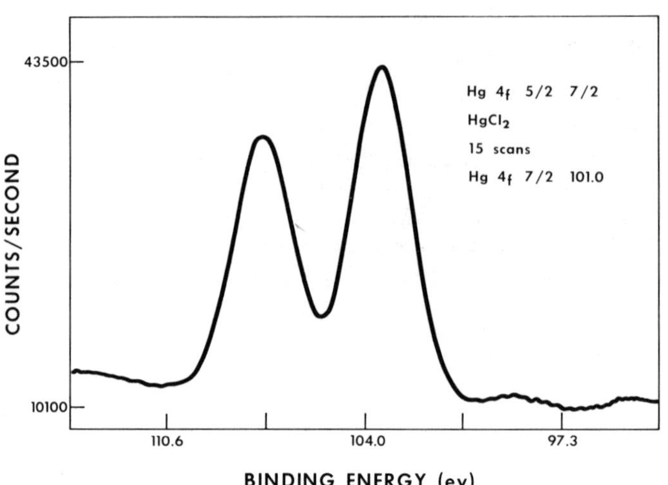

Figure 27. Mercury 4f electron spectra from mercury II chloride. (Millard and Friedman, 1974).

	Sp3/2	Hg 4f7/2
Ovalbumin	163.0	
Lysozyme	162.6	
Papain	162.8	
Mercury Papain	162.8	100.5
PCMB Ovalbumin	163.0	100.5
Mercury Dibenzyl Mercaptide	162.4	100.6
$HgCl_2$		101.0
$HgIISO_4$		100.6
HgICl		100.25
HgS	161.0	99.6

Figure 28. Mercury and sulfur binding energies. (Millard and Friedman, 1974).

	Sp3/2	Hg 4f7/2	Clp3/2	
BSA $HgCl_2$	162.4	100.1	200.3	196.7
Lysozyme $HgCl_2$	162.6	100.4	200.5	196.7
Hemoglobin $HgCl_2$	162.5	100.5	200.7	197.1
Histidine HCl				196.7
Cysteine HCl				196.7
$HgCl_2$		101.0	198.9	

Figure 29. Sulfur, mercury and chlorine binding energies. (Millard and Friedman, 1974).

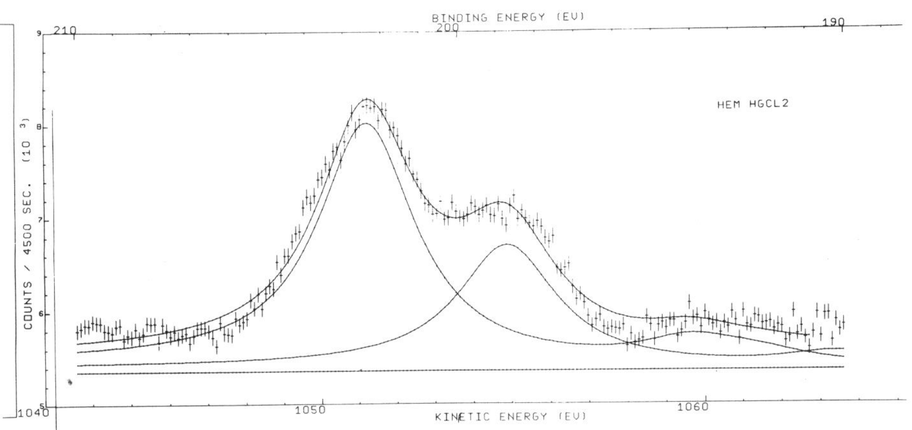

Figure 30. Chlorine 2p electron spectra from hemoglobin. (Millard and Friedman, 1974a).

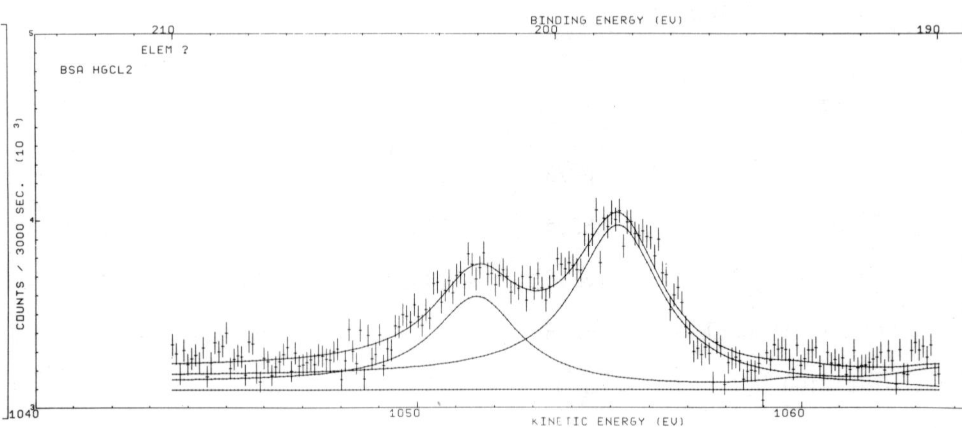

Figure 31. Chlorine 2p electron spectra from bovine serum albumin. (Millard and Friedman, 1974a).

ELEMENTAL ANALYSIS BY XPS

Quantitative Determination of Nitrogen and Sulfur Content

Compound	Element	Experimental Weight %	Calculated Weight %
Cysteine* (1)	Nitrogen	8.2	8.2
	Sulfur	18.0	19.3
Cysteine* (2)	Nitrogen	8.5	8.5
	Sulfur	20.0	19.3
Cystine (1)	Nitrogen	13.0	11.7
	Sulfur	26.3	26.7

*Cysteine-hydrochloride-monohydrate

Elemental Analysis

Seed	% N(XPS)	% N(anal)	% S(XPS)	% S(anal)
Barker Barley	1.5 $\pm$ 0.2	1.5 $\pm$ 0.2	.05 $\pm$ .02	.01 $\pm$ .01
Ramona Wheat	1.6 $\pm$ 0.2	1.7 $\pm$ 0.2	.05 $\pm$ .02	.06 $\pm$ .03
Rapida Oats	2.2 $\pm$ 0.2	1.9 $\pm$ 0.2	.1 $\pm$ .03	.02 $\pm$ .02
Light Red Kidney Bean	3.2 $\pm$ 0.3	4.1 $\pm$ 0.3	.08 $\pm$ .04	.13 $\pm$ .06

Seed Protein Distribution (Light Red Kidney Bean)

Sample	% N(XPS)	X N(anal)	% Basic A.A.(XPS)	% Basic A.A.(anal)
Endoplasm	3.1 $\pm$ 0.3	4.0 $\pm$ 0.3	17 $\pm$ 5	17.4
Embryo and Endoplasm	3.1 $\pm$ 0.3	4.2 $\pm$ 0.3	18 $\pm$ 5	

Figure 32. (Klein and Kramer, 1971).

gained by inspection of the data in Figure 33. The ratio of the nitrogen and sulfur atoms in several proteins was estimated using area ratios and values for the sensitivity of atoms from Wagner's paper. The lysine groups of bovine serum albumin were trifluro-acetylated according to the procedure described by Goldberger (1967). The fluroine one s line spectra from a sample of the trifluroacety-lated protein is shown in Figure 34. From the area under the fluroine line spectra the number of acetylated lysine groups can be estimated. The lysine group in casein was modified by treatment with ethyl vinyl sulfone (Friedman & Finley, 1974). The sulfur 2p line spectra from the modified protein is shown in Figure 35. The intense line around 168 eV is due to the sulfone sulfur and represents the amount of lysine modification. The line at lower binding energy $\sim$ 163 eV is due to the remaining sulfur in the molecule. Limited success was achieved in detecting the presence of nitrogen in modified functional groups. Bovine serum albumin was reduced and the sulfhydryl groups modified with p-nitrostyrene (Masri et. al., 1972). The nitrogen 1S line spectra from the modified protein is shown in Figure 36. A very weak line at higher binding energy is present at 404.8 eV and is due to the aromatic nitro group from the modified protein.

	N/S Ratio	
	Calculated	Observed(XPS)
Lysozyme	17.1	17.7
BSA	19.2	20.4
Conalbumin	20.8	22
Wool	10.9	8.9

Figure 33. Atom ratios in proteins estimated from XPS data.

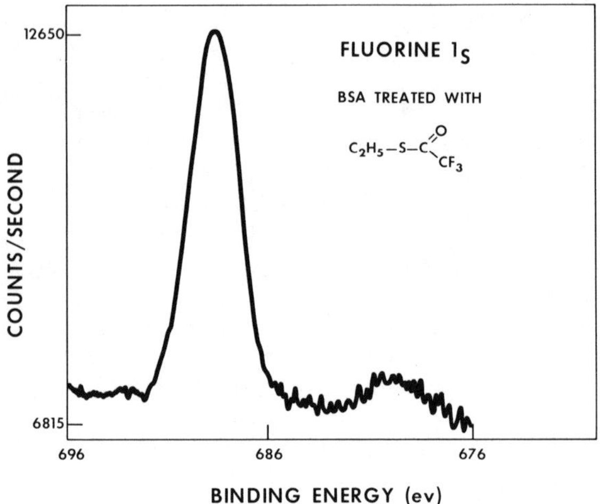

Figure 34. Fluorine is electron spectra from trifluoroacetylated bovine serium albumin. (Millard and Masri, 1974).

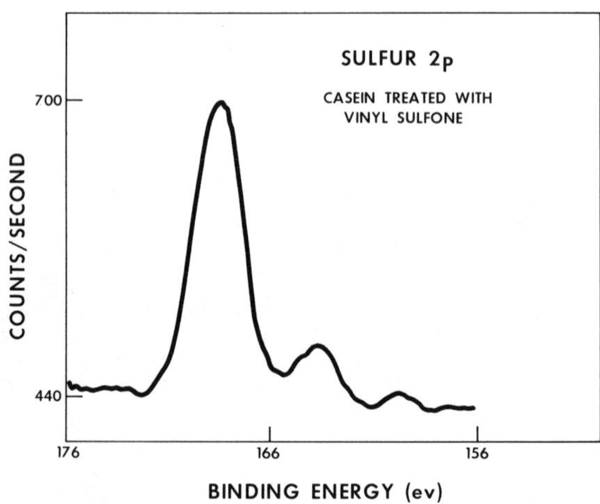

Figure 35. Sulfur 2p electron spectra from caesin treated with vinyl sulfone.

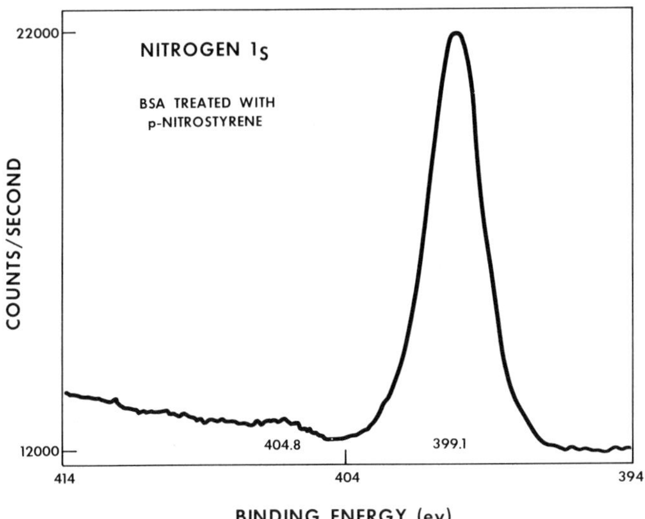

Figure 36. Nitrogen 1s electron spectra from reduced bovine serium albumin treated with p-nitrostyrene. (Millard and Masri, 1974).

ACKNOWLEDGEMENTS

The author thanks W. P. Klein, Laboratory of Chemical Biodynamics, Lawrence Radiation Laboratory, University of California for the figures from the Ph.D thesis of L. N. Kramer.

REFERENCES

Adman, E. T., Sieker, L. C. and Jensen, L. H. (1973). The structure of bacterial ferredoxin. J. Biol. Chem. 248:3987-3996.

Baddiley, J., Hancock, I. C. and Sherwood, P. M. A. (1973). XPS studies of magnesium ions bound to the cell walls of the gran-positive bacteria. Nature 243:43-45.

Baker, A. D., Brundle, C. R. and Thompson, M. (1972). Electron spectroscopy. Chemical Society Reviews 1:355. The Chemical Society London.

Bodmer, A., (1971). The Photoelectron spectrometer. Abstracts of the ESCA symposium laboratory of inorganic chemistry. ETH Zurich, Switzerland, Oct. 4-6.

Brundle, C. R. (1971). Some recent advances in photoelectron spectroscopy. Appl. Spectroscopy 25:8-23.

Carlson, T. A., Cheng, K. L. and Carver, J. C. (1973). X-ray photoelectron spectra of ethylene diaminetetracetic acid and its metal complexes. Inorg. Chem. 12:1702-1704.

Carter, C. W., Freer, S. T., Xuong, Ng. H, Alden, R. A., and Kraut, J. (1972). Structure of the iron sulfur cluster in the chromatium iron protein at 2.25 Å resolution. Cold Spring Harbor Symp. Quant. Biol. 36: 381.

Clark, D. T. (1972). Electron spectroscopy for chemical analysis. Ann. Reports Progr. Chem. 68:Sec B Organic Chemistry 91. The Chemical Society London.

Friedman, M., and Finley, J. W. (1974). Chemical modification of wool, soluble proteins, and cereal grains by ethyl vinyl sulfone., submitted for publication.

Goldberger, R. F., (1967). Trifluroacetylation of ε amino groups. Methods in Enzymology 11:317. Enzyme Structure. Hirs, C. H. W. ed. Academic Press, N.Y., N.Y.

Grunthaner, F. J. (1974). Electronic structure, surface reactivity and site analysis of transition metal complexes and metalloproteins by x-ray photoelectron spectroscopy. Ph.D. Thesis, California Institute of Technology, Pasadena, California.

Hayes, R. G. and Zeller, M. V. (1973). X-ray photoelectron spectroscopic studies on the electronic structures of porphyrin and phthalocyanine compounds. J. Am. Chem. Soc. 95:3855.

Hercules, D. M. (1970). Electron spectroscopy. Anal. Chem. 42:20A.

Hercules, D. M. (1972). Electron spectroscopy II. X-ray photoexcitation. Anal. Chem. 44:106R-112R.

Ibers, J. A., Collman, J. P., Brock, C. P., Dolcetti, G., Farnham, P. H., Lester, J. E., and Reed, C. A. (1973). A bent vs linear nitrosyl paradox. Infrared and X-ray photoelectron spectra of $CoCl_2NOL_2$ and crystal structure with $L = P(CH_3)[C_6H_5]_2$. Inorg. Chem. 12:1304-1313.

Jensen, L. H., Watenpaugh, K. D., Sieker, L. C. and Herriott, J. R. (1972). The structure of a non-heme iron protein:rubredoxin at 1.5. Å resolution. Cold Spring Harbor Symp. Quant. Biol. 36:359.

Jolly, W. L. and Hollander, J. M. (1970). X-ray photoelectron spectroscopy. Accounts Chem. Res. 3:193-199.

Klein, M. P. and Kramer, L. N. (1971). Application of extended huckel theory to X-ray photoelectron spectra of transition metal complexes. Chem. Phys. LeHers 8:183-186.

Kramer, L. N., (1971). Investigations of bonding, structure and quantative analysis in biological systems by means of X-ray photoelectron spectroscopy. Ph.D. Thesis, University of California, Berkeley, California.

Kramer, L. N. and Klein, M. P. (1971). Estimation of protein quality and quantity by X-ray photoelectron spectroscopy. Abstracts of the 161st Natl. Meeting of ACS, Los Angeles, California, AGFD paper 18.

Leibfritz, D. (1972). X-ray photoelectron spectroscopy of iron containing proteins--the valency of iron in ferredoxins. Angew. Chem. Internat. Edit. 11:232.

Lindberg, B. J. (1971). ESCA 23rd Int. IUPSC Congress Boston, 33-60.

Lucchesi, C. A. and Lester, J. E. (1973). Electron Spectroscopy instrumentation. J. Chem. Ed. 50:A205, A269.

Masri, M. S., Windle, J. J. and Friedman, M. (1972). p-nitrostyrene: New alkylating agents for sulfhydryl groups in reduced soluble proteins and keratins. Biochem. Biophys. Res. Commun. 47:1408-1413.

McDowell, C. A., Frost, D. C. and Ishitani, A. (1972). X-ray photoelectron spectroscopy of copper compounds. Mol. Phys., 24, 861-877.

Millard, M. M., and Masri, M. S., 1974. Detection and estimation of chemical modifications in proteins by XPS. Submitted for for publication.

Millard, M. M., and Friedman, M., (1974). X-ray photoelectron spectroscopy studies of mercuric and methylmercuric derivatives of proteins., submitted for publication.

Millard, M. M., and Friedman, M., (1974a). Estimation of organically bound chlorine and chloride ion in proteins by X-ray photoelectron spectroscopy., submitted for publication.

Palmberg, P. W. (1973). Quantitative analysis of solid surfaces by auger electron spectroscopy. Anal. Chem. 45:549A.

Shirley, D. A., Ed. (1972). Electron spectroscopy, North Holland Publishing Co.

Siegbahn, K., Nordling, C., Fahlman, A., Nordberg, R., Hamrin, K., Hedman, J., Hohansson, G., Bergmark, T., Karlson, S. -E., Lindgren, I., and Lindberg, B. (1967). "ESCA, Atomic, Molecular and Solid State Structure Studied by Means of Electron Spectroscopy," Almqvist and Wiksells AB, Stockholm.

Wagner, C. D. (1972). Sensitivity of detection of the elements by photoelectron spectrometry. Anal. Chem. 44:1050-1053.

Wesser, U., Jung, G., Ottnad, M. and Bohnenkamp, W. (1972). X-ray photoelectron spectra of bovine erthrocuprein. FEBS Letters. 25:346.

Wesser, U., Jung, G., Ottnad, M., Bohenkamp, W. and Bremser, W. (1973). X-ray photoelectron spectroscopic studies of copper and zinc amino acid complexes and superoxide dimutase. Biochim. Biophys. Acta. 295:77-86.

26

METAL ANALYSIS IN BIOLOGICAL MATERIAL BY ENERGY DISPERSIVE X-RAY FLUORESCENCE SPECTROSCOPY

F. William Reuter[1] and William L. Raynolds[2]

[1]Western Regional Research Laboratory, Agricultural Research Service, U.S. Department of Agriculture, Berkeley, California 94710, U.S.A.

[2]Department of Nutritional Sciences, University of California, Berkeley, California 94720

INTRODUCTION

The problem of metal analysis is rapidly moving from the laboratory of specialized chemists to the general purpose lab of the analyst. Metals in animal and plant products need to be routinely determined. Foods must list mineral content. Previously much of trace metal analysis has been directed toward single elements. Iron, iodine, sodium, potassium and Mg are routinely studied. The knowledge, however, that 26 minerals occur naturally in plant and animal material often contaminated by metallic pollutants such as Pb, Cd, Hg encourages the analyst to rely more on instruments than intuition for detection of the various minerals in a given sample (Mertz and Cornatzer, 1971).

Many instruments are in various stages of development from bench top one of a kind that require constant maintenance and skilled technical and theoretical experience to relatively standardized machines and techniques that can be run with easily trained personnel. The person who wants mineral information as a diagnostic tool or as a technique of quality control is interested in a system that has been fully prepared by the manufacturer. The fact that no one sells instruments that are completely computer controlled indicates that the decision of spectrum interpretation is still too complicated to trust to a machine. Nonetheless there are 6 techniques for quantitative elemental analysis that do not require foreknowledge for sample composition: Emission Spectroscopy, Spark Source Mass Spectroscopy, Neutron Activation Analysis, Electron Spectroscopy for Chemical Analysis and X-ray Fluorescence Spectroscopy. Of them x-ray fluorescence offers the most promise as a technique for routine metal analysis in biological material. It is non-destructive, it will process liquids or solids, the raw data is relatively uncomplicated, it can be automated, and set up time is minimal. Currently its principle drawback is dection limits which do not exceed 1 ppm for commercial instruments but some experimental instruments have demonstrated these can be reduced a factor of 10.

THEORY

The physical principle underlying x-ray fluorescence spectroscopy is when an x-ray incident on an unknown atom makes a K- or L-shell vacancy, an electron from the next shell will fill the vacancy with the corresponding emission of an x-ray photon whose energy equals the energy difference between the 2 shells. Since the binding energy of atomic electrons is determined by the nuclear change, the energy of emitted fluorescent x-rays for an element is unique to that element. In general there are two prominent x-rays in the energy region we will consider whose energy increases with atomic number. Any spectron is hence relatively simple and the combination of possible interferences few.

The intensity of any fluorescent x-ray line from a sample is to first order directly proportional to the amount of the element present. Since second order effects can be greater than 10 percent, any general treatment must take them into account.

For an incident photon along ℓ_1, causing a fluorescence in dx, leaving along ℓ_2 and passing through air to a detector, the probability of detecting a trace fluorescent x-ray is from an element in dl_1 is

$$dP = e^{-\mu_i \rho \ell_1} n\sigma e^{-\mu_f \rho \ell_2} e^{-\mu_{air}\rho_{air}\ell_{air}} \frac{\Omega}{4\pi} \varepsilon \, d\ell_1$$

where (Figure 1)

- μ_i is the absorption coefficient of the sample to incident x-rays.
- ρ is the density of the sample.
- ℓ_1 is the distance traveled into the sample.
- n is the density of fluorescing atoms in the matrix.
- σ is the probability of exciting a fluorescent x-ray.
- μ_f is the absorption coefficient of the sample to the fluorescent x-ray.
- ℓ_2 is the distance traveled out of the sample.
- $\exp(-\mu_{air}\rho_{air}\ell_{air})$ is the probability of air absorption.

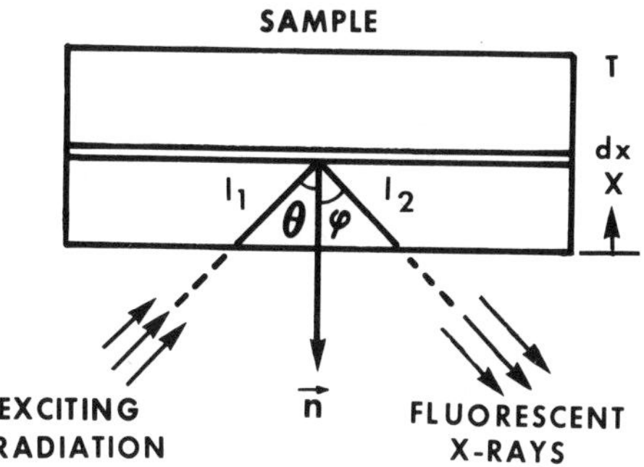

Figure 1. X-ray fluorescence from a thin sample.

Ω is the solid angle subtended by the detector.
ε is the detector effeciency.

Converting to units of x and integrating over the sample gives the probability of detecting an x-ray from the whole sample.

$$P = \frac{\Omega\varepsilon}{4\pi} \exp(-\mu_{air}\rho_{air}\ell_{air}) \frac{n\sigma}{\mathrm{Cos}\theta} \int_0^T \exp[-\rho x(\frac{\mu_i}{\mathrm{Cos}\theta} + \frac{\mu_f}{\mathrm{Cos}\phi})]dx$$

where T is the sample thickness.

Integrating gives

$$P = \frac{\Omega\varepsilon}{4\pi} \exp(-\mu_{air}\rho_{air}\ell_{air}) \frac{n\sigma}{\mathrm{Cos}\theta} \frac{1 - \exp(-\mu_c\rho T)}{\mu_c\rho} \tag{1}$$

where

$$\mu_c = \mu_i/\mathrm{Cos}\theta + \mu_f/\mathrm{Cos}\phi$$

Equation (1) can be simplified by combining all the parameters that depend on the instrument such as ϕ, θ, Ω, and the incident beam strength into a constant k and by combining those parameters which determine the overall excitation probability into σ_i such that

$$\sigma_i = \varepsilon\sigma \exp(-\mu_{air}\rho_{air}\ell_{air}) \tag{2}$$

and

$$C_i = kn_i\sigma_i[1 - \exp(-\mu_c\rho T)]/\mu_c\rho \qquad (3)$$

Where i indicates the element under investigation and C_i is the x-ray counts measured from that element.

Equation (3) can be simplified to read

$$C_i = kn_i\sigma_i T/F_i$$

where

$$F_i = \rho\mu_c T/[1 - \exp(-\mu_c\rho T)] \qquad (4)$$

F_i is governed by the whole sample. It varies from 1 for energetic x-rays which are not absorbed by the sample to larger values for x-rays which are absorbed as shown in Figure 2. It is discussed later under Matrix Absorption.

The constants σ_i and k must be determined for a particular instrument. If a thin film elemental standard is used for k then F equals 1 and σ_{std} can be set to 1 so that

$$k = C_{std}/(nT)_{std}$$

and

$$n_i T = C_i F_i/k\gamma_i$$

Where γ_i is σ_i/σ_{std} a relative cross section normalized to the standard element.

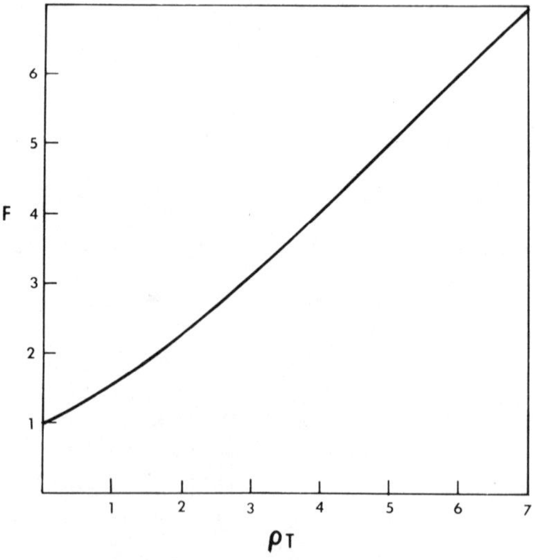

Figure 2. The change in the matrix absorption coefficient as a function of sample density X thickness where μ_c is 1.

It should be noted that $n_i T$ is the unknown which is measured. It is sometimes called the areal density and has the units weight per unit area of sample. Ppm is obtained by dividing by the areal density of the entire sample.

$$ppm = C_i F_i / k\gamma_i W \qquad (5)$$

Where W is the areal density of the sample.
If the area of the sample is known, the total weight of the unknown can be measured. If the density is known, sample thickness T can be measured.

The simplicity of equation (5) is important for several reasons. Once the γ_i have been determined, they need not be determined again. F can be determined on each sample directly so that calibration curves do not have to be made for each sample type. For samples of less than 1000 ppm, the matrix coefficient will be constant. Thin samples have a higher signal to noise ratio than thicker ones. Thin samples of biological material are on the order of 1 mm thick with areal densities of 30-40 mg/cm^2. They are self supporting and can be handled easily.

INSTRUMENTATION

We would like to point out here that we will direct nearly all our remarks to the techniques of energy dispensive x-ray fluorescence spectroscopy. Wavelength dispensive spectroscopy is used routinely in the metallurgical analysis of alloys and catalysts (Birks 1959, Liebhafsky et al., 1960). For unknown elemental analysis however the simplicity, lower cost and simultaneous spectrum accumulation of energy dispersive machines have demonstrated their usefullness (Gilfrich 1973).

The last few years have seen large advances in instrumentation for energy dispersive fluorescence analysis. As with any instrumentation it is the advancement in transducers in this case x-ray sources and x-ray detectors which have provided the foundation for the improvements. The Lawrence Berkeley Laboratory has spearheaded many of the innovations which some manufacturers have been quick to adopt.

The progress in energy resolution of lithium-drifted silicon detectors has made it possible to separate the x-ray lines of adjacent elements. Signal to noise was further increased by the development of the guard ring detector (Jaklevic and Goulding, 1972a) and the promise of 0.1 ppm detection limits in biological material was anticipated (Goulding et al., 1972). It appears though that at this time manufacturers have chosen not to use these electronic sophistocations and instead claim that a larger detector with a beam collimated to strike only the center will accomplish the same results.

The x-ray sources for most of the early work were radioactive isotopes (Rhodes, 1971). They are easily moved and continue to be used in mobile (Kneip and Laurer, 1973) and research situations where source-detector configurations are constantly changing. If one is willing to handle a hot source, they can be as potent as an x-ray tube (Baglan et al., 1973). However in cases where the sample is normally brought to the source, x-ray tubes have been shown to be as stable as sources in long counting situations (Dyer et al., 1972). The bi-enegetic spectrum of an isotope source has also been duplicated with a transmission targeted x-ray tube (Jaklevic et al., 1972b). Here again manufacturers have short cut this new x-ray tube technology by filtering the beam of a conventional tube to produce similar results. This is accomplished by sending the beam from say a Rh x-ray tube through a thin layer a Rh foil. Energies other than those of the principle Rh lines are selectively absorbed in the foil.

A variable exciting line will also increase sesitivity. This has provided the rationale for development of secondary radiators with x-ray tubes (Porter, 1973). Such a system begins to look a little like a diffraction x-ray spectrometer with collimators and a large water cooled tube. A pulsed tube has also been demonstrated which allows an increase in count rate and optimization of electronics for maximum resolution. (Jaklevic et al., 1972c).

The sound x-ray technique of 90° incident-exit geometry continues to be pointed out (Kneip and Laurer, 1972, Baglan et al., 1973) as well as the use of thin samples which enhance signal to noise ratio (Giauque and Jaklevic, 1972, Rhodes, 1971). The *in situ* measurement of matrix effects has reduced the need for large numbers of standards for calibration curves (Giauque et al., 1973). This was theoretically described in detail in the previous section. The ease of preparing biological samples into thin specimens allows one to take advantage of these developments. Several manufacturers have built automatic sample changers and have added dedicated computers to perform all or part of the data reduction (Wood et al., 1973).

To date a system with 0.1 ppm detection limits has not been put on the market. However such a system is within the state of the art of present technology. Such a system coupled with a computerized data analysis system capable of performing data reduction without an operator assisted peak search would be a major step toward automated nutritional evaluation of foods and medical diagnostic services.

In dried food stuffs an instrument with Mo, Rh or Ag exciting lines and 1-5 ppm detection limits in the region of Cu can generally detect the presence of K, Ca, Mn, Fe, Ni, Cu, Zn, Br,

Rb, Sr, and Pb, though Mn, Ni, Cu and Pb are very near the limits of quantitation. In blood such a system shows K, Fe, Cu and Zn (Ong et al., 1973, Baglan et al., 1973). An increase in sensitivity in this region to 0.1 ppm would permit better accuracy for the elements near current detection limits and in foods allow detection of such nutritionally important elements as Cr and Se and the pollutant Hg. Some research instruments are near this limit (Giauque et al., 1973). If measurements are performed in vacuum, P, S, and Cl are easily detected. Unfortunately the poor fluorescence yield and low energy lines of Na and Mg will make it difficult to include these elements in a broad spectrum technique. For highest sensitivity it appears as though a diffraction instrument married to the detector and electronics of an energy dispersive machine will be the most sensitive for individual elements. However the fact that the x-rays of the majority of the nutritionally important trace elements had pollutants fall in the 3 to 11 keV region will provide a place for energy dispersive machines with maximum sensitivity in that range.

INSTRUMENT DESIGN

An instrument is shown schematically in Figure 3[a]. X-rays from a Rh targeted tube are passed through .030" of Rh foil. The collimated beam strikes the sample. Fluorescent x-rays are collimated, and detected in a solid state lithium-drifted detector (175 ev FWHM at 5.9 KeV, 1000 cps). The pulses are stored in a 512 multi-channel analyser.

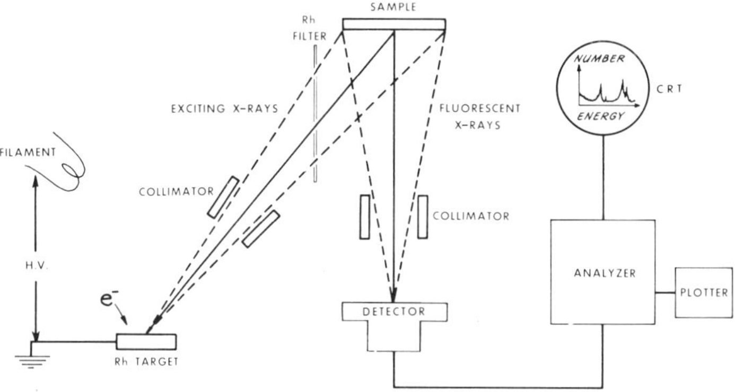

Figure 3. Energy dispersive x-ray fluorescence spectrometer.

[a]Finnigan Corporation, Sunnyvale, California.

The rhodium targeted tube was chosen because it has an exitation energy in a region of the periodic table where biological trace elements and trace pollutants are not localized. The obvious exception is Cd. By the same reasoning the sample chamber is lined with 0.01" indium foil an uncommon trace pollutant which can be obtained in easily worked sheets of high purity. The collimators are indium and antimony. In spite of these precautions there persists an Fe impurity equivalent to about 5 ppm.

Radiation damage is commonly cited as a counter example of the nondestructiveness of x-ray fluorescence spectroscopy. The instrument, however, delivers 30R/min at the sample at standard operating conditions of 40 KV and 1 ma. These conditions result in a 500 R dose to the sample in a 1000 sec. measurement. This is about the range of LD_{50} for whole animals but it is below the LD_{50} of single celled plants and animals (Comar, 1955). In a simple test of destructiveness on single cell plants yeast was prepared as a standard sample and was irradiated for 2000 sec. or a dose of 1000R. When grown in a nutrient medium the irradiated yeast produced the same amount of gas as a control indicating it was not affected by the dose.

SAMPLE PREPARATION

Exactly 300 mg of solids are spread out uniformally in a 1.25" dia. stainless steel die and pressed to 15,000psi. The resulting thin pellet will generally support itself and can be stored for future reference in a plastic "petrislide"[b]. This sampling method for solids is preferred because it requires minimal sample preparation without contamination. Analysis of solids has been used as described on flaky substances, powders, peanut skins, bran, ground wool (Masri and Friedman, 1974), and other materials. We have encountered no contamination from the die showing Fe, Cr, or Ni.

Solids which are too concentrated, contain too much of a constituent for analysis by the above technique or ones that are too powdery to press into a pellet are mixed with Whatman cellulose (chromatographic grade, 200 mesh) then pressed into a pellet for analysis. The cellulose acts as a binder as well as diluent free of most impurities except about 5 ppm Fe and 1 ppm Cu. Millipore filters can be used to collect a precipitate which can be analysed wet or dry.

A liquid approach has been worked out in which a piece of filter paper is saturated with the liquid and placed between two sheets of mylar. The mylar prevents evaporation and accompanying migration of solutes. The reproducibility of this technique is less than with solids. For most biological samples which contain water, however, some concentration can be achieved by drying.

This will increase the solids concentration about 5-10 times. Such procedures rarely introduce impurities and can be routinely performed.

When material can be obtained in large quantities, ashing may concentrate the elements 10 to 100 times. However, concentration by this means is laborious and elements can be lost. One ashing technique we have tried is low temperature ashing in which the sample is placed in a low temperature oxygen plasma[c]. Oxygen radicals produced in the plasma combine with the solid hydrocarbons to produce gaseous CO_2 and H_2O. One hundred grams of flour were ashed by this technique into a light fluffy ash which could be easily compacted. The concentration factor was 185. K, Ca, Mn, Fe, Cu, Zn and Br were observed in the flour sample and K, Ca, Mn, Fe, Cu, Zn, Co, Ni, Rb, Sr and Mo were observed in the ashed material. Br and possibly Se, Pb and Hg are lost in this technique.

INSTRUMENT CALIBRATION

Equation (5) requires that several instrument constants be determined. They are discussed below.

k is the instrument standard relating counts of Cu to $\mu g/cm^2$ of Cu. It is measured daily to account for instrument drift. The thin film standard was made according to a method described by Luke (1968). 10 $\mu g/cm^2$ of Cu was precipitated with Na diethyldithiocarbamate and the precipitate collected on a millipore filter 0.45μ hole size. The filter was placed between two sheets of mylar 0.15 x 10^{-3}in. thick secured by two metal interlocking rings. The standard was kindly measured by Robert Giauque of the Lawrence Berkeley Laboratory and determined to be 10.2$\mu g/cm^2$. Daily instrument drift is $\pm$ 0.9%.

Relative elemental detection efficiencies σ_i are determined with cellulose standards made by mixing 10 g of powdered cellulose with 10 ml of salt solution and dried to give a mixture of 200 ppm metal. The numerical values are normalized to copper the metal in the standard. The relative efficiencies are governed as shown in equation (2) by atomic parameters of x-ray excitation and emission, air absorption and detector effieciency. A curve of the relative efficiencies of K-shell x-rays for this instrument is shown in Figure 4. The curve for L-shell x-rays is similar in shape. The fact that Cu is 0.94 instead of 1.0 gives some indication

---

[b]Millipore Corporation, Bedford, Massachusetts.

[c]International Plasma, Hayward, California.

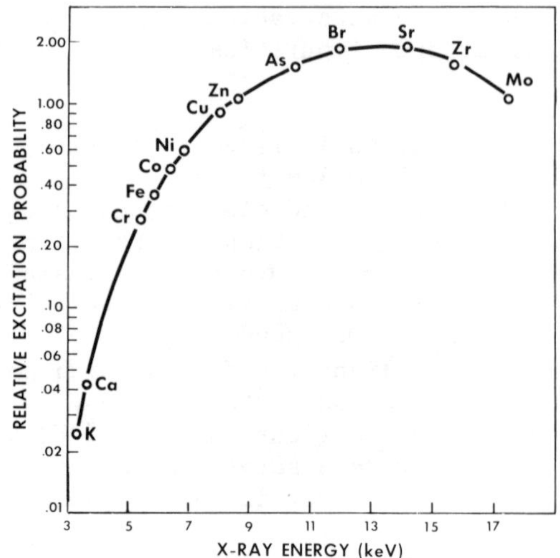

Figure 4. The K-shell relative excitation probability as a function of K-shell x-ray energy.

of the relative error in this standardization technique. A more accurate technique of vapor depositing precisely known amounts of metals on aluminum is mentioned by Giauque et al. (1973).

The elemental signal C is obtained by taking a background point on either side of the peak, assuming a straight line background under the peak and integrating a fraction of the peak area. Since the data is stored digitally in the memory, this calculation can be performed with a small dedicated computer, by hand, or submitting the whole job to a larger computer. Once the background points have been chosen, data reduction is very straight forward. The selection of background points for each data reduction are valid for only one sample type. A more general background correction scheme would have greater applicability but would have to be chosen in light of larger computing requirements.

Since K and L-shell spectral energies overlap, there are cases where two elements interfere. Then it is necessary to use the second x-ray peak which has a constant intensity ratio to the primary peak. The most common interference in biological material is the Pb Lα and the As Kα. In this case the Pb Kβ which overlaps the Kr Kα can be measured. The Pb signal can then be subtracted from the As to give the true As signal. Similarly the Hg Lβ and Br Kα interference can be corrected using the Hg Lα which overlaps Ge Kα, an uncommon trace mineral. Finally the only other common interference of interest is the Fe Kβ and the Co Kα.

Cobalt is usually less than the detection limits in any case, but even when it is in higher concentrations, the iron concentration may mask it. The ubiquity of iron in most samples limits Co detection to only those amples in which cobalt is greater than 1 or 2% of the iron concentration.

The matrix absorption correction F is obtained by measuring the sample and a pure element as shown in Figure 5. The signal I is corrected by subtracting the signal from the sample, so that only true attenuation is measured. F is calculated from equation (4) by noting that

$$I/I_o = \exp(-\mu_c \rho T)$$

so that

$$F = \ln(I_o/I)/(1 - I/I_o) \qquad (6)$$

For a typical biological sample, F varies with x-ray energy. Though we currently measure F separately for each element, it is possible to combine several elements into the elemental disk and measure values simultaneously.

Another approach is to assume that $\mu \propto E^{-n}$ where $n \sim 3$ (Compton and Allison, 1935). Then measure one F to determine the constant of proportionality and compute the rest. We have done this and found that for cellulose $n \sim 2.6$ (Figure 6). Such a technique is only valid where μ is determined by a light element matrix and does not have any absorption edges in the region of interest.

In general, F is constant for each element until the elemental concentration exceeds 1000 ppm. Figure 7 shows a function of F_{Cu} vs. concentration of Cu in cellulose. The highest concentrations were

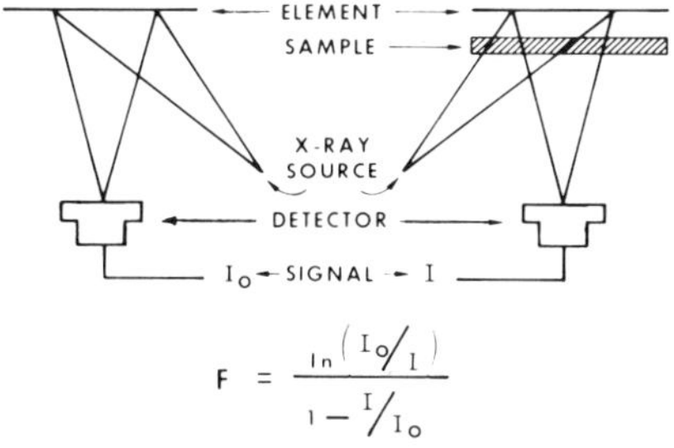

$$F = \frac{\ln(I_o/I)}{1 - I/I_o}$$

Figure 5. The experimental configuration to obtain the matrix absorption correction. F is calculated from equation (6).

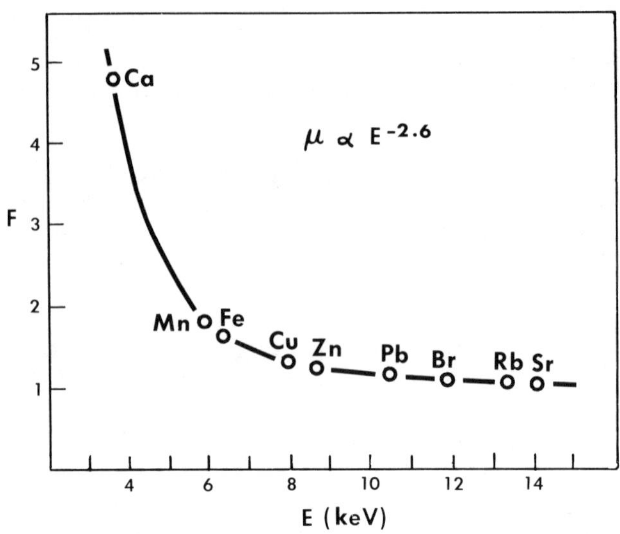

Figure 6. F obtained from one measurement at Fe and calculating the others assuming μ is proportional to $E^{-2.6}$ where E is x-ray energy.

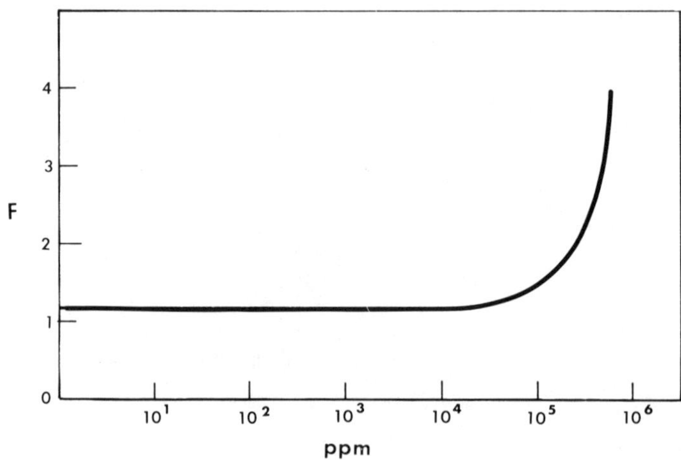

Figure 7. F as a function of concentration of Cu in a light element matrix. The results of the calculation are shown in Figure (8).

were obtained with pure Cu salts. Figure 8 shows the percent difference in the measured and expected concentrations as a function of concentration. When F is greater than 4 the assumptions of equation (5) are no longer valid. Even though a signal is evident from K and Ca, these elements are not accurately determined by this technique.

DETECTION LIMITS

Figure 9 shows the detection limits in biological material for a standard run. The curve was produced assuming a signal of three times the standard deviation of the background for a 1000 second measurement of a cellulose sample. The shape of the curve with increasing atomic number is explained as follows. The light element x-rays are strongly absorbed by the air between the sample and detector. Sensitivity increases with increasing atomic number from two factors: 1) air attenuation decreases, and 2) the probability of x-ray emission increases as the fluorescent energy approaches the excitation energy. On the low energy side of the Rh excitation energy the background becomes very large from Compton backscattering of the Rh exciting lines and so sensitivity is decreased. From Rh to gadolinium x-rays are excited by x-ray tube Bremsstrahlung. This excitation mode does not have the low signal to noise ratio that single line excitation does because part of the reflected exciting radiation is at the line of interest. In this region sensitivity decreases with atomic number because the amount of radiation above the absorption edge of the element of interest decreases. In practice a higher sensitivity is obtained in this

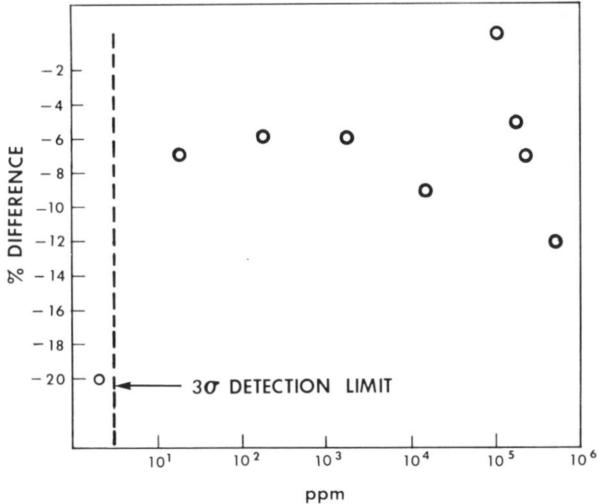

Figure 8. The % difference in measured and prepared concentrations of Cu in a light element matrix as a function of concentration. Measurement time is 1000 seconds or less.

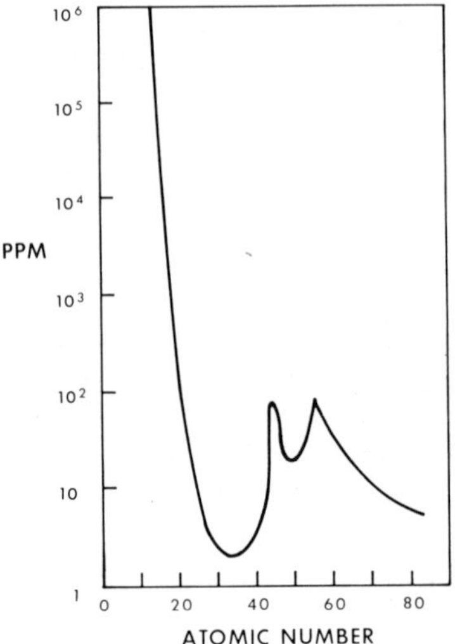

Figure 9. Detection limits in biological material.

region by increasing the x-ray tube energy to the maximum voltage and by looking at this region of the spectrum separately. L-shell x-rays of elements heavier than gadolinium are detected. These start at energies similar to those of calcium and so are attenuated by the air. Sensitivity increases with increasing atomic number as the fluorescent energy approaches the excitation energy.

EXAMPLES

When all the constants for a given instrument have been determined, the ability to measure the concentration of the elements in a biological material are determined solely by the ability of the researcher to make a homogeneous sample and whether the concentrations in the sample exceed the detection limit of the instrument. The x-ray energies are high enough that any biological matrix material is accounted for by equation (5) whether cellulose, starch, water or blood plasma. This is one of the strong points of x-ray fluorescence. We have found the National Bureau of Standards orchard leaves Standard Reference Material 1571 to be a suitable material to verify this instrumental technique. There is little dispute over the values and the material contains a wide range of elements. Table I shows the results of a measurement of orchard leaves from this laboratory.

TABLE I. TRACE ELEMENT CONCENTRATION IN N.B.S. ORCHARD LEAVES

Element	Concentration (ppm) X-ray Fluorescence	N.B.S.
Cr	<18.3[a]	2.3[b]
Mn	101.±10.	91.±4.
Fe	319.±32.	300.±20.
Co	<5.7[a]	0.2[b]
Ni	<3.9[a]	1.3±0.2
Cu	13.8±1.4	12.±1.
Zn	27.4±2.7	25.±3.
As	12.3±1.2	14.±2.
Br	8.9±0.9	10.[b]
Rb	11.5±1.2	12.±1.
Sr	42.2±4.2	37.[b]
Pb	47.1±4.7	45.±3.

[a]Not detected, concentration less than the detection limit shown.

[b]No standard deviations reported.

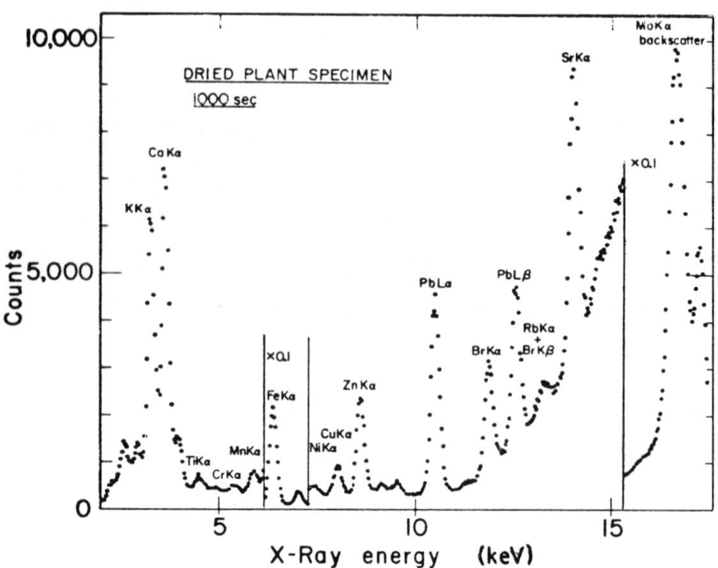

Figure 10. X-ray spectrum of plant material courtesy of Giauque and Jaklevic (1972).

An example of a raw data spectrum of plant material that is contaminated with Pb is shown in Figure 10 (Giauque and Jaklevic, 1972). The peaks from 13 elements are clearly evident. The detection limits of this instrument are of the same shape and about a factor of 10 better than the limits shown in Figure 9.

ACKNOWLEDGEMENTS

The authors express their gratitude to Chuchson Yokota for technical assistance and to Julia Almestad for preparing the graphs and illustrations for publication.

References

Baglan, R. J., Brill, A. B. and Patton, J. A. (1973). Applications of Non-dispersive X-ray Fluorescence Techniques for In Vitro Studies. IEEE Trans. Nucl. Sci. Eng. NS-20, No.1:379-388.

Birks, S. L. (1959). Chemical Analysis, Vol.XI, X-ray Spectrochemical Analysis. Interscience Publishers.

Comar, C. L. (1955). Radioisotopes in Biology and Agriculture. McGraw-Hill.

Compton, A. H. and Allison, S. K. (1935). X-rays in Theory and Experiment. D. Van Norstrand Co.

Dyer, G. R., Gedcke, D. A. and Harris, T. R. (1972). Fluorescence Analysis Using an Si(Li) X-ray Analysis System with Low Power X-ray Tubes and Radioisotopes in Adv. in X-ray Anal. 15:228-239.

Giauque, R. D. and Jaklevic, J. M. (1972). Rapid Quantitative Analysis by X-ray Spectroscopy in Adv. in X-ray Anal. 15:164-175.

Giauque, R. D., Goulding, F. S., Jaklevic, J. M., and Pehl, R. H. (1973). Trace Element Determination with Semiconductor Detector X-ray Spectrometers. Anal. Chem. 45:671-681.

Gilfrich, J. V. (1973). Applications of X-ray Analysis to Environmental and Biomedical Studies in Adv. in X-ray Anal. 16:1-9.

Goulding, F. S., Jaklevic, J. M., Jarrett, B. V., and Landis, D. A. (1972). Detector Background and Sensitivity of Semiconductor X-ray Fluorescence Spectrometers in Adv. in X-ray Anal. 15:470-482.

Jaklevic, J. M. and Goulding, F. S. (1972a). Semiconductor Detector X-ray Fluorescence Spectrometry Applied to Environmental and Biological Systems. IEEE Trans. Nucl. Sci. Eng. NS-19, No.3:384-391.

Jaklevic, J. M., Giauque, R. D., Malone, D. F., and Searles, W. L. (1972b). Small X-ray Tubes for Energy Dispersive Analysis Using Semiconductor Detectors in Adv. in X-ray Anal. 15:266-275.

Jaklevic, J. M., Goulding, F. S., and Landis, D. A. (1972c). High Rate X-ray Fluorescence Analysis. IEEE Trans. Nucl. Sci. Eng. NS-19, No.3:392-395.

Kneip, T. J. and Laurer, G. R. (1972). Isotope Excited X-ray Fluorescence. Anal. Chem. 44:57A-68A.

Liebhafsky, H. A., Pfeiffer, H. G., Winslow, E. H., and Zemany, P. D. (1960). X-ray Absorption and Emission in Analytical Chemistry. John Wiley and Sons.

Luke, C. L. (1968). Determination of Trace Elements in Inorganic and Organic Materials by X-ray Fluorescence Spectroscopy. Anal. Chem. Acta. 41:237-250.

Masri, M. S., and Friedman, M. (1974). This volume.

Mertz, W., and Cornatzer, W. E. (1971). Newer Trace Elements in Nutrition. Marcel Dekker.

Ong, P. S., Lund, P. K., Litton, C. E., and Mitchell, B. A. (1973). An Energy Dispersive System for the Analysis of Trace Elements in Human Blood Serum in Adv. in X-ray Anal. 16:124-133.

Porter, D. E. (1973). High Intensity Excitation Sources for X-ray Energy Spectroscopy. X-ray Spectrometry. 2:85-89.

Rhodes, J. R. (1971). Design and Application of X-ray Emission Analzers Using Radioisotope X-ray or Gamma Ray Sources in Energy Dispersion X-ray Analysis, Series ASTM Special Technical Publication 485, Philadelphia.

Wood, W. G., Mathiesen, J. M., and Mgebroff, J. S. (1973). X-ray Energy Analysis of particulate Matter on Filter Paper in Adv. in X-ray Anal. 16:134-144.

27

THE APPLICATION OF PERTURBED DIRECTIONAL CORRELATION OF GAMMA RAYS TO THE STUDY OF PROTEIN-METAL INTERACTIONS

G. Graf, J. C. Glass, and L. L. Richer

Departments of Biochemistry and Physics

North Dakota State University

Fargo, North Dakota 58102

Several branches of nuclear spectroscopy--magnetic resonance and the Mössbauer effect in particular--have been the subject of considerable interest to obtain structural and motional information about biological macromolecules. Recently, a third technique--perturbed gamma-gamma directional correlation--is being tested for the study of rotational correlation times, internal rotations and conformational changes in biological macromolecules. When two gamma rays from a radioactive nucleus are emitted in a cascade and detected with a coincidence spectrometer, the coincidence counting rate depends strongly on the angle between their directions of propagation. This directional correlation can be perturbed by the interaction of the nucleus with fields existing in macromolecules. A study of such perturbed directional correlation (PDC) has the potential to provide information on protein-metal interactions. The use of a radioactive nucleus as a rotational tracer to label macromolecules offers the possibility of obtaining information on protein structure with the sensitivity and instrumental simplicity of radioactive tracer techniques. Work to date has centered on measurement of internal electric field gradients of metalloproteins, molecular correlation times and the correlation time of water associated with proteins. This paper is a brief review of the theory and experimental technique of perturbed directional correlation of gamma rays in comparison to Mössbauer spectroscopy and the methods of magnetic resonance. Particular emphasis will be placed on the isotopes which can be used in PDC especially those which may occupy sites in biological macromolecules.

Magnetic resonance, a well established phenomenon for structural studies in organic chemistry, has already demonstrated its great potential in the field of biological systems. The technique was extremely useful for characterizing conformational changes in proteins and in elucidating some mechanisms of enzymatic catalysis. In recent years the techniques of pulsed and Fourier transform NMR have been developed and were particularly useful for the study of water associated with macromolecules.

Mössbauer spectroscopy represents a less generally accepted resonance technique in biological investigations. So far its application has been limited to hemoproteins, heme prosthetic groups and iron-sulfur proteins. The Mössbauer method measures small changes in the nuclear energy levels of a suitable nuclide, produced by its electronic surroundings. These changes may be attributed to the three possible interactions: isomer shifts, nuclear quadrupole interaction, and magnetic field interaction. By measuring these, the Mössbauer nuclide serves as an extremely small probe which does not perturb the electronic configuration as it yields information about the surroundings. The method is limited only by the number of nuclides which are compatible with biological systems.

The magnetic resonance methods have the potential for studying proteins in their native solution states, but these techniques are limited to the measurement of a single electronic phenomenon and they generally require the insertion of a labeled compound which may perturb the normal state of the macromolecule. In contrast, Mössbauer spectroscopy has the capability of detecting all electronic interactions in the vicinity of a nuclide, the spectrum contains no interfering signals from other atomic species and the Mössbauer nuclide does not perturb the normal configuration of the protein. However, Mössbauer samples must be in a solid state--usually at liquid nitrogen or even at liquid helium temperatures--to minimize recoil effects. Under these conditions protein conformations and electronic states may be widely different from the native states. Thus, it is customary in Mössbauer studies of proteins to correlate the interpretation of data with results obtained with complementary techniques such as magnetic resonance and magnetic susceptibility.

There appears to exist a strategy for studying proteins at levels of increasing sophistication. The initial study usually employs conventional biochemical techniques. Then, fluorescence spectroscopy, proton magnetic resonance and x-ray crystallography reveal further structures which are compatible with the results of the previous investigations. Electron paramagnetic resonance, and Mössbauer spectroscopy form the next level of investigation and provide information about the electronic and magnetic environment produced by the tertiary structure. The employment of perturbed directional correlation of gamma rays for the study of macromolecules represents such a complementary technique which can overcome some of the difficulties encountered in other nuclear methods. In the case of PDC the number of nuclides which may be used to investigate internal fields is greatly increased, the experiments can be conducted at room temperature and very small ion-probe concentrations--in the order of 10^{-12}M--are usually quite sufficient. In most cases, the investigation and data analysis is also more straightforward than in Mössbauer spectroscopy.

BASIC THEORY

The theory of gamma-gamma directional correlation without the presence of perturbing fields, was first developed by Hamilton (1940) and later extended by Racah (1951) and Fano (1953). The effect of extranuclear fields on gamma-gamma directional correlations was first examined by Alder (1952; 1953; 1963), as well as Coester (1954), Paul and Brunner (1963) for static magnetic and electric fields and the problem of time dependent fields was treated by Abragam and Pound (1953). Very extensive summaries on the early experimental and theoretical work dealing with unperturbed and perturbed gamma-gamma directional correlations may be found in the survey articles by Biedenharn (1962), Devons and Goldfarb (1957), Frauenfelder and Steffen (1966), Frauenfelder (1961), and Karlsson (1964).

From electromagnetic and nuclear theory the probability of emission of photons by a radioactive nucleus depends on the angle between the nuclear spin direction and the direction of emission. In the usual experimental situation, however, the emitting nuclei in a radioactive sample are randomly oriented and so the observed distribution of emitted photons is isotropic. To observe an anisotropic distribution it is thus necessary to examine photons emitted by nuclei with a common spin orientation. This may be achieved if we consider a gamma-gamma cascade decay, in which a nucleus emits two or more genetically related radiations. The succeeding radiation then shows a definite angular correlation with respect to the first one. For this, the general term is angular correlation which comprises directional correlation and polarization correlation. In directional correlation measurements only the directions of the two radiations are observed. In polarization correlation one determines also the linear or circular polarization of the radiations (Frauenfelder and Steffen, 1966). In this paper the discussion is restricted to directional correlation, referred to as perturbed directional correlation (PDC) in the presence of extranuclear fields.

Consider a nucleus which is formed in an excited level having a spin I_i, as shown in Figure 1. Suppose that the nucleus decays to some final state, not necessarily stable, which has a spin I_f and that the decay process involves the successive emission of two radiations R_1 and R_2. If we observe the photon R_1 in some fixed direction $\vec{k}_1$, then we are effectively picking out nuclei that statistically have similar spin orientations and the second radiation R_2 will usually have a definite angular distribution with respect to the direction $\vec{k}_1$. We will suppose for the moment that the average lifetime τ_N of the nuclear intermediate level is very small so that the orientation is not disturbed between the times of emission of R_1 and R_2. Then the angular distribution of R_2 with respect to the direction of R_1 is independent of the time interval between the emission of R_1 and R_2.

In practice, measurements are made to obtain the directional correlation function $W(\theta)$, which is defined as the relative probability that radiation R_2 is emitted into solid angle $d\Omega$ at some angle θ with respect to $\vec{k}_1$. The most straightforward method of obtaining $W(\theta)$ has utilized first order perturbation theory and the density matrix formalism to give (Sakai et al., 1963)

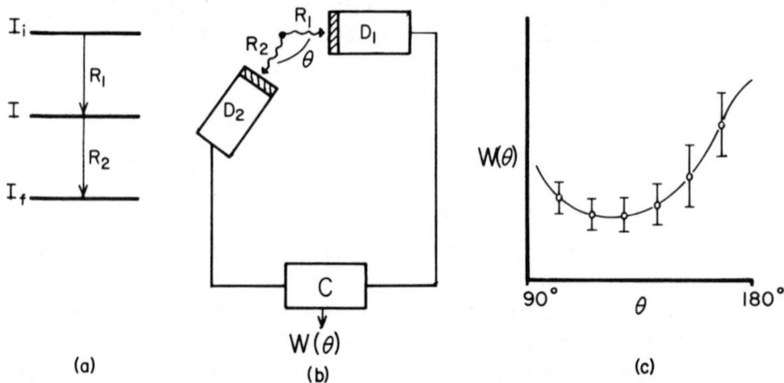

Figure 1. Examination of a gamma-gamma cascade (a), with a coincidence spectrometer (b), to measure the probability of seeing radiation R_2 at some angle θ with respect to the direction of radiation R_1 (c). The two detectors, one fixed and one movable, are placed at a relative angle θ. Each detector D_1 and D_2 forms one side of a coincidence system C which is electronically adjusted to look at either radiation R_1 or R_2. Coincidence measurements are made at various angles to determine the directional correlation function $W(\theta)$.

$$W(\theta) = 1 + \sum_{k_{even}=2}^{\infty} A_{kk}(0)\, P_k\,(\cos\,\theta)\ , \tag{1}$$

where the $P_k(\cos\,\theta)$ are ordinary Legendre polynomials and the A_{kk} are weighing factors which depend on I_i, I, I_f and the multipolarities of R_1 and R_2. In practice, for the nuclei we will be considering, the coefficient $A_{44} << A_{22}$ and the directional correlation function become simply

$$W(\theta) = 1 + A_{22}(0)\, P_2(\cos\,\theta)\ , \tag{2}$$

where the coefficient $A_{22}(0)$ again refers to a nucleus that is not disturbed during the time period between the emissions of R_1 and R_2. In Eq.(2), however, the radiation source and detectors are assumed to be infinitesimal in size and in a typical experiment one obtains the correlation function

$$W(\theta) = 1 + \frac{A_{22}(0)\ P_2(\cos\,\theta)}{Q_2(R_1)Q_2(R_2)}\ . \tag{3}$$

where the $Q_2(R_1)$ and $Q_2(R_2)$ are correction factors which depend on source and detector size and shape, energies of R_1 and R_2, as well as source to detector distance. These have been calculated numerically and may be found in the works of Yates (1963) and Rose (1953). The coefficient $A_{22}'(0)$ is most easily found by measuring the anisotropy A, defined from

$$A(0) \equiv \frac{W(\pi) - W(\pi/2)}{W(\pi/2)}\ , \tag{4}$$

which is related to $A_{22}'(0)$ by

$$A_{22}'(0) = \frac{2A(0)}{3+A(0)}\ . \tag{5}$$

The known values of $A_{22}(0)$ for gamma-gamma cascades of interest in this article are shown in Table I.

In all of the above discussion it has been assumed that the decaying nuclei are free from perturbing forces and that the intermediate level lifetime τ_N is very small. For a typical experiment, however, τ_N may range from .1 nsec to 600 nsec and the nucleus will have time to interact with any perturbing force fields present. In such a case the final level of the transition involving radiation R_1 will not be the same as the initial level of the transition involving R_2 and the directional correlation function becomes dependent on the time period between emissions of R_1 and R_2. This time-dependent directional correlation function has the general form

$$W(\theta,t) = 1 + A_{22}(0)\, G_{22}(t)\, P_2(\cos\theta) , \tag{6}$$

where we are again considering $A_{44}(0) \ll A_{22}(0)$. The development of Eq. (6) may be found in the review article by Frauenfelder and Steffen (1966). The function $G_{22}(t)$, known as the perturbation factor, is dependent on the spin of the intermediate level and parameters involved in the particular kind of perturbation which occurs. If the intermediate level has a g-factor then magnetic interactions are possible. For a static magnetic field, or a field which varies little over τ_N, the perturbation factor has the form

$$G_{22}(t) = \sum_n C_n \cos n\omega_B t , \tag{7}$$

where ω_B is the Larmor precession frequency, given by

$$\omega_B = - \frac{2\pi g B}{h} \mu_N . \tag{8}$$

Measurement of ω_B can thus infer values of internal magnetic fields of materials, provided that g is known. ($\mu_N = 5.05 \times 10^{-24}$ erg/gauss is the nuclear magneton.)

In the case of rapidly fluctuating magnetic hyperfine fields, as in spin-lattice relaxation, the time τ_c between field fluctuations can be much less than τ_N. In this case the perturbation factor has the simple form

$$G_{22}(t) = e^{-\lambda_2 t} , \tag{9}$$

with

$$\lambda_2 = 4\pi J\ (J+1)\ \frac{A^2}{h}\ \tau_c\ . \qquad (10)$$

The hyperfine interaction is assumed to have the form $H = A\vec{I}\cdot\vec{J}$, where $\vec{I}$ is the nuclear intermediate level spin and $\vec{J}$ is the electronic angular momentum. For known $\vec{J}$ and coupling constant A, measurement of $G_{22}(t)$ infers a value for relaxation or vibration time τ_c.

When τ_c is the same order of magnitude as τ_N, $G_2(t) \sim e^{-\lambda_2 t} \cos \omega_B t$. This case, for various values of τ_c/τ_N, has been investigated extensively by Scherer (1970).

For nuclei with excited state spin $I \geq 1$ it is possible for an intermediate level to have an electric quadrupole moment and electric quadrupole interactions may be observed whose strength depends on the quadrupole moment Q as well as the magnitude and symmetry of the electric field gradient at the nuclear site. In the case of static electric quadrupole interactions, the perturbation factor has the form, for $A_{44} << A_{22}$,

$$G_{22}(t) = \sum_n s_{2n} \cos n\omega_e\ t\ , \qquad (11)$$

where it has been assumed that the medium containing the cascade decaying nuclei is polycrystalline and that the field gradient at the nuclear site is axially symmetric. Values of s_{2n}, for various intermediate level spins, are tabulated in the previously mentioned review article by Frauenfelder and Steffen. The characteristic interaction frequency ω_e is related to the quadrupole interaction frequency ω_e by

$$\omega_e = 3\omega_Q \text{ for integral I}$$

$$\omega_e = 6\omega_Q \text{ for half integral I}\ , \qquad (12)$$

and

$$\omega_Q = \frac{2\pi e\ |QV_{zz}|}{4I\ (2I-1)\ h}\ , \qquad (13)$$

where V_{zz} is the value of the axially symmetric electric field gradient at the nuclear site, and e is electron charge. In cases where a nuclear probe occupies a site having axial symmetry, as in tetragonal or hexagonal structures, a measurement of $G_{22}(t)$ infers a value for $|V_{zz}|$ if the quadrupole moment Q is known.

In the case of orthorhombic or rhombohedral fields $G_{22}(t)$ must be calculated numerically, but the basic effect of asymmetric fields is to induce a pseudoperiodic behavior of $G_{22}(t)$. With a principal axis transformation the field may be defined in terms of the asymmetry parameter η, given by

$$\eta = \frac{V_{xx} - V_{yy}}{V_{zz}} , \tag{14}$$

such that $0 \leq \eta \leq 1$. For rhombohedral fields η is in the range of .05 to .2, while for orthorhombic fields $.2 < \eta < .4$. The behavior of $G_{22}(t)$ changes rapidly with increasing η and a measurement of the time-differential behavior of $G_{22}(t)$ can yield both the magnitude and the symmetry of the field gradient at the nuclear site. Time-differential behavior of $G_{22}(t)$ for $0 \leq \eta \leq 1$ has been calculated for various intermediate level spins by Matthias et al. (1962; 1963). Perturbation factors for I=1 are shown in Figure 2.

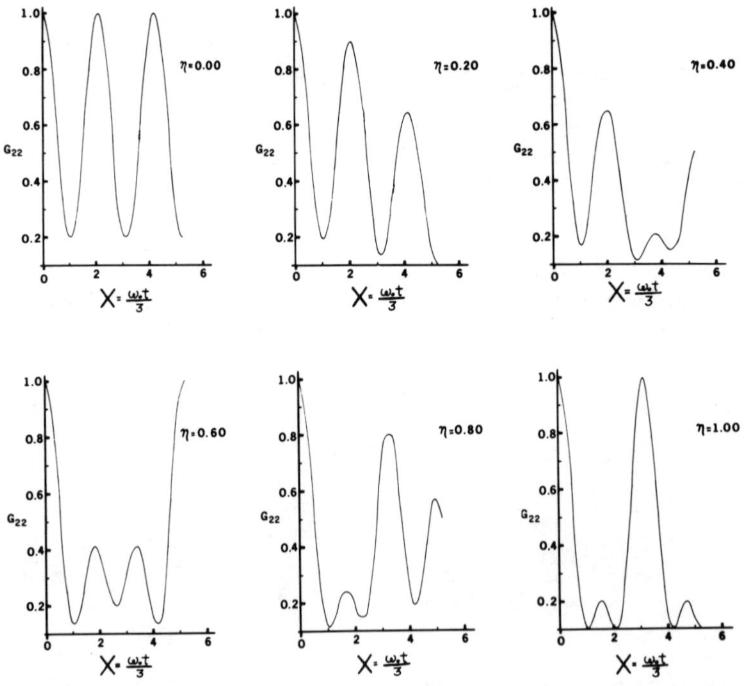

Figure 2. Time-differential perturbation factors for an intermediate level spin I=1 and various values of the asymmetry parameter.

In many cases it is possible that the ion containing a cascade decaying nucleus occupies two or more inequivalent sites in a solid, and the perturbation factor will thus be a function of more than one characteristic interaction frequency. In both the magnetic and electric case this situation can be most easily analyzed by examining the Fourier transform of the perturbation factor. Inspection of the power spectrum of the Fourier transform yields the characteristic interaction frequencies. Some details on such an approach are given in the recent work by Raghavan et al. (1973). In solids which have a large dislocation density the number of characteristic interaction frequencies is quite large and the perturbation factor must be averaged over some probability distribution for the field gradient at the site of a cascade decaying nucleus. In the case of random dislocations, a normal probability distribution may be assumed, having the form

$$P(\omega)\, d\omega = \frac{1}{\sqrt{2\pi}\sigma} \exp\left[- \{(\omega-\omega_e)/\sqrt{2}\sigma\}\right]^2 d\omega\ , \qquad (15)$$

where we define $\delta = \sigma/\omega_e$ as the relative width of the distribution. The observable perturbation factor is then

$$g_{22}(\omega,t) = \frac{\int_{-\infty}^{\infty} G_{22}(\omega,t)\, P(\omega)\, d\omega}{\int_{-\infty}^{\infty} P(\omega)\, d\omega}\ . \qquad (16)$$

As an example, for I=1, $G_{22} = .6 + .4 \cos \omega_e t$. For a normal distribution of field gradient, from Eq. (16), the modified form of the perturbation factor is given by $g_{22} = .6 + (.4 \cos \omega_e t) \exp(-\omega_e{}^2\, \delta^2\, t^2/2)$. In general, the presence of a distribution of field gradients leads to a damping of the oscillatory perturbation factors for a static electric quadrupole interaction. Such effects have been observed (Glass and Kliwer, 1968) for relative widths as high as $\delta = 0.25$. A theoretical fit to their data is shown in Figure 3.

For solids and viscous liquids, in which ionic contributions to the field gradient are large, it is possible to observe time-dependent electric quadrupole interactions. As in the magnetic case, the perturbation is given by Eq. (9), but for a time varying electric field gradient

$$\lambda_2 = \frac{72\pi^2}{80}\, \frac{\tau_c}{h^2}\, (eQ)^2\, |V^2{}_{zz}|\, \frac{4I\,(I+1)-7}{I^2(2I-1)^2}\ , \qquad (17)$$

from which is inferred a value for the product of the correlation time τ_c and rms value of the field gradient. Observation of perturbation factors of this kind have been useful in examination of correlation times in liquids (Steffen, 1956) and, more recently, have verified the random motion of one fluorine ion among the faces of the fluorine octahedra in $(NH_4)_3HfF_7$ (Andrade et al., 1970).

In the interpretation of a possible time dependent interaction care must be taken to also include effects due to decays which occur preceeding the gamma-gamma cascade. These effects due to preceeding decays, or after effects, are primarily due to nuclear recoil, or electron shell rearrangement. Nuclear recoil is particularly important if α-decay preceeds a gamma-gamma cascade, since the daughter nucleus will generally recoil out of a lattice site in the environment to be studied and a time dependent interaction, due to relative motion of the daughter nucleus and the lattice, will be observable. Recoil effects must also be considered if the gamma-gamma cascade is preceeded by a gamma transition involving an energy of 1 MeV or more.

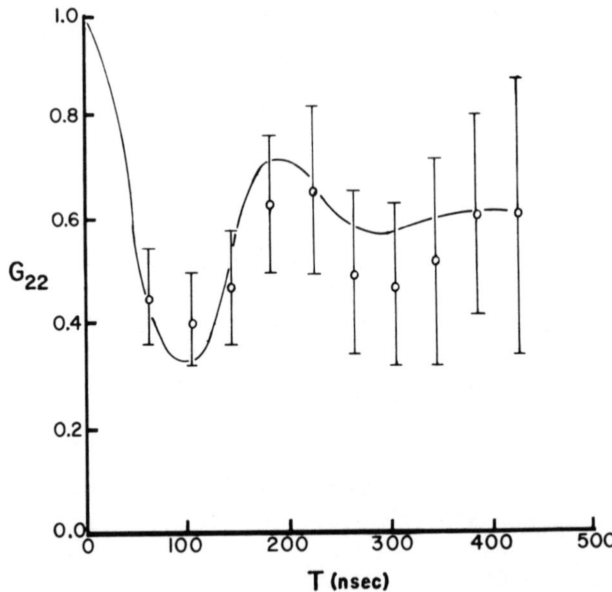

Figure 3. Time-differential perturbation factor for interaction of the 68 keV level in ^{44}Sc with a distribution of axially symmetric electric field gradient at the ^{44}Ti ion site in $BaTiO_3$. Relative width of distribution $\delta = 0.25$.

In many cases of interest the gamma-gamma cascade decaying nucleus is formed by electron capture decay of some parent nuclide. The recoil energy of the daughter is given (Wu and Moszkowski, 1966) by

$$E_R = 140.2 \frac{E^2}{M} \text{ electron volts} , \qquad (18)$$

with E the disintegration energy in units of electron rest energy and M the mass of the recoiling daughter atom in atomic mass units. In the case of nuclides of interest in this paper the recoil energy is generally less than 2 eV and recoil effects may be neglected. Effects due to electron shell rearrangement following a (primarily) K-capture process must be seriously considered, particularly if the gamma-gamma cascade occurs shortly after this process. In the inner shells the electron hole propagates outwards predominantly through x-ray emission. In the outer shells, however, the propagation is primarily due to emission of Auger electrons and the availability of electrons near the unstable ion becomes critical to the total time for electron shell rearrangement. In a metallic-like environment, with a large supply of conduction electrons, hole propagation in the outer shells takes place within a few vibration times of the environment. For lesser conductivities, electron shell rearrangements may take 10^{-10} to 10^{-7} sec to occur and a time dependent electric quadrupole interaction can take place due to the field gradient caused by the propagating electron hole. For good insulators, electron shell arrangement can take microseconds to occur and it is even possible to observe static interactions due to the field of an electron hole that is stationary during the intermediate level lifetime τ_N. Such an effect has been pointed out in the recent work by Klemme et al. (1973). For good insulators, ^{44}Sc is the only gamma-gamma cascade decaying nuclide which can be used without fear of after effects, since the gamma-gamma cascade is initiated by the decay of an isomeric state having a lifetime of 50 nsec.

In some solids it is possible to have magnetic and electric perturbations present at the same time. Calculation of perturbation factors for the case where the symmetry axis of the electric field gradient is parallel to the magnetic field was originally treated by Alder et al. (1953). The general case of non-parallel fields has been calculated numerically by Alder et al. (1963) and the computer program used for the calculations is available from Alder et al. or the authors of this paper. The problem of calculating time-differential perturbation factors for non-coaxial static electric and magnetic fields has been most recently treated in the work of Boström et al. (1970) for intermediate level spins of 3/2 and 5/2. In general, the time-differential perturbation factors for combined static electric and magnetic interactions are complex oscillatory functions and details will not be given in this paper.

In some cases, interpretation of observed time-differential perturbation factors may be quite complicated, particularly when time resolution is poor for the time-differential coincidence system used. In such a case, much information can be obtained by examining the time-integrated perturbation factor, defined by

$$\overline{G_{22}(t)} = \frac{\int_0^t G_{22}(t)\, e^{-t/\tau_N}\, dt}{\int_0^t e^{-t/\tau_N}\, dt} , \tag{19}$$

where t is again the time between emission of R_1 and R_2. The quantity $\overline{G_{22}(t)}$ may be obtained by performing a weighted average of $G_{22}(t)$ between t = 0 and t. If $t>4\tau_N$ the value of $\overline{G_{22}(t)}$ closely approximates the total time-integrated perturbation factor given by

$$\overline{G_{22}(\infty)} = \frac{\int_0^\infty G_{22}(t)\, e^{-t/\tau_N}\, dt}{\int_0^\infty e^{-t/\tau_N}\, dt} = \frac{1}{\tau_N} \int_0^\infty G_{22}(t)\, e^{-t/\tau_N}\, dt . \tag{20}$$

For a time dependent interaction,

$$\overline{G_{22}(\infty)} = \frac{1}{\tau_N} \int_0^\infty e^{-\lambda_2 t}\, e^{-t/\tau_N}\, dt = \frac{1}{1+\lambda_2\tau_N} , \tag{21}$$

from which λ_2 may be exacted directly. For very large $\lambda_2\tau_N$, however, $\overline{G_{22}(\infty)}$ approaches zero value and the angular distribution of R_2 with respect to the direction of R_1 can become isotropic. In the case of a static electric interaction, on the other hand,

$$\overline{G_{22}(\infty)} = \frac{1}{\tau_N} \int_0^\infty s_{2n}\, e^{-t/\tau_N} \cos(n\omega_e t)\, dt .$$

$$= \sum_n s_{2n} \frac{1}{1 + (n\omega_e\tau_N)^2} . \tag{22}$$

If the characteristic period $2\pi/\omega_e$ is much less than the time resolution of the coincidence system used, or if it is much less than τ_N, then $\overline{G_{22}(\infty)}$ approaches the limiting value of s_{0n}. This number has a value .6 for I=1, .2 for I=3/2, .37 for I=2, and .2 for I=5/2. Measurement of $\overline{G_{22}(\infty)}$ can thus distinguish between a strong static electric interaction and a time dependent interaction. The values of s_{0n}, known as the "hard core values", are insensitive to any distribution in characteristic interaction frequencies, but vary rapidly with a change of asymmetry parameter η between $\eta = 0$ and $\eta = .2$ for intermediate level spins of I=1, 2, 5/2, 3, and 7/2. For these spins then, one can distinguish between the presence of axially symmetric or asymmetric fields by measuring $\overline{G_{22}(\infty)}$.

EXPERIMENTAL CONDITIONS AND INSTRUMENTATION

As in Mössbauer spectroscopy, the number of nuclides available for use in PDC is somewhat limited. In choosing a nuclide for a particular experiment, consideration should be given to the following factors:

1. The parent lifetime should be long, preferably months or years, although PDC work has been done with parent lifetimes as short as 7 hours. Short parent lifetimes necessitate the use of 4-detector time-differential coincidence systems and good statistics are difficult to obtain.

2. The daughter nuclide should have a strong gamma-gamma cascade branch of decay. A good working number to use is that at least 40% of the parent nuclide-decays lead to the desired gamma-gamma cascade.

3. The intermediate level of the gamma-gamma cascade should have a lifetime from one to a few hundred nanoseconds. This range is compatible with the perturbing field strengths commonly observed in solids and liquids.

4. The energies of the cascade gamma rays should be between 100 keV and 1 MeV. For very high energies, after effects may be observable. For energies below 100 keV, detector time resolution becomes relatively poor.

5. The magnetic dipole moment μ and the electric quadrupole moment Q of the intermediate level should be known. This is rarely true.

6. The recoil energy in the decay of the parent nuclide should be minimal.

7. The parent nuclide and ground state of the daughter nuclide should decay by β-emission or electron capture. Positron emission is not desirable because annihilation in the source produces prompt gamma pairs which have an extremely high anisotropy within one time

resolution of zero time delay between emission of the cascade gammas.

8. The ions containing both parent and daughter nuclides should be chemically compatible with the medium to be studied and should occupy definite lattice sites.

There is no known nuclide which satisfies all of the above requirements. A survey of the Nuclear Data Sheets and Table of Isotopes by Lederer et al. (1968) shows 29 nuclides, however, which satisfy some of the above requirements and thus have possible value in PDC experiments. These nuclides and some of their significant properties, are given in Table I.

In any PDC experiments, a correlation function is measured as a function of time delay between emission of two radiations. This necessitates electronically defining the exact time at which radiation R_2 is emitted after R_1. The basic circuit used to do this is shown in Figure 4.

The basic function of the time pickoff unit is to modify the detector pulse so as to define the time of occurance of the detected event. In various ways, the unit puts out a logic pulse whose leading edge carries the time information.

In any PDC experiment, a basic concern is time resolution, which depends both on the type of time pickoff device and the detector being used. The two phenomena which contribute to poor time resolution are walk and jitter. The contribution of walk is particularly important for scintillation detectors and fixed threshold discriminators and is caused by the spread in rise time of detector pulses for a given event (Nutt, 1967; Compton and Johnson, 1967).

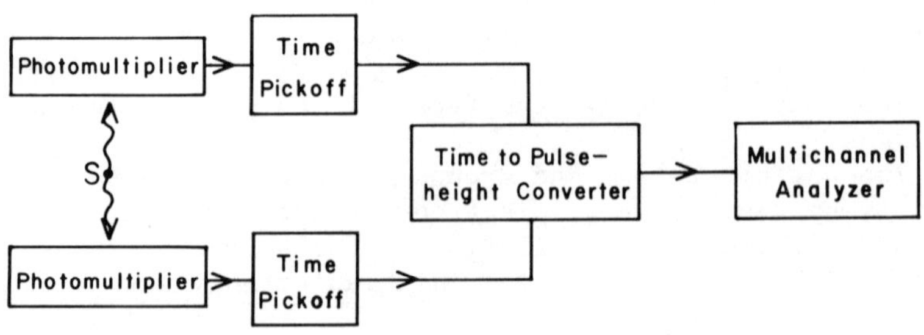

Figure 4. Basic time-differential coincidence system.

TABLE I

Isotopes for Gamma-Gamma Directional Correlations

Most data on moments is taken from the review of Lindgren (1964). Other specific references are given in the table.

Isotope	Parent	Parent Lifetime	Level Lifetime(ns)	Level Spin	Cascade Energy (keV)	Remarks
$^{44}_{21}Sc$	$^{44}_{22}Ti$	49y.	153	1	78-68	$\mu = .7\mu_N$ $Q = .109b$ (Ref. 1) $A_{22} = 0.042$ (Ref. 2)
$^{57}_{26}Fe$	$^{57}_{27}Co$	270d.	100	3/2	122-14	$\mu = -0.16\mu_N$ $Q = + .15b$ $A_{22} = -.024$ (Ref. 3)
$_{48}Cd$	$^{111}_{49}In$	2.8d.	85	5/2	173-247	$\mu = -.8\mu_N$ $Q = +.9b$ $A = -.21$ (Ref. 4)
$^{120}_{50}Sn$	$^{120}_{51}Sb$	5.8d.	5.5	5	200-90	$\mu = -.3\mu_N$
$^{133}_{55}Cs$	$^{133}_{56}Ba$	7.2y.	6	5/2	356-81	calculated $\mu = 3.2\mu_N$ $Q = 0.51b$ (Ref. 5) $A_{22} = .036$ (Ref. 6)
$^{161}_{66}Dy$	$^{161}_{65}Tb$	6.9d.	27	5/2	49-26	$\mu = .53\mu_N$

References: (1) P. Doshi, J. C. Glass, and M. Novotny, Phys. Rev. B7(1973); (2) M. R. Haas, and J. C. Glass, Phys. Rev. B4(1971)147; (3) B. M. Virabyan, S. S. Durgar'yan, I. L. Eganyan, A. G. Malayan, and K. I. Pyuskyulyan, Isv. Akad. Nauk Arm. SSR, Fiz. 7(1972)93; (4) P. Lehman, and J. Miller, J. Phys. Radium 17(1956)526; (5) D. C. Choudbury and J. N. Friedman, Phys. Rev. C (to be published); (6) F. T. Avignone, G. D. Frey, and L. D. Hendrick, Phys. Rev. C1(1970)635.

Table I (Continued)

Isotope	Parent	Parent Lifetime	Level Lifetime (ns)	Level Spin	Cascade Energy (keV)	Remarks
$^{181}_{73}Ta$	$^{181}_{72}Hf$	43d.	15.4	5/2	133-482	$\mu = 3.3\mu_N$ $A = 0.37$ (Ref. 7)
$^{199}_{80}Hg$	$^{199}_{81}Ti$	7.4h.	2.3	5/2	333-153	$\mu = 1.04\mu_N$ $A = -0.42$ (Ref. 8)
$^{204}_{82}Pb$	$^{204}_{83}Bi$	11.2h.	260	4	1200-375	$\mu = .22\mu_N$
$^{237}_{93}Np$	$^{237}_{92}U$	6.75d.	60	5/2	208-59	$\mu = 2.1\mu_N$
$^{48}_{23}V$	$^{48}_{24}Cr$	23h.	11	2	116-310	
$^{77}_{34}Sc$	$^{77}_{35}Br$	57h.	9.3	3/2	580-248	Relatively weak cascade
$^{83}_{36}Kr$	$^{83}_{37}Rb$	83d.	150	7/2	32-10	
$^{86}_{39}Y$	$^{86}_{40}Zr$	17h.	29	2	28-243	

References: (7) P. Da R. Andrade, A. Vasquez, J. D. Rogers, and E. R. Fiaga, Phys. Rev. B1(1970) 2912; (8) L. Grodyens, R. W. Bauer, and H. H. Wilson, Phys. Rev. 124(1961)1897.

Table I (Continued)

Isotope	Parent	Parent Lifetime	Level Lifetime (ns)	Level Spin	Cascade Energy (keV)	Remarks
$^{100}_{45}Rh$	$^{100}_{46}Pd$	4d.	240	2	126-75	
$^{131}_{55}Cs$	$^{131}_{56}Ba$	12d.	9.6	7/2	294-78	Weak cascade
$^{157}_{64}Gd$	$^{157}_{63}Eu$	15h.	460	5/2	413-64	
$^{168}_{68}Er$	$^{168}_{69}Tm$	85d.	110	.4	448-198	$\mu = 1.80\mu_N$ (Ref. 9) $A_{22} = .007$
$^{169}_{69}Tm$	$^{169}_{70}Yb$	32d.	660	7/2	63-197	
$^{187}_{75}Re$	$^{187}_{74}W$	24h.	570	9/2	499-72	
$^{197}_{80}Hg$	$^{197}_{80}Hg$	24h.	7	5/2	165-134	Hg isomer decay
$^{238}_{93}Np$	$^{242}_{95}Am$	152y.	50	6	42-136	Note alpha decay of parent. Recoil energy very large.

Reference: (9) B. T. Kim, W. K. Chu, and J. C. Glass, Phys. Rev. C (in press).

Using various means, however, walk can be made quite small and is a minor factor in the ultimate time resolution achievable. The phenomena of jitter is caused (Hyman *et al.*,1964; Bertolaccini *et al.*, 1967) by the statistical time variation of the detector signal as seen by the time pickoff device. In the case of solid state detectors this is contributed to strongly by system noise and severely limits the ultimate time resolution achievable with these detectors. In the case of scintillation detectors, noise is not usually important, but jitter is still present due to statistics of emission and collection of photoelectrons. This can be improved somewhat by using scintillators which have a good light output per unit incident energy (Present *et al.*, 1964; Bertolini *et al.*, 1966), such as NaI(Ti), Naton 136, Pilot B, NE 102, NE 213, or NE 218. The major improvement in jitter elimination, however, has come (Hyman, 1965; McDonald and Gedcke, 1967) with the development of photomultiplier tubes having low noise, high quantum efficiency, fast anode pulse risetime and small transit time spread (time it takes a photoelectron to go from the photocathode to the first dynode). Bialkali cathode tubes, such as the RCA 8575, have been used with NaI(Tl) scintillators to achieve time resolutions as low as 5 nsec for a gamma ray energy of 14 keV.

With minimal jitter due to proper detector selection, time resolution depends primarily on walk; which in turn depends on the type of time pickoff device used. The two basic systems we will consider here are the slow crossover system, and leading edge timing.

The slow crossover (McGuire and Palmer, 1967), or crossover pickoff, system is shown in Figure 5. It is the most economical system to use and, having good walk characteristics, is useful over a wide range of detector pulse heights. It is also, however, quite sensitive to jitter and over a narrow dynamic range has a time resolution 11-times poorer than that obtainable with leading edge timing. This system is best used for gamma ray energies greater than 100 keV.

A leading edge timing (Gedcke and McDonald, 1968) time-differential coincidence system is shown in Figure 6. Modifications of this system appear in the type of fast discriminator used. Leading edge timing offers good time resolution over a narrow dynamic range, but with fixed threshold discrimination has poor walk characteristics and so should be used only over a narrow spread of pulse heights, obtainable for gamma energies over 100 keV. For energies less than 100 keV time resolution has been significantly improved by the advent of the constant fraction height trigger (Braunsfurth and Korner, 1965), which gives minimal walk over a wide dynamic range. This has been further improved with introduction of so-called snap-off timing discrimination, which is essentially a combination of the constant fraction method with snap-off diode integration of the anode pulse from the detector.

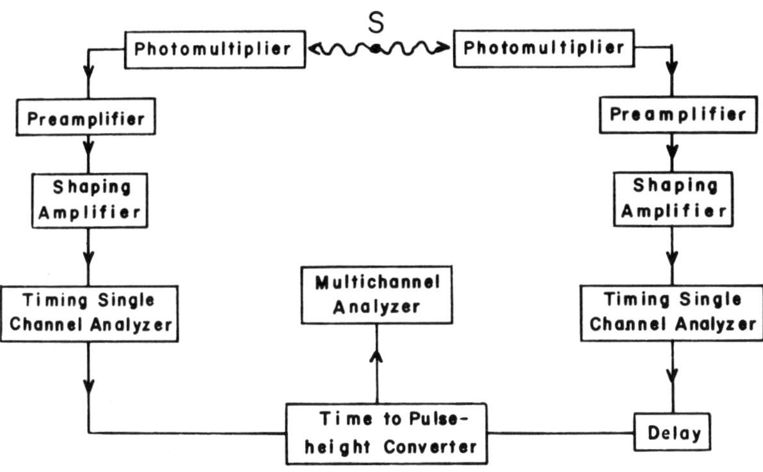

Figure 5. Slow crossover (crossover pickoff) coincidence system.

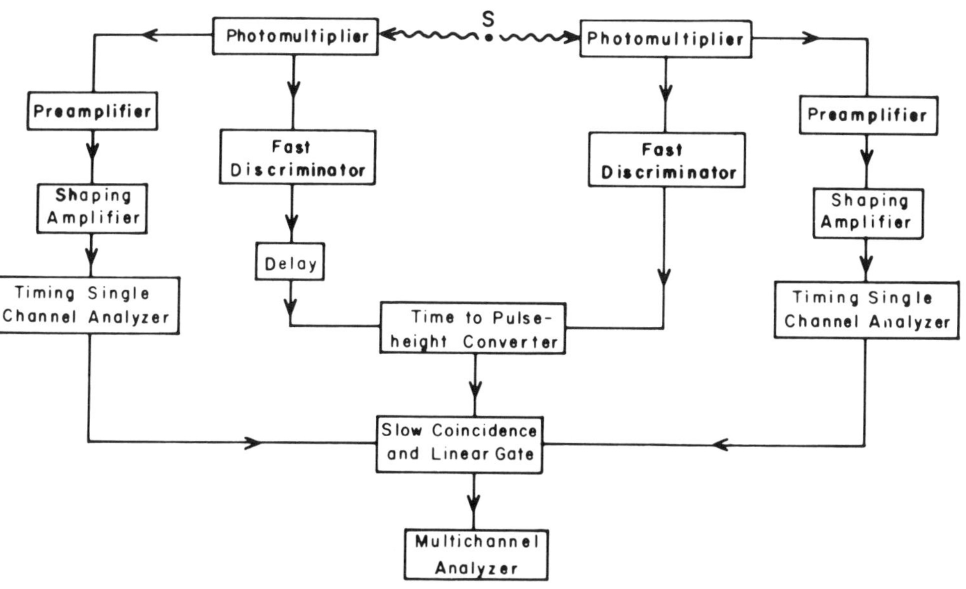

Figure 6. Leading edge timing--time-differential coincidence system.

In PDC experiments involving very short parent nuclide lifetimes it is extremely difficult to obtain good statistics with a 2-detector system. Because of this, 3-detector systems, with one detector movable, have been developed which improve statistics by a factor of 2 for a given running time. Such systems have been well described (Bodenstedt *et al.*, 1972).and have been used to work with short-lived parent nuclides such as ^{111}In and ^{199}Tl. As an extension of this, four-detector systems have been built (Wagner and Forker, 1969) which will simultaneously give 4 independent measurements of a time-differential directional correlation function.

Finally, it is sometimes convenient to automate the data-cycling and detector positioning of a coincidence system, particularly when detector position must be changed every few minutes because of observations involving a very short parent nuclide lifetime. Such systems (Netz and Bodenstedt, 1973) have been constructed and, with the availability of an on-line computer, can be automatically controlled by a data acquisition program.

APPLICATION TO THE STUDY OF PROTEIN-METAL INTERACTIONS

Several techniques have recently been applied to investigate rotational correlation times, internal motions, and conformational changes of protein molecules. The depolarization of fluorescence of small chromophores can be used to measure the rotational correlation time of a chromophore bound to a macromolecule (Stryer, 1965; Yguerabide *et al.*, 1970). The orientation and motion of stable free radicals bound to macromolecules are easily monitored by electron spin resonance (Stryer and Griffith, 1965; Hamilton and McConnell, 1968). Halide ions have been used as chemical probes for nuclear magnetic resonance studies of proteins labeled with metal atoms (Stengle and Baldeschwieler, 1966; Haugland *et al.*, 1967). By using these techniques, information on local characteristics of the macromolecules near the labeling site is usually obtainable. In most cases, the labels can be inserted into the protein molecules at the regions of interest by using selective and specific chemical modifications. Labels have been tied chemically to substrate and inhibitor molecules which can interact subsequently with the active regions of enzymes or antibodies.

The potential of perturbed directional correlation studies to yield motional and structural information about protein molecules has also been recognized recently. Various authors calculated the PDC expected from nuclei attached to molecules that reorient slowly (Marshall and Meares, 1972), anisotropically (Marshall *et al.*, 1972), with internal rotation (*ibid.*), by random jumps caused by strong collisions (Lynden-Bell, 1971), or in one dimension (Shirley, 1971). As the theoretical understanding of the effects of molecular dynamics and structure on PDC became more complete, the method gained considerable potential as a labeling technique (Glass and Graf, 1970; Meares, 1972; Richer, 1973).

The first work in this field (Leipert _et al_., 1968) employed the γ-ray cascade in ^{111}In following the decay of $^{111}In^{3+}$ bound to sites in bovine serum albumin (BSA). Cadmium metal (12.75 percent ^{111}Cd) was irradiated with 10 MeV protons in the Berkeley 88-inch cyclotron, producing 2.8 day ^{111}In by the (p,n) reaction. After chemical separation from the cadmium target the ^{111}In tracer was added as indium(III) ion to aqueous solutions containing BSA. The PDC measurements were made with a four-detector delayed coincidence spectrometer using 0.2 ml liquid samples. For the unperturbed directional correlation of the ^{111}Cd cascade following ^{111}decay, the coefficient $A_{22}G_2(t)$ should be independent of the delay time t and equal to 0.20. The plot of the anisotropy A(t) against time (t) for this cascade in the presence of BSA at pH 5.7 showed that the directional correlation was strongly perturbed. When the pH of the solution was reduced to 2.8, A(t) approached more closely the unperturbed behavior. Because BSA undergoes a conformational transition between pH 5.7 and 2.8 (Steinhardt and Beychok, 1964), the changes in the directional correlation may reflect changes in the effective rotational correlation time at the binding sites of ^{111}In or they may simply reflect a change in the effective binding constant. At pH 5.7 the protein molecule may be more rigid, giving a larger value for the rotational correlation time, τ_c, and a larger perturbation in the directional correlation. The shape of the delayed coincidence plot suggests that only time-independent quadrupole interactions are involved in this case. Because the γ-ray cascade observed in the daughter ^{111}Cd nuclei follows electron capture decay of ^{111}In on the binding sites, the possibility that after effects following decay might lead to detachment of cadmium ions from these sites, became a major concern in the experiment. To avoid detachment the cadmium ions have to achieve a stable chemical configuration during the lifetime of the 419 keV state of ^{111}Cd ($t_{1/2}$ = 1.2 x 10^{-10}s). The displacement of $^{111}In^{3+}$ by EDTA and dialysis of the BSA-indium(III) ion solution, and analogous qualitative experiments with polyglutamic acid suggested that the radioactive tracer was firmly bound to the macromolecule and reflected the effective molecular rotational correlation time at the binding site.

To eliminate uncertainties associated with the electron capture decay of ^{111}In, it was also desirable to use the isomeric ^{111m}Cd state as the parent nucleus. Meares _et al_. (1969) showed that $^{111m}Cd^{2+}$ binds into the Zn^{2+} position in carbonic anhydrase (EC 4.2.1.1) isolated from bovine red blood cells. The perturbation factor function for this case resembled that of a polycrystalline sample indicating that the $^{111m}Cd^{2+}$ binding site was essentially immobilized. Excited ^{111}Cd was prepared by the reaction

^{110}Cd (n,p)^{111}Cd using the University of California, Berkeley, reactor. Because of the short lifetime ($t_{1/2}$ = 49 min) of ^{111}Cd metastable state, three freshly irradiated samples of cadmium were required for each measurement. The directional correlation experiments were made with a four-detector fast-slow gamma-ray coincidence spectrometer using NaI(Tl) crystal detectors. Native bovine carbonic anhydrase was obtained commercially. It was demonstrated to be enzymatically active by the method of Wilbur and Anderson (1948). Apo-carbonic anhydrase was prepared by dialyzing native carbonic anhydrase against o-phenanthroline. The anisotropy of $^{111m}Cd^{2+}$ in buffered (pH 6.7) sodium chloride solution was only weakly perturbed, characteristic of the free ion in solution. The anisotropy for buffered $^{111m}Cd^{2+}$ in the presence of native carbonic anhydrase was again characteristic of the free ion in solution although its lower value may indicate that a small fraction of the ions is bound to the native enzyme. The anisotropy plot for $^{111m}Cd^{2+}$ in the presence of apo-carbonic anhydrase in the same buffer showed that the correlation was strongly perturbed. The shape of the plot suggests that mainly time-independent quadrupole interactions are responsible for the perturbation of the directional correlation. Under these conditions $\tau_c > (e^2qQ)^{-1}$ and the nuclear spin system is effectively immobilized, i.e. the $^{111m}Cd^{2+}$ is rigidly bound to the apoenzyme. The qualitative dependence of the perturbed directional correlation of the ^{111m}Cd cascade on the molecular rotational correlation time at the metal binding site was further illustrated by measuring the anisotropy at various temperatures for a series of $^{111m}Cd^{2+}$ solutions containing 1M N-benzyliminodiacetic acid. Since the correlation time for the motion of the metal-N-benzyliminodiacetic acid complex is a function of temperature, the relationship could be easily established. For frozen solution the finite hard-core value of $G_{22}(t)$ for large t indicates that the condition $\tau_c >> (e^2qQ)^{-1}$ is satisfied.

As more sophisticated experiments became possible, it was desirable to explore theoretically the sensitivity of gamma-ray correlation patterns to molecular orientation and dynamics. Shirley (1971) examined the influence of molecular geometry, orientation and dynamics on gamma-ray directional correlation patterns from solute macromolecules labeled with rotational tracers such as ^{111m}Cd under certain model conditions. A nucleus of spin 5/2 acted upon by an axially symmetric electric field gradient and bound to a rod-like macromolecule was considered. Under static conditions (with no molecular rotation) the time dependent correlation pattern is quite sensitive to molecular orientation and, for oriented molecules, also to the angle between the axis of the field gradient tensor and the molecular axis. Macromolecules were considered whose shapes can be characterized by a single preferred axis (e.g. rods or disks). An equation could be derived that relates the correlation pattern to molecular orientation and local properties of the tracer atom. The sensitivity to molecular geometry and orientation could be well illustrated with explicit calculations for the few cases that can be constructed with the macromolecules oriented parallel or perpendicular to E and the electric field gradient axis parallel or perpendicular to the molecular symmetry axis. As the molecules in a given experimental situation become oriented the observed correlation pattern will vary continuously from the random situation exhibiting a considerable diagnostic value in studying the static properties and orientation of macromolecules. When molecular rotation is allowed, a classical model can be applied if the rotation is sufficiently slow. This model was used to calculate relaxation curves for several geometrical configurations with the restriction that the macromolecules rotate about their long axes. Thus, the perturbation factor could be made sensitive to the dynamic behavior of the molecules and it is reasonable to expect that under favorable conditions it may be possible to study molecular dynamics. These features of rotational tracers are especially interesting because of the potential of *in vivo* studies and possible medical applications. Although the results were obtained for a nuclear level with spin 5/2, similar calculations could be done with any other spin for which $I \geq 1$, the condition required for quadrupole interactions.

Most proteins in their native form in solution fall into the general category of the prolate ellipsoidal shape. If such a molecule undergoes rotational diffusion in a continuous medium, the rotational motions about the different axes of the ellipsoid occur at different rates (Woessner, 1962; Huntress, 1968). Details of molecular rotational motion in liquids are not completely understood. For example, do the molecules undergo rotational motion by random diffusion through infinitesimal increments or through larger jumps? Furthermore, how are the rates of molecular rotational motion related to macroscopic fluid properties or to the molecular tranlational coefficient? An intriquing question is, how are the internal rotations related to each other or to the rotation of the molecule as a whole? The structure of a protein molecule is in a dynamic state that is capable to respond to cooperative forces at various levels of organization such as between residues, regions, subunits, or molecular superstructures. Marshall *et al*. (1972) analyzed the effect of molecular shape and flexibility on gamma-ray directional correlation. They considered the PDC for a gamma-ray cascading nucleus rigidly bound to an asymmetric-top molecule in solution, using a diffusional model for reorientation. The calculations are supplemented by several theoretical plots of time-integrated anisotropy for symmetric-top molecules of fixed shape and varying size and for several choices of the characteristic angle of attachment between the molecular symmetry axis and the principal axis of the electric field gradient of the nucleus. It can be shown that the case of a tracer rigidly bound through a single flexible link to a spherical molecule is formally identical to the case of a tracer rigidly bound to a symmetric-top molecule. The calculations and plotted results illustrate that PDC presents a particularly attractive means to study specific sites of proteins in solution, since the observed parameter (time-integrated anisotropy) is quite insensitive to molecular shape but changes markedly in the presence of local flexibility and internal rotations at the site of attachment of the tracer to the macromolecule. These changes contain information on both the rate of the motion as well as the geometry of the attachment. The appeal of such studies is greatly enhanced by the fact that the sensitivity approaches 10^{-12}M for these measurements.

An alterntive to the diffusional model of molecular motion is the strong collision model. Lynden-Bell (1971) used the strong collision approximation, in which reorientation through any angle is equally probable, to calculate the time-dependent directional correlations and their integrals for a nucleus whose intermediate spin is 5/2. The results are compared with the experimental observation of the indium-cadmium and the cadmium-cadmium cascades where the nucleus is bound to various proteins (cf. Leipert _et al_., 1968 and Meares _et al_., 1969). The calculations were confined to situations that are macroscopically isotropic, i.e. there are no external fields and no preferred molecular orientations within the sample. The maximum difference between the diffusional and strong collision model appears to be in the slow motion limit, but even in this limit the difference is small and would not be obvious from a qualitative comparison of the experimental curves. Although it is probable that the diffusion model provides a better description of the motion of large, approximately spherical, molecules, the effects of strong collisions on nuclear spin systems are much easier to calculate.

Using time-dependent perturbation theory, Abragam and Pound (1953) have calculated the effect of rapid molecular motion on the angular correlation. This treatment is useful only when the molecular rotational correlation time is much shorter than the nuclear mean lifetime, τ_N. Considering that one can readily imagine situations for which molecular motions are slow compared to the nuclear mean lifetime, (e.g. for the 247-keV state of ^{111}Cd $\tau_N = 1.22\text{x}10^{-7}$ sec), such as macromolecules in solution, smaller molecules in very viscous solutions, solids near the melting point, etc., Marshall and Meares (1972) extended the theory of perturbed directional correlations of gamma radiation to include the possibility of adiabatic variation of the intermediate state interaction Hamiltonian as it might arise from slow rotational diffusion of the molecule to which the emitting ion is bound. They showed that the adiabatic variation in the interaction Hamiltonian introduces a time dependence into the angles which expresses the orientation of the molecular frame. Thus, terms are obtained which include the rotational diffusion constant for molecular reorientation in the intermediate state. The adiabatic approximation should hold for rotational correlation times of 10^{-7} sec or longer. One immediate application is the determination of the rotational correlation time as a function of temperature for metal ions in a frozen matrix. Barrett _et al._ (1970) studied the perturbed directional correlation of $^{111}Cd^{2+}$ in frozen aqueous solutions of $InCl_3$ and $In(NO_3)_3$ in the temperature range -140 to -20°C. Combination of these experimental results with the theory indicates that the rotational correlation time in this case is less than 10^{-6} sec at temperature above -120°C. Since the theory is independent of the nuclear mean lifetime, it should prove interesting to extend the results to cover rotational correlation times which are of the order of the nuclear lifetime. Again, the possible influence of after effects following the electron capture decay of ^{111}In makes the quantitative interpretation of the results somewhat uncertain. In this respect, the use of ^{111m}Cd as a source should yield more reliable data. Since the instrumentation required for the measurements is simple, information concerning local environments of macromolecules, ions, and complexes in highly viscous media, such as the cellular environment, should be easily obtained.

The technique of PDC was used by Glass _et al._ (1972) to study the internal field of viscous liquids. The isotope chosen for this work was ^{133}Ba, which decays by electron capture to ^{133}Cs (Figure 7). The 356-81 keV gamma-ray cascade decay of ^{133}Cs involves an intermediate level having a spin of 5/2 and a half life of 6.3 nsec. This isotope was chosen primarily for its compatibility with water. For a normal water environment the perturbation factor for the ^{133}Cs cascade should be unity because the correlation time in normal water is very much less than the intermediate lifetime. From the experimental perturbation factor for the gamma cascade decay of ^{133}Cs in pure glycerol (Figure 8) and in water condensed in fine capillaries a correlation time of the order of magnitude of 6 nsec could be estimated. The perturbation factors obtained for the viscous samples were oscillatory. This oscillatory behavior could not be explained on the basis of a pure static electric quadrupole interaction. The data may well indicate the presence of a free-ion hyperfine interaction in ^{133}Cs. It is presumed that other non-interacting viscous liquids with suitable correlation times can be found and this convenient method to measure free-ion hyperfine interaction may be extended to other gamma-gamma cascade decaying nuclei (Glass _et al._, 1973).

In view of the increased interest in the interaction between electric field gradients and nuclear quadrupole moments, Haas and Shirley (1973) reported the results of a comprehensive study of PDC. Although their work was done on metals, low-molecular weight insulating solids and simple solutions, the scope of the technique was greatly expanded and the amount of quantitative information required

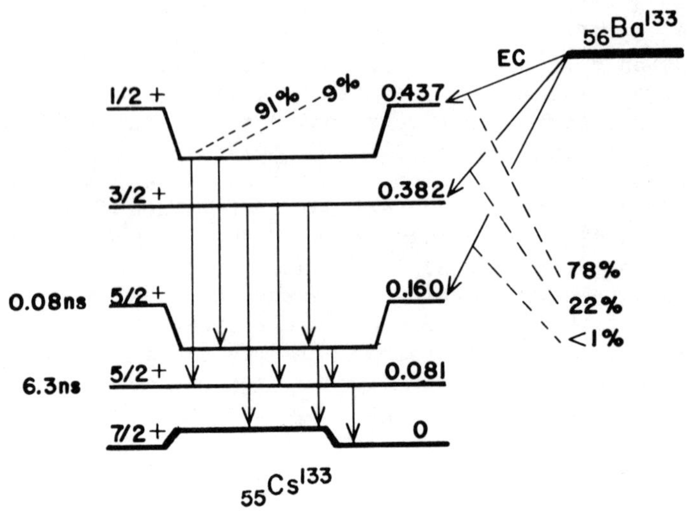

Figure 7. Abridged decay scheme for ^{133}Ba. The 356 keV-81 keV gamma-gamma cascade in ^{133}Cs may be used for PDC.

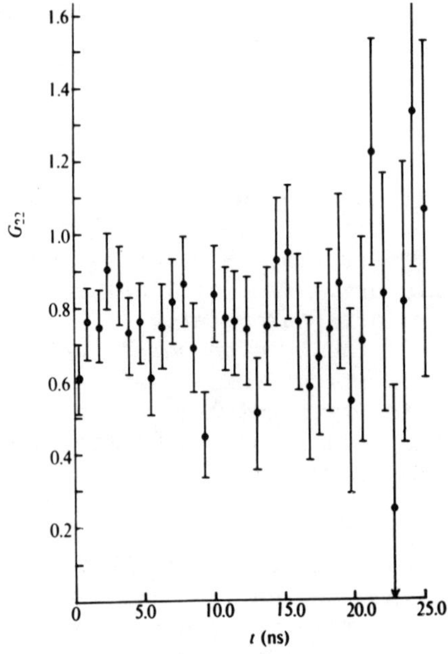

Figure 8. Experimental perturbation factor for 356-81 keV gamma cascade decay of ^{133}Cs in pure glycerol.

for future biochemical applications, considerably increased. A total of fourteen gamma-ray cascades, following electron capture decay (EC), isomeric transition (IT, with no elemental transformation) or beta-decay (β^-, with elemental transformation) were employed: ^{44}Ti-^{44}Sc (EC), ^{99}Rh-^{99}Ru (EC), ^{100}Pd-^{100}Rh (EC), ^{111}Ag-^{111}Cd (β^-), ^{111m}Cd-^{111}Cd (IT), ^{111}In-^{111}Cd (EC), ^{115}Cd-^{115}In (β^-), ^{117}Cd-^{117}In (β^-), ^{131m}Te-^{131}I (β^-), ^{181}Hf-^{181}Ta (β^-), ^{187}W-^{187}Re (β^-), ^{199m}Hg-^{199}Hg (IT), ^{204m}Pb-^{204}Pb (IT), and ^{204}Bi-^{204}Bi (EC). Several nuclear spins were confirmed and the quadrupole moments of ten excited nuclear states were determined or estimated from the data. Quadrupole coupling constants were determined for the excited states of several nuclei for a variety of isotope-lattice combinations. The nuclei ^{111}Cd, ^{115}In, ^{117}In, ^{189}Hg, and ^{204}Pb were used to determine a total of fifty quadrupole coupling constants in insulating inorganic and organic solids. It was demonstrated that good determinations of quadrupole coupling constants could be made following isomeric transitions and beta-decay. However, it was not possible to derive a coupling constant from a gamma-ray cascade preceded directly by electron capture decay, presumably because the sudden creation of a K-hole and the Auger and "shake-off" events that follow, destroy the chemical integrity of the species under study. Relaxation times were determined also for a number of liquid samples such as simple ionic or molecular solutions. Measurements on liquids or on gases appear to be of particular interest for studying molecular motion. Molecular behavior in gases is of the simplest kind and it would be a valuable check of the perturbed directional correlation theory if a systematic study could be made on the gaseous state. The simplest situation would be presented by a diatomic molecule with a nucleus decaying through an isomeric transition. Preliminary attempts have been made to study the isolated linear molecule dimethylcadmium at a pressure low enough to make the time between collisions long compared to the lifetime of the intermediate state.

Meares and Westmoreland (1972) in a paper presented at the 1971 Cold Spring Harbor Symposium on Quantitative Biology corroborated the usefulness of the perturbed directional correlation technique to study motional characteristics of labeled macromolecules and to investigate in vivo systems at concentrations of the tracer as low as 10^{-12}M. The ^{111}In used in their work was prepared in the Oak Ridge cyclotron by the ^{111}Cd(p,n)^{111}In reaction on enriched cadmium foil. The directional correlation measurements were made with a four-detector coincidence spectrometer using a NaI(Tl) crystal detector. In one series of experiments conclusive evidence was given that the $^{111}In^{3+}$ tracer was incorporated into the native ribonuclease (EC 3.1.4.22) of bovine pancreatic origin, with only a slight chemical modification and without noticeable change in the chemical behavior of the enzyme. Since there are no sites in the native molecule which would specifically bind indium(III) ions, the disulfide bridge connecting Cys 65 and Cys 72 was reduced and the metal ion inserted into the reduced disulfide bridge according to the procedure of Sperling *et al*. (1969). The plots of $G_{22}(t)$ for the $^{111}In^{3+}$ derivative of ribonuclease and for native ribonuclease to which $^{111}In^{3+}$ has been added were determined. The directional correlation for $^{111}In^{3+}$ in the presence of the native ribonuclease is only weakly perturbed, but strong perturbation could be observed when the tracer was bound to the partially reduced ribonuclease. This was an indication that the tracer is in an environment with a much longer rotational correlation time. Since in any type of labeling experiment it is important to consider how much the label perturbed the system, the temperature denaturation curves of the native and of the labeled enzyme were compared. The transition temperature for the thermal denaturation of the labeled ribonuclease was 0.7°C less than the transition temperature for the native sample, a difference within the experimental error. Since the rotational correlation time at the tracer binding site should be sensitive to the denaturation of the protein molecule, it was of interest to follow the time-integrated perturbation factor as a function of guanidinium hydrochloride concentration. The data suggested that the perturbation factor was quite sensitive to different stages of the denaturation process. In another series of experiments, by the same authors, the tracer was incorporated *in vivo* into animal tissues. The decay characteristics of ^{111}In make it an excellent isotope for medical studies. Injection of radioactive indium compounds leads to the accumulation of radioactivity in various tissues. Little is known, however, about the chemical behavior and fate of these compounds *in vivo*. In the PDC experiments ^{111}In-labeled compounds had been injected into the tail vein of a white mouse. The mouse was then placed in the barrel of a slightly modified 60-ml plastic syringe, centered among the detectors, and PDC measurements were made for periods of 13.4 min. The uniformly low values of the time-integrated perturbation factor observed with indium chloride, indium citrate and indium nitrilotriacetate was an indication that the indium ions have been displaced from these ligands and bound to protein in the tissue. The small values of the perturbation factor reflected the long rotational correlation times characteristic of macromolecules. In fact, Stern *et al*. (1967) suggested that the indium is almost quantitatively bound to transferrin in the blood. On the other hand, the EDTA complex of indium is extremely stable and the distinctly higher values of the perturbation factor found after the injection of the complex indicated that it remained intact *in vivo*. This is supported by the fact that the radioactivity is excreted rapidly in the urine.

The lack of versatility to attach radioactive rotational tracers selectively to specific sites on macromolecules has been a severe limitation of the technique. Meares *et al*. (1972) described the use of a small bifunctional organic molecule, 1-p-nitrophenylethylenediaminetetracetic acid (NO_2Ph-EDTA) that chelates indium(III) ions so that they are not released into aqueous buffer solutions at neutral pH and bind to the combining site of rabbit antibody to dinitrophenyl groups (anti-Dnp). The average association constant between anti-Dnp and the labeled hapten could be estimated from PDC data obtained with rabbit antiserum. These measurements were compared to the results of equilibrium dialysis. It appears that the chemical selectivity available with other labeling techniques may be achieved also for perturbed directional correlation since NO_2Ph-EDTA is a potential intermediate to several other labeling compounds. The ^{111}In used in these experiments was prepared in the Oak Ridge cyclotron by the ^{111}Cd (p,n)^{111}In reaction on enriched cadmium oxide. The PDC measurements were made with the four-detector coincidence spectrometer with a NaI(Tl) crystal detector. The samples for the experiments were prepared by addition of 100-μl aliquots containing about 10μCi of ^{111}In as NO_2Ph-EDTA chelate to 100 μl of antiserum. The initial concentration of the chelate in the sample was less than 0.2 μM. These experiments illustrated several characteristics of PDC that makes it unique as a labeling method. About 10^{10} molecules containing ^{111}In are required for a PDC experiment, so it is possible to make measurements on very small quantities of the macromolecules of interest. Quantitative PDC measurements are best made on a point-source of radiation, so quite small volumes of sample can be studied. With the apparatus used in these experiments the measurement could be completed in four hours, but this time requirement can be greatly reduced by using more instrumentation. Thus measurements can be made on relatively unstable systems.

Glass and Graf (1970) reported the use of perturbed directional correlation to study internal fields of water associated with hydrated protein molecules. For this measurement the nuclide ^{133}Cs was used which is formed by the electron capture decay of ^{133}Ba (Figure 7). The time-dependent anisotropy of the cascade was measured by the use of a two-detector time-differential coincidence system. A 100 μCi sample of ^{133}Ba was dissolved in a solution of bovine carbonic anhydrase. Portions of this solution were frozen at liquid nitrogen temperature and the frozen samples lyophilized to constant weight. The freeze-dried samples were adjusted to approximately 5 percent and 20 percent water content by exposure to 0.2 and 0.6 relative vapor pressures, respectively. The samples were placed in sealed thin glass cylinders which approximated the geometry of a linear source. Measurement of specific activities by the Wilbur-Anderson method (1948) showed no change with respect to the original enzyme preparation. First, the coefficient A_{22} was measured by performing an unperturbed directional correlation experiment. The $^{133}Ba^{2+}$ was suspended in 6M HCl to minimize hydrolysis effects. Brownian motion in the solution eliminates the effects of perturbing fields and the unperturbed coefficient A_{22} may be measured. Then, the time dependence of the anisotropy was determined in the perturbed experiments. The results are shown in Figure 9. The time-dependent perturbation factors for the carbonic anhydrase samples are shown in

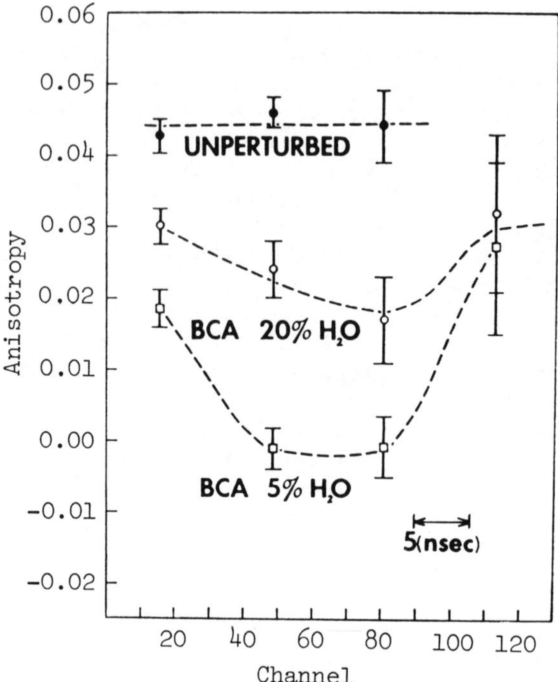

Figure 9. Time dependence of anisotropy for 356-81 keV gamma cascade decay of ^{133}Cs in hydrated bovine carbonic anhydrase (BCA). Unperturbed anisotropy is from ^{133}Cs in aqueous 6N HCl solution. Data has been integrated over the 10 nsec time resolution of the coincidence system.

Figure 10. For the 5 percent hydrated sample the behavior of the perturbation factor shows the presence of a static electric quadrupole interaction. Because static fields are not expected in an aqueous environment it seems likely that the $^{133}Ba^{2+}$ occupies definite sites of axial symmetry in the protein molecule. For the 20 percent hydrated carbonic anhydrase sample the behavior of the experimental perturbation factor indicates the presence of a static interaction, although the comparison with the perturbation factor for the 5 percent hydrated sample suggests also the presence of a time-dependent interaction which is seen by a certain fraction of $^{133}Ba^{2+}$ while the remainder occupy sites on the protein molecule and see only a pure static electric quadrupole interaction. If a fraction n_1 of the $^{133}Ba^{2+}$ is attached to the carbonic anhydrase molecules, the fraction n_2 must be in the water associated with the molecules. Then, the total perturbation factor is given by

$$G_{22}(t) = n_1 G_{22}{}^{s}(t) + n_2\, e^{-\lambda_2 t} \tag{1}$$

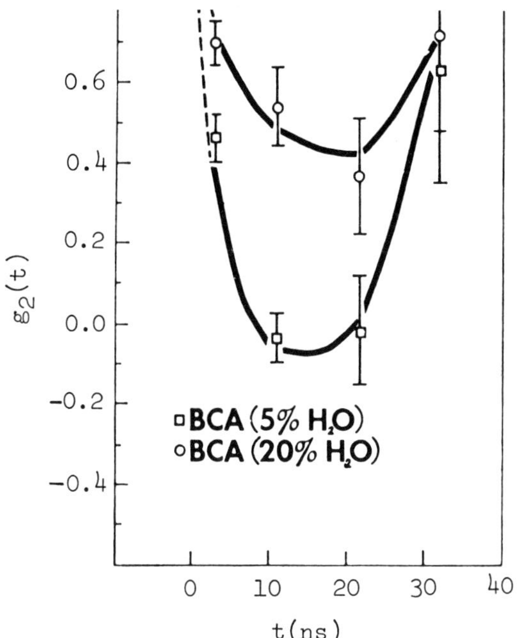

Figure 10. Experimental perturbation factors with theoretical fit for 356-81 keV gamma cascade decay of ^{133}Cs in hydrated bovine carbonic anhydrase (BCA).

The static perturbation factor $G_{22}{}^{s}(t)$ is known from the 5 percent hydrated source. Eq.(1) was solved numerically for several values of n_1, n_2, and λ_2. A best fit of the data gives $n_1 = 0.45$, $n_2 = 0.55$, and $\lambda_2 = (8.20 \pm 1.00 \times 10^6\ sec^{-1})$. If the water is normal with a correlation time $\tau_c \simeq 10^{-11}$ sec, then time dependent interactions should not be observable. If the water is ordered, with a rotational correlation time much larger than 10^{-11} sec, time-dependent interactions become possible due either to fluctuating fields in the water or to the propagation of the electron hole following the electron capture decay of ^{133}Ba. It is suggested that the fraction n_2 of the $^{133}Ba^{2+}$ is in the water associated with the carbonic anhydrase molecule. A relaxation time for this environment may be obtained from λ_2. It is found that $\tau = 1/\lambda_2 = (1.22 \pm 0.15) \times 10^{-7}$ sec, which indicates that the water associated with carbonic anhydrase molecules is highly ordered.

Richer et al. (1973a) measured the time-differential anisotropy for the cascade of ^{133}Cs (Figure 7) implanted at several sites of bovine carbonic anhydrase. The commercially available enzyme was further purified by preparative polyacrylamide gel disc electrophoresis and the most anionic band (commonly designated as BCA-B) was used for the preparation of the source. In order to achieve incorporation of the tracer, it was necessary first to demetallate the enzyme. The removal of zinc was accomplished by dialysis against o-phenanthroline and then the demetallated sample was dialyzed against 0.5 μCi ^{133}Ba until equilibrium was reached. Atomic absorption spectroscopy indicated a 10–15 percent incorporation of the tracer in the sample, which was enclosed in a glass capillary to serve as a source for the measurement of the time dependence of the correlation function. In these experiments a time-differential coincidence system was used with two NaI(Tl) crystal detectors, with crossover pickoff timing on the 356 keV start side and constant fraction timing on the 81 keV side. The time resolution achieved with the combination was approximately 3 nsec. The anisotropy appears to be time dependent (Figure 11). The corresponding time-differential (Figure 12) and time-integrated (Figure 13) perturbation factor strongly suggests the form $G_{22} = e^{-\lambda_2 t}$ which would be expected for a time-dependent interaction. A time-dependent interaction may be due to rapidly fluctuating fields or to the propagation of the electron hole after the electron capture of ^{133}Ba. An inspection of the work of Eley (1968) on the conductivity of biopolymers shows that carbonic anhydrase is a sufficiently good conductor so as to rule out possible "after effects".

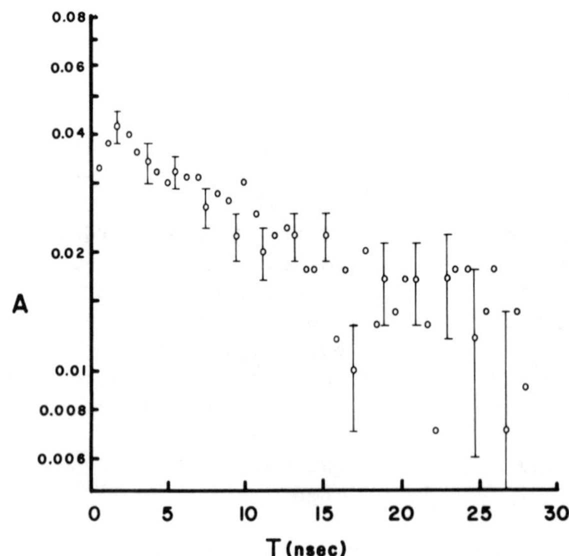

Figure 11. Time-dependence of the anisotropy of the 356 keV–81 keV cascade of ^{133}Cs in dehydrated BCA-B.

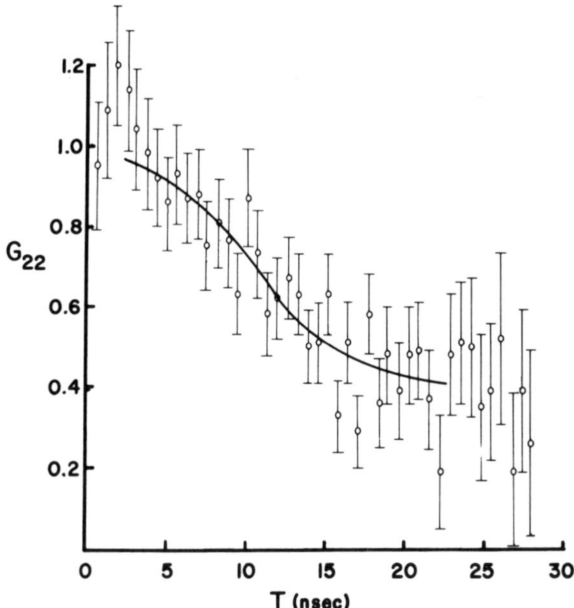

Figure 12. Time-differential perturbation factor for the 356 keV-81 keV cascade of ^{133}Cs in dehydrated BCA-B.

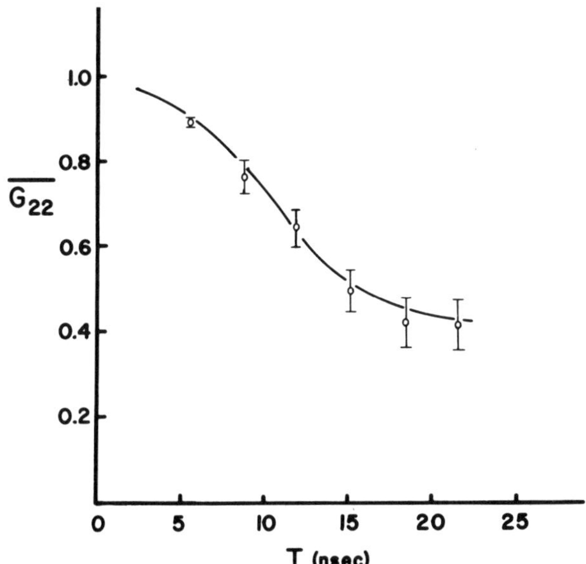

Figure 13. Perturbation factor integrated over 3 nsec time resolution of the coincidence system, for the 356 keV-81 keV cascade of ^{133}Cs in dehydrated BCA-B.

For strongly fluctuating local fields one must consider both electric and magnetic interaction. Tentatively, the data were interpreted in terms of fluctuating magnetic fields. The possibility of time-dependent electric quadrupole interactions, however, could not be ruled out definitely. Sites occupied by the $^{133}Cs^{2+}$ could not be identified with any great certainty. Since demetallation was necessary prior to the incorporation of the tracer, it is highly probable that the $^{133}Ba^{2+}$ is attached somewhere in the active region of the molecule. At the pH used in the metal incorporation step, the likely sites of attachment are the His at the active site or one or more Glu and Asp residues known to be in the active region.

Richer et al. (1973b) reported also the use of PDC to study the internal field at the active site of cobalt carbonic anhydrase. Electrophoretically homogeneous samples were prepared again by preparative polyacrylamide gel disc electrophoresis. The samples were demetallated by dialysis against o-phenanthroline and the zinc ion at the active site replaced by $^{57}Co^{2+}$ with no appreciable loss of specific activity. Time-differential measurements were made for the 122 keV-14 keV cascade in ^{57}Fe (Figure 14) bound to the active site of the protein. The results indicated that the anisotropy rapidly goes to zero (Figure 15). This behavior can occur only as a consequence of a large nuclear quadrupole interaction. The relaxation constant was calculated as $\lambda_2 = 7.0 \times 10^7$ sec and it is related to the nuclear quadrupole interaction and to the field gradient at the

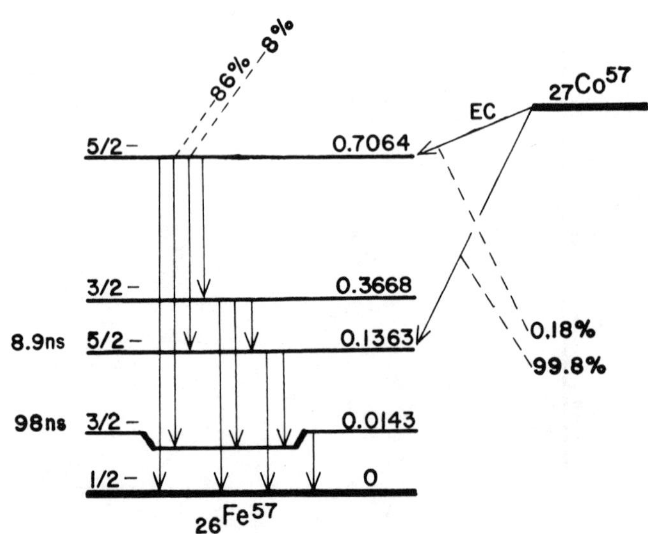

Figure 14. Abridged decay scheme for ^{57}Co. The 122 keV-14 keV gamma-gamma cascade in ^{57}Fe may be used for PDC experiments.

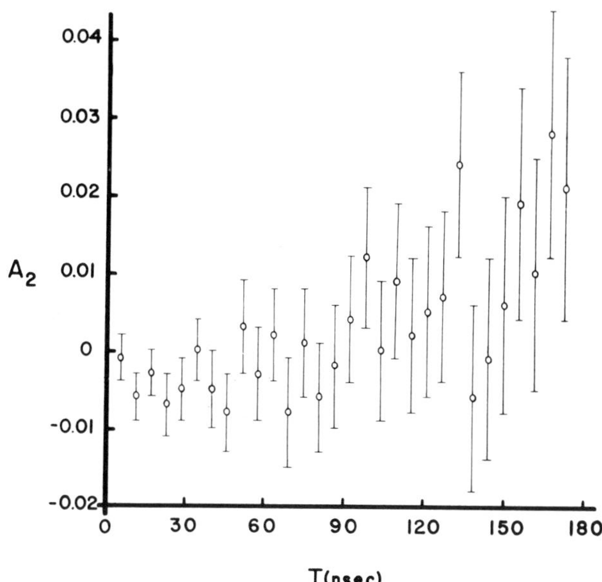

Figure 15. Time-differential A_2 coefficient for the 122 keV-14 keV cascade of ^{57}Fe at the active site of bovine carbonic anhydrase.

decaying nuclide. For the ^{57}Fe the quadrupole moment has been reported to be 0.29 barn (Bearden and Dunbar, 1970). Several authors (Hower et al., 1971: Chiqnell et al., 1972; Mushak and Coleman, 1972) reported the rotational correlation time of the bovine carbonic anhydrase molecule as approximately 2×10^{-8} sec. Using these values and assuming axial symmetry, a minimum value of 10^{20} V/m^2 could be established for the electric field gradient at the metal binding site of bovine carbonic anhydrase. Reports of other investigations of the active site of this protein have suggested an axially symmetric distorted tetrahedral coordination for the cobalt and the presence of a strong quadrupole interaction. The perturbed directional correlation study produced a good approximation of the electric field gradient which participates in this interaction.

CONCLUDING REMARKS

It is apparent from the above discussion that the perturbed directional correlation of gamma-rays provides an excellent technique to study proteins and protein-metal interactions. In the past five years the pioneering work centered mainly on rotational characteristics of proteins, structure of water associated with protein molecules and internal field strengths and field symmetries at or around the active site of enzymes. Considering that over one-third of the known enzymes and many molecular superstructures require metal ions for their function, the field seems to be wide open for this technique. A large number of metallo-enzymes contain transition metal ions at well defined binding sites. In many cases convenient cascading isotopes can be found for these ions and built-in rotational tracers can be easily provided. The potential of ^{62}Zn is a case in question (Meares, 1973). If the metal is directly involved in the catalytic process, the active site can be studied specifically. In some other cases the rotational tracer can be incorporated into a suitable labeling compound, hapten, and possibly, a substrate or inhibitor molecule. Both structural and dynamic information is readily available from these experiments. The technique could be extended to multi-enzyme systems, allosteric effects and enzyme-membrane interactions. The extreme sensitivity of the measurements (10^{-12}M) and the negligibly small denaturation effects owing to the low tracer concentrations, render the technique particularly suitable for *in vivo* experiments. The perturbed directional correlation of gamma rays should be of exceptional value in biomedical research since it expands the scope of the already established radioactive tracer techniques into the dimensions of nuclear spectroscopy. A rapid progress in this field is expected when the advantages of instrumental simplicity and the straightforward nature of the theory are recognized.

ACKNOWLEDGEMENTS

This work was supported in part by the North Dakota Agricultural Experiment Station (Journal Article No. 439). The authors wish to thank Mr. E. Handel, graduate student, for technical assistance and Mrs. C. O'Leary for the preparation of the manuscript.

REFERENCES

Abragam, A. and Pound, R. V. (1953). "Influence of Electric and Magnetic Fields on Angular Correlations", Phys. Rev. 92: 943-952.

Alder, K. (1952). "Theory of Directional Correlation", Helv. Phys. Acta 25: 235-258

Alder, K., Albers-Schonberg, H., Heer, E. and Novey, T. B. (1953). "Measurements of the Nuclear Moments of Excited States by Angular Correlation Methods", Helv. Phys. Acta 26: 761-784.

Alder, K., Matthias, E., Schneider, W. and Steffen, R. M. (1963). "Influence of a Combined Magnetic Dipole and Electric Quadrupole Interaction on Angular Correlations", Phys. Rev. 129: 1199-1213.

Andrade, P. Da R., Vasquez, A., Rogers, J. D. and Fraga, E. R. (1970). "Nuclear Relaxation in $(NH_4)_3HfF_7$ Studied by Gamma-Gamma Angular Correlation", Phys. Rev. B1: 2912-2917.

Barrett, J. S., Cameron, J. A., Gasdner, P. R., Keszthelyi, L. and Prestwich, W. V. (1970). "Perturbed Angular Correlation Study of the Environment of ^{111}Cd Nuclei in Ice", J. Chem. Phys. 53: 759-763.

Bearden, A. J. and Dunham, W. R. (1970). in Structure and Bonding, eds. Hemmerich, P., Jorgensen, C. K. and Neilands, J. B. (Springer-Verlag, New York), Vol. 8, pp. 1-52.

Bertolaccini, M., Bussolatti, C., Cova, S., Donati, S. and Svelto, V. (1967). "Statistical Behavior of the Scintillation Counter", Nucl. Instrum. Methods 51: 325-328.

Bertolini, G., Cocchi, M., Mandl, V. and Rota, A. (1966). "Time Resolution Measurements with Fast Photomultipliers", IEEE Trans. Nucl. Sci. NS-13(3): 119-126.

Biedenharn, L. (1962). in Nuclear Spectroscopy, Part B, ed. Ajzenberg-Selove, F. (Academic Press, New York), pp. 732-810.

Bodenstedt, E., Ortabase, V. and Ellis, W. H. (1972). "Perturbed Angular Correlation Studies of the Quadrupole Interactions of ^{111}Cd in Different Metallic Environments and the Electric Quadrupole Moment of the 247 keV State", Phys. Rev. B6: 2209-2222.

Boström, L., Karlsson, E. and Zetterlund, S. (1970). "Calculation of Differential Angular Correlation Factors for Non-coaxial Electric and Magnetic Fields", Physica Scripta 2: 65-69.

Bozek, E., Golczewski, J., Hrynkiewicz, A. Z., Polak, G., Rybicka, M., Styczen, B. and Styczen, J. (1971). "The Perturbed Angular Correlations for the (596-42) keV Cascade in Cu", in Angular Correlations in Nuclear Disintegration, eds. Van Krugten, H. and Van Nooijen, B. (Rotterdam University Press, The Netherlands), pp. 571-575.

Braunsfurth, J. and Korner, H. J. (1965). "Time Resolution Properties of NaI Scintillators", Nucl. Instrum. Methods 34: 202-212.

Chiqnell, C. F., Starkweather, D. K. and Erlich, R. H. (1972). "The Interaction of Some Spin-labeled Sulfonamides with Bovine Erythrocyte Carbonic Anhydrase B", Biochim. Biophys. Acta 271: 6-15.

Coester, F. (1954). "Influence of Extranuclear Fields on Angular Correlations", Phys. Rev. 93: 1304-1308.

Compton, P. D. Jr. and Johnson, W. A. (1967). "Determination of Charge Sensitivity of a Tunnel Diode Voltage Threshold Discriminator Used In Time Difference Measurements", IEEE Trans. Nucl. Sci. NS-14(1): 116-125.

Devons, S. and Goldfarb, L. (1957). in Encyclopedia of Physics 42, ed. Flügge, S. (Springer-Verlag, Berlin), pp. 362-554.

Eley, D. D. (1968). in Organic Semiconducting Polymers, ed. Katon, J. E. (Marcel Dekker, New York), pp. 259-294.

Fano, U. (1953). "Geometrical Characterization of Nuclear States and Theory of Angular Correlations; National Bureau of Standards Report 1214", Phys. Rev. 90: 577-579.

Frauenfelder, H. (1961). in Methods of Experimental Physics 5, eds. Yuan, L. and Wu, C. (Academic Press, New York), pp. 214-264.

Frauenfelder, H. and Steffen, R. M. (1966). in Alpha-, Beta- and Gamma-Ray Spectroscopy, ed. Siegbahn, K. (North-Holland, Amsterdam), pp. 997-1198.

Gedcke, D. A. and McDonald, W. J. (1968). "Design of the Constant Fraction of Pulse Height Trigger for Optimum Time Resolution", Nucl. Instrum. Methods 58: 253-260.

Glass, J. C. and Kliwer, J. K. (1968). "Static Quadrupole Interaction of Scandium-44 in Barium Titanate", Nucl. Phys. A115: 234-240.

Glass, J. C. and Graf, G. (1970). "Directional Correlation Studies of Anomalous Water in Carbonic Anhydrase", Nature 226: 635-636.

Glass, J., Gupta, J., Graf, G. and Richer, L. (1972). "Directional Correlation Study of Capillary Grown Anomalous Water", Nature, Physical Science 235: 14-15.

Glass, J., Graf, G., Richer, L. and Klee, M. (1973). "Directional Correlation Measurements of ^{133}Cs in Viscous Liquids", (in preparation for Phys. Rev.).

Goodwin, D. A., Meares, C. F. and Song, C. H. (1973). "The Study of ^{111}In Labeled Compounds in Mice Using Perturbed Angular Correlations of Gamma Radiation", Radiology (in press).

Haas, H. and Shirley, D. A. (1973). "Nuclear Quadrupole Interaction Studies by Perturbed Angular Correlations", J. Chem. Phys. 58: 3339-3355.

Hamilton, D. R. (1940). "Directional Correlation of Successive Quanta", Phys. Rev. 58: 122-131.

Haugland, R. P., Stryer, L., Stengle, T. R. and Baldeschwieler, J. D. (1967). "Nuclear Magnetic Resonance Studies of Antibody-Hapten Interactions Using a Chloride Ion Probe", Biochemistry 6: 498-502.

Hower, J. F., Henkens, R. W. and Chestnut, D. B. (1971). "A Spin Label Investigation of the Active Site of an Enzyme", J.A.C.S. 93: 6665-6671.

Huntress, W. T. Jr. (1968). "Effects of Anisotropic Molecular Rotational Diffusion on Nuclear Magnetic Relaxation in Liquids", J. Chem. Phys. 48: 3524-3533.

Hyman, L. G. Schwarcz, R. M. and Schluter, R. A. (1964). "Study of High Speed Photomultiplier Systems", Rev. Sci. Instrum. 35: 393-406.

Hyman, L. G. (1965). "Time Resolution of Photomultiplier Systems", Rev. Sci. Instrum. 36: 193-197.

Karlsson, E. (1964). Perturbed Angular Correlations, eds. Karlsson, E., Matthias, E. and Siegbahn, K. (North-Holland, Amsterdam).

Klemme, B., Herzog, P., Schafer, G., Folle, R. and Netz, B. (1973). "The Quadrupole Moment of the 535 keV State of ^{133}La", Phys. Letters 45B: 38-39.

Lederer, C. M., Hollander, J. M. and Perlman, J. eds. (1968). in Table of Isotopes (John Wiley and Son, New York),pp. 191, 281.

Leipert, T., Baldeschwieler, J. D. and Shirley, D. A. (1968). "Applications of Gamma-Ray Angular Correlations to the Study of Biological Macromolecules in Solution", Nature 220: 907-909.

Lynden-Bell, R. M. (1971). "Perturbation of the Angular Correlation of Gamma Rays by Molecular Motion", Mol. Phys. 21: 891-900.

Marshall, A. G. and Meares, C. F. (1972). "Effect of Slow Rotational Diffusion on Angular Correlations", J. Chem. Phys. 56: 1226-1229.

Marshall, A. G., Werbelow, L. G. and Meares, C. F. (1972). "Effect of Molecular Shape and Flexibility on Gamma-Ray Directional Correlations", J. Chem. Phys. 57: 364-370.

Matthias, E. Schneider, W. and Steffen, R. M. (1962). "Nuclear Level Splitting Caused by a Combined Electric Quadrupole and Magnetic Dipole Interaction", Phys. Rev. 125: 261-268.

Matthias, E., Schneider, W. and Steffen, R. M. (1963). "Some Remarks on Static Electric Quadrupole Perturbations in Angular Correlations", Phys. Letters 4: 41-43.

McDonald, W. J. and Gedcke, D. A. (1967). "Time Resolution Studies on Large Photomultipliers", Nucl. Instrum. Methods 55: 1-14.

McQuire, R. L. and Palmer, R. C. (1967). "A Fluorescence Emission Detector with 100 Picosecond Time Resolution", IEEE Trans. Nucl. Sci. NS-14(1): 217-221.

Meares, C. F., Bryant, R. G., Baldeschwieler, J. D. and Shirley, D. A. (1969). "Study of Carbonic Anhydrase Using Perturbed Angular Correlations of Gamma Radiation", Proc. Nat. Acad. Sci. USA 64: 1155-1161.

Meares, C. F. (1972). "The Application of Perturbed Angular Correlation Studies to Biological Macromolecules", Ph.D. Thesis, Stanford University.

Meares, C. F., Sundberg, M. W. and Baldeschwieler, J. D. (1972). "Perturbed Angular Correlation Study of a Haptenic Molecule", Proc. Nat. Acad. Sci. USA 69: 3718-3722.

Meares, C. F. and Westmoreland, D. G. (1972). "The Study of Biological Macromolecules Using Perturbed Angular Correlations of Gamma Radiation", Cold Spring Harbor Symp. Quant. Biol. 36: 511-516.

Meares, C. F. (1973). "Perturbed Angular Correlations and Protein-Labeling with Metal Ions", 166th ACS National Meeting Abstr., Chicago, Illinois.

Mushak, P. and Coleman, J. E. (1972). "Electron Spin Resonance Studies of Spin-labeled Carbonic Anhydrase", J. Biol. Chem. 247: 373-380.

Netz, G. and Bodenstedt, E. (1973). "Measurement of the Electric Quadrupole Moment of the 482 keV State of Tantalum-181 by the Time-Differential Perturbed Angular Correlation Technique", Nucl. Phys. A208: 503-508.

Nutt, R. (1967). "Optimum Design of Sensitive Tunnel Diode Discriminators", IEEE Trans. Nucl. Sci. NS-14(1): 110-115.

Paul, H. and Brunner, W. (1963). "Contribution to the Theory of Gamma-Gamma Angular Correlations Under Combined Magnetic and Inhomogeneous Electric Fields. II The Interaction Operator", Ann. Phys. 9: 323-329.

Pound, R. V. and Wertheim, G. K. (1956). "Directional Correlation and Electric Quadrupole Moments of Mercury Isotopes", Phys. Rev. 102: 396-399.

Present, G., Schwarzschild, A., Spirn, I. and Wotherspoon, N. (1964). "Fast Delayed Coincidence Technique: The XP1020 Photomultiplier and Limits of Resolving Times Due to Scintillation Characteristics", Nucl. Instrum. Methods 31: 71-76.

Racah, G. (1951). "Directional Correlation of Successive Nuclear Radiations", Phys. Rev. 84: 910-912.

Raghaven, R. S., Raghaven, P., Kaufmann, E. N., Krien, K. and Naumann, R. A. (1973). "Time-Differential Perturbed Angular Correlation Measurements of the Hyperfine Field at Mercury in Iron", Phys. Rev. B7: 4132-4137.

Richer, L. L. (1973). "Gamma-Ray Perturbed Directional Correlation Studies of Bovine Carbonic Anhydrase", Ph.D. Thesis, North Dakota State University.

Richer, L., Glass, J. and Graf, G. (1973a). "Gamma-Gamma Directional Correlation Study of Bovine Carbonic Anhydrase", 165th ACS National Meeting Abstr., Dallas, Texas.

Richer, L., Glass, J. and Graf, G. (1973b). "Gamma-Gamma Directional Correlation Study of Cobalt-Carbonic Anhydrase", 166th ACS National Meeting Abstr., Chicago, Illinois.

Rose, M. (1953). "Analysis of Angular Correlation and Angular Distribution Data", Phys. Rev. 91: 610-615.

Sakai, M., Nayawa, H., Ikegami, H. and Yamazaki, T. (1963). "Proceedings of the International Conference on the Role of Atomic Electrons in Nuclear Transformations", (Warsaw, Poland).

Scherer, C. (1970). "Gamma-Gamma Angular Correlations: A Model for Statistical Perturbation with any Correlation Time", Nucl. Phys. A157: 81-92.

Shirley, D. A. (1970). "Estimates of Correlation Times of Dissolved Complexes from Rotational Tracer Experiments", J. Chem. Phys. 53: 465-466.

Shirley, D. A. (1971). "Influence of Molecular Geometry, Orientation, and Dynamics on Angular Correlation Patterns from Rotationally Labeled Macromolecules", J. Chem. Phys. 55: 1512-1521.

Sperling, R., Burstein, Y. and Steinberg, T. Z. (1969). "Selective Reduction and Mercuration of Cystine IV-V in Bovine Pancreatic Ribonuclease", Biochemistry 8: 3810.

Steffen, R. M. (1956). "Influence of the Time-Dependent Quadrupole Interaction on the Directional Correlation of the ^{111}Cd Gamma-Rays", Phys. Rev. 103: 116-125.

Steinhardt, J. and Beychok, S. (1964). in The Proteins, ed. Neurath, H. (Academic Press, New York), Vol. II, pp. 213-216.

Stengle, T. R. and Baldeschwieler, J. D. (1966). "Halide Ions as Chemical Probes for NMR Studies of Proteins", Proc. Nat. Acad. Sci. USA 55: 1020-1022.

Stern, H. S., Goodwin, D. A., Scheffel, U., Wagner, H. N. and Kramer, H. H. (1967). "^{113m}In for Blood-Pool and Brain Scanning", Nucleonics 25: 62.

Stryer, L. (1965). "The Interaction of a Naphthalene Dye with Apomyoglobin and Apohemoglobin. A Fluorescent Probe of Non-polar Bonding Sites", J. Mol. Biol. 13: 482-495.

Stryer, L. and Griffith, O. H. (1965). "A Spin-Labeled Hapten", Proc. Nat. Acad. Sci. USA 54: 1785-1791.

Wagner, H. F. and Forker, M. (1969). "An Investigation of Systematic Error in Time-Differential Angular Correlation Measurements Performed with a Four-Detector Apparatus", Nucl. Instrum. Methods 69: 197-208.

Wertheim, G. K. and Pound, R. V. (1956). "Time-Dependent Directional Correlation of 1.1hr. ^{104}Pb", Phys. Rev. 102: 185-189.

Wilbur, K. M. and Anderson, N. G. (1948). "Electrometric and Colorimetric Determination of Carbonic Anhydrase", J. Biol. Chem. 176: 147-154.

Woessner, D. E. (1962). "Nuclear Spin Relaxation in Ellipsoids Undergoing Rotational Brownian Motion", J. Chem. Phys. 37: 647-654.

Wu, C. S. and Moszkowski, S. A. (1966). Beta Decay (Interscience Publishers, New York), p. 196.

Yates, M. (1963). "Angular Correlation Attenuation Coefficients for Photopeaks Measured with NaI Crystals", Nucl. Instrum. Methods 23: 152-154.

Yates, M. (1964). in Perturbed Angular Correlations, eds. Karlsson, E., Matthias, E. and Siegbahn, K. (North-Holland, Amsterdam), pp. 453-466.

Yguerabide, J., Epstein, H. F. and Stryer, L. (1970). "Segmental Flexibility in an Antibody Molecule", J. Mol. Biol. 51: 573-590.

SUBJECT INDEX